GREAT EVENTS
FROM
HISTORY II

GREAT EVENTS FROM HISTORY II

Science and Technology Series

Volume 5
1969-1991

Edited by

FRANK N. MAGILL

SALEM PRESS

Pasadena, California   Englewood Cliffs, New Jersey

Library of Congress Cataloging-in-Publication Data
Great events from history II. Science and technology
series / edited by Frank N. Magill.
    p.   cm.
   Includes bibliographical references and index.
   1. Science—History—20th century. 2. Technology—
History—20th century. I. Magill, Frank Northen, 1907-

Q125.G825 1991
509′.04—dc20
ISBN 0-89356-637-3 (set)   91-23313
ISBN 0-89356-642-x (volume 5)   CIP

92-0369

LIST OF EVENTS IN VOLUME V

GREAT EVENTS
FROM
HISTORY II

THE FIRST HUMANS LAND ON THE MOON

Category of event: Space and aviation
Time: July 20, 1969
Locale: Sea of Tranquillity, the Moon

For one moment the world stood as one, looked up at the Moon, and knew it would be changed forever by the presence of two humans walking on that landscape

Principal personages:

NEIL A. ARMSTRONG (1930-   ), the Apollo 11 commander who flew many pioneering high-speed aircraft, including the X-15

MICHAEL COLLINS (1930-   ), an Air Force colonel who was the Apollo 11 command module pilot

EDWIN E. ("BUZZ") ALDRIN, JR. (1930-   ), an Air Force colonel and Apollo 11 lunar module pilot who flew as pilot on the last Gemini mission in 1966

Summary of Event

". . . I believe that this nation should commit itself to achieving the goal, before this decade is out, of landing a man on the moon and returning him safely to the earth. No single space project in this period will be more impressive to mankind, or more important for the long-range exploration of space; and none will be so difficult or expensive to accomplish." This was the challenge made to the National Aeronautics and Space Administration (NASA) and the American aerospace community by President John F. Kennedy on May 25, 1961.

The manned lunar landing project began in the infancy of NASA on June 18, 1959. It was to be the culmination of years of technology development and testing. Although it would be an ambitious undertaking, requiring launch vehicles many times larger than any in existence, it could be realized by following a systematic plan of building on existing hardware until the technology caught up with the vision.

The first step was Project Mercury, designed to place a manned spacecraft into orbit for twenty-four hours. The basic skills of safely getting a human up and back, as well as being kept alive in the hostile environment of space, materialized. The Gemini program developed the techniques necessary for accomplishing the Moon landing, including rendezvous and docking of two orbiting spacecraft, and extra-vehicular activity—EVA (space walk). The two-man Gemini spacecraft was a vast improvement over its forerunner. It incorporated a modular system of components that permitted relatively easy servicing and replacement.

The United States' program to land a human on the Moon was named Apollo—after the Greek god of music, prophecy, medicine, light, and progress—because the "image of the god Apollo riding his chariot across the sun gave the best representation of the grand scale of the proposed program."

The command module (CM) was a conical spacecraft that carried the three astronauts to and from the Moon. The service module (SM) contained the engines used for maneuvers, oxygen and water, the electrical power system, and the large communications antenna. The lunar module (LM), a two-stage, spiderlike vehicle, was designed to carry two astronauts to the lunar surface and provide a shelter for them during their stay. The descent stage had a single, throttleable engine that would slow the craft and permit a soft landing on its four spindly legs. The ascent stage contained the pressurized crew compartment, a system of small attitude control engines, and a single, fixed-thrust engine to lift the astronauts and their cargo back into lunar orbit.

The first manned flights in the Apollo program were to be launched atop the Saturn 1B. First would come a few flights using only the command and service modules (CSM), followed by dual launches of the CSM and LM to test rendezvous and docking.

On January 27, 1967, during a ground test of the spacecraft and launch vehicle, a fire broke out inside the command module, which was pressurized with pure oxygen. The fire rapidly spread, generating toxic fumes and eventually rupturing the pressure vessel of the spacecraft. The astronaut crew, Virgil I. ("Gus") Grissom, Edward H. White, and Roger B. Chaffee, died from a combination of the toxic fumes and the heat of the fire.

The manned orbital flight of Apollo 7, after nearly two years of waiting, was quickly followed by the circumlunar Apollo 8 mission. Apollo 9 tested the CSM and LM in Earth's orbit, while Apollo 10 did the same in lunar orbit. By mid-1969, everything was ready for Apollo 11. Apollo 11 began its journey at 9:32 A.M. eastern daylight time on July 16, 1969, atop the Saturn 5, the height of which equaled the length of a football field. It rode a flame three times its length and climbed slowly into the warm Florida air. Twelve minutes later, the spacecraft and its crew were safely in Earth's orbit. Two and one-half hours later, the third stage propelled its cargo toward the Moon. The CSM separated from the stage, turned 180 degrees, and pulled out the LM. The spacecraft were placed in a "barbecue" mode, rotating around the long axis at a rate of three revolutions per minute, permitting even heating of the spacecraft by the sun. Seventy-six hours after launch, the spacecraft entered lunar orbit. Final checkout of the landing craft began almost immediately. A final sleep/rest period commenced on the morning of the fifth day. At 7:00 A.M., the crew awakened and ate their breakfast. Later, the LM's systems were powered up, and the crew donned their pressure suits. At 1:44 P.M., the LM undocked from the CSM and began its descent to the lunar surface. Michael Collins inside the CM *Columbia*, told his companions in the LM *Eagle* to "take care." Neil A. Armstrong replied that they would see him later.

A 29.8-second burn of the descent engine put the LM into a 106 by 16 kilometer orbit. Forty minutes later, the engine ignited again to begin powered descent. *Eagle* was 15 kilometers above the lunar surface, 554 kilometers from the landing target, flying face down, landing gear forward. At an altitude of 12 kilometers, the LM rolled face up and the crew could no longer see the lunar surface. At 2.4 kilometers above the surface on its final approach, *Eagle* was pitched upright and Armstrong began to look for the landing site. At 1,500 meters above the surface, the hovering phase began and Armstrong had to decide where to land. Several alarms sounded in the cockpit—the computer was being overloaded with data. A probe attached to one of *Eagle*'s legs contacted the surface and a blue light illuminated in the cabin.

The LM settled onto the lunar surface in the southwestern part of Mare Tranquillitatis at 4:17 P.M. on the evening of July 20, 1969. Immediately, the crew began procedures for an emergency lift-off. Things went well, and the two astronauts were given a "go" for the moonwalk. They ate their first lunch on the Moon but bypassed a rest period because they were anxious to proceed with the EVA. Airpack donning and equipment checkout took longer than anticipated. After a bit of trouble bleeding off the remnants of their cabin air, the hatch at the front of the LM opened and the EVA began.

Armstrong, being the mission commander, was the first out of the LM. He pulled a lanyard, which opened a compartment on the descent stage, containing a television camera. Armstrong then climbed down the ladder attached to the front leg of the LM and stepped from the foot pad. It was 10:56 P.M. "That's one small step for a man, one giant leap for mankind," he said. Cautiously, Armstrong began moving around on the surface. He took photographs and collected a small sample of soil and rocks, which he placed in a pocket on the leg of his spacesuit.

Eighteen minutes later, Edwin E. ("Buzz") Aldrin, Jr., joined him on the surface. "Beautiful view. Magnificent desolation." The two astronauts walked, hopped, and loped around the landing site collecting 21 kilograms of bulk and documented soil samples. They exposed a sheet of aluminum to the solar wind and deployed a scientific package consisting of a passive seismometer and the Laser Ranging Retroreflector, an array of mirrored cubes that would bounce a laser beam from Earth back to the ground station.

After two and a half hours on the surface, they climbed back aboard *Eagle*. At 2:54 P.M. on July 21, the ascent stage's engine ignited and carried them back to *Columbia*. They entered the command module with their cargo of moon rocks, sealed the connecting hatch, and jettisoned *Eagle*. At fifty-five minutes past midnight on July 22, the Service Propulsion System engine ignited and Apollo 11 headed home. On July 24, the CM separated from the SM and began its fiery reentry. Later, the three main parachutes opened, slowing the CM for a soft water landing. At 12:50 P.M., the goal set by President Kennedy eight years earlier was met.

Impact of Event

The first manned lunar landing was perhaps modern man's greatest accomplishment. There would be five more landings and ten other astronauts would stay longer, travel farther, deploy more experiments, and collect more samples. Someone had to be first, and it was this distinction that separated the flight of Apollo 11 from the rest. Two humans traveled where no human had gone before, putting their lives in the hands of the technology that had carried them so far.

The next flight, Apollo 12, would show that its predecessor had not been a fluke. Its lunar module made a pinpoint landing, within 2 kilometers of its target site. Apollo 13 almost proved to be as unlucky as its number might indicate: An oxygen tank exploded in the service module on the way to the Moon. A little luck and a lot of skill on the part of the crew and ground controllers brought the flight to a safe conclusion.

An investigation into the Apollo 13 accident delayed the next flight for nearly a year. Apollo 14 took two astronauts, including Alan B. Shepard, to a region of the Moon called Fra Mauro. Shepard, America's first astronaut in space, became the first (and, so far, only) lunar golfer when he attached a specially made club head to the end of a sample return container handle and swung at a genuine golf ball.

The final three lunar landing missions were designed to provide a maximum amount of scientific investigation. They utilized the first lunar "dune buggies," battery-powered roving vehicles that could extend the distance traveled from the LM to 10 kilometers.

After Apollo 17, NASA swung into the Skylab program, a project designed to place a laboratory into Earth's orbit, where it could be manned for periods up to three months. Skylab was not a true space station, although many have called it one. It could not be manned permanently and could not be resupplied with its most important consumable: air.

On July 24, 1975, six years after Apollo 11 splashed down, America's last flight of an expendable manned spacecraft concluded. The Apollo-Soyuz Test Project was a joint Soviet-American venture. The space race had given way to détente.

The United States would not journey into space for almost six years. A new era in space exploration was beginning: the age of reusable spacecraft. The Space Transportation System would make space accessible to ordinary people, even a teacher. Satellites would be carried into orbit and gently deployed. If they were disabled, astronauts could go out and fix them or bring them back to the ground for repairs. In the twenty years since Apollo 11, manned spaceflight had become almost routine, at least to those not involved with the missions. Space travel, however, would reveal that death was not partial to any one type of traveler, even a teacher.

Bibliography

Aldrin, Edwin E., "Buzz," Jr., and Malcolm McConnell. *Men from Earth*. New York: Bantam Books, 1989. As one of the first men on the Moon, Aldrin tells the story of America's journey to the lunar surface from a perspective unavailable to most. His description of the Apollo 11 landing reads like a James Michener novel. A briefly annotated but extensive bibliography as well as photographs are included.

Aldrin, Edwin E. "Buzz," Jr., with Wayne Warga. *Return to Earth*. New York: Random House, 1973. For years, the Apollo 11 crew trained to become the crew to make the first landing on the Moon. Unfortunately, NASA forgot to teach them how to be ordinary human beings after they returned to Earth. Aldrin chronicles his career from his glory days as an astronaut to his post-Apollo battle with alcohol and depression.

Brooks, Courtney G., James M. Grimwood, and Loyd S. Swenson, Jr. *Chariots for Apollo: A History of Manned Lunar Spacecraft*. NASA SP-4205. Washington, D.C.: Scientific and Technical Information Branch, 1979. One of the titles in The NASA History Series, this is the "official" record of America's voyage to the Moon. There are many black-and-white photographs of the flights leading to the Apollo 11 mission. Detailed source notes and an extensively annotated bibliography provide a wealth of information for the researcher.

Collins, Michael. *Carrying the Fire: An Astronaut's Journeys*. New York: Farrar, Straus & Giroux, 1974. Collins writes an intriguing look at life as a test pilot, taking the reader into the mysterious world of space travel and behind the scenes at the Manned Spacecraft Center in Houston. Captures the spirit of adventure that intrigues nonastronauts.

_____. *Liftoff: The Story of America's Adventure in Space*. New York: Grove Press, 1988. An evenly written tale of accomplishments and failures. Takes a personal look at space travel from a perspective inaccessible to the majority of readers.

Columbia Broadcasting System. *10:56:20 P.M. EDT, 7/20/69: The Historic Conquest of the Moon as Reported to the American People by CBS News over the CBS Television Network*. New York: Columbia Broadcasting System, 1970. An interesting look at the televising of the most significant event in the history of exploration. It is filled with glimpses into the personal lives of the participants. This moment-by-moment chronicle gives a true feeling of what the world thought about humanity's greatest adventure.

Compton, William David. *Where No Man Has Gone Before: A History of Apollo Lunar Exploration Missions*. NASA SP-4214. Washington, D.C.: Government Printing Office, 1989. This title in The NASA History Series tells of the astronauts chosen to participate in humanity's greatest adventure and introduces the people behind the scenes who made it all possible. The work has many black-and-white photographs and drawings and a completely annotated source listing.

MacKinnon, Douglas, and Joseph Baldanza. *Footprints: The Twelve Men Who Walked on the Moon Reflect on Their Flights, Their Lives, and the Future*. Washington, D.C.: Acropolis Books, 1989. One of the best books on the Apollo 11 mission. The authors interviewed the twelve men who had walked on the Moon. The questions ranged from the technical to the sublime, but each interview revealed more about the inner feelings of the astronauts than any prior essay.

U.S. Manned Spacecraft Center, Houston, Tex. Mission Evaluation Team. *Apollo 11 Mission Report*. NASA SP-238. Springfield, Va.: National Technical Information Service, 1971. This official report on Apollo 11 contains a wealth of data on the flight. There is a mission summary and sections that detail the flight, spacecraft, crew activities, orbital and surface experiments, in-flight demonstrations, biomedical evaluations, and mission ground support. A definitive study of the Apollo 11 flight.

Russell R. Tobias

Cross-References

Sputnik 1, The First Artificial Satellite, Is Launched (1957), p. 1545; The United States Launches Its First Orbiting Satellite, Explorer 1 (1958), p. 1583; Luna 2 Becomes the First Man-Made Object to Impact on the Moon (1959), p. 1614; Gagarin Becomes the First Human to Orbit Earth (1961), p. 1693; Shepard Is the First United States Astronaut in Space (1961), p. 1698; Glenn Is the First American to Orbit Earth (1962), p. 1723; The First Space Walk Is Conducted from Voskhod 2 (1965), p. 1787; The Lunar Orbiter 1 Sends Back Photographs of the Moon's Surface (1966), p. 1825; Apollo 12 Retrieves Surveyor 3 Parts from the Lunar Surface (1969), p. 1913.

APOLLO 12 RETRIEVES SURVEYOR 3 PARTS FROM THE LUNAR SURFACE

Category of event: Space and aviation
Time: November 19-20, 1969
Locale: Ocean of Storms, the Moon

A lunar landing made with pinpoint accuracy paved the way for future scientific exploration of the Moon

Principal personages:
CHARLES ("PETE") CONRAD (1930-), a Navy captain who was the Apollo 12 commander
RICHARD F. GORDON (1920-), a Navy captain and Apollo 12 command module pilot
ALAN L. BEAN (1932-), a Navy captain and Apollo 12 lunar module pilot

Summary of Event

Four months after the triumph of Apollo 11, the mission planners at the National Aeronautics and Space Administration (NASA) were about to show that an astronaut crew not only could repeat the lunar landing mission but also could do so with controlled precision. The flight to the Ocean of Storms would begin in a storm on Earth.

The successful completion of the Apollo 11 mission was a relief to all who had worked so hard to accomplish the task. It was even more so for the Apollo 12 team, who had their launch pushed back to November, giving them two additional months to prepare. Some of that preparation time would be extremely useful, since the mission planners had selected the Surveyor 3 site as their target.

Surveyor 3 landed on April 20, 1967, and transmitted 6,326 pictures and large amounts of scientific data for two weeks. Even though the scientists on the Apollo Site Selection Committee had rejected unanimously its location as unsuitable, Surveyor offered a unique opportunity. Future lunar landings needed to be made at specific locations, and a landing by Apollo 12's Lunar Module (LM) within 1 kilometer of a target would demonstrate this capability. In addition, by retrieving certain removable components from Surveyor, the effects of their exposure to two years on the Moon's surface could be studied.

Apollo 12 would set up the first operational Apollo Lunar Surface Experiments Package (ALSEP), make extensive geological observations and sample collection, and conduct photographic observations from lunar orbit.

Launch day—November 14, 1969—dawned gray, cold, and wet. Weather radar showed rain showers 130 kilometers north of the Cape Kennedy launch complex and heading south. Forecasters, however, believed that the area would be dry at the scheduled lift-off time. An hour and twenty-two minutes before launch, a pump replenishing liquid oxygen in the Saturn 5 launch vehicle's tanks failed. Contrary to the strictest launch rules, the decision was made to fly with only the backup pump operating. The astronaut crew had trained extensively for the specific conditions at the planned landing site and if the day's launch window was missed, it would be another month before it reopened.

At seven-tenths of a second past 11:22 A.M., eastern standard time, Apollo 12 lifted off into the thick overcast. Just before the vehicle disappeared, two streaks of lightning flashed toward Launch Complex 39. At the same time, Mission Commander Charles ("Pete") Conrad in the Command Module *Yankee Clipper* radioed that they had lost momentarily the inertial guidance system. Apparently, as the vehicle passed through the electrically charged clouds, the lightning discharged down the ionized (and therefore, electrically conductive) exhaust plume. This overloaded the fuel cells, which provide power to the spacecraft, and caused them to shut down automatically. The crew reset the circuit breakers and the flight continued normally.

Once safely in Earth orbit, the crew began the task of ensuring that all systems were operating properly. The inertial guidance platform was realigned and the Saturn's third stage placed the spacecraft on its way to the Moon. Several minutes later, *Yankee Clipper* extracted the Lunar Module *Intrepid* from the stage and the two proceeded to their destination. A single midcourse maneuver was performed to place the craft on a fuel-saving hybrid trajectory, one which would not bring them automatically back to Earth.

The journey to the Moon was uneventful, as was lunar orbit insertion. While in orbit, the astronauts performed photographic tasks, including taking pictures of Apollo 13's target landing site in the Fra Mauro region. On the fourth day, Conrad and Bean powered up the LM and began their descent to the Ocean of Storms.

At 1:54:36 A.M. EST on November 20, Conrad set *Intrepid* down on the surface. As the dust settled, Conrad and Alan L. Bean began to describe the landing site, but they could not see Surveyor. Had there been windows in the back of the LM, they would have spotted the dormant spacecraft a mere 163 meters behind them.

Five and a half hours after landing, Conrad made his way out of *Intrepid*'s front hatch and down the ladder to the surface. On his way down, he pulled a lanyard, releasing the door on a storage bay in the lower half of the LM. A color television camera in the bay transmitted his descent to the surface. "Man, that [step] may have been a small one for Neil, but that's a long one for me," Conrad exclaimed as he stepped from the lowest rung of the ladder to the foot pad. Conrad, one of the shorter astronauts, was 15 centimeters shorter than Armstrong. The lowest ladder rung was about 75 centimeters above the foot pad. As he looked around, Conrad spotted the Surveyor spacecraft behind the LM.

In the event an emergency lift-off from the Moon became necessary, a small contingency sample was collected by Conrad while Bean prepared his descent to the surface. Soon, they were both on the Moon, preparing to deploy the ALSEP.

The two spent nearly four hours on the lunar surface during the first walk, collecting 16.7 kilograms of soil and rock samples, taking photographs, and setting up their experiments' package. The next day, armed with new plans formulated by geologists in Houston for collecting samples, the pair began their second EVA. During the next four hours, they covered a little more than a kilometer and collected 17.6 kilograms of samples. Half way through the second EVA, the astronauts approached the Surveyor III spacecraft and began photographing it in its untouched state. Later, television pictures taken of the same area by Surveyor's camera immediately after it landed could be compared with the photographs. With near reverence for the only human artifact ever encountered on the Moon, Conrad and Bean examined it and noted that much of its originally white surface had turned brown. They attributed the change to a layer of dust on the exposed surfaces that was easily brushed away to reveal the original white paint beneath. They removed Surveyor's television camera and cut off electrical cabling and structural tubing for study by scientists back in Houston. Then they removed the trenching scoop, collected a few more samples, and headed back toward the LM.

At the close of the second moonwalk, Conrad and Bean climbed back into *Intrepid* and prepared for the journey home. They were covered in a layer of moon dust and despite their efforts to brush it off, it stuck with them. Before rejoining Richard F. Gordon in *Yankee Clipper*, they attempted to vacuum the dust, with little success. They brought a considerable amount with them into the Command Module and it remained in the spacecraft until they returned to Earth.

After undocking *Intrepid* and sending it to crash into the Moon to help calibrate the Passive Seismometer they had left at the landing site, the crew fired *Yankee Clipper*'s engine to bring them back home. Early on the morning of November 24, Apollo 12 splashed down some 600 kilometers east of Pago Pago.

Impact of Event

The flight of Apollo 12 is overshadowed generally by its predecessor Apollo 11 and by its successor Apollo 13. Apollo 11 fulfilled the goal set by President Kennedy to land a man on the Moon by the end of the decade and return him safely to Earth. From the beginning, Apollo 12 was planned as a backup to Apollo 11. If something were to go wrong, the Apollo 12 crew still could achieve the goal by the deadline. Once Apollo 11 was successfully completed, the crew had to shift their attention to an even more difficult goal, that of an accurate landing on the Moon. They were to land within a kilometer of a targeted site. This was equivalent to firing a rifle to hit a dime on a moving train 32 kilometers away.

Subsequent lunar missions would use the knowledge gained by Apollo 12 both in the targeting of landings and in the science that would be accomplished. None would require the accuracy of Apollo 12's landing feat, but the techniques used by ground controllers and by the astronauts would permit future crews to look for the most desirable site for their exploration. Geologists in Houston had always pushed for the inclusion of a geologist in a lunar landing crew. They believed that a geologist was best suited for finding the most significant samples to bring back. They also believed that a geologist could best describe the location of the sample before it was retrieved, thereby making it easier to understand the geological processes it took to place the sample where it had been.

Conrad and his fellow test-pilot astronauts argued that it took all of the piloting skills he had to accomplish the landing and that pilots could be trained to be relatively good field geologists. In addition, they argued that even a trained field geologist would have trouble doing much field geology on the Moon. With its smaller diameter, and therefore closer horizon, in combination with the unusual lighting conditions, the visual clues used on Earth did not work on the surface of the Moon. Once the pilots had a handle on landing techniques, a geologist could be brought along as the Lunar Module Pilot, since the Commander actually lands the LM. The geologists would have to wait for Apollo 17 through 20 to practice their trade up close. Nevertheless, public interest in lunar landings waned after Apollo 13 and Congressional cutbacks forced NASA to cancel the last three planned missions. A geologist, Harrison Schmitt, did make it to the Moon. He was Lunar Module Pilot on the Apollo 17 mission, the last flight of Apollo.

In the meantime, the seven pilot-astronauts who explored the lunar surface after Apollo 12 proved that they were capable of being trained to find important geological examples. The morphology and stratigraphy of the vicinity could be understood to the point that they would know where to find these samples. They spent more time collecting and documenting the samples than doing on the spot geology. In doing so, they were able to bring back numerous examples for the geologists to study.

The crew of Apollo 12 could not claim to be the first or the last to explore the Moon. They did not walk the farthest nor drive a lunar dune buggy. They did not spend the most time on the surface nor did they bring back the most rocks. They do, however, still hold the record for landing the closest to their target.

Bibliography

Bilstein, Roger E. *Stages to Saturn: A Technological History of the Apollo/Saturn Launch Vehicles.* NASA SP-4206. Washington, D.C.: Government Printing Office, 1980. This book chronicles the Apollo program literally from the ground up. Details of each Saturn flight are presented in chronological order. Fully illustrated and sporting an impressive forty-page bibliography, the book provides details on Apollo available nowhere else.

Brooks, Courtney G., James M. Grimwood, and Loyd S. Swenson, Jr. *Chariots for Apollo: A History of Manned Lunar Spacecraft.* NASA SP-4205. Washington, D.C.: Scientific and Technical Information Branch, 1979. The "official" record of America's voyage to the Moon. Contains many black-and-white photographs of the flights leading to the Apollo 11 mission. Detailed source notes and an extensively annotated bibliography provide much information for the researcher.

Compton, William David. *Where No Man Has Gone Before: A History of Apollo Lunar Exploration Missions.* NASA SP-4214. Washington, D.C.: Government Printing Office, 1989. Tells of the astronauts chosen to participate in humankind's greatest adventure and it introduces the people behind the scenes who made it all possible. Contains many black-and-white photographs and drawings and a completely annotated source listing, which is almost as valuable as the manuscript.

MacKinnon, Douglas, and Jospeh Baldanza. *Footprints: The Twelve Men Who Walked on the Moon Reflect on Their Flights, Their Lives, and the Future.* Washington, D.C.: Acropolis Books, 1989. One of the best books on the Apollo 11 mission. The authors interviewed the twelve men who had walked on the Moon. Each interview revealed more about the inner feelings of the astronauts than any prior publication.

National Aeronautics and Space Administration. *Analysis of Surveyor 3: Material and Photographs Returned by Apollo 12.* Washington, D.C.: National Technical Information Service, 1972. This book provides the researcher with information about what exposure to the lunar "atmosphere" does to man-made equipment. This information was useful in designing equipment for replacement on the lunar surface by future crews. Copies of the black-and-white photographs taken by astronauts Conrad and Bean are included.

U.S. Manned Spacecraft Center, Houston, Tex. *Apollo 12 Preliminary Science Report.* NASA SP-235. Springfield, Va.: Scientific and Technical Information, 1970. Topics covered include: landing site selection criteria; mission description; and a summary of scientific, photographic, and geologic results. Contains many photographs, line drawings, and data tables. Each section has its own comprehensive reference list. Provides many facts and figures for the researcher.

Russell R. Tobias

Cross-References

Goddard Launches the First Liquid Fuel Propelled Rocket (1926), p. 810; Sputnik 1, the First Artificial Satellite, Is Launched (1957), p. 1545; The United States Launches Its First Orbiting Satellite, Explorer 1 (1958), p. 1583; Luna 2 Becomes the First Man-Made Object to Impact on the Moon (1959), p. 1614; Gagarin Becomes the First Human to Orbit Earth (1961), p. 1693; Shepard Is the First United States Astronaut in Space (1961), p. 1698; Glenn is the First American to Orbit Earth (1962), p. 1723; The Lunar Orbiter 1 Sends Back Photographs of the Moon's Surface (1966), p. 1825; The First Humans Land on the Moon (1969), p. 1907.

BARTON AND HASSEL SHARE THE NOBEL PRIZE FOR DETERMINING THE THREE-DIMENSIONAL SHAPES OF ORGANIC COMPOUNDS

Category of event: Chemistry
Time: December 10, 1969
Locale: Stockholm, Sweden

Barton and Hassel studied the exact shape of molecules made from carbon atoms, facilitating greater understanding and predictions concerning these ubiquitous natural materials

Principal personages:

SIR DEREK H. R. BARTON (1918-), an English organic chemist noted for his insightful views on the exact shape of complicated naturally occurring compounds; the cowinner of the 1969 Nobel Prize in Chemistry

ODD HASSEL (1897-1981), a Norwegian physical chemist whose detailed studies of simple carbon-ring molecules laid the foundation for understanding molecules; cowinner of the 1969 Nobel Prize in Chemistry

WILLIAM HENRY PERKIN (1860-1929), an English organic chemist who attempted the earliest syntheses of cyclic compounds containing less than five-carbon atoms

ADOLF VON BAEYER (1835-1917), a German organic chemist who was one of the giants of nineteenth century organic chemistry; winner of the 1905 Nobel Prize in Chemistry

Summary of Event

Of all the varied arrangements of carbon atoms in the world of naturally occurring molecules, the ring has provided the greatest challenge and fascination for chemists. Organic chemists (whose chief concern is the way in which atoms are assembled in molecules and the reactions by which those molecules are formed and broken apart) and physical chemists attempted to understand the most basic and quantitative level of how molecules arrange and interact with one another.

One of the most famous German chemists of the late nineteenth century, Adolf von Baeyer, thought deeply about the three geometry of such systems. He wondered why one finds five- and six-carbon rings in great abundance in nature and never any other size. Baeyer devised a clever explanation.

It was well established by 1885 that the carbon atoms in the most commonly found ring systems formed bonds having angles of about 109.5 degrees. If a three-carbon ring were to form, it would have to be a triangle and since Euclid's time it had been known that such a geometric figure lay in a single plane and had interior angles of 60 degrees.

It seemed clear to Baeyer that such bonds would place a great strain on the carbon atoms forming them and would contribute greatly to the instability of the resulting molecule. By a similar argument, a four-carbon ring with bond angles of 90 degrees would also be highly strained and reactive. In contrast, five- and six-membered rings would have interior angles of 108 and 120 degrees, respectively, angles in good agreement with normal expectations for carbon-to-carbon bonds. At the same time, if the number of carbon atoms were to grow larger, so would the angles, straining the bonds, just as the smaller angles did.

Baeyer's theory seemed to be a neat and reasonable explanation of then-current observations. In any experimental science, however, facts continue to be discovered and theories either must be able to include them in their scope or they must give way to new and more general explanations. Theories are meant to be guides for understanding and making predictions.

Early work in Baeyer's laboratory seemed to support his speculations. William Henry Perkin undertook the synthesis of small ring compounds. (Synthesis for an organic chemist means making more complicated molecules from simpler ones. It is necessary that the exact structure of the starting material be known and that the chemical conversion be well understood.) Using a chain of carbon atoms, Perkin linked the first and last to form a ring, or cycle. He found that small rings could be prepared, but that they were markedly less stable than the common five- and six-membered examples. It was to be nearly a half century before the fundamental error in Baeyer's theory would be widely appreciated.

Only five years after its publication, however, the most basic assumption of Baeyer's theory was being questioned. As early as 1890, Ulrich Sachse noted that there was no fundamental reason why the six-carbon and larger rings must lie in a single plane and that the strain would be eliminated if a slight puckering was introduced. By building models of the six-membered cyclohexane ring, Sachse showed that two such forms should be strain-free: one shaped like a lawnchair and a second, which he called a bed or bathtub form. The state of theory and technique in organic chemistry at that time, however, doomed attempts to obtain any convincing evidence of the existence of the two forms, and Sachse's idea was left for nearly thirty years.

In 1918, Ernst Mohr added the crucial thought that if two six-membered rings were linked together by sharing two common carbon atoms, two distinct forms could be expected. Most important, these could be interconverted only at a great cost of energy. By contrast, he suggested, the two forms of the simple six-carbon ring might be interconverted easily by the thermal energy available at room temperature. Then, in 1925, Walter Hückel succeeded in showing that two forms of the fused ring system did, in fact, exist.

These early and penetrating thoughts and experiments led directly to the revolutionary work of Sir Derek H. R. Barton and Odd Hassel. They were awarded the 1969 Nobel Prize in Chemistry for demonstrating the unity and utility of these ideas. Hassel, a professor at the University of Oslo, studied the six-carbon ring directly by introducing one or more heavy atoms such as chlorine in place of hydrogen. In this way, Hassel was able to show, using X-ray and electron measurements, that the two forms suggested by Sachse actually existed. More important, Hassel showed that one form constitutes about 99 percent of the mixture because of the enhanced stability resulting from its greater energy efficiency. The two forms are interconverted about 1 million times a second at room temperature.

These important results were being obtained in the midst of World War II, but because Hassel refused to publish his work in German journals, the Western scientific community was deprived of his most significant ideas and experimental findings for many years. His precise and fundamental studies of the simplest six-carbon rings were noteworthy.

Barton carried out his early work at the Imperial College (University of London) and at Harvard University. He studied Hassel's work as it became available, and has acknowledged his debt to his colleague. Barton was honored for applying the detailed understanding of the six-membered ring to more complicated systems and their reactions.

It is especially important to appreciate the relationship between Hassel and Barton's work. They did not work together in the usual sense; in fact, the war prevented all communication. In his important 1950 publication concerning the conformations of steroids, Barton acknowledges Hassel and cites many of Hassel's articles. The example of how science is built with each advance dependent upon earlier work could not be clearer.

The term conformation was not universally accepted and much less widely understood in 1950. Barton took pains in several of his earliest articles to define the word carefully and clearly. Chemists had accepted, since the classical work of Louis Pasteur, that, under appropriate circumstances, the same set of atoms could form more than one compound that differed only in the exact arrangement of those atoms in space. Now, they were being asked to accept that even a single molecular species could assume various spatial arrangements that differed significantly in energy.

Impact of Event

It is not an overstatement to maintain that Hassel and Barton caused a revolution in the way organic chemistry is understood and practiced in the last half of the twentieth century. The results of their studies have, in the words of one eminent chemist, "lifted out of the chemical literature the flat formulas of the 1930's and 1940's and [sic] replaced them with the three-dimensional structures by which we view molecules."

Except for numerical details, Hassel's concept of the dynamic structure of six-carbon rings explains very well the chemistry of such systems. Furthermore, it laid the foundation for theoretical and experimental work carried out since 1943, that cyclohexane is the most studied and probably the best understood structure in modern organic chemistry. For example, the usual textbook representations fail to make clear that the pairs of hydrogen atoms connected to each of the six-carbon atoms fall into two distinct sets. Hassel showed convincingly that half occupy positions roughly perpendicular to the average plane of the carbon atoms, while the others deviate only slightly from that plane. Accepting this does not make its usefulness clear necessarily, but it was soon discovered that great energy differences are observed also when the attached atom(s) are something other than hydrogen. In this way, it is now possible to fix effectively a given ring in a particular spatial arrangement and study its chemistry in fine detail.

This breadth and depth of understanding is of fundamental importance because the six-carbon ring forms the basis of an almost incomprehensible array of compounds that occur in nature. Perhaps the most widely known is cholesterol. Cholesterol is known to make cell walls stronger and to increase their utility in keeping the materials and processes of life functioning in an orderly fashion. Cholesterol also provides the basic structure from which it is possible for the human body to synthesize the huge variety of steroids that serve as carriers of information in our chemical systems. Furthermore, this central molecule represents the starting material for the preparation of natural and human-made drugs so vital in the prevention and treatment of disease.

At the same time, it requires only the short step of changing an oxygen atom for a carbon atom to include an equally incalculable number of sugars and other carbohydrates in this picture of three-dimensional six-atom rings. In fact, this type of speculation was introduced by Jacob Böeseken, while Mohr and Hückel were making their contributions.

This new view of molecular structure, given extensive experimental validation by Hassel and developed so enthusiastically by Barton, has shown itself capable of explaining a large number of previously inexplicable experimental observations. As with any useful theory, conformational analysis is now called upon to predict the outcome of new experiments. The success of these leaps into the future has made this approach one of the most powerful tools in the possession of modern chemists.

Bibliography

Andersen, Per, Otto Bastiansen, and Sven Furberg, eds. *Selected Topics in Structure Chemistry: A Collection of Papers Dedicated to Professor Odd Hassel on His 70th Birthday, 17 May 1967.* Oslo: Universitetsforlaget, 1967. This book contains a short sketch of Hassel and a study of his career titled "Forty-five Years of Achievement." Both pieces are written by people who knew Hassel well. A good photograph and an extensive bibliography of Hassel's publications are included.
Barton, Sir Derek H. R. "The Principles of Conformational Analysis." In *Les Prix Nobel en 1969.* Stockholm: Imprimerie Royal, 1970. This article is very technical, but also contains carefully presented definitions and historical background. Accompanied by excellent references, photograph and career summary, and a most important introductory speech. Hassel's acceptance speech is included also, but is

very difficult for the nonspecialist.

Eliel, Ernest L. "Nobel Laureates in Economics, Chemistry, and Physics." *Science* 166 (November 7, 1969): 718-720. This article, written by a chemist who works in the same general field, goes well beyond the usual congratulatory message between colleagues. It offers genuine and readable insight on very difficult concepts, coupled with excellent personal appreciations.

Finley, K. Thomas. "The Synthesis of Carbocyclic Compounds: A Historical Survey." *Journal of Chemical Education* 42 (October, 1965): 536-540. This article places emphasis on the ring strain theory of Baeyer and the work and personality of the scientists who brought it to its modern form. Offers good background for understanding conformations.

Wasson, Tyler, ed. *Nobel Prize Winners.* New York: H. W. Wilson, 1987. This book offers biographical sketches of Nobel Prize winners, placing emphasis on their work. Includes excellent photographs, but only limited references.

K. Thomas Finley
Patricia J. Siegel

Cross-References

Ochoa Creates Synthetic RNA (1950's), p. 1363; Watson and Crick Develop the Double-Helix Model for DNA (1951), p. 1406; Cohen and Boyer Develop Recombinant DNA Technology (1973), p. 1987.

THE FLOPPY DISK IS INTRODUCED FOR STORING DATA USED BY COMPUTERS

Category of event: Applied science
Time: 1970
Locale: San Jose, California

Engineers at IBM Corporation's San Jose laboratory developed the disk drive, an inexpensive means of storing data in computer applications

Principal personages:
ANDREW D. BOOTH (1918-   ), an English inventor who developed paper disks as a storage medium during World War II
REYNOLD B. JOHNSON, a design engineer at IBM's research facility who oversaw development of magnetic disk storage devices
ALAN SHUGART, an engineer at IBM's research laboratory who first developed the floppy disk as a means of mass storage for mainframe computers

Summary of Event

When the IBM Corporation decided to concentrate on the development of computers for business use in the 1950's, it faced a problem that had troubled the earliest computer designers: how to store data reliably and inexpensively both during immediate processing and for later use. In the early days of electro-mechanical computation (the early 1940's), the English inventor Andrew D. Booth produced spinning paper disks on which he stored data by means of punched holes, only to abandon the idea because of the insurmountable engineering problems he foresaw. To work reliably, such disks would have to rotate quickly and in a stable fashion; in order to record enough data to be efficient and therefore economically feasible, the read/write head—the device that would encode and decode data on the spinning disk—would have to hover very close to the disc surface without actually touching it. The project was abandoned when these requirements were deemed impossible.

An alternative technology—punched cards and the punched card reader—soon became the industry standard for storing data. The origin of punched cards dates back to 1801, when the French inventor Joseph-Marie Jacquard invented an automatic weaving loom for which patterns were stored in pasteboard cards. Charles Babbage, an early computing pioneer, refined this storage medium for his "analytical engine." Then, as now, data were represented on a card by punched holes that corresponded to a letter or command. The virtue of such a device was its relative simplicity and reliability: At any given location on it, a card either had or did not have a punch. Two disadvantages—slow speeds (one might have to sift through a long sequence of cards to find what the computer needed) as well as low density of data storage—spurred the industry to look for other inexpensive, yet faster, means of retaining information for computer use.

When the Ampex Corporation demonstrated its first magnetic audio tape recorder after World War II, computer designers were quick to see the possibilities such a device could offer. Shortly after Ampex's introduction, in August of 1949, the BINAC (Binary Automatic Computer) was introduced with a storage device that consisted of what appeared to be a large tape recorder. BINAC's makers soon introduced a more advanced machine, the UNIVAC (Universal Automatic Computer), which used a strong metal tape instead of the plastic tape used by BINAC, which was subject to distortion by stretching or by breaking altogether. Unfortunately, the disadvantages of metal tape soon became apparent: It was considerably heavier, its edges were razor sharp and thus dangerous, and it created considerable friction and wear on the read/write head. Later advances offset these problems to some degree. The formulation of plastic tape eventually produced sturdy media, and IBM introduced a transport mechanism that used a vacuum to spool out lengths of tape not immediately passed by the read/write head in order to increase transport speed. As a consequence, magnetic tape became (and remains) a viable medium for storage computer use. Punched cards, however, have become a part of history.

Now that magnetism had demonstrated its feasibility as a means of storing electronic data, however, alternate ideas were proposed, including rapidly rotating drums whose construction allowed engineers to read data from them quickly. Whereas a tape might have to be fast forwarded nearly to its end to locate a specific piece of data, a drum rotating at high speed had a short period of rotation. To achieve maximum flexibility, drums rotated at speeds up to 12,500 revolutions per minute and could store more than 1 million bits (or approximately 125 kilobytes) of data. To be sure, the physical demands of such a device moving at that speed were great; even faster speeds (necessary if increased capacity and transfer rates were to be achieved) presented difficult practical engineering problems.

In 1952, Reynold B. Johnson was hired by IBM's San Jose laboratory to develop a means of transferring data from previously recorded magnetic media to punched cards. After reading a paper written by inventor Jacob Rabinow, who proposed a stack of independently rotating disks as a means of storing data, Johnson and his staff began development in earnest. By May, 1955, IBM announced the production of the hard disk unit. It was part of the RAMAC 350 whose very name, "Random Access Method of Accounting and Control," emphasized one very important feature of the hard disk device: its ability to access data randomly, not like a tape that moved either forward or backward to a given point. The hard disk unit consisted of fifty platters, each two feet in diameter, mounted permanently on a spindle rotating at 1,200 revolutions per minute; a read/write head was moved to the appropriate disk and then positioned on the one side of the disk that held data in order to read the information required. To speed things even more, the next version of the device, similar in design, employed one hundred read/write heads—one for each of its fifty double-sided disks. Computer designers and users now had what they had long awaited: flexible, mass data storage at a reasonable price (in terms of mainframe computers), even if the size of IBM's first commercial unit earned it the nickname "jukebox."

The development of the floppy disk drive was a logical evolutionary outgrowth of the introduction of the hard disk drive, although it was not until the late 1960's that the floppy took shape as a result of the efforts by Alan Shugart (it was announced by IBM as a ready product in 1970). First created as a means of restoring the operating system of mainframe computers that, perhaps because of a power outage, needed to be restarted, the floppy seemed in some ways to be a step back, for it operated more slowly than a hard disk drive and did not store as much data. Initially, it consisted of a single thin plastic disk of eight inches in diameter and was developed without the protective envelope in which it is now universally seen. The addition of that jacket gave the floppy its single greatest advantage over the hard disk: portability with reliability. Another advantage soon proved itself: The floppy is resilient to damage. In a hard-disk drive, the read/write heads must hover thousandths of a centimeter over the disk surface in order to attain maximum performance. Should even a small particle of dust get in the way or should the drive unit be bumped too hard, the head may "crash" into the surface of the disk and ruin its magnetic coating; the result is a permanent loss of data. Because the floppy operates with the read/write head in contact with the flexible plastic disk surface, individual particles of dust or other contaminants are not nearly as likely to cause a disaster.

As a result of its advantages, the floppy disk was the logical choice for mass storage in personal computers (PCs), which were developed a few years after the floppy disk's introduction. Because the conditions under which it operates are less stringent than those of the hard disk (which, to achieve the density of data storage of which it is capable, must be assembled in a "clean-room" environment), the floppy disk drive is relatively inexpensive to produce. Continuing advancements in floppy disk design, technology, and fabrication have permitted ever larger amounts of data storage in increasingly smaller formats (some disks have been as small as two inches in diameter). The floppy is an important storage device in an age when hard disk drives for PCs have become rather inexpensive and is attested by the fact that manufacturers continually are developing new floppy formats and improved storage densities.

Impact of Event

While still of use in the world of mainframe computer operation, the floppy disk has had its greatest impact in the realm of the personal computer. There is little doubt that personal computing would be very different today were it not for the availability of inexpensive floppy disk drives. While the hard disk had been in existence for some years before the first floppy was announced and had reached a high level of sophistication, hard disk drives were quite expensive to produce. As a result, they were initially built in a fixed format of multiple platters in order to keep costs (relative to data storage capacity) reasonably low. Later developments in hard-drive technology allowed for removable disk packs, still made of units containing multiple platters (usually more than ten), but these devices required large, washing machine-sized drives which, among other things, could create a partial vacuum in the actual disk chamber. It is clear that such technology was unsuitable for the office or home environment and that only large research organizations would have a great need for it to sustain the necessary expenditure. To be sure, alternative storage media have been used, especially in the home computer market, where cassette tape recorders have been employed. (In fact, when IBM introduced its PC in 1981, the machine provided as standard equipment a connection only for a cassette recorder as a storage device; a floppy disk was an option, though an option few did not take.) The awkwardness of such alternative devices—their slow operating speed and the sequential nature of the storage medium—presented clear obstacles to the acceptance of the personal computer as a basic information appliance. The floppy drive gives computer users relatively fast storage at low cost.

Floppy disks provide more than merely economical data storage. Since they are built to be removable (unlike hard drives), they represent a very basic means of transferring data between machines of like architecture. Indeed, prior to the popularization of local area networks (LAN) for the interchange of computer data, the floppy was known as a "sneaker" network: One merely carried the disk by foot to its destination computer. While most personal computers sold for serious use today include a hard drive of some capacity, floppy disks are the primary means of distributing new software to users. Packaged in a simple mailer, even the very flexible floppy has shown itself to be quite resilient to the wear and tear of postal delivery. More recently, the 3.5-inch disk improves upon the design of the original 8-inch and 5.25-inch floppies by protecting the disk medium within a hard plastic shell and by protecting the area where the read/write heads are positioned by means of a sliding metal door. Given this additional protection, greater data densities are possible and have been realized.

Bibliography

Horn, Carin E., and James L. Poirot. *Computer Literacy: Problem Solving with Computers.* 2d ed. Austin, Tex.: Swift Publications, 1985. Useful primarily for its extensive and well-executed glossary and bibliography. Otherwise, a useful general introduction to personal computers.

Kean, David W. *IBM San Jose: A Quarter Century of Innovation.* San Jose, Calif.: International Business Machines Corporation, 1977. Provides somewhat technical discussion of design and implementation of the floppy disk drive, which, inasmuch as it was developed after the hard-disk drive, shares the same basic operating principles. Well documented.

Logsdon, Tom. *Computers Today and Tomorrow: The Microcomputer Explosion.* Rockville, Md.: Computer Science Press, 1985. Along with a useful glossary and bibliography, explains the rudiments of computer hardware function, especially disk drives.

Sanders, Donald H. *Computers Today.* New York: McGraw-Hill, 1983. Textbook that is eminently accessible and thorough. Covers the basics (what they are and what they do) as well as functions of hardware, software; includes discussion of the impact—both virtues and vices—of computers in society.

Sobel, Robert. *IBM vs. Japan: The Struggle for the Future.* New York: Stein & Day, 1985. While concentrating on the similarities and differences in corporate styles in IBM and in Japanese organizations, this text provides a useful and revealing early history of the origins of the IBM personal computer series.

Time-Life Books. *Understanding Computers: Memory and Storage.* Alexandria, Va.: Time-Life Books, 1987. Provides a very clear discussion of how modern mass storage devices work (including disk drives, tape drives and optical media). Good illustrations. Contains a bibliography, as well as a glossary of terms related to storage and memory functions.

Joseph T. Malloy

Cross-References

UNIVAC I Becomes the First Commercial Electronic Computer and the First to Use Magnetic Tape (1951), p. 1396; Bubble Memory Devices Are Created for Use in Computers (1969), p. 1886; Apple II Becomes the First Successful Preassembled Personal Computer (1977), p. 2073; The IBM Personal Computer, Using DOS, Is Introduced (1981), p. 2169; IBM Introduces a Personal Computer with a Standard Hard Disk Drive (1983), p. 2240; Optical Disks for the Storage of Computer Data Are Introduced (1984), p. 2262.

LUNOKHOD 1 LANDS ON THE MOON

Category of event: Space and aviation
Time: November 10, 1970-October 1, 1971
Locale: Mare Imbrium, the Moon

The Soviet remote controlled lunar rover, Lunokhod 1, was the first self-propelled craft to explore another world

> *Principal personages:*
> G. N. BABAKIN (1914-1971), the chief designer of interplanetary spacecraft
> VALENTIN P. GLUSHKO (1906-   ), the chief designer of rocket engines
> A. M. ISAYEV (1908-1971), the chief designer of spacecraft propulsion systems
> SERGEI P. KOROLEV (1907-1966), the chief designer of the Soviet space program from 1957 to 1966
> VASILI P. MISHIN (1917-   ), the chief designer of the Soviet space program from 1966 to 1974
> MIKHAIL S. RYAZANSKIY (1909-1987), the chief designer of rocket guidance systems

Summary of Event

The Lunokhod 1 remote-controlled lunar exploration vehicle was carried to the Moon aboard the Luna 17 spacecraft. Luna 17 was launched from the Baikonur Cosmodrome, near Tyuratam, Kazakhstan, in the Soviet Union, by a Proton booster at thirteen hours, forty-four minutes, Greenwich mean time, on November 10, 1970. The 5,700-kilogram Luna 17 spacecraft was placed initially in a temporary 192-by-237-kilometer Earth parking orbit. Prior to completing the first orbit, the Proton's fourth stage injected the Luna 17/Lunokhod 1 spacecraft into a translunar trajectory.

Two midcourse corrections were performed on November 12 and 14, 1970, by Luna 17's rocket motor while en route to the Moon. Luna 17 entered an 84-kilometer circular retrograde orbit around the Moon on November 15, 1970. The next day, the spacecraft lowered its perilune (lowest point of an orbit around the moon) to 19 kilometers. The following day, Luna 17 fired its retro-rocket and began its descent to the lunar surface. A radar altimeter cut off the braking rocket at an altitude of 3,000 meters, then refired it again at 740 meters. The descent engine was cut off a final time at 20 meters altitude, and the descent was completed with the craft's vernier engines.

The spacecraft and its Lunokhod 1 cargo made a successful soft landing at three hours, forty-seven minutes, Greenwich mean time, on November 17, 1970. Luna 17 touched down on the inner slope of a shallow 150-meter-wide crater on Mare Imbrium at lunar coordinates 38 degrees north, 35 degrees south. The lander was banked and pitched 4 degrees as it sat on the surface.

After landing, two sets of ramps—one forward and one aft of the Lunokhod—were lowered to the surface. The first television photographs were returned at five hours, thirty-one minutes, Greenwich mean time, and showed the ramps were clear of obstacles. At six hours, twenty-eight minutes, Greenwich mean time, the rover was driven slowly down one of the ramps onto the lunar surface.

The word "Lunokhod" means "moon walker." The 1.35-meter high, 756-kilogram rover looked like a polished titanium alloy insect and consisted of a bathtub-shaped instrument compartment 2.15 meters long, mounted on eight electrically driven aluminum wire mesh wheels set in two rows 1.6 meters apart with a wheel base of 2.21 meters. Each of Lunokhod's eight 51-centimeter-diameter wheels had a two-speed electric drive motor and independent suspension. If any drive motor froze up in the lunar dust, a small explosive charge would sever the drive shaft and the wheel would freewheel. The rover was capable of moving even if two wheels on each side were disabled. Various items from the Lunokhod, including the drive motors, had been tested previously in lunar orbit by the Luna 12 and 14 spacecraft.

Lunokhod was capable of traveling on a 45-degree slope. If this angle was exceeded, the rover's brakes engaged automatically until ground control could steer the vehicle away from danger. The rover's magnesium alloy instrument compartment contained a special polonium 210 radioisotopic heat source to keep internal equipment from freezing in the -150 degree Celsius lunar night. During the day, excess heat was radiated away and internal temperature was sustained in the 15 to 20 degree Celsius range.

Lunokhod used two forward-looking television cameras, which returned a photograph every twenty seconds for navigation while moving on the lunar surface. These cameras had a field of view of 50 degrees and were capable of sending stereo views of closeup objects. A second photographic system consisted of four 80-by-205-millimeter, 1.3-kilogram telephotometers, which operated much like a scanning facsimile photograph transmitter. These units returned high resolution 500-by-6,000-pixel panoramas of the lunar surface.

Lunokhod used several different methods to analyze the lunar soil. The first involved photographing the rover's tracks to see how deep the wheels sank into the lunar dust. Another passive method used a trailing ninth wheel, which acted as an odometer. The revolutions of the odometer wheel were checked against the revolutions of the drive wheels to measure the slip between the wheels and the loose Moon dust. The bearing strength of the lunar surface was tested by a penetrometer consisting of a flat metal crusiform, which was stamped into the surface and its depth photographed by a television camera. Chemical analysis of the lunar soil was accomplished with the Röntgen Isotopic Fluorescence Method of Analysis (RIFMA) instrument. This apparatus focused a beam of electrons onto the surface to be analyzed, inducing it to emit X rays whose spectrum could be measured to show the concentrations of aluminum, silicon, magnesium, potassium, calcium, iron, and titanium.

An X-ray telescope operating in the 2 to 10 angstrom range and cosmic-ray detectors also were carried by the Lunokhod. Cosmic-ray data from the Moon were used in conjunction with data returned simultaneously from Soviet Venus and Mars spacecraft. A 3.7-kilogram French laser retroreflector consisting of ten separate 14-centimeter prisms was mounted above the forward television cameras. This instrument had a reflective capacity three times that of the Apollo laser reflectors. Lunokhod was parked during the lunar night in a position that aimed the laser reflectors at Earth. The French used the 102-centimeter telescope at Pic du Midi Observatory, while the Soviets used the 262-centimeter telescope at the Crimean Astrophysical Observatory to fire bursts of red ruby laser light at the Lunokhod for lunar ranging experiments.

Lunokhod was remotely operated by a five-person team located at the Soviet Deep Space Communications Center. The team consisted of a commander, driver, navigator, systems engineer, and radio operator.

Control of the Lunokhod was directed through the Soviet deep space tracking antenna at Yevpatoria in the Crimea. Because the Soviets did not have a worldwide tracking network like the National Aeronautics and Space Administration (NASA), the Lunokhod could be operated and navigated only while the Moon was visible from the Crimea. Typically, the Lunokhod was active only four hours per day. Navigating the Lunokhod on the Moon's surface by remote control was complicated by the slow scan rate of the forward-looking television cameras and the three-second round trip time for signals between Earth and the Moon. By the time controllers received the picture, Lunokhod already would be several meters ahead of that position.

After landing, Lunokhod traveled in a southeasterly direction, then looped back toward the initial starting area. After arriving back at the lander and inspecting it, Lunokhod traveled north. Travel distance during each lunar day varied from several hundred meters to 2 kilometers. Steep slopes, boulder fields, and craters of various sizes were encountered. At one point, angles were so steep that Lunokhod's brakes were applied while a complicated series of maneuvers were mapped to back the rover away from danger.

Roving operations on the Moon by Lunokhod had to be suspended for three days near each local lunar noon because high sun elevations erased shadows and made navigation by television difficult.

The Lunokhod was designed for a ninety-day lifetime, but was active on the Moon for eleven months. Although the RIFMA instrument failed after seven months, Lunokhod survived for 322 Earth days until September, 1971. Attempts to revive the rover after the lunar night on October 1, 1971, failed after the radioisotope heat source exhausted its fuel and the lander froze in the night.

Impact of Event

The Soviet Lunokhod 1 was the first roving unmanned exploration vehicle to travel on another world. In contrast to other unmanned spacecraft, which had returned data about one small area of the Moon, Lunokhod traveled over a wide area while taking measurements.

The Lunokhod rover was not a true autonomous robot but was entirely directed by remote control from Zvedniyi Gorodok, the Soviet manned space center, near Moscow. The chief designers were G. N. Babakin, Valentin P. Glushko, A. M. Isayev, Sergei P. Korolev, Vasili P. Mishin, and Mikhail S. Ryazanskiy. The rover was very similar in operation to modern radio-controlled model cars and provided the Soviets with experience in teleoperations (the science of remote exploration). The fact that control of the rover was carried out from the Soviet manned space center showed the link between the Lunokhod program and the Soviet manned lunar landing program. In the late 1960's, it was obvious to the Soviets that their manned lunar landing program was experiencing lengthy delays, while the American program was quickly approaching its goal of a manned lunar landing by the end of the decade.

During its eleven-month lifetime, Lunokhod 1 returned twenty thousand separate photographs of the lunar surface, including two hundred panoramas. The rover traveled 10.5 kilometers over the lunar terrain and examined 80,000 square meters of surface. Over that distance, five hundred penetrometer and twenty-five RIFMA soil analyses were performed. The RIFMA soil analysis experiment confirmed that the lunar soil was made primarily of volcanic basalt. On December 12, 1970, the RIFMA apparatus detected radiation from a solar flare, which the Soviets claimed would have been lethal to an astronaut on the Moon.

Sensors continuously reported Lunokhod's internal pressure and temperature, pitch and roll attitude, and temperatures of the solar cells, wheels, and motors. The readings showed that the solar cells can reach 121 degrees Celsius. The sun-side wheels reached 100 degrees Celsius, whereas the shadow-side wheels were only 19 degrees Celsius. Temperatures recorded by the Lunokhod during a lunar eclipse on February 10, 1971, quickly fell from 138 degrees Celsius to −100 degrees Celsius.

In addition to lunar science, the Lunokhod 1 mission provided the Soviets with valuable engineering data on lubricating moving parts in a vacuum. An unusual problem to overcome was the tendency of metallic parts to freeze together, a process called vacuum welding. The thermal stresses induced by the lunar day and night cycles were also studied for data needed to improve future space machinery. Initial fears that Lunokhod's wire mesh wheels would clog with lunar dust attracted by static electricity were proved to be unfounded. During operation, Moon dust fell off the wheels like dry sand.

Joint laser ranging experiments were also carried out in collaboration with the French. Precise measurement of the travel time of the laser light reflected from Lunokhod helped fix the Earth-Moon distance to within 30 centimeters. Lunokhod 1 also became the target of international politics and propaganda. During Lunokhod's first lunar night, the Soviets and the French both attempted laser shots at the rover's laser reflector. The Soviets were successful, while the French were not. Open speculation accused the Soviets of supplying the French the wrong lunar coordinates, which would assure the Soviets the honor of being first to locate the Lunokhod.

The Soviets argued that remote exploration of the Moon by Lunokhod-type rovers was much cheaper and safer than human exploration. The Lunokhod mission came only seven months after the failed Apollo 13 mission. The Soviets cited the Apollo 13 problems as part of the justification for unmanned exploration of the Moon.

Bibliography

Gatland, Kenneth. *Robot Explorers.* London: Blandford Press, 1972. A chronology of Soviet and American lunar and planetary space exploration programs. Contains numerous color illustrations providing insights into the design and functions of American and Soviet lunar and planetary spacecraft. Recommended for general audiences.

McDougall, Walter A. . . . *The Heavens and the Earth.* New York: Basic Books, 1985. A political history of the space age. Well researched. Describes and analyzes the decisions by the leaders of both the United States and the Soviet Union and their effects on the respective space programs. Suitable for general audiences.

Short, Nicholas M. *Planetary Geology.* Englewood Cliffs, N.J.: Prentice-Hall, 1975. Summarizes the accomplishments and scientific results of both American and Soviet lunar and planetary space programs. Stresses the chemical nature of the Moon and inner planets, their geological similarities and differences, and their origins. College-level reading, illustrated with many diagrams and photographs taken by the space missions discussed.

Smolders, Peter. *Soviets in Space.* New York: Taplinger, 1974. A well-illustrated narrative on all aspects of the Soviet space program. Suitable for wide audiences, it concentrates on the successful portions of the Soviet space program. Contains numerous diagrams and photographs.

Turnill, Reginald. *Spaceflight Directory.* London: Frederick Warne, 1978. A lavishly illustrated summary of spaceflight activities by all nations. Lists chronologies of major manned and unmanned space missions. Suitable for readers at high school and college level.

Wislon, Andrew. *Solar System Log.* London: Jane's, 1987. A compilation of all manned and unmanned lunar and planetary space flights up to mid-1985 by all space-faring nations. A well-illustrated chronology of the history, spacecraft, mission, and discoveries of all deep space exploration missions. Suitable for all readers.

Robert Reeves

Cross-References

Sputnik 1, the First Artificial Satellite, Is Launched (1957), p. 1545; Luna 2 Becomes the First Man-Made Object to Impact on the Moon (1959), p. 1614; Luna 3 Provides the First Views of the Far Side of the Moon (1959), p. 1619; The

Lunar Orbiter 1 Sends Back Photographs of the Moon's Surface (1966), p. 1825; The First Humans Land on the Moon (1969), p. 1907; Apollo 12 Retrieves Surveyor 3 Parts from the Lunar Surface (1969), p. 1913; Mars 2 Is the First Spacecraft to Impact on Mars (1971), p. 1950.

DIRECT TRANSOCEANIC DIALING BEGINS

Category of event: Applied science
Time: 1971
Locale: United States to Europe

The first use of direct distance dialing (DDD) began from the United States to Europe

Principal personage:
ALMON BROWN STROWGER, the inventor of the automatic dial telephone

Summary of Event

Direct distance dialing between the United States and Europe is a relatively recent development. Direct dialing for local calls that pass through a single exchange has been available since the 1920's, but overseas toll calls continued to be handled manually by overseas operators until 1963. Once direct distance dialing technology was available for overseas telephone calls, those operators used the dial system until it became available to the general public in 1971. In fact, all toll calls required operator assistance until the 1950's. By 1960, direct distance dialing was available to 54 percent of Bell Telephone customers.

Until 1930, placing an overseas call was a time-consuming undertaking, requiring that the caller place a request with the overseas operator, then hang up to wait for the call to be put through. Once a circuit was available and a connection had been established, the operator would call back, ensure that both parties were on the line, then hang up. Over the years, as the number of available circuits increased, the amount of time required to make a connection decreased, and most calls could be placed while the caller remained on the line. The Bell Company showcased this improved service in a popular exhibit at the 1939 World's Fair in New York City.

Some direct toll dialing was available prior to World War II, but deployment of the technology was interrupted by the war, and it was not until 1951 that direct distance toll dialing began in the continental United States.

Touch-tone dialing was introduced in 1963. By 1969, the entire telephone network of the United States was operating on automatic systems with only a few, mostly rural, exchanges still using manual operators for toll calling.

Overseas telephone service took years to develop. While the first undersea telegraph cable had been laid and put into service in 1858, the first transatlantic telephone cable was not put into service until 1956, nearly one hundred years later. Part of the reason was the effort by AT&T to develop radiotelephony as a cost effective alternative to cable for overseas calls. Eventually, radiotelephony was all but abandoned as it was proved to be unreliable. Transoceanic radio telephone conversations were sent via shortwave, which was susceptible to interference and noise resulting from constantly changing weather patterns. Also, shortwave had other disadvantages: Each conversation required two frequencies—one for the caller, the other for the receiver. This was a major problem because of the growing demand for access to the overseas telephone network. In order to provide enough circuit capacity, hundreds of shortwave frequencies would have to be used. This was a problem because, unlike cable, the electromagnetic spectrum over which shortwave is broadcast is finite, and demand for access to the shortwave band was increasing as more and more countries sought to establish their own shortwave services during the first half of the twentieth century.

One reason why direct overseas toll dialing took so long to develop was the limited availability of circuits. The first transatlantic submarine telephone cable had only sixty-four voice circuits, which limited simultaneous calls to that number. If direct distance dialing had been available in 1956, chances are that demand would have continually outstripped supply, and telephone customers would have quickly become frustrated by busy signals. Overseas operators were able to intercept that problem by taking requests, then calling back once the connection was made. In effect, overseas operators were hand holders, taking the bother out of standing in line for overseas circuit access. Soon after the first undersea telephone cable was placed in operation, another submarine cable with 128 circuits was deployed. During the 1960's, several more cables added circuit capacity. By 1971, the year customers began overseas direct distance toll dialing, another cable was put in place with forty-two hundred voice circuits adding to those already in service.

Another telephone delivery system that first made its appearance in 1962 added to circuit capacity. Communication satellites were placed in orbit with increasing frequency during the late 1960's and early 1970's. The first was the famed AT&T satellite Telstar. It received telephone calls from ground stations, amplified them, then relayed them back to ground stations on the opposite side of the ocean over the equivalent of six hundred voice circuits. Eventually, the number of satellite circuits numbered in the hundreds of thousands. As the number of circuits increased, cost per call decreased substantially, thus encouraging more growth in the volume of voice traffic. In the 1990's, communications satellites make it possible for callers from the continental United States to place calls to any location in the world, even those areas of the world with poor telecommunication infrastructure. In fact, some countries have established domestic telephone services using satellites as the primary means of distribution as the laying of cable would be too costly or geographically prohibitive.

Impact of Event

Direct distance dialing offers convenience to telephone customers everywhere. In 1901, the inventor of the automatic dial telephone, Almon Brown Strowger, sought to eliminate the telephone operator from the connection process for security reasons, but users soon discovered another advantage. Direct dialing was faster, especially during times of peak usage. Still, the number of long-distance circuits was fairly small in relation to the number of circuits running through the central exchange in even the smallest communities. Until 1970, the number of overseas circuits was even smaller. Once direct distance dialing became available, the convenience of making overseas calls led to more of them, which increased system usage and drove down the cost of making an overseas call.

In the years since the first direct toll calls were made, overseas voice traffic has increased exponentially. As telecommunications infrastructure has improved around the world, more and more calls are placed from Europe, Asia, Africa, and virtually every corner of the world. Developing technology also made it possible for the technical quality of circuitry to be improved significantly, which proved to be a critical factor when demand for digital transmission began to grow in the 1970's. As computers evolved, more and more data traffic was transmitted overseas, representing 15 percent of all overseas transmissions by 1980 and a significantly larger proportion in the 1990's. The volume of overseas telephone transmission of voice and data combined expanded enormously in the 1980's.

Most direct distance transoceanic dialing originated in the United States. More telephone calls are made within the New York City area than in all of England and France combined in any given year. Increasing globalization of industry and commerce has had a positive impact on the volume of overseas calling, and improvements in technology have expanded the capability of circuitry to transmit new forms of data that require high capacity and fast transmission speeds.

The appearance of direct distance dialing represented a major step forward in the evolution of transoceanic telecommunications technology. It made the world seem smaller—anywhere in the world, communication was a few taps of the touch-tone pad. Connect time was fast, transmission quality was high, and a new era of high-speed, high-quality worldwide telephone communication had begun.

Bibliography

Brooks, John. *Telephone: The First Hundred Years.* New York: Harper & Row, 1976. This excellent corporate history of the Bell system includes many anecdotes and colorful stories about the early years of telephony, giving life and context to what could otherwise be described as a highly technical description of the birth and development of one of the world's most remarkable companies. Includes a discussion of the evolution of transoceanic services offered by AT&T.

Danielian, N. R. *AT&T: The Story of Industrial Conquest.* New York: Vanguard Press, 1939. A good look at some of the personalities involved in the development of the world's largest telephone network. Provides a strong backdrop for gaining an understanding of how and why decisions were made regarding the adoption of technical innovations at AT&T through the years.

Millman, S., ed. *A History of Engineering and Science in the Bell System: Communications Sciences (1925-1980).* New York: The Laboratories, 1984. Discusses the various foundations of applied physics in communications technology, including television, radio, lightwave transmission, and digital communications. Index.

_____. *A History of Engineering and Science in the Bell System: Physical*

Sciences (1925-1980). New York: The Laboratories, 1983. This book contains a very good overview of the problems (and solutions) encountered in the construction of the first cables designed for transoceanic service. Discusses such arcane subjects as the theory of electronic switching and digital transmission. Index.

O'Neill, E. F., ed. *A History of Engineering and Science in the Bell System: Transmission Technology (1925-1975)*. Short Hills, N.J.: AT&T Bell Laboratories, 1985. This large volume contains a good discussion of the development of direct distance dialing (DDD) by the Bell System and its deployment throughout the United States and eventually in its overseas operation.

Ress, Etta Schneider. *Signals to Satellites in Today's World*. Mankato, Minn.: Creative Educational Society, 1965. This heavily illustrated book is a layperson's overview of the evolution of communication technology over the centuries up to the mid-1960's. It includes a good discussion of the history of transoceanic cable from the middle of the nineteenth century to the deployment of the first telephone cable in 1956. Illustrated.

Smits, F. M., ed. *A History of Engineering and Science in the Bell System: Electronics Technology (1925-1975)*. Indianapolis: The Laboratories, 1985. This book looks at developments in electronics that have been incorporated into the design of long-distance cable and related equipment, including the transistor, integrated circuits, and optical devices. Index.

Michael S. Ameigh

Cross-References

Strowger Invents the Automatic Dial Telephone (1891), p. 11; Marconi Receives the First Transatlantic Telegraphic Radio Transmission (1901), p. 128; The First Transcontinental Telephone Call Is Made (1915), p. 595; Transatlantic Radiotelephony Is First Demonstrated (1915), p. 615; The Principles of Shortwave Radio Communication Are Discovered (1919), p. 669; The First Transatlantic Telephone Cable Is Put Into Operation (1956), p. 1502; Telstar, the First Commercial Communications Satellite, Relays Live Transatlantic Television Pictures (1962), p. 1728; The First Commercial Test of Fiber-Optic Telecommunications Is Conducted, (1977), p. 2078.

THE MICROPROCESSOR "COMPUTER ON A CHIP" IS INTRODUCED

Category of event: Applied science
Time: 1971
Locale: Santa Clara, California

Developers at Intel Corporation produced the first microprocessor, combining the basic logic circuits of a computer onto a single chip

Principal personages:
ROBERT NORTON NOYCE (1927-1990), a semiconductor engineer who was coinventor of the integrated circuit and cofounder of Intel Corporation
GORDON E. MOORE (1929-   ), a cofounder of Intel Corporation
WILLIAM SHOCKLEY (1910-   ), a coinventor of the transistor, who was one of the recipients of the 1956 Nobel Prize in Physics
MARCIAN EDWARD HOFF, JR. (1937-   ), an Intel engineer who was generally regarded as the inventor of the world's first microprocessor chip
JACK ST. CLAIR KILBY (1923-   ), an assistant vice president of Texas Instruments who shared credit for the invention of the integrated circuit

Summary of Event

The microelectronics industry began shortly after World War II with the invention of the transistor. During the war, it was discovered while radar was being developed that certain crystalline substances, such as germanium and silicon, possessed unique electrical properties that made them excellent signal detectors. This class of materials became known as semiconductors, because they were neither conductors nor insulators. Immediately after the war, Bell Telephone Laboratories began to conduct research on semiconductors in the hope that they might yield some benefits for communications. The Bell physicists learned to control the electrical properties of semiconductor crystals by "doping" them with minute impurities. When two thin wires for current were attached to this material, a crude device was obtained that could amplify voice. The transistor, as this device was called, was developed late in 1947. The transistor duplicated many functions of vacuum tubes, but was smaller, required less power, and generated less heat. The three Bell Laboratories scientists who headed its development—William Shockley, Walter H. Brattain, and John Bardeen—won the 1956 Nobel Prize in Physics.

Shockley left Bell Laboratories and returned to Palo Alto, California, where he formed his own company, Shockley Semiconductor Laboratories, as a subsidiary of Beckman Instruments. Palo Alto is the home of Stanford University, which, in 1954, set aside 655 acres of its land as a high-technology industrial park, known as Stanford Research Park. One of the first small companies to lease a site there was Hewlett-

Packard. Many others followed, and the surrounding area of Santa Clara County gave rise in the 1960's and 1970's to a booming community of electronics firms known as "Silicon Valley." On the strength of his prestige, Shockley recruited eight young scientists from the eastern United States to work for him. One was Robert Norton Noyce, an Iowa-bred physicist with a doctorate from Massachusetts Institute of Technology. After working at Philco's transistor division, Noyce came to Shockley's company in 1956. The "Shockley Eight," as they became known in the industry, soon found themselves at odds with their boss over issues of research and development. Seven of the dissenting scientists negotiated with industrialist Sherman Fairchild, and they convinced the remaining holdout, Noyce, to join them as their leader. The Shockley Eight defected in 1957 to form a new company, Fairchild Semiconductor, in nearby Mountain View. Shockley's company never recovered from the loss of these scientists and soon went out of business.

At the same time, an established Dallas company, Texas Instruments, began its own production of transistors. By 1957, the Soviets had launched Sputnik, Earth's first artificial satellite. The American aerospace industry struggled to catch up, and government contracts beckoned to the companies that could find a reliable manufacturing process for transistors. Texas Instruments was one of the first companies to obtain defense contracts for the manufacture of transistors.

Research efforts at Fairchild Semiconductor and Texas Instruments now focused on putting several transistors on one piece, or chip, of silicon. The first step involved making miniaturized electrical circuits. Jack St. Clair Kilby, a researcher at Texas Instruments, succeeded in making a circuit on a chip, consisting of tiny resistors, transistors, and capacitors, all connected with gold wires. He and his company filed for a patent on this "integrated circuit" in February, 1959. Noyce and his associates at Fairchild Semiconductor followed in July of that year with an integrated circuit manufactured by a "planar process," which involved laying down several layers of semiconductor, isolated by layers of insulating material. Although Kilby and Noyce are generally recognized as coinventors of the integrated circuit, Kilby has the sole honor of membership in the National Inventors Hall of Fame for his efforts.

A significant development in semiconductor technology was the introduction of photolithography in the mid- and late 1950's. This process enabled manufacturers to etch the semiconductor by projecting a photo reduction of an actual circuit diagram. The development of the integrated circuit, the planar manufacturing process, and photolithography were all crucial steps in the evolution of microelectronics. They reduced the need for labor-intensive attachment of individual components to a circuit board, greatly facilitating mass production.

In the early 1960's, Fairchild Semiconductor and Texas Instruments led the semiconductor industry, in great measure because of contracts for missile guidance system components. By the time International Business Machines (IBM) introduced its 360 line of computers in 1963, computer manufacturers were using "solid state" (semiconductor) components in their logic circuits. Computer memory had developed from vacuum tubes to "magnetic core," which consisted of thousands of tiny ferrite rings through which wires were run. Federal regulations were mandated that required new television sets be able to receive ultrahigh frequency broadcast (UHF), which was not practical with vacuum tubes. Semiconductor industry leaders saw enormous potential markets in computer memory and television components.

By 1968, Fairchild Semiconductor had grown to a point where many of its key Silicon Valley managers had major philosophical differences with the East Coast management of their parent company. This led to a major exodus of top-level management and engineers. Many started their own companies. Noyce, Gordon E. Moore, and Andrew Grove left Fairchild to form a new company in Santa Clara called Intel with $2 million from venture capitalist Arthur Rock. Intel's main business was the manufacture of computer memory integrated circuit chips. By 1970, they were able to develop and bring to market a random-access memory (RAM) chip that was purchased in mass quantities by several major computer manufacturers, providing large profits for Intel.

In 1969, Marcian Edward Hoff, Jr., an Intel research and development engineer, met with engineers from Busicom, a Japanese firm. Busicom wanted Intel to design a set of integrated circuits for their desktop calculators, but Hoff told them their specifications were too complex. Nevertheless, Hoff began to think about the possibility of incorporating all the logic circuits of a computer central processing unit (CPU) into one chip. He began to design a chip called a microprocessor, which, when combined with a chip to hold the program and one to hold the data, would become a small, general purpose computer. Noyce encouraged Hoff and his associates to continue his work on the microprocessor, and Busicom contracted with Intel to produce the chip. Frederico Faggin was hired from Fairchild, and he did the chip layout and circuit drawings. In January, 1971, the Intel team had their first working microprocessor, called the 4004. The following year, Intel made a higher capacity microprocessor, the 8008, for Computer Terminals Corporation. That company contracted with Texas Instruments to produce a chip with the same specifications as the 8008, which they produced in June, 1972. Other manufacturers soon produced their own microprocessors.

In *Silicon Valley Fever*, Hoff admitted: "If we had not made the 4004 in 1971, someone else would have invented the microprocessor in a year or two." Indeed, Texas Instruments also claims ownership of the inventions of the microprocessor and the microcomputer. Furthermore, according to a 1990 article in *Computerworld*, a California engineer named Gilbert Hyatt received a patent on July 17, 1990, for a single-chip microcomputer. The patent had been filed in December, 1970, for work dating back to 1968. Regardless of who actually invented the microprocessor, its appearance in the early 1970's was truly a great event.

Impact of Event

The evolution of integrated circuits produced a corresponding evolution of computer technology. Profits from the sales of integrated circuit memory chips to computer companies allowed the semiconductor makers to research and develop new integrated circuits with more power. In 1964, Moore, the director of research at Fairchild Semiconductors and one of the original Shockley Eight, noted that the number of elements in the most advanced integrated circuits had doubled each year since 1959 and made a prediction that the trend would continue which became known as Moore's law. Coupled with Moore's law is the "learning curve," which describes the 20 to 30 percent decline in an integrated circuit's price for each doubling of its market lifetime as production efficiency improves.

The integrated circuit industry became a fiercely competitive arena where companies such as Texas Instruments, Intel, Motorola, and Fairchild Semiconductors contended. Many other new companies were created in the process. After leaving Fairchild, Shockley's original team eventually produced more than thirty corporate descendants in Silicon Valley.

By 1975, microcomputer kits were sold to hobbyists for less than a thousand dollars. More powerful computers made by Apple, Commodore, and Radio Shack soon appeared in stores for the same price. Intel was assured a dominating role in the industry when its 8088 microprocessor was used as the basis of the IBM Personal Computer (PC), introduced in 1981. IBM's prestige as the world's largest computer company legitimized the PC for business use. Also, it created a standard that served to unify a growing number of divergent microcomputer architectures.

The learning curve of microprocessor technology made available for a few dollars computers the size of a fingernail that would have filled a room and cost hundreds of thousands of dollars a decade before. Computers had existed previously behind closed doors as the mysterious domain of a few scientists and technicians, but the microprocessor allowed almost anyone to own a computer, often without even being aware of it. The microprocessor found its way into automobiles, microwave ovens, wristwatches, telephones, and an abundance of other ordinary items encountered in everyday life.

On the cutting edge of technology, ever more powerful computers were developed, often using a parallel architecture of several CPUs. The declining cost of microelectronic technology made it possible for more and more companies to invest in and make use of a global network of communications satellites. In 1987, American semiconductor manufacturers, concerned by increasing competition from Japan and other countries in the world market, formed a consortium called the Semiconductor Manufacturing Technology Institute, or Sematech. The following year, Noyce accepted the position of chief executive officer. Sematech's efforts at bolstering the declining role of the United States as a world leader in high technology suffered a sudden setback when Noyce died in 1990.

Afer two and a half decades, Moore's law was still an accurate measure of integrated circuit development. Noyce and others admitted that chip densities have an upper limit imposed by the laws of physics but believed it would not be reached until the beginning of the twenty-first century. The future of this technology is difficult to predict, but it is known with certainty that computer technology will continue to develop rapidly.

Bibliography

Hanson, Dirk. *The New Alchemists: Silicon Valley and the Microelectronics Revolution*. Boston: Little, Brown, 1982. In this excellent book on semiconductor technology, Hanson chronicles its evolution from the days of Thomas Alva Edison. The history, methods, and culture of the semiconductor industry is richly described, along with its impact on communications, the military, consumer products, and larger social issues such as freedom and unemployment. Some of Hanson's predictions of the future are dated, but many have occurred sooner than he expected.

Harrington, Maura J. "Micro Chip Patent Rewrites History." *Computerworld* 24 (September 3, 1990): 4. This short article contains the startling report that twenty years after filing with the U.S. Patent Office, Gilbert Hyatt was granted a patent for the first microcomputer chip. The fact that his claim predates others may mean Hyatt will get a royalty on every microprocessor chip ever made.

Large, Peter. *The Micro Revolution Revisited*. London: Frances Pinter, 1984. An English view of the impact of microprocessors on human society. Contains a description of how chips are made and their history.

Noyce, Robert N. "Microelectronics." *Scientific American* 237 (September, 1977): 63-69. In his article, Noyce cites Moore's law as the basis for predicting the rapid growth and proliferation of computing power. The revolutionary impact on society that Noyce predicted continues to unfold.

Osborne, Adam. *An Introduction to Microcomputers*. Vol. 1, *Basic Concepts*. Rev. ed. Berkeley, Calif.: Osborne/McGraw-Hill, 1980. For those who desire to learn about how microprocessors work, Osborne provides a good introduction. The book is filled with illustrations and diagrams that help explain the components and functions of microprocessors and their related devices to the nontechnical reader.

_____. *Running Wild: The Next Industrial Revolution*. Berkeley, Calif.: Osborne/McGraw-Hill, 1979. Industry pioneer Osborne's account of the microprocessor's impact on technology, society, and the future. Unfortunately, the unusual insight evident in this book was not enough to ensure the success of the company Osborne founded in 1980 to manufacture the Osborne 1 portable computer. After two years of huge initial sales, production and cash flow problems caused Osborne Computer to go bankrupt.

Rogers, Everett M., and Judith K. Larsen. *Silicon Valley Fever*. New York: Basic Books, 1984. A thorough chronicle of Silicon Valley, the hub of America's semiconductor industry, with an emphasis on the companies that form its history, economics, and culture. Chapter 2, "The Rise of Silicon Valley," and chapter 6, "Winning at the Game: Intel," are particularly informative. Everett and Larsen spent two years interviewing industry leaders, and the resulting work contains many personal accounts that offer an intimate look at key events.

Charles E. Sutphen

Cross-References

Fleming Files a Patent for the First Vacuum Tube (1904), p. 255; Shockley, Bardeen, and Brattain Discover the Transistor (1947), p. 1304; Bardeen, Cooper, and Schrieffer Explain Superconductivity (1957), p. 1533; Easki Demonstrates Electron Tunneling in Semiconductors (1957), p. 1551; The Floppy Disk Is Introduced for Storing Data Used by Computers (1970), p. 1923; Texas Instruments Introduces the First Commercial Pocket Calculator (1972), p. 1971; Apple II Becomes the First Successful Preassembled Personal Computer (1977), p. 2073; The IBM Personal Computer, Using DOS, Is Introduced (1981), p. 2169; IBM Introduces a Personal Computer with a Standard Hard Disk Drive (1983), p. 2240; Optical Disks for the Storage of Computer Data Are Introduced (1984), p. 2262.

MARINER 9 IS THE FIRST KNOWN SPACECRAFT
TO ORBIT ANOTHER PLANET

Category of event: Space and aviation
Time: 1971-1972
Locale: Mars

Between September, 1971, and October, 1972, Mariner 9 sent back to Earth a wealth of close-up photographs of Mars and many other data essential to interplanetary exploration

Principal personages:
GEOFFREY ARTHUR BRIGGS (1941-   ), the task group leader of mission analysis
ELLIOTT CHARLES LEVINTHAL (1922-   ), the task group leader of data processing
CONWAY B. LEOVY (1933-   ), the principal investigator of atmosphere phenomena
JOHN FRANCIS MCCAULEY (1932-   ), the principal investigator of geology
JAMES BARNEY POLLACK (1938-   ), the principal investigator of satellite astronomy
CARL SAGAN (1934-   ), the principal investigator of visible surface features
BRADFORD ADELBERT SMITH (1931-   ), the task group leader of mission operations
GÉRARD HENRI DE VAUCOULEURS (1918-   ), the principal investigator of geodesy and cartography

Summary of Event

On September 22, 1971, an unmanned American spacecraft designated Mariner 9 successfully achieved orbit around the planet Mars at an altitude of 1,398 kilometers. Over the next thirteen months, television relays aboard the satellite sent back almost three thousand high-quality, close-up photographs of Mars' surface that, along with other data recorded by sophisticated scientific equipment on board the Mariner, revolutionized scientific knowledge about Mars, about planetary development and the evolution of the solar system, and about Earth. The information gathered by the Mariner 9 mission represented an indispensable step in humankind's exploration of outer space. It also produced a completely unexpected dividend in the form of increased cooperation between space scientists in the United States and their counterparts in the Soviet Union, one of the small steps toward easing international tensions and reducing the threat of nuclear war.

It is not surprising that Mars has received more attention from space scientists than any other planetary body. Since the inception of modern astronomy in the sixteenth century, both scientists and the writers of speculative fiction have identified Mars as the most likely of the planets to bear Earth-like life. Beginning with H. G. Wells and Edgar Rice Burroughs, modern science-fiction writers have made the red planet the abode of a host of fantastic sentient creatures. Their imaginary Martians were hardly more startling than the assertions by prominent astronomers such as Percival Lowell in the earlier part of the twentieth century that an ancient race of Martians were engaged in a massive irrigation project on their dying world, their canals clearly visible through the most powerful Earth telescopes on particularly clear nights.

The Mariner 9 mission was the fifth and most ambitious of the Mariner missions designed to gather information concerning Mars. Three earlier missions, Mariner 4 in 1965 and Mariners 6 and 7 in 1969, had achieved close flybys of Mars. All three spacecraft sent back photographs of parts of Mars' surface and limited data concerning Martian surface conditions. The information gained by the three missions was enormously interesting, but incomplete. Consequently, American space scientists began planning a space mission that they hoped would answer many of the perplexing questions concerning Mars.

National Aeronautics and Space Administration (NASA) officials authorized the Mariner 8 and 9 projects in 1968. NASA scientists planned the missions to take place during the next "window" period (a "window" is what astronomers call a period of time during which two heavenly bodies are in the most favorable relative positions to facilitate travel between them) in 1971. NASA scientists planned six objectives to be performed by the Mariners over a period of three months after they achieved orbit. These objectives included the relay of photographs of almost the entire Martian surface to Earth, along with data concerning Martian atmospheric content and surface temperatures. To design experiments to achieve the objectives, NASA officials selected scientists to form several different teams and task forces and appointed a principal investigator to head each one. The principal investigators included Conway B. Leovy, John Francis McCauley, James Barney Pollack, Carl Sagan, and Gérard Henri de Vaucouleurs. The task group leaders were Geoffrey Arthur Briggs, Elliott Charles Levinthal, and Bradford Adelbert Smith: These team and task force leaders coordinated their efforts with those of the scientists and engineers responsible for designing the spacecraft and launching them into Martian orbit.

The specifics of the plan that eventually evolved called for two virtually identical spacecraft to be inserted into Martian orbit, one with a periapsis (closest approach to the planet) of 1,250 kilometers with an inclination of 80 degrees to Mars' equator, the second with a periapsis of 850 kilometers and a 50 degree inclination. NASA scientists expected three thousand photographs from the two spacecraft, which would permit virtually the entire surface of the planet to be mapped. The other scientific equipment aboard the Mariners would permit much more sensitive analyses of the surface and high-altitude atmospheric conditions on Mars than were possible with the earlier flyby missions.

The design of the two spacecraft was very similar to that of Mariners 6 and 7. The major difference was the large rocket engines necessary to insert Mariners 8 and 9 into Martian orbit. At the same time, other NASA scientists and engineers built new facilities to track the mission and record and interpret the data sent back by the spacecraft. Mission controllers also evolved plans for emergency changes in the mission plans if something should go wrong, or in the event of unexpected discoveries. Subsequent events proved the wisdom of those contingency plans. NASA personnel moved Mariners 8 and 9 from the Jet Propulsion Laboratory in California, where they were built, to the Kennedy Space Center in Florida early in 1971. Technicians at the Kennedy Space Center launched Mariner 8 aboard a Centaur rocket on May 9, 1971, only to witness its crash into the Atlantic Ocean 563 kilometers northwest of Puerto Rico, caused by a malfunction in the rocket's guidance system. Project scientists immediately began implementing contingency plans that resulted eventually in Mariner 9 accomplishing almost all of the goals originally set for the two spacecraft. Scientists chose a new orbit with an 11.98-hour orbital period, a periapsis of 1,250 kilometers, and a 65 degree inclination. This orbit permitted Mariner 9 to observe and photograph almost as much of the Martian surface as would have been possible had both Mariners achieved Martian orbit.

The Kennedy Space Center personnel successfully launched Mariner 9 on May 30, 1971. Six days later, launch mission control scientists made a correction in the spacecraft's trajectory that delivered it into orbit around Mars 167 days later. The hoped-for results of the mission, however, were not immediately forthcoming. More than three months after Mariner's launch and almost two months before it achieved Martian orbit, astronomers detected a problem on the surface of Mars that might prevent Mariner 9 from achieving most of its objectives. The problem was a gigantic dust storm that eventually grew to the point that the entire planet was shrouded by a huge cloud of dust. The storm was not unprecedented. Astronomers had observed similar but smaller storms on Mars during other close approaches in preceding years, but it was sheer bad luck that such an enormous storm should occur just as Mariner 9 approached orbit around the red planet. When the spacecraft sent back its first photographs of Mars as it approached orbit on November 8, 12, and 13, no details of the planet's surface could be discerned through the dust except several dark spots that were identified eventually as high mountain peaks. After Mariner 9 achieved orbit, the photographs it relayed to Earth showed virtually nothing of Mars' surface, only the dust storm itself. The other equipment aboard the spacecraft was able to perform measurements of the Martian atmosphere, but not to the degree originally planned.

Fortunately for the Mariner 9 mission, NASA ground control was able to control equipment aboard the spacecraft. They delayed the photography of the planet's surface and other tests until the dust storm abated. Space scientists in the Soviet Union were not so fortunate. On November 13, 1971, officials in the Soviet Union announced that their space program five months earlier had launched two spacecraft toward Mars with even more ambitious objectives than those of Mariner 9. The Soviet spacecraft—designated Mars 2 and Mars 3—both contained soft-landing components. After achieving orbit, the landing components were to separate from the orbital craft and parachute to the Martian surface. The landing craft would then relay photographs of the Martian surface and data concerning surface conditions to Earth via the orbiters. In addition, the Soviet orbiters were equipped with cameras and other instruments similar to those aboard Mariner 9.

Mars 2 achieved Martian orbit on November 27 and separated into its two components. The landing component of Mars 2 became the first known human-made object to land on Mars, but it apparently crash-landed and never transmitted the hoped-for data. Mars 3 arrived in Martian orbit three days later and successfully separated its landing component. The lander soft-landed on Mars but ceased transmitting signals less than two minutes later for unknown reasons. The Mars 2 and Mars 3 orbiters were designed to take all their pictures automatically immediately after arriving in orbit, and thus transmitted nothing to their designers but pictures of the raging dust storm with no surface features of the planet visible. Thus, the Soviet Mars missions of 1971 failed to accomplish most of their objectives. The more flexible Mariner 9, however, exceeded the most optimistic expectations of its designers.

Mariner 9 continued to send information to Earth until October 27, 1972, 516 days after leaving Earth. On that date, its supply of attitude control gas depleted, it tumbled out of control. Significantly, United States space scientists shared the information with their Soviet counterparts under an agreement reached by the governments of the two nations on October 20, 1971, while Mariner 9, Mars 2, and Mars 3 were still en route to Mars.

Impact of Event

The Mariner 9 mission represents a significant step in humankind's painfully slow attempt to reach the stars. It achieved a number of firsts in space exploration, revolutionized scientific understanding of Mars and of planetary evolution, and furnished information essential to the success of subsequent American and Soviet missions to Mars. Mariner 9 also occasioned an important bridging of the ideological chasm between the United States and the Soviet Union.

Mariner 9 was the first known spacecraft to orbit another planet. The photographs it relayed to Earth allowed the first accurate mapping of the surface of another planet. While waiting for the dust storm to clear, it relayed the first close-up photographs of Mars' two satellites. After the dust storm cleared, the cameras aboard the spacecraft photographed almost the entire surface of Mars. When the nearly three thousand photographs arrived on Earth, they forced a complete reevaluation of Mars by NASA scientists. In addition to evidence of recent volcanism and photographs of a huge canyon that dwarfs the Grand Canyon in Arizona, the pictures relayed to Earth by Mariner 9 provided positive indications that some form of liquid (presumably water) once flowed freely on the Martian surface. The flowing liquid carved typical river valley channels in many areas of Mars.

Other equipment aboard Mariner 9 detected water vapor and oxygen in the Martian atmosphere in much greater quantities than predicted by NASA experts. Other instruments determined that the Martian atmosphere had at some time in the past been denser than is presently the case. These data, together with the evidence for free-flowing liquid on the Martian surface, led NASA scientists to conclude that Mars at some time in the past had a much more Earth-like atmosphere, very possibly conducive to the evolution of life. This conclusion raises a chilling question: What happened on Mars to transform it into the barren and apparently lifeless planet revealed by Mariner 9's cameras and by subsequent Viking space probes? Scientists wondered if something similar could happen to Earth.

Mariner 9's success, coupled with the failure of the Soviet spacecraft, revealed the necessity of flexibility in the programming of subsequent space probes. Equally important, the pictures Mariner 9 transmitted to Earth permitted NASA scientists to choose the most favorable locations for the Viking missions that soft-landed scientific equipment on Mars several years later. Thus, Mariner 9 became an indispensable step in the continuing human efforts to explore the planets of the solar system. The data transmitted by Mariner 9 were also of great importance to the Soviet space program and of great importance to world peace. The information-sharing agreement reached during Mariner's voyage has since been expanded and extended to areas other than space exploration.

Finally, the success of the Mariner 9 mission greatly stimulated public interest in the space program in the United States. That interest, and the congressional support it generated in the United States, accelerated the NASA program of unmanned planetary exploration.

Bibliography

Carr, Michael H. *The Surface of Mars.* New Haven, Conn.: Yale University Press, 1981. Many photographs, drawings, and maps. A detailed exposition of everything known concerning surface conditions on Mars. Much of the information contained is too technical to be of interest to a general audience, but will be fascinating for all who desire in-depth knowledge about Mars.

Ezell, Edward Clinton, and Linda Neuman Ezell. *On Mars: Explorations of the Red Planet, 1958-1978.* NASA SP-4212. Washington, D.C.: Government Printing Office, 1984. A summary of everything learned about Mars through 1978 from the various space missions of the United States and the Soviet Union. Contains numerous photographs and technical drawings of Mars and the space vehicles sent there from Earth. Concentrates on United States accomplishments, but evaluates Soviet contributions to Martian exploration. Suitable for all audiences.

Firsoff, Valdemar Axel. *The New Face of Mars.* Hornchurch, England: Ian Henry, 1980. Contains a short history of astronomical observations of Mars and conjectures based on these observations. Shows how those conjectures have evolved with the ever-increasing knowledge about Mars. Contains some photographs. Suitable for readers from high school level through adult.

Greeley, Ronald. *Planetary Landscapes.* London: George Allen and Unwin, 1985. Discusses the origin, surface features, and evolution of all the inner planets in the light of discoveries made through space exploration. Especially valuable in putting planetary features of Mars into the context of the other planets of the solar system. Contains many maps and photographs. Suitable for high school and college readers.

Hartmann, William K., and Odell Raper. *The New Mars: The Discoveries of Mariner 9*. NASA SP-337. Washington, D.C.: Government Printing Office, 1974. Contains an account of speculations concerning Mars before Mariner 9, and a brief account of earlier Mars space missions. Valuable for a sophisticated analysis of the data sent to Earth by Mariner 9 in nontechnical language. Includes many Mariner 9 photographs.

Mariner 9 Television Team and the Planetology Program Principal Investigators. *Mars As Viewed by Mariner 9*. NASA SP-329. Washington, D.C.: Government Printing Office, 1974. Contains a brief account of the Mariner 9 mission and most of the photographs sent back to Earth during the mission. Photographs are accompanied by expert analysis. Recommended for all audiences.

Sheldon, Charles H., II. *United States and Soviet Progress in Space: Summary Data Through 1971 and a Forward Look*. Washington, D.C.: Government Printing Office, 1972. An overview of the space exploration efforts of the United States and the Soviet Union, with particular reference to the Mariner 9 and Mars 2 and 3 missions. Contains speculation about the nature and scope of future missions, and the possibility of United States-Soviet cooperation in space exploration. Suitable for readers from high school through adult.

Paul Madden

Cross-References

Mariner 2 Becomes the First Spacecraft to Study Venus (1962), p. 1734; Mars 2 Is the First Spacecraft to Impact on Mars (1971), p. 1950; Pioneer 10 Is Launched (1972), p. 1956; Mariner 10 Is the First Mission to Use Gravitational Pull of One Planet to Help It Reach Another (1973), p. 2003; Soviet Venera Spacecraft Transmit the First Pictures from the Surface of Venus (1975), p. 2042; Viking Spacecraft Send Photographs to Earth from the Surface of Mars (1976), p. 2052; Voyager 1 and 2 Explore the Planets (1977), p. 2082; *Columbia*'s Second Flight Proves the Practicality of the Space Shuttle (1981), p. 2180; The First Permanently Manned Space Station Is Launched (1986), p. 2316.

MARS 2 IS THE FIRST SPACECRAFT
TO IMPACT ON MARS

Category of event: Space and aviation
Time: May 19, 1971-March, 1972
Locale: Mars

The Soviet planetary probe Mars 2 was the first human-made object to impact the surface of the planet Mars

Principal personages:

G. N. BABAKIN (1914-1971), the chief designer of interplanetary spacecraft

VLADIMIR N. CHELOMEI (1914-1984), the designer of the Proton launch vehicle

VALENTIN P. GLUSHKO (1906-   ), the chief designer of rocket engines

A. M. ISAYEV (1908-1971), the chief designer of the spacecraft propulsion system

ALEKSANDR KONOPATOV, the chief designer of upper-stage rocket engines

VASILI P. MISHIN (1917-   ), the chief designer of the Soviet space program from 1966 to 1974

NIKOLAI A. PILYUGIN (1908-1982), a designer of spacecraft guidance systems

MIKHAIL S. RYAZANSKIY (1909-1987), a designer of rocket and spacecraft control systems

Summary of Event

The 1971 Mars planetary launch window saw the successful flight of a new generation of Soviet Mars spacecraft. Mars 2 was launched at sixteen hours, twenty-two minutes, forty-nine seconds Greenwich mean time, on May 19, 1971, by a Proton booster from the Baikonur Cosmodrome in Kazakhstan, Soviet Union. This marked the first successful use of the giant 1,002,000-kilogram-thrust Proton booster in a Soviet planetary launch. Vladimir N. Chelomei was the designer of the Proton launch vehicle. After circling Earth once in a temporary parking orbit, Mars 2 was successfully injected into a planetary transfer orbit at seventeen hours, fifty-nine minutes on May 19, 1971. Its twin, the Mars 3 spacecraft, was launched May 28, 1971, and a third Soviet Mars launch failed on May 10, 1971, when booster failure left Cosmos 419 trapped in Earth orbit.

The new Mars probes weighed 4,650 kilograms, more than half of this weight being fuel for a rocket burn to enter Mars orbit. The probes consisted of two sections, the main spacecraft bus that would enter orbit around Mars and a 450-kilogram Mars landing capsule. The 4.1-meter-tall spacecraft were built around a 1.8-meter-diameter cylindrical body, which housed the fuel tanks containing nitric acid and amine-based propellants for the 9.86- to 18.89-kilonewton-thrust KTDU-425A course correction and Mars orbit insertion engine. A pressurized 2.3-meter-diameter toroidal flare at the base of the spacecraft bus carried the command, communications, and navigation equipment. Atop each Mars craft was a 1.2-meter-diameter sterilized landing capsule, nested under a 2.9-meter-diameter conical heat shield. Extending from the sides of the spacecraft were two 2.3- by 1.4-meter solar panels, spanning a total of 5.9 meters. Radiator panels for radiating excess heat were mounted on the back side of the solar panels. A 2.5-meter-diameter high-gain dish antenna extended from the side of the craft.

During the flight to Mars, studies of the solar wind and cosmic rays were carried out. While en route to Mars, Mars 2 made three course corrections. The first was performed on June 17, when the craft was 7 million kilometers from Earth. The second was performed on November 20, a week before arriving at the planet. Before ejecting its landing capsule and entering Mars orbit on November 27, the spacecraft performed its third course correction using onboard sensors to measure Mars optically and to compute the duration of the engine burn.

The Mars 2 mission proceeded according to plan until arrival at Mars on November 27, 1971, after a flight of 192 days. As the Mars 2 spacecraft approached the planet, Mars was enduring the most severe planetwide dust storm in recent history. The entire surface of the planet was obscured by layers of dust many kilometers thick, blown by several-hundred-kilometer-per-hour winds.

The Mars 2 orbiter only had enough fuel to enter Mars orbit without the 635-kilogram lander and heat shield combination. The spacecraft did not have the flexibility to wait safely in Mars orbit with the lander until the end of the gigantic Martian dust storm. Four and one-half hours before the spacecraft entered orbit around Mars, the landing capsule and its 185-kilogram protective shell were ejected. A solid-propellant rocket motor shifted its path to intersect the planet while the main spacecraft entered an elliptical orbit around Mars.

The landing capsule was called an Automatic Mars Station (AMS). It used a hybrid aerodynamic and rocket-braking system to attempt a soft landing. The extremely thin Martian atmosphere complicated the descent. If the lander entered at too shallow an angle, it would not slow down and would escape into space again. If the entry angle were too steep, it would descend too fast for the parachute to operate properly.

The Mars 2 landing capsule entered the Martian atmosphere at a velocity of 6 kilometers per second. After aerodynamic braking using the craft's heat shield, a decelerometer controlled the release of a drogue parachute, which was ejected while still descending at supersonic speed. This small chute pulled out the main parachute canopy, which remained in a reefed condition until slowing to Mach 1. At this velocity, the 15-meter-diameter main canopy was fully opened and the heat shield fell away. Because the Martian atmosphere is so thin at the surface, a landing using a parachute alone was not possible. While still falling at 90 meters per second, and at an altitude of 30 meters, a radar altimeter triggered a 10,000-kilogram-thrust retrorocket attached to the parachute lines, which fired for one second, cushioning the lander's impact. After this landing, a smaller rocket then pulled the parachute off to one side so it would not cover the landing capsule. Touchdown occurred three minutes after atmospheric entry at Martian coordinates 45 degrees south, 58 degrees east, in an area 500 kilometers southwest of Hellas.

The landing capsule was spherical with an offset center of gravity to make it roll into an upright position on the surface. The upper half of the lander's shell was designed to split open into four petals to stabilize the spacecraft on the surface. The Mars 2 lander was thought to have landed at a velocity of 20 meters per second with an impact of 500 g's. Although the lander was designed to survive an impact shock of 1,000 g's, there was no contact with the capsule after landing. It is assumed that the extensive planet-wide dust storm enveloping Mars at the time of landing overwhelmed the lander and caused it to crash. Had it landed successfully, the capsule would have used atmospheric temperature and pressure sensors, a mass spectrometer for chemical analysis of the atmosphere, a wind anemometer, devices to measure the chemical and physical properties of the Martian soil, and panoramic television equipment to study the landing site. Stereoscopic views of the landing site were to be relayed by dual television cameras. Signals from the lander were to be relayed to Earth by the Mars 2 spacecraft orbiting the planet. No biological or life detection experiments were carried by the lander.

After the loss of the lander, the Mars 2 mission concentrated on Mars science from orbit. The spacecraft bus fired its retrorocket at twenty-three hours, nineteen minutes Greenwich mean time, on November 27, 1971, and entered an initial Mars orbit of 1,380 by 25,000 kilometers, inclined 48.9 degrees, with a period of eighteen hours.

The spacecraft investigated surface temperature and water vapor content in the Martian atmosphere. Using infrared sensors, Mars 2 could determine the thickness of the carbon dioxide atmosphere over various areas and thus build up a surface relief map. Surface reflectivity and the density of the atmosphere were also measured.

The Mars 2 orbiter also carried a wide-angle as well as a 4-degree-field-of-view telephoto camera. Twelve exposures of the Martian terrain were to be taken on photographic film and developed automatically for transmission to Earth using a 1,000-by-1,000-pixel facsimile scan system. Because the photographic sequence had been programmed in advance and could not be delayed, the global dust storm on Mars at the time Mars 2 arrived prevented successful surface photography. The Soviets later claimed that planetary imaging was only a "subsidiary role" in the mission. When the Mars 2 mission ended in March, 1972, the spacecraft had completed 362 orbits of the planet.

Impact of Event

After seven failed Mars exploration attempts in the 1960's, the Soviets designed a new generation of planetary exploration spacecraft weighing four times that of previous planetary probes. Extensive tests in Earth orbit of this new spacecraft were carried out by the Cosmos 379 and Cosmos 382 launches on November 24, and December 2, 1970.

Soviet accounts of the Mars 2 mission indicate that the design of the second-generation Mars probes was heavily influenced by prior experience with Venera and Luna missions to Venus and the Moon. The new spacecraft design was accomplished in a relatively short period of time by a new design team whose average age was under thirty. The chief designers were G. N. Babakin, Valentin P. Glushko, A. M. Isayev, Aleksandr Konopatov, and Vasili P. Mishin. The designers included Chelomei, Nikolai A. Pilyugin, and Mikhail S. Ryazanskiy. This indicated the Mars effort was being carried out by a new planetary team independent of the Venus and lunar exploration groups, and a new Soviet commitment to Martian exploration was under way.

The Mars 2 and 3 missions were carried out concurrently with the American Mariner 9 mission to Mars. Cold War tensions between the United States and the Soviet Union had eased by this time, and the heavy political emphasis placed on earlier space missions was less evident. In a cooperative move, a special teletype hot line was set up between the American Jet Propulsion Laboratory (JPL) and the Soviet Coordinating and Computing Center to share data returned by the Mars probes and Mariner 9. Although the Mars 2 lander was a failure, it was the first human-made object to land on the surface of Mars. It carried a commemorative Soviet emblem to the Martian surface.

The giant Martian dust storm of 1971 was studied by the Mars 2 orbiter. Soviet space scientists concluded that the winds were not constant but occurred only in the initial phases of the storm. The fine dust blown to altitudes as high as 10 kilometers then took months to settle. Spectroscopic analysis of these dust clouds showed they were 60 percent silicon, ranging from 2 to 15 micrometers in diameter. The Soviets deduced further that, during the dust storm, the surface temperature of Mars dropped between 20 and 30 degrees Celsius because of blocked sunlight, while the atmosphere warmed up as it absorbed solar heat. When the dust storm ended, Martian surface temperatures ranged from 13 degrees Celsius in the Southern Hemisphere's summer, to −93 degrees Celsius in the Northern Hemisphere's winter. The northern polar cap had cooled down to −110 degrees Celsius.

Studies of the effects of the global Martian dust storm were instrumental in understanding the concept of "nuclear winter" as it applied to the aftermath of global nuclear war on Earth. Analysis of the cooling effects of the dust suspended in the atmosphere of Mars led to theories about global cooling and plant life extinction on Earth from similar effects caused by dust and smoke particles created by many simultaneous nuclear detonations.

Measurements by the Mars 2 orbiter indicated the planet's atmosphere is primarily carbon dioxide with a surface pressure of only 5.5 to 6 millibars, or 0.5 percent that on Earth. Water vapor content of the Mars atmosphere measured two thousand times less than Earth's.

Observations showed that at 100 kilometers altitude, carbon dioxide is broken up by solar ultraviolet radiation into carbon monoxide molecules and oxygen atoms. Water vapor at high altitudes is broken down into atomic hydrogen and oxygen. Traces of atomic oxygen were found at altitudes of 1,127 to 1,287 kilometers in concentrations of 100 atoms per cubic centimeter.

Mars 2 showed that Mars possesses a very weak magnetic field and has an ionosphere only one-tenth as dense as Earth's. Three featureless pictures of the Martian surface returned by the Mars 2 orbiter were shown on Moscow television on January 22, 1972.

Bibliography

Gatland, Kenneth. *Robot Explorers.* London: Blandford Press, 1972. A chronology of Soviet and American lunar and planetary space exploration programs. Contains numerous color illustrations providing insights into the design and functions of American and Soviet lunar and planetary spacecraft. Descriptive narrative provides detailed results of all Soviet and American lunar and planetary exploration spacecraft and their missions. Suitable for general audiences.

McDougall, Walter A. . . . *The Heavens and the Earth.* New York: Basic Books, 1985. A political history of the space age. Well researched and heavily footnoted. Heavy emphasis on the key political and technological leaders of the time. Suitable for all audiences.

Short, Nicholas M. *Planetary Geology.* Englewood Cliffs, N.J.: Prentice-Hall, 1975. Summarizes the accomplishments and scientific results of both American and Soviet lunar and planetary space programs. Stresses the chemical nature of the moon and inner planets, their geological similarities and differences, and their origins. College-level reading. Illustrated with many diagrams and photographs.

Smolders, Peter. *Soviets in Space.* New York: Taplinger, 1974. A well-illustrated narrative on all aspects of the Soviet space program. Concentrates on the successful portions of the Soviet space program as reported by the Soviet Union. Contains numerous diagrams and photographs illustrating the technical details of Soviet spacecraft and their missions.

Turnill, Reginald. *Spaceflight Directory.* London: Frederick Warne, 1978. A lavishly illustrated summary of spaceflight activities by all nations. Lists chronologies of major manned and unmanned space missions. Technical narrative describes worldwide space activities by nation and program. Suitable for readers at high school and college level.

U.S. Congress. Senate Committee on Commerce, Science, and Transportation. *Soviet Space Programs: 1976-80.* Part 2. Washington D.C.: Government Printing Office, 1985. Comprehensive descriptions of all phases of unmanned Soviet space programs. Provides a detailed overview of the technical development of Soviet unmanned space activities, scientific investigations, and results. Recommended for general audiences. Standard reference for data on Soviet space programs.

Wilson, Andrew. *Solar System Log.* London: Jane's, 1987. A compilation of all manned and unmanned lunar and planetary spaceflights up to mid-1985 by all space-faring nations. A well-illustrated chronology of the history, spacecraft, mission, and discoveries of all deep space exploration missions. Suitable for all readers.

Robert Reeves

Cross-References

Goddard Launches the First Liquid Fuel Propelled Rocket (1926), p. 810; The First Rocket with More than One Stage Is Created (1949), p. 1342; Sputnik 1, the First Artificial Satellite, Is Launched (1957), p. 1545; Parker Predicts the Existence of the Solar Wind (1958), p. 1577; Luna 2 Becomes the First Man-Made Object to Impact on the Moon (1959), p. 1614; Venera 3 Is the First Spacecraft to Impact on Another Planet (1965), p. 1797; Manabe and Wetherald Warn of the Greenhouse Effect and Global Warming (1967), p. 1840; Mariner 9 Is the First Known Spacecraft to Orbit Another Planet (1971), p. 1944; Soviet Venera Spacecraft Transmit the First Pictures from the Surface of Venus (1975), p. 2042; Viking Spacecraft Send Photographs to Earth from the Surface of Mars (1976), p. 2052.

PIONEER 10 IS LAUNCHED

Category of event: Space and aviation
Time: March, 1972
Locale: Cape Canaveral, Florida

Pioneer 10 initiated exploration of the outer planets, which led to a better understanding of the evolution of the solar system

Principal personages:
CHARLES F. HALL (1920-   ), an American physicist who was the Pioneer 6-13 project manager
RICHARD O. FIMMEL, an American engineer and Pioneer project manager succeeding Hall
JAMES A. VAN ALLEN (1914-   ), an American physicist who was a principal science investigator for the Pioneer project
JOHN A. SIMPSON (1916-   ), an American physicist and a principal science investigator for the Pioneer project
EDWARD J. SMITH (1927-   ), an American physicist and a principal science investigator
RICHARD B. MILLER, an American physicist and Deep Space Network manager for the Pioneer project

Summary of Event

In the opening days of space exploration, the search for an understanding of the evolution of the solar system began with studies of the inner planets, Mercury, Venus, Earth, and Mars, followed by studies of the outer planets, Jupiter, Saturn, Uranus, Neptune, and Pluto. There is an enormous difference between the outer and inner planets. For example, Jupiter is large enough to hold fourteen hundred Earths, and its volume is almost two and one-half times more than that of all the other planets together. It is a huge, rapidly spinning sphere of cold gases, hydrogen, helium, methane, water, and ammonia, and more complex, but as yet unknown, chemicals. Jupiter has a diameter eleven times that of Earth and spins on its axis twice as fast as Earth. In the 1960's, Congress authorized the first mission to explore the outer planets. The mission was designated as Pioneer 10. The previous Pioneer missions—1 through 9—were designed to explore space in the vicinity of Earth.

The Ames Research Center was assigned responsibility for project management and for mission and spacecraft design, with the understanding that the spacecraft would be built under a system contract with U.S. industry. Responsibility for deep-space communications, Earth to spacecraft and spacecraft to Earth, was assigned to the Deep Space Network managed by the Jet Propulsion Laboratory (JPL). Key personnel included Charles F. Hall, Richard O. Fimmel, John A. Simpson, Edward J.

Smith, and Richard B. Miller. Ames also contracted with JPL for radio navigation support.

The Pioneer 10 spacecraft was launched from Cape Canaveral on March 3, 1972, aboard an Atlas-Centaur launch vehicle that incorporated a solid-propellant third stage. At the time, Pioneer 10 attained the highest injection energy ever achieved, as is attested by the fact that the spacecraft required only eleven hours to cross the lunar orbit. After a twenty-one-month flight, the spacecraft arrived at its radius of closest approach to Jupiter at a distance of approximately 2.8 R_j (Jovicentric Jupiter radius) on December 4, 1973.

Spacecraft had already traveled to Venus and Mars, but this Pioneer spacecraft had to face traversing the asteroid belt, which lies between Mars and Jupiter. The asteroid belt was known to contain many thousands of small rocky bodies, possibly including untold numbers of very small particles. There were doubts that spacecraft could safely cross this region; yet, it was important to study the outer giants, because to reach the more distant planets requires full use of Earth's orbital velocity around the sun and a spacecraft trajectory directly through the heart of the asteroid belt. The gravity and orbital motion of Jupiter are then used to urge a spacecraft flying by that planet to the high velocities needed to reach other distant planets— Saturn, Uranus, and Neptune—within reasonable time and with reasonable scientific payloads. Pioneer 10 then had to face the unknown radiation environment of Jupiter.

The spacecraft is spin-stabilized, having a spin rate of 4.8 revolutions per minute. The spacecraft spin axis is parallel to the axis of the 2.74-meter-diameter high-gain antenna reflector and is kept pointed toward the earth in order to maximize the communication bit rate. The maximum bit rate used during the Jupiter encounter was 1,024 bits per second. Spacecraft spin axis precession maneuvers are required periodically in order to maintain Earth pointing and were performed approximately six days prior to and two days after the Jupiter flyby. Electrical power for the experiments and spacecraft subsystems is supplied by four radioisotope thermoelectric generators. The generators are located at the end of two long booms. Inspection of in-flight data indicates that these batteries have produced negligible interference with any of the Pioneer 10 experiments. During the encounter, the spacecraft approached Jupiter in the midmorning sector of the sunlit hemisphere and exited near Jupiter's dawn meridian.

The first spacecraft to fly by Jupiter also achieved many other firsts: namely, first spacecraft images of the large Galilean satellites; first spacecraft to go beyond the last known planet of our solar system (June 13, 1983); and first spacecraft to carry a message for the benefit of an extraterrestrial civilization. In the field of space communications, each day a new distance record is being achieved in communication between humans on Earth and a spacecraft in interstellar space. As of July, 1990, the spacecraft was 7.5 billion kilometers from the sun. On this occasion, the radio signal requires seven hours to reach Earth from the spacecraft. It should be noted that radio communications were discovered in the early part of the twentieth century by Guglielmo Marconi in Bologna, Italy. His early experiments were over a distance of approximately 50 meters, and the progress in technology has made it possible now to communicate to the far-reaching edge of the solar system.

Scientists gave high priority to extending the exploration of the outer space frontiers. In the early 1970's, scientific descriptions of the outer parts of the solar system still differed enormously because details were lacking from observations made from Earth. There were many unknowns about the far reaches of interplanetary space beyond the orbit of Mars and about the very large and distant planets of the outer solar system, where most of the matter and the angular momentum of the total system are concentrated.

The science instruments were designed to gather new knowledge on interplanetary space between Earth and Jupiter, and beyond; certain instruments were to characterize the environment of Jupiter and its Galilean satellites. Science instruments were installed at the deep-space stations, which provided data to characterize the atmospheres of Jupiter and its satellites, the gravitational field of these bodies, detection of gravitational waves, and tests of Albert Einstein's theory of relativity. These phenomena affect the radio signal from the spacecraft; this signal also is used to carry the communication of data from the spacecraft to Earth. Discoveries made by Pioneer 10 had far-reaching consequences for future space exploration and for an understanding of the solar system and of vast magnetospheres containing energetic plasmas that cannot be duplicated in laboratories on Earth. The myth of a hazardous asteroid belt was dispelled; the anticipated concentration of small particles did not exist. Spacecraft could reach Jupiter safely and use the Jovian gravitational slingshot to hurtle them to more distant planets.

The Pioneer missions made discoveries about the distribution of particles causing the zodiacal light and the Gegenschein, glows in the night sky observed from Earth. They also mapped the background of starlight. At Jupiter, the Pioneers explored the giant planet's magnetosphere and found that it is disk-shaped and bigger than the sun. The magnetic field of Jupiter—ten thousand times stronger than Earth's field—is opposite in direction to Earth's field; its dipole moment is offset and tilted so as to cause a wobbling of the huge magnetosphere as the planet spins on its axis.

Also discovered was a ring current within the magnetosphere and radiation belts whose trapped electrons have an intensity ten thousand times Earth's trapped electrons, and its protons, one thousand times Earth's. Jupiter's magnetosphere was revealed as the source of high-energy electrons observed everywhere that spacecraft have traveled in the heliosphere.

The ratio of helium to hydrogen in the Jovian atmosphere was found to be 0.14, fairly close to the 0.11 ratio of the sun. Close-up pictures were obtained of the belts and zones, of the Great Red Spot, and of the polar regions. A new understanding was obtained of the weather patterns on a giant, rapidly rotating planet with no solid surface. The heat balance was measured and showed that Jupiter emits 1.7 times the heat the planet receives from the sun, indicating that heat is still being generated in its vast interior.

Impact of Event

Pioneer 10 continued to provide rich scientific data years after traveling beyond the last known planet of the solar system. It has been searching for Planet X; namely, a planet beyond Pluto. According to research by John Anderson, a scientist at JPL, something is disturbing the orbits of Uranus and Neptune. Data suggest that the disturbances are real and that they amount to about 4,300 kilometers for Neptune and 2,700 kilometers for Uranus. This magnitude would require a planet with a mass one to four times that of Earth. Theories place Planet X at about 100 astronomical units from the sun, in an orbit with a period of about one thousand years that would take it as close to the sun as 40 astronomical units. Anderson stated: "If there is a planet out there, it could very well be detected by the two Pioneers. We want to continue to [maintain communications with] them for as long as possible." This investigation is made possible by analyses of the observables generated by the Deep Space Network while it is in communication with the spacecraft. These observables are the radial velocity of the spacecraft relative to Earth through the Doppler effect on the radio signal and the distance between the spacecraft and Earth made possible by instruments of the Deep Space Stations. The Doppler effect is measured at each of the Deep Space Stations and provides a means for the detection of a certain class of gravitational waves predicted by the theory of relativity. As the spacecraft travels into interplanetary space, it will define the boundaries of the sun's effect on the interplanetary medium. All these experiments, as well as the continuing measurements of particles and fields, were expected to continue until 2000.

Since Pioneer 10 passed the orbit of the last known planet, one of its notable achievements was to provide indications that it was closer to the edge of the heliosphere—the area of the sun's influence—than scientists had believed, as Pioneer investigator Darrell L. Judge, of the University of Southern California, reported. These indications of a smaller heliosphere were provided by the ultraviolet photometer. "Various pieces of evidence independent of each other suggest to many of us that we may be getting reasonably close to the inner edge of the heliosphere," Judge stated. A "shock" boundary, where hot particles hurtling out from the sun at tremendous velocity interact with the "cool interstellar breeze" moving into the solar system, may be located at about 50 astronomical units.

James A. Van Allen outlined how scientists will know when Pioneer 10 has escaped the sun's influence and penetrated the interstellar medium, which he said would be a "classical achievement of historic quality": The solar wind will cease and cosmic-ray intensity will be constant.

Bibliography

Dyal, P., and R. O. Fimmel. "Exploring Beyond The Planets: The Pioneers 10 and 11 Missions." *Journal of the British Interplanetary Society* 37 (October, 1984): 468-479. Describes the main goals of the mission for the 1990's.

Fimmel, Richard O., William Swindell, and Eric Burgess. *Pioneer Odyssey: Encounter with a Giant*. NASA SP-349. Washington, D.C.: Government Printing

Office, 1974. A discussion of the Pioneer 10 mission to Jupiter and its results. Jupiter and the spacecraft design are described briefly, and each scientific experiment aboard Pioneer 10 is discussed in detail.

Fimmel, Richard O., James Van Allen, and Eric Burgess. *Pioneer: First to Jupiter, Saturn, and Beyond*. NASA SP-446. Washington, D.C.: Government Printing Office, 1980. The continued mission of Pioneer 10 into the outer solar system necessitated an update of the earlier publication by Fimmel, Swindell, and Burgess (cited above). Results from further analyses of data from the encounter with Jupiter and interplanetary results are included.

French, Bevan M., and Stephen P. Maran. *A Meeting with the Universe*. NASA EP-177. Washington, D.C.: Government Printing Office, 1981. Describes what was learned about the universe by going into space. It is a history of space exploration by NASA, universities, other government agencies, and industries. A novel experiment in writing about science for nontechnical readers.

Mead, Gilbert D. "Pioneer 10 Mission: Jupiter Encounter." *Journal of Geophysical Research* 79 (September 1, 1974): 25. Contains comprehensive reports of almost all the Pioneer 10 experiments. Interesting reading.

Montoya, Earl J., and Richard O. Fimmel. *Space Pioneers and Where They Are Now*. NASA EP-264. Washington, D.C.: Government Printing Office, 1987. As part of the modern exploration of space frontiers, this volume describes the accomplishments of the Pioneer missions, including Pioneer 10, and the scientific objectives of that mission in the future as it heads out of the solar system.

N. A. Renzetti

Cross-References

Mariner 2 Becomes the First Spacecraft to Study Venus (1962), p. 1734; Mariner 9 Is the First Known Spacecraft to Orbit Another Planet (1971), p. 1944; Mars 2 Is the First Spacecraft to Impact on Mars (1971), p. 1950; Soviet Venera Spacecraft Transmit the First Pictures from the Surface of Venus (1975), p. 2042; Viking Spacecraft Send Photographs to Earth from the Surface of Mars (1976), p. 2052; Voyager 1 and 2 Explore the Planets (1977), p. 2082.

HOUNSFIELD INTRODUCES A CAT SCANNER THAT CAN SEE CLEARLY INTO THE BODY

Category of event: Medicine
Time: April, 1972
Locale: Atkinson Morley's Hospital, Wimbledon, England

Hounsfield developed computerized axial tomography (CAT) to produce detailed pictures of the body, providing an invaluable tool for medical diagnosis

Principal personages:
    GODFREY NEWBOLD HOUNSFIELD (1919-   ), an English electronics engineer who in 1972 developed the first practical CAT scanner
    ALLAN M. CORMACK (1924-   ), a South African-born American physicist, who in the 1960's independently developed a mathematical basis for CAT scanning
    JAMES AMBROSE, an English radiologist at Atkinson Morley's Hospital, who participated in the clinical studies associated with the first CAT scanner

Summary of Event

In 1979, the Nobel Prize in Physiology or Medicine was awarded to Godfrey Newbold Hounsfield, an electronics engineer at England's Electrical and Musical Instruments (EMI) Limited, and Tufts University physics professor Allan M. Cormack, for their contributions to development of the X-ray technique called computerized axial tomography (CAT). Medical experts lauded CAT as a revolutionary new procedure, possessed of amazing capabilities. The Nobel committee stated that "no other method of X-ray diagnosis within such a short time has led to such remarkable advances in research and in a multitude of applications." At the same time, consumer advocacy groups attacked CAT as an overly expensive plaything for physicians.

CAT is a technique that collects X-ray data and uses a sophisticated computer to assemble it into a three-dimensional image of an opaque, solid mass, such as a human body. The importance of CAT is eloquently stated by Richard A. Robb, in his introduction to *Three-Dimensional Biomedical Imaging* (1985). That is, "to be able to see into the body has been and is a primary capability desired and necessary to study and elucidate the basic processes of life, and to diagnose disease conditions that perturb and endanger the normal function of these biological processes."

It was not until the development of CAT that this became possible in a sophisticated fashion, by merging use of X rays with modern computer technology. This merger led to another name for CAT, computer-assisted tomography. CAT is a technique of medical radiology, an area of medicine that began after Wilhelm Röntgen's

1895 discovery of the high-energy electromagnetic radiations he named X rays.

Röntgen and others soon produced useful X-ray images of parts of the human body, and physicians were quick to learn that these images were valuable diagnostic aids. Medical X-ray analysis, however, developed slowly. For many years, it was confined to the familiar X-ray analysis in which broad, high-energy beams of X rays pass through a body part, hit a covered sheet of photographic film, and develop it because of their energy.

This process is useful because bones absorb much more X-ray energy than do the soft tissues that surround them. Therefore, when conventional X-ray films are developed, the bones appear as prominent white areas (dark areas of X-ray negatives). The conventional method is useful for the identification of bone fractures and has some value in diagnosing health problems in the soft tissues. Distinction between soft tissues is poor, however, and differences in normal and diseased soft tissue cannot be assessed.

In the late 1950's and early 1960's, Cormack, at the Department of Physics of Tufts University, pioneered a mathematical method for obtaining detailed X-ray absorption patterns in opaque samples meant to model biological samples. Cormack had become interested in X rays in 1956, while serving as a radiologic physicist at the Groote Schuur Hospital (Cape Town, South Africa). His studies at Groote Schuur, and later at Tufts University (Medford, Massachusetts), used narrow X-ray beams, passed through samples at many different angles. Because the technique probed test samples from many different points of reference, it became possible—with use of proper mathematics—to reconstruct the interior structure of a thin slice of the object studied.

Cormack published his data but received almost no recognition because computers that could analyze the data in an effective fashion had not yet been developed. X-ray tomography, however—that is, the procedure of using X rays to produce detailed images of thin sections of solid objects (from the Greek, *tomos*, or section)— had been born. It remained for Godfrey Hounsfield independently, and reportedly with no knowledge of Cormack's work, to design the first practical CAT scanner.

Hounsfield, at EMI's Central Research Laboratory (Middlesex, England), like Cormack, realized that X-ray tomography was the most practical approach to developing a medical body imager. It could be used to divide any three-dimensional object into a series of thin slices that could be reconstructed into images by use of appropriate computers. Hounsfield developed another mathematical approach to the method. He estimated that the technique would allow the very accurate reconstruction of images of thin body sections with a sensitivity "two orders of magnitude" above that of the X-ray methodology then in use. Moreover, he proposed that his method would enable researchers and physicians to distinguish between normal and diseased tissue. Hounsfield was correct about that.

The prototype instrument that Hounsfield developed was quite slow, requiring nine days to scan an object. Soon, he modified the scanner so that its use took only nine hours, and he obtained successful tomograms of preserved human brains and the fresh brains of cattle. The further development of the CAT scanner then proceeded quickly and yielded an instrument that required 4.5 minutes to gather tomographic data and 20 minutes to produce the tomographic image.

In late 1971, the first clinical CAT scanner was installed at Atkinson Morley's Hospital in Wimbledon. By early 1972, the first patient, a woman with a suspected brain tumor, was examined, and the resultant tomogram identified a dark, circular cyst in her brain. Additional data collection from other patients soon validated the technique. Hounsfield and EMI patented the CAT scanner in 1972, and the findings were reported at that year's annual meeting of the British Institute of Radiology.

Hounsfield published a detailed description of the instrument in 1973. Hounsfield's clinical collaborator, James Ambrose, published on the clinical aspects of the technique. Neurologists all around the world were ecstatic about the new tool that allowed them to locate tissue abnormalities with great precision.

The CAT scanner consisted of an X-ray generator, a scanner unit composed of an X-ray tube and a detector in a circular chamber about which they could be rotated, a computer that could process all the data obtained, and a cathode-ray tube on which tomograms were viewed. To produce tomograms, the patient was placed on a couch, head inside the scanner chamber, and the emitter-detector was rotated 1 degree at a time. At each position, 160 readings were taken, converted to electrical signals, and fed into the computer. In the 180 degrees traversed, 28,800 readings were taken and processed. The computer then converted the data into a tomogram (a cross-sectional representation of the brain that shows the differences in tissue density). A Polaroid picture of the tomogram was then taken and interpreted by the physician in charge.

Other scientists produced important discoveries related to the CAT scanner. Hounsfield, they noted, obtained his ultimate success because only he was able to achieve the necessary synthesis of a mathematical solution to image reconstruction, recognition of the clinical need for the instrument, and the engineering technology required to produce the CAT scanner.

Impact of Event

Many neurologists agree that CAT is the most important method developed in the twentieth century to facilitate diagnosis of disorders of the brain. Even the first scanners could distinguish between brain tumors and blood clots and help physicians to diagnose a variety of brain-related birth defects. In addition, the scanners are believed to have saved many lives by allowing physicians to avoid the dangerous exploratory brain surgery once required in many cases and replacing more dangerous techniques (for example, pneumoencephalography, which required a physician to puncture the head for diagnostic purposes).

By 1975, improvements in the scanner, including quicker reaction time and more complex emitter-detector systems, allowed EMI to introduce full-body CAT scanners to the world market. Then, it became possible to examine other parts of the body—including the lungs, the heart, and the abdominal organs—for cardiovascular problems, tumors, and other structural health disorders. The technique became so ubiquitous that many departments of radiology changed their names to departments of medical imaging.

Use of CAT scanners has not been accepted without problems. Part of the reason for this is their great cost—ranging from about $300,000 for early models to $1 million for modern instruments—and resultant claims by consumer advocacy groups that the scanners are unnecessarily expensive toys for physicians. In fact, as a result of this kind of feeling, the Carter administration enacted 1978 legislation that made CAT scanners subject to issuance of "certificates of need" for which hospitals had to apply before purchasing them.

Despite such problems, CAT scanners have become important everyday diagnostic tools in many areas of medicine. Furthermore, continuation of the efforts of Hounsfield and others has led to more improvements of CAT scanners and to use of nonradiologic, nuclear magnetic resonance (NMR) imaging in such diagnosis.

Hounsfield has been widely recognized for his endeavors in both CAT and NMR. His awards, in addition to the Nobel Prize, include the Barclay Prize of the British Institute of Radiology (1974), the Albert Lasker Award (1975), and the Gairdner Foundation Award (1976). Hounsfield, who became head of the Medical Systems Division at EMI, is also a fellow of the British Royal College of Radiologists and an honorary fellow of the Royal College of Physicians. In addition, he has received more than a dozen honorary doctorates from renowned universities worldwide.

In that light, it has been of interest to some that neither Hounsfield nor Cormack had doctorates at the time of his great discovery. Cormack also achieved acclaim, serving as chairman of the Tufts University Physics Department (1968-1976) and reaching Tufts' highest professorial rank, university professor (1980). Tufts also awarded Cormack an honorary doctor of science degree (1980), and he was elected to the American Academy of Arts and Sciences.

Bibliography

Ambrose, James. "Computerized Transverse Axial Scanning (Tomography): Part 2, Clinical Application." *British Journal of Radiology* 46 (October, 1980): 1023-1047. Describes CAT scanner applications, indicating that data are used to construct images that allow qualitative and quantitative examination. Notes that tomograms are examined in the same way as conventional radiographs and that lesions appear as altered density of soft tissues, interpreted in the light of known pathological changes.

Assessing Computed Tomography. Rockville, Md.: National Center for Care Technology, 1981. Part of a monograph series designed for technical exploration of societal implications of health care technology. Component articles address evaluation of computed body tomography, brain imaging, computed tomography of the head and spine, a neurologic overview of the technique, and contributions of the technique to neurosurgery.

Di Chiro, Giovanni, and Rodney A. Brooks. "The 1979 Nobel Prize in Physiology or Medicine." *Science* 206 (November 30, 1979): 1060-1062. Describes the development, fundamental operation, and evolution of the CAT scanner. Also identifies the preeminence of Hounsfield's contributions, without overlooking other contributors to the field. Discusses some factors contributing to the success of Hounsfield's endeavors. Includes bibliographical material.

Hounsfield, Godfrey N. "Computed Medical Imaging: Nobel Lecture, December 8, 1979." *Journal of Computer Assisted Tomography* 4 (October, 1980): 665-674. A Nobel acceptance lecture; describes aspects of development of the CAT scanner, focusing on early tests, principles of the technique, improvement of the system, its accuracy, and the relationship between resolution and picture noise. Discusses future improvements; includes important diagrams and tomograms.

_____. "Computerized Transverse Axial Scanning (Tomography): Part 1, Description of System." *British Journal of Radiology* 46 (October, 1980): 1016-1022. Describes the original CAT scanner. Explains the system wherein multiple X-ray readings are converted, by computer, to a series of pictures of cranial slices. Notes that the system is one hundred times as sensitive as conventional X-ray systems and able to display variations in soft tissues of similar density.

Robb, Richard A. *Three-Dimensional Biomedical Imaging*. Vol. 1. Boca Raton, Fla.: CRC Press, 1985. A technical book that provides useful information and important illustrations for the interested reader. A series of articles on major aspects of biomedical imaging. Chapters 3 to 5 cover many aspects of CAT scanning, renamed X-ray computed tomography (CT) here. Includes technical description of the basic principles of the method and the implementation and expansion of its applications; covers some advanced CT systems and their use.

S., B. M. [staff writer]. ". . . and Cormack, Hounsfield for Medicine." *Physics Today* 32 (December, 1979): 19-20. Describes the development of the CAT scanner and includes some biographical information for both laureates. Briefly describes aspects of the operation, theory, and economics of the scanner. Mentions an early solution of the mathematical aspect of tomography by Johann Radon.

Sanford S. Singer

Cross-References

Röntgen Wins the Nobel Prize for the Discovery of X Rays (1901), p. 118; Salomon Develops Mammography (1913), p. 562; Berger Develops the Electroencephalogram (EEG) (1929), p. 890; Moniz Develops Prefrontal Lobotomy (1935), p. 1060; Cerletti and Bini Develop Electroconvulsive Therapy for Treating Schizophrenia (1937), p. 1086; X Rays from a Synchrotron Are First Used in Medical Diagnosis and Treatment (1949), p. 1336; Donald Is the First to Use Ultrasound to Examine Unborn Children (1958), p. 1562.

GELL-MANN FORMULATES THE THEORY OF QUANTUM CHROMODYNAMICS (QCD)

Category of event: Physics
Time: September, 1972
Locale: Pasadena, California

Gell-Mann developed the theory of quantum chromodynamics to describe the characteristics of elementary particles (quarks)

Principal personages:

MURRAY GELL-MANN (1929-   ), an American physicist who pioneered the discovery and classification of subatomic particles, hypothesized the existence of quarks and postulated the theory of quantum chromodynamics, and was the winner of the 1969 Nobel Prize in Physics

HARALD FRITZSCH (1943-   ), a German physicist who assisted Gell-Mann in developing the theory of quantum chromodynamics

WILLIAM BARDEEN, an American physicist who assisted Gell-Mann and Fritzsch in developing the theory of quantum chromodynamics

Summary of Event

From antiquity, it had been speculated by philosophers and scientists that everything was made up of successively smaller particles, but that beyond a certain size, the particles became no smaller. These fundamental particles, of which everything was made, the Greeks called *atomos*, meaning indivisible. The English version of *atomos* became "atom." That idea held for thousands of years, moving from mere philosophical speculation to confirmed scientific detail by the beginning of the nineteenth century.

In the seventeenth century, Robert Boyle was instrumental in bridging the gap between the Greek philosophy and actual scientific manipulation of elementary atomic theory. John Dalton expanded on the idea and forged atomic theory into a full-blown experimental science based on elemental activity. It was not until the beginning of the twentieth century that it was discovered that atoms were not, in fact, indivisible, but rather that atoms were actually made up of even smaller parts.

In 1904, the first suggestion was made that the atom incorporated tiny subparticles called electrons and that they orbited a central core. In 1910, Ernest Rutherford, an English physicist, discovered the atom's core, and it was called the nucleus. Three years later, one of Rutherford's students, Niels Bohr, a Danish physicist, qualified the nature of the electron orbital about the nucleus. By 1927, the atom's structure had been largely solved. A new science called quantum mechanics defined the atom's internal structure as consisting of tiny electrons "orbiting" at a distance from the nucleus, which contained an assortment of relatively heavier protons and neutrons. By 1930, a concentrated effort was launched by a newly emerging branch of physics (particle physics) to probe even deeper into the atom's secrets. Much evidence existed that there were smaller particles yet to be discovered within the atom's core.

The first particle accelerator (atom smasher) was put into experimental use in 1932. The purpose of the accelerator was literally to cause one atom to collide with another at extremely high speeds and break up into their elementary parts. Physicists then record the particles as they fly off in the collision. It was during a series of these particle accelerator experiments in the early 1960's that California Institute of Technology physicist Murray Gell-Mann developed a series of brilliant postulations regarding the results of these particle accelerator experiments. By late 1963, Gell-Mann had enough evidence to publish his theory that the nucleus of protons and neutrons was made up of even smaller particles. In reference to a passage in James Joyce's book *Finnegans Wake* (1939), Gell-Mann called these small pieces of protons and neutrons "quarks." He said he chose this name as "a gag . . . a reaction against pretentious scientific language." He published the first discussion of quarks in February, 1964. In 1969, he was awarded the Nobel Prize for his subatomic classification schemes.

Gell-Mann postulated no less than six different kinds of quarks called "flavors" (up, down, bottom, top, strange, and charm), and, according to Gell-Mann, each of the quark flavors comes in three "colors" (red, green, or blue). The assignment of "colors" to quarks gave rise to a whole new branch of quantum physics called "quantum chromodynamics."

It was discovered that a proton or neutron is made up of three quarks, one of each color. A proton consists of two up quarks and one down quark, while a neutron consists of two down quarks and one up quark. The quarks in both proton and neutron are bound tightly together.

The assignment of flavors and their names and the colors and their names were made in continuation of Gell-Mann's rebellion against pretentious scientific names. Such convention seemed to be popular in 1960's physics; however, the names assigned by the use of the bizarre nomenclature stuck permanently. Nevertheless, it does give rise to some confusion. Obviously, there are no actual "flavors" in quantum mechanics (much less flavors defined as up, down, and so on), and likewise, at the subatomic level, there are no actual "colors." These terms—flavors and colors—define the specific quantum characteristic of the elementary particle. Through classification and subclassification in the quantum chromodynamic nomenclature, the particles can be classed according to their characteristics and behavior.

In 1972, quarks had been assigned their flavors already, but the colors had not been apportioned yet. Additional work on the particle accelerators demonstrated that there were some unexplained discrepancies in experiments using the Gell-Mann quark theory. Gell-Mann collected the data and conferred with his colleagues Harald Fritzsch and William Bardeen. They realized that their original definition of quarks still had a missing dimension in addition to flavor, which they called color.

Gell-Mann chose the dimension of color, because outside the destruction of the accelerator, the quark particles are fixed together into consolidated particles called "hadrons," where their color constituents cancel so that they are composite "white." Only when they are torn apart by the collisions in the accelerator does their composite white color decay into their individual colors—the fundamental colors red, green, and blue. Gell-Mann, Fritzsch, and Bardeen united the color concepts with the other quark ideas into a single formulation that united all the aspects of nuclear particles. Gell-Mann called the theory "quantum chromodynamics" (QCD). He presented this QCD theory in September, 1972.

In the theory of QCD, the multicolored quarks are held together by a binding force called "gluons." This aspect of a binding force is not only critical to any discussion of QCD but also fundamental to all of nature and drives the community of particle physics even today. Gluons make up what is called the "strong force." Of the four forces of nature, there is the gravitational force that works over astronomical distances and is comparatively very weak. The electromagnetic force exists between atoms and molecules. The weak force is responsible for all radioactivity. The strong force exists in the form of gluons that tie quarks together.

Quantum chromodynamics describes the forces holding quarks inside packets called hadrons. QCD explains that the quarks are so tightly bound within the hadrons that the force of the gluon holding the quarks in its hadron increases with distance. The force is so strong that if one attempted to liberate a quark from a hadron, one would only create another hadron in the process. One fundamental particle of nature is not addressed by QCD. Subatomic entities called "leptons" do not enter into the QCD theory because QCD is concerned with gluons and quarks. The six different kinds of leptons make up the electron family.

Impact of Event

Quantum chromodynamics is a work of science that stands as a benchmark hypothesis on the landscape of physical theory. It fulfills a long-term dream of physicists to have a complete theory of one of the four fundamental forces of nature: the strong nuclear force and how it interacts with elementary particles at the atomic core.

As a prelude to QCD, Richard P. Feynman, Julian Schwinger, and Shin'ichirō Tomonaga formulated what was known as quantum electrodynamics (QED) in the 1940's. QED became known as a "field theory" of quantum mechanics. In terms most basic to particle physics, it is also called a "relativistic quantum field theory." It unifies all that is known about the interaction of light with electrons, and it became the field theory that summarized the experimental and theoretical fundamentals of the electromagnetic force. Quantum chromodynamics was formulated about twenty years later. QCD is a relativistic quantum field theory; it consolidates the experimental and theoretical information of the strong forces within the atomic core.

Immediately, QCD clarified a mixture of perplexing observations that had been compiled from numerous accelerator experiments. It enabled a clear understanding of some previously undefined observations. Futhermore, it enabled prediction in the sense that it so completely described the workings within the nucleus that physicists were able to predict some events before the experiments took place, the ultimate validity of any theory. QCD is an exceptionally difficult array of complex mathematics that string probabilistic events together in a bewildering assortment of mathematical events. Because of this degree of difficulty, it becomes an intricate and enigmatic task to relate the data streaming in from particle accelerators to the field theory itself. Supercomputers have been employed to handle such processing, and all the final possible results from QCD have yet to be compiled.

It is the dream of physicists to unite the field theories into a single "grand unified field theory." Already, there is a single theory that unifies the weak and electromagnetic force field theories, called the "electro-weak unified field theory." The final goal is to unify all four field theories into a single grand unified theory of nature in which the success of defining precisely the interactions of the nucleus can be applied to all of nature.

Bibliography

Crease, Robert P., and Charles C. Mann. *The Second Creation*. New York: Macmillan, 1986. Crease and Mann follow the development of twentieth century physics from its nineteenth century roots to the most enigmatic mysteries of the late 1980's. It microscopically examines characters and personalities as well as the issues of physics. Gell-Mann's approach to QCD is examined in detail and how it fits into the other experimental questions of its time. His relationship to the ongoing work at research centers is detailed. Highly readable; probably the most complete book ever written on the personalities and work of twentieth century particle physicists.

Hawking, Stephen W. *A Brief History of Time*. New York: Bantam Books, 1988. In this eminently readable work, a prominent physicist examines the universe from his view of creation to the late 1980's. Hawking examines the far-flung reaches of space and time from black holes to the interior of the atom and discusses the elementary particles of the atomic nucleus. Written for a wide audience; illustrated.

Pagels, Heinz R. *The Cosmic Code*. New York: Simon & Schuster, 1982. This book describes quantum physics as "the language of nature." Pagels, a physicist, embarks on a literary quest to explain some of the most profoundly difficult topics in quantum physics in a clear manner for the general reader. Pagels opens up the interior of the atom for a clear inside view. Illustrated.

_____. *Perfect Symmetry*. New York: Simon & Schuster, 1985. In this follow-up book to *The Cosmic Code* (cited above), Pagels delves deeper by providing a layperson's perspective. Pagels supports a lively discussion of the grand unified field theories. He also touches on the frontiers of atomic physics and speculates on where the unified field theories will ultimately lead. Very readable and has become a classic of the genre.

Sutton, Christine. *The Particle Connection*. New York: Simon & Schuster, 1984. Sutton, a physicist turned reporter, explains the particle accelerator. She discusses how the machine is used and the nature of the particle chase at CERN, the European particle accelerator laboratory. Illustrated; written so that the student with a reasonable grounding in science can grasp its message.

Trefil, James S. *The Unexpected Vista*. New York: Charles Scribner's Sons, 1983. This book seeks to explain some of the most exciting concepts of physics to the general reader. In a wide-ranging exposé of contemporary ideas, Trefil explains concepts from magnets to a clear discussion of the ultimate theory of grand unification. For a wide audience; illustrated.

Dennis Chamberland

Cross-References

Thomson Wins the Nobel Prize for the Discovery of the Electron (1906), p. 356; Bohr Writes a Trilogy on Atomic and Molecular Structure (1912), p. 507; Rutherford Presents His Theory of the Atom (1912), p. 527; Rutherford Discovers the Proton (1914), p. 590; Einstein Completes His Theory of General Relativity (1915), p. 625; Lawrence Develops the Cyclotron (1931), p. 953; Chadwick Discovers the Neutron (1932), p. 973; Cockcroft and Walton Split the Atom with a Particle Accelerator (1932), p. 978; Hofstadter Discovers That Protons and Neutrons Each Have a Structure (1951), p. 1384; The Liquid Bubble Chamber Is Developed (1953), p. 1470; Georgi and Glashow Develop the First Grand Unified Theory (1974), p. 2014.

TEXAS INSTRUMENTS INTRODUCES
THE FIRST COMMERCIAL POCKET CALCULATOR

Category of event: Applied science
Time: September, 1972
Locale: Dallas, Texas

The production of a portable, reliable, hand-held calculator by Texas Instruments brought to reality the dream of mathematicians that had gone unfulfilled for more than five thousand years

> *Principal personages:*
> JACK ST. CLAIR KILBY (1923-), the inventor of the first monolithic integrated circuit (popularly known as the semiconductor microchip)
> JERRY D. MERRYMAN (1932-), the project manager of the Texas Instruments team that invented the first portable calculator
> JAMES VAN TASSEL (1929-), an inventor and expert on semiconductor components

Summary of Event

In the earliest accounts of civilizations that developed number systems to record mathematical calculations, evidence has been found of efforts to fashion a device that would permit people to perform these calculations with reduced effort and increased accuracy. The ancient Babylonians are regarded as the inventors of the first abacus (or counting board, from the Greek *abakos*, meaning "board" or "tablet"). It was originally little more than a row of shallow grooves with pebbles or bone fragments as counters, but it enabled people five thousand years ago to calculate much more efficiently than the often awkward and complex number systems.

The introduction of Hindu-Arab numerals into Europe during the eighth and ninth centuries suggested that further advances were possible, but the next step in mechanical calculation did not occur until John Napier, a Scottish baron, placed figures on rods in the early seventeenth century. This concept led to the first slide rule, a pair of circles numbered by William Oughtred of Cambridge, which made it possible to perform rough but rapid multiplication and division. Oughtred's invention in 1623 was paralleled by the work of a German professor, Wilhelm Schickard, who built a "calculating clock" in the same year, but because the record of his work was lost until 1935, the French mathematician Blaise Pascal generally was thought to have built the first mechanical calculator, the "Pascaline," in 1645.

Other versions of mechanical calculators were built in subsequent centuries, but none was rapid or compact enough to be useful beyond specific laboratory or mercantile situations. The dream of such a machine continued to fascinate scientists and mathematicians, but the crucial development that made it possible did not occur until the middle of the twentieth century, when Jack St. Clair Kilby of Texas Instruments invented the silicon microchip (or integrated circuit) in 1958. The chip is a sliver of germanium hardly more than a centimeter long that could do the work of the much larger transistor. Kilby had been familiar with the transistor (invented by Bell Laboratories in 1947) from his work at Centralab Inc., a radio and television parts manufacturer in Milwaukee, Wisconsin. In joining Texas Instruments, a pioneering company that had manufactured the first silicon transistor in 1954, he had the opportunity to pursue his ideas about putting all the individual components of a circuit on an integrated, single, compact base. Patrick Haggerty, then president of Texas Instruments, had written in 1964 that "integrated electronics" would "remove limitations" that determined the size of instruments, and he recognized that Kilby's invention of the microchip made the creation of a portable, hand-held calculator a practical possibility. He challenged Kilby to put together a team to design a calculator that would be as powerful as the large, electromechanical models in use at the time but small enough to fit into a coat pocket. Working with Jerry D. Merryman and James Van Tassel, Kilby began to work on the project in October, 1965.

Five elements had to be designed. The logic designs, which enabled the machine to perform the actual calculations; the outer keyboard, which signaled the assigned problem to the functioning circuit; the power supply, which drove the entire operation; the readout, which provided the answer to the problem; and the outer enclosure, which contained the works, all required specific creative solutions to what were basically new problems. Kilby recalls that once a particular size for the unit had been determined (something that could be easily held in the hand), project manager Merryman was able to develop the initial logic designs in three days. Merryman had been able to accomplish everything required without exceeding the relatively meager power supply that would be available from the low-output batteries necessary to ensure portability.

Van Tassel contributed his experience with semiconductor components to solve the problems of packaging the integrated circuit. Most available integrated circuits in use had fourteen or sixteen leads, and Van Tassel was obliged to work with 120 leads, the minimum required to handle the basic functions Kilby believed were crucial for success. Van Tassel had to determine the coefficient of expansion of the separate chips and to devise a way to make reliable contacts among the chips so that the integrated circuit was attached securely at all points. In addition, the chips that were available were inconsistent in quality and had to be individually probed. Those found to satisfy the standards had to be individually matched for common workable areas. The display to record the results required a thermal printer that would work on the low power source, and it had to include a microencapsulated ink source that was available when the paper readout was pressed against a heated digit. After the result was recorded in this manner, the paper had to be advanced for the next calculation. Kilby, Merryman, and Van Tassel filed for a patent on their work in 1967.

Although this relatively small, working prototype of the minicalculator made the transistor-operated design of the much larger desk calculator obsolete, the cost of setting up new production lines and the necessities of developing a market made it impractical to begin production immediately. Instead, Texas Instruments and Canon,

Inc., of Tokyo formed a joint venture, which led to the introduction of the Canon Pocketronic Printing Calculator in Japan in April, 1970, and in the United States that fall. Built entirely of Texas Instruments parts, this four-function machine with three metal oxide semiconductor (MOS) circuits was similar to the prototype designed in 1967, but it weighed about one-third less because of a plastic case replacing the almost solid aluminum case in the original. It used nickel-cadmium batteries, which were lighter in weight and reliable enough to replace the silver-zinc batteries in the original. The calculator was priced at four hundred dollars, weighed 740 grams, and measured 101 millimeters wide by 208 millimeters long by 49 millimeters high. It could perform twelve-digit calculations and worked up to four decimal places.

In September, 1972, Texas Instruments put the Datamath, its first commercial hand-held calculator using a single MOS chip, on the retail market. It weighed 340 grams and measured 75 by 137 by 42 millimeters. The Datamath was priced at $120 and included a full-floating decimal point unit that could appear anywhere among the numbers on its eight-digit light-emitting-diode (LED) display. It came with a re-chargeable battery that could also be connected to a standard AC outlet. It used algebraic forms for its four basic functions—addition, subtraction, multiplication, and division—and enabled the user to press the keys as the problem progressed. At the heart of the calculator, the integrated semiconductor circuit (the silicon chip) contained all the necessary electronics for performing all functions, and it had ten digit keys, seven function keys, and a decimal location key. When the battery needed charging, all eight decimal points were activated. Beyond its standard computation capability, the Datamath would turn off the display except for the character in the first position after fifteen seconds if the action was interrupted, thereby conserving power until another keyboard entry was made. One of its more sophisticated features was a constant/chain switch to offer a selection between a function permitting multi-plication or division by a constant or making available the last calculation for use in further calculation in a chain of entries. This limited memory element anticipated the evolution of the calculator/computer combination with much more extensive memory storage. The Datamath had a total of 111 parts, 43 of which were electronic.

Impact of Event

In recognition of the significance of the work by the Texas Instruments team of Kilby, Merryman, and Van Tassel, in 1975 the Smithsonian Institution accepted the world's first miniature electronic calculator for its permanent collection, noting that it was the forerunner of more than 100 million pocket calculators then in use. By the 1990's, more than 50 million portable units were being sold each year in the United States, and many units selling for less than ten dollars could do much more than the CS-104 made by Sharp Corporation of Tokyo in 1964, a 55-pound model that sold for twenty-five hundred dollars. Kilby's dream had been realized.

Beyond the immediate practicality of the device, however, this invention, which heralded the age of the low-cost consumer-portable calculator and was instrumental in launching the still-growing electronic calculator industry, also revolutionized the manner in which the human race related to the world of numbers. Instead of gaining familiarity with figures by painstakingly adding long columns of numbers, children were now able to combine the basic principles of addition and subtraction with the instant responses to problems the simple calculators provided. Thus, they may have become less skilled with numbers, but they were able to overcome the fear of numbers that held many people back from even a preliminary understanding of mathematics. Similarly, while classes in many fields of engineering and industrial education were occupied previously by the laborious tasks of working out routine problems through numerous steps, the calculator permitted students to race through multi-stepped problems—the classic extensive number crunching required in many graduate laboratory projects in particular—and permitted more classroom time for more creative and analytical thinking.

Prior to 1970, most calculating machines were of such dimensions that professional mathematicians and engineers were tied to their desks or carried slide rules when they were compelled to work in the field. By 1975, Keuffel & Esser, the largest slide rule manufacturer in the world, was producing its last model, and mechanical engineers found that problems that had previously taken a week could now be solved in an hour. As Irene Kim, a news editor for *Mechanical Engineering* magazine, stated, four major aspects of engineering could be handled by calculators, including "running programs on calculators; using preprogrammed calculators as control devices; using calculators to help write programs on a computer; and basic calculating which may or may not include programming." The portability of the hand-held calculator made it ideal for use in remote locations, such as those a petroleum engineer might have to explore, while its rapidity and reliability made it an indispensable instrument for construction engineers, architects, and real estate agents, who could figure the volume of a room and other building dimensions almost instantly and then produce cost estimates by referring to previous programs almost on the spot. Beyond all the uses that the mechanical engineering professions required, the nonprofessional segment of the population that has begun to depend on remote-control devices to operate television monitors and VCR machinery and small cellular phones for access to clients, customers, and colleagues has also developed new habits of communication and apprehension dependent upon Texas Instruments' pioneering work.

Bibliography

Augarten, Stan. *Bit by Bit: An Illustrated History of Computers.* New York: Ticknor & Fields, 1984. Includes an excellent history of calculating machines and a chapter on Kilby's work. Written very lucidly; well illustrated with diagrams and color photographs, as well as black-and-white photographs and portraits. Excellent notes and bibliography.
Braun, Ernest, and Stuart Macdonald. *Revolution in Miniature: The History and Impact of Semiconductor Electronics.* Cambridge, England: Cambridge University Press, 1978. Though not concentrating specifically on the use of the calculator,

this thorough and clearly written volume covers the entire field of semiconductor electronics, explaining the place of the calculator in terms of historic developments. Most of the material is for the well read or specialist with some knowledge of scientific terminology.

Editors of *Electronics. An Age of Innovation: The World of Electronics, 1930-2000.* New York: McGraw-Hill, 1981. One of a series of general science/textbook publications, it is written clearly with the general reader in mind. Many color illustrations, explanatory illustrations, diagrams, tables, glossary, and index.

Kim, Irene. "Functions at the Fingertips." *Mechanical Engineering* 112 (January, 1990). A very informative and accessible essay about the work of Kilby, Merryman, and Van Tassel, combined with a survey of the multiple uses of hand-held calculators by mechanical engineers in the last decade of the twentieth century.

Reid, T. R. *The Chip: How Two Americans Invented the Microchip and Launched a Revolution.* New York: Simon & Schuster, 1984. A thorough, detailed history of the invention and development of the microchip, combining history and technology on both the level of the general reader and the scientist.

Leon Lewis

Cross-References

Shockley, Bardeen, and Brattain Discover the Transistor (1947), p. 1304; UNIVAC I Becomes the First Commercial Electronic Computer and the First to Use Magnetic Tape (1951), p. 1396; Sony Develops the Pocket-Sized Transistor Radio (1957), p. 1528; Easki Demonstrates Electron Tunneling in Semiconductors (1957), p. 1551; Bubble Memory Devices Are Created for Use in Computers (1969), p. 1886; The Microprocessor "Computer on a Chip" Is Introduced (1971), p. 1938.

JANOWSKY PUBLISHES A CHOLINERGIC-ADRENERGIC HYPOTHESIS OF MANIA AND DEPRESSION

Category of event: Medicine
Time: September 23, 1972
Locale: Vanderbilt University, Nashville, Tennessee

Janowsky and coworkers proposed a cholinergic-adrenergic hypothesis of mania and depression, which facilitated the understanding of manic depression and the design of treatment methodology

Principal personages:

DAVID STEFFAN JANOWSKY (1939-   ), an American psychiatrist who developed the cholinergic-adrenergic hypothesis of mania and depression

JOHN MARCELL DAVIS (1933-   ), an American psychiatrist who collaborated with Janowsky in the development of the cholinergic-adrenergic hypothesis

Summary of Event

Hospital beds and mental institutions all over the United States are filled with the victims of serious mental illness. There are so many people affected by mental illness that large numbers of its "milder" victims have been released from these overcrowded institutions. Many of them wander the streets of the cities, homeless and in severely mentally impaired states. Americans read about mental illness in the newspapers and hear about it on radio and television.

What is mental illness? How is it cured? These important questions have been asked for many thousands of years. Less is known about their answers than is desired. Mental illness, however, is often divided into two basic kinds: the "organic" and the "functional" types. Organic mental illness results from an injury or a known disease (for example, diabetes) that alters the structure of the brain, changes its ability to function correctly, or affects some other part of the nervous system. Cure of this type of mental illness depends upon surgery and other methods that cure the causative disease.

The basis for functional mental illness—often called affective disorder—is more subtle and, therefore, has evaded clear understanding more easily. It is defined as being caused by operational flaws of mental function. The most severe types of functional mental illness are collectively called insanity. One of these, manic-depressive psychosis, will be considered. The manic-depressive person alternates very rapidly between an excessively happy (manic) state and a severely depressed (depressive) state. Consequently, such people are not capable of coping with the world around them. Hypotheses concerning such affective disorders, now termed the manic-depressive psychosis, date back to the father of medicine, Greek Hippocrates of Cos (460-377 B.C.), who coined the term "melancholia" to describe severe depression. Hippocrates suggested that melancholia was caused by the accumulation of "black bile and phlegm, which darkened the spirit and made it become melancholy." Modern understanding of the phenomenon begins with consideration of the function and dysfunction of the nervous system in which it occurs.

The human nervous system is composed of a 1.4- to 1.8-kilogram central computer—the brain—made up of cells called neurons and a network of neuron wires—the nerves—that communicate signals to the rest of the body via nerve impulses. When nerve impulses pass through the nervous system correctly, they allow one to recognize and to respond appropriately to the world. Dysfunction of nerve impulse generation and passage through the nerves is believed to lead to functional mental illness. Therefore, effective passage and correct control of nerve impulses is essential to a normal mental state.

Understanding of normal and pathologic nervous system function requires the explanation of the terms synaptic gap and neurotransmitter action. To begin with, the neurons are separated from each other by tiny spaces, about twenty-millionths of a centimeter wide, the synaptic gaps. The passage of nerve impulses through nerves requires them to cross thousands of synaptic gaps in their travel. Nerve impulse transport across synaptic gaps is mediated by the biochemicals called neurotransmitters. The best known of the neurotransmitters is acetylcholine, which acts in "cholinergic" nerves. Dysfunction of cholinergic nerves, via disruption of acetylcholine action, is thought to be a major component of mental disease. This idea arose, in part, from observation of impaired mental function in people exposed to small amounts of insecticides and nerve gases that act by disrupting acetylcholine production and use.

Other neurotransmitters associated with mental disease include catecholamines and indoleamines. The main catecholamine neurotransmitters are the hormone epinephrine and its close cousins norepinephrine and dopamine, produced by the adrenal glands. Catecholamines control nerve impulse transmission by "adrenergic" portions of the nervous system. The indoleamines, especially serotonin, function in neurons related to sleep and sensory perception. They are believed to be associated with symptoms of affective disorders that include sleep and sensory dysfunction.

The current theories of depression and mania arose from the catecholamine (actually, norepinephrine) hypothesis of affective disorders. This hypothesis was proposed in 1965, by Joseph Schildkraut, in *The American Journal of Psychiatry*, and others. The hypothesis focused mostly on the adrenal catecholamine, norepinephrine. It was proposed that the depressive state arose because of suboptimum norepinephrine production or utilization (decreased noradrenergic activity) and that the manic state arose from excess norepinephrine production or utilization (increased noradrenergic activity).

Acceptance of the catecholamine hypothesis led to the examination of norepinephrine levels in normal and mental disease states; efforts to use these levels to explain how existing drugs, electric shock, and other known treatments affected mania and depression; attempts to choose new drugs for therapeutic use on the basis of their effects on the norepinephrine levels; and study of other catecholamines and related "biogenic amines." These efforts led to the modification of the catecholamine hypothesis. First, dopamine (a catecholamine cousin of norepinephrine) was implicated in the function of the central nervous system. Then, it was noticed that several of the major tranquilizers (for example, reserpine) decreased both the dopamine and the norepinephrine levels. Consequently, it was suggested that the catecholamine hypothesis should include dopamine. In fact, low central nervous system levels of dopamine soon became even more intimately associated with depression than low norepinephrine levels.

The biogenic indoleamine, serotonin, was also implicated in depression because it, too, was depleted by major tranquilizers like reserpine. Then, it was shown that the action of "tricyclic antidepressants" was related mostly to serotonin levels. Because of this, an indoleamine (or serotonin) hypothesis of affective disorders was born.

In 1972, David Steffan Janowsky, John Marcell Davis, and coworkers at Vanderbilt University's Psychiatry Department proposed a cholinergic-adrenergic hypothesis of mania and depression, in *The Lancet*. The hypothesis focused upon the cholinergic neurotransmitter, acetylcholine. This, at once, expanded the conceptual basis of manic-depressive psychosis. Janowsky's theory, unlike others before it, recognized the importance of interaction between the various systems involved in nervous transmission and proposed that the affective state of any individual represents a balance between the noradrenergic and the cholinergic activity. Janowsky and his colleagues proposed that depression was a disease of "relative cholinergic predominance," while mania was defined as being a disease of relative "adrenergic predominance." They also suggested that manic-depressive illness was caused by compensatory overreaction of the central autonomic nervous system.

Impact of Event

Manic-depressive psychosis is a worldwide affective disorder that has a lifetime prevalence rate of between 1 percent and 2 percent. Those people afflicted with the disease exhibit severe emotional disturbance and mood swings that make it difficult for them to function within the framework of reality. Their symptoms include greatly disordered thought processes, thought disturbance, delusions and hallucinations, and feelings of grandeur and its reverse, severe depression. The chronic nature of these symptoms—occurring episodically—places manic depressives at serious social and financial risk, and many of them end up in mental institutions.

While there is still no sure treatment for the disease, current explanations of the phenomenon and its effective treatment began with the catecholamine hypothesis of affective disorders proposed in 1965. This hypothesis focused mostly on the adrenal catecholamine, norepinephrine, and supposed that the depressed state arises from suboptimum norepinephrine production or utilization, while manic states arises from its excess. Acceptance of the hypothesis generated endeavors that included the use of norepinephrine levels to explain how existing therapeutic drugs and other known treatments affected the disease, and the choice of new drugs for therapeutic use on the basis of their effects on norepinephrine levels. Soon, other catecholamines, related biogenic amines, and the nerve transmission process were implicated as well. Generally, individual researchers believed that a single phenomenon or system was responsible for manic depression.

Janowsky did not agree with this view. In his often-cited 1972 article (in *The Lancet*) he stated, "In most investigations and reviews, a single-monoamine hypothesis of affect has been proposed. We postulate that the cholinergic nervous system is also involved in the regulation of affect." His cholinergic-adrenergic hypothesis of mania and depression was the next logical step in defining the disease process and its possible treatment. That is, recognizing the importance of interaction among the several systems participant in the function of the nervous system and proposing that the affective state of an individual represents a balance between these systems.

It seems probable that this concept helped to spur on the discovery and delineation of the many concepts and treatment modalities now utilized by psychiatric practitioners to treat manic depression. Such methodology, including psychoendocrine concepts, tricyclic antidepressants, lithium, and inhibitors of monoamine oxidase inhibitors, is well described in *Depression and Mania* (1988), edited by Anastasios Georgotas and Robert Cancro. There, the legacy of Janowsky's efforts is seen, in the realization of the interactive nature of the various treatments, despite lack of overall agreement on their relative importance.

Bibliography

Cohen, Seth, and David Dunner. "Bipolar Affective Disorder: Review and Update Depression." In *Modern Perspectives in the Psychiatry of the Affective Disorders*, edited by John G. Howells. New York: Brunner/Mazel, 1988. The article reviews manic-depressive psychosis, describes its clinical appearance, and outlines the methods used for its differential diagnosis. Other aspects of the disease that are covered include its epidemiology, genetics, outcome, and treatment with lithium. The authors provide forty-one useful references on all aspects of the disease they discuss.

Georgotas, Anastasios, and Robert Cancro, eds. *Depression and Mania*. New York: Elsevier, 1988. Forty-two sophisticated articles deal with aspects of mania and depression, including historical issues, epidemiology, diagnosis, disease etiology, medications and other treatments, ways to assess the disease, laboratory tests, and how to differentiate manic depression from related psychoses. The book is quite technical, but it contains much information of use to average readers.

Janowsky, David S., et al. "Neurochemistry of Depression and Mania." In *Depression and Mania*, edited by Anastasios Georgotas and Robert Cancro. New York: Elsevier, 1988. The article reviews development of modern concepts of the neurochemistry of these mental diseases, including norepinephrine and the catecholamine hypothesis; acetylcholine and the cholinergic-adrenergic hypothesis; and serotonin, dopamine, neuropeptides, and other neurotransmitters. Interaction of these systems is stressed throughout the text. More than 160 references on historical and modern topics are included.

Janowsky, David S., Dominick Addario, and S. Craig Risch. *Psychopharmacology Case Studies.* 2d ed. New York: Guilford Press, 1987. Sixty-four case studies are presented. They outline practical use of current psychopharmacologic theory to treat mania, depression, and other mental diseases. Some topics included are treatment of affective disorders, complications of regular medical drug use, side effects of antipsychotic drugs, antipsychotic drug maintenance, working with the elderly, and iatrogenic psychologic effects.

Janowsky, David S., M. Khaled El-Yousef, John M. Davis, and H. Joseph Sekerke. "A Cholinergic-Adrenergic Hypothesis of Mania and Depression." *The Lancet* 2 (September, 1972): 632-635. Janowsky's hypothesis focuses on cholinergic neurotransmission but recognizes the importance of the other systems involved. It proposes that affective state represents a balance between noradrenergic and cholinergic activity. Depression and mania are viewed as diseases of relative cholinergic and adrenergic predominance, respectively. Manic depression is deemed to be caused by overreaction of part of the nervous system.

Schildkraut, Joseph J. "The Catecholamine Hypothesis of Affective Disorders: A Review of Supporting Evidence." *American Journal of Psychiatry* 2 (November, 1965): 509-522. Schildkraut describes the clinical basis for the catecholamine (norepinephrine) hypothesis of affective illness. It is pointed out that depression and elation are associated with "catecholamine" deficiency and excess, respectively, at important brain sites. Evidence for the hypothesis, including effects of therapeutic drugs, is also cited.

Stryer, Lubert. *Biochemistry.* 2d ed. San Francisco: W. H. Freeman, 1981. Chapter 37 of this excellent biochemistry text gives a good summary of the concepts of nerve transmission and neurotransmitters. Topics presented include acetylcholine as a neurotransmitter, acetylcholine-related drugs and poisons, and catecholamines and other neurotransmitters. A number of useful illustrative diagrams and references are included.

Sussman, Norman, and Robert Cancro. "Differential Diagnosis of Manic-Depressive and Schizophrenic Illnesses." In *Depression and Mania*, edited by Anastasios Georgotas and Robert Cancro. New York: Elsevier, 1988. The article reviews various aspects of diagnosis of these psychoses, including clinical presentation of manic syndrome, associated thought disorders, delusions and hallucinations, mood disturbance, hyperactivity, and comparison. The issues delineated present a useful beginning for those who wish information in the area. Seventy references are given for those who wish more complete coverage of specific issues.

Sanford S. Singer

Cross-References

Sherrington Delivers *The Integrative Action of the Nervous System* (1904), p. 243;

Berger Develops the Electroencephalogram (EEG) (1929), p. 890; Moniz Develops Prefrontal Lobotomy (1935), p. 1060; Cerletti and Bini Develop Electroconvulsive Therapy for Treating Schizophrenia (1937), p. 1086; Wilkins Discovers Reserpine, the First Tranquilizer (1950's), p. 1353; Sperry Discovers That Each Side of the Brain Can Function Independently (1960's), p. 1635.

THE UNITED STATES GOVERNMENT BANS DDT USE TO PROTECT THE ENVIRONMENT

Category of event: Earth science
Time: December 31, 1972
Locale: United States

The United States Environmental Protection Agency banned the use of the pesticide DDT in order to protect human health and the environment

Principal personages:

WILLIAM D. RUCKELSHAUS (1932-   ), an American attorney, politician, and bureaucrat who made the decision to ban the domestic use of DDT except in cases of extreme emergency

RACHEL CARSON (1907-1964), an American environmentalist, marine biologist, and author whose 1962 book *Silent Spring* brought the scientific debate over the safety of widespread DDT use into the public eye

PAUL HERMANN MÜLLER (1899-1965), a Swiss chemist whose experiments in chlorinated hydrocarbons as the basis for insecticides ended in 1939 with the discovery of DDT

Summary of Event

On June 14, 1972, the Chief Administrator of the United States Environmental Protection Agency (EPA), William D. Ruckelshaus, announced that the pesticide use of the chemical known as DDT would be prohibited in the United States, except in cases of emergency, effective December 31, 1972. The ban ended nearly three decades of domestic use of the chemical, which had become controversial in the late 1950's and 1960's because of its potential to harm human health and the environment.

DDT is an abbreviation for the chemical name dichloro-diphenyl-trichloroethane. It was accidentally discovered by German chemist Othmar Ziedler in 1874, although it was not until the 1930's, when the Swiss chemical firm J. R. Geigy began searching for an economical pesticide to control potato beetles, that its insecticidal properties were discovered. The project was directed by Paul Hermann Müller, who won the 1948 Nobel Prize in Physiology or Medicine for applying DDT to public health crises. In 1942, Geigy representatives brought samples of a 5 percent solution of DDT to the United States Department of Agriculture (USDA) in the form of a pesticide spray and powder, trade name "Gesarol."

While domestic use of DDT did not begin until near the end of World War II, the United States military used the material to control disease-carrying insects such as typhoid-carrying lice in Italy and malaria-carrying mosquitoes in the South Pacific. By the end of the war, when production exceeded War Production Board requirements, the surplus was released for domestic civilian use. While DDT gained its early popularity in the realm of public health, its domestic use in the control of disease had dropped dramatically by 1972 with the advent of safer alternatives and, unfortunately, developing insect resistance to the poison. From 1946 to 1956, however, the United States Public Health Service carried out insect eradication programs in areas prone to mosquito infestation such as the New Jersey shore, the Mississippi River delta, and the Texas coast, where mosquito-borne diseases such as malaria and encephalitis remained in at least endemic levels. Aerial spraying of infested waterways reduced mosquito populations by upward of 90 percent.

The most important economic use of the substance was as an agricultural insecticide to control pests such as the boll weevil and the cabbage worm. For example, through 1972, the cotton industry applied pesticides to well over two-thirds of its acreage, accounting for 80 percent of all domestic DDT use. The USDA used DDT to control tree-destroying insects such as the gypsy moth, which was responsible for the destruction and defoliation of hundreds of thousands of acres of U.S. forests. Between 1945 and 1972, the USDA applied pesticides, mostly DDT, to an estimated 30 million acres of woodlands. Despite its obvious economic and public health benefits, concerns regarding DDT's health hazards had surfaced as early as the mid-1940's. These escalated in the wake of the growing environmental movement of the 1960's. Most evidence pointed to the fact that DDT posed a more significant threat to wildlife than to human health, although concerns over chronic effects increased with the 1962 publication of Rachel Carson's *Silent Spring*, the catalyst for the new American environmental movement in general and DDT opponents in particular.

The earliest medical tests conducted during the war showed that humans suffered few if any immediate adverse effects from DDT exposure and ingestion of low doses, although sufficiently high amounts could kill. Deaths from DDT poisoning fell into two categories: accidental ingestion of relatively high quantities, and suicide. According to the Food and Drug Administration (FDA), DDT posed no threat to human health when used in sufficiently low quantities—enough to kill pests, yet harmless to humans. In three decades of domestic use, no deaths were attributed to contact with or ingestion of low level residual amounts of DDT.

The United States Committee on Medical Research concluded a 1948 DDT study with the findings that DDT accumulated in the body and was stored in fatty tissues. In addition, DDT was regularly excreted in mother's milk and, therefore, could be passed on to nursing infants. The study showed that in the short term, this accumulation was a factor in minor kidney and liver damage, but since the study was conducted over a short period of time, it could not conclude whether this accumulation resulted in any long-term ailments or caused cancer. Tests of DDT's potency as a carcinogen conducted by the FDA and the National Cancer Institute remained inconclusive regarding cancer in humans, although liver cancer did occur in laboratory rats exposed to low levels of DDT over long periods of time. The ambiguity of studies of chronic health effects and cancer in humans were the main ingredients in Carson's *Silent Spring*. The novel told the story of an ecology destroyed by indiscriminate chemical usage and polarized anti- and pro-DDT coalitions, moving health and environmental disputes from the scientific literature into courtrooms and legislatures. DDT had most assuredly saved hundreds of thousands of human lives from disease and starvation, but it was becoming clear that careless application of the material was harming the environment. While fears over cancer and uncertainty over other chronic effects caused some public alarm, fears of widespread environmental damage were the major cause of DDT's eventual demise.

Scientists and laypersons alike began to notice the ecological effects of widespread DDT use as early as the late 1940's. One of the earliest signs was the high proportion of robins and other birds killed in areas where the chemical was used to control tree disease. Fish deaths were widely reported in New York's gypsy moth eradication campaigns. From the late 1940's into the early 1960's—corresponding to the highest use of DDT—the average thickness of some bird species' eggshells, most notably the bald eagle, peregrine falcon, and California brown pelican, decreased. This resulted in higher than normal mortality rates among offspring and posed the very real danger of extinction. When this information was combined with Carson's argument that DDT residues had spread as far as the polar regions, where no DDT spraying had occurred, citizen environmental groups, including the Environmental Defense Fund, pushed for legislation that would greatly curtail or ban DDT use.

Beginning in 1962, public hearings in Wisconsin, Michigan, and Arizona ushered in the eventual banning or restriction of DDT use in twenty-seven states. By 1970, the USDA canceled nearly all uses of DDT, including those on tobacco, shade trees, and aquatic areas. By 1971, federal responsibility for pesticide use was transferred to the EPA, which announced in January its intent to ban all nonemergency DDT uses. Public hearings beginning in August and lasting until March, 1972, included testimony from public health officials, chemical manufacturers, farmers, the USDA, and the Environmental Defense Fund. Faced with growing evidence of the dangers of DDT misuse and the fact that, in most economically sensitive applications, DDT use had already been severely curtailed, EPA Chief Administrator Ruckelshaus canceled all remaining crop uses of the chemical effective December 31, 1972. From that point forward, DDT could be used only in cases judged by the EPA to be public health emergencies.

Impact of Event

Since the 1972 ban on all crop uses of DDT came at a time when farmers and pesticide manufacturers were searching for, and implementing, alternatives to the chemical, the economic impact of the ban was minimal. The greatest impact of the ban itself could be seen in the effects of curtailed DDT use on the overall ecology. Sociologically, the ban was the culmination of a series of events, beginning with the publication of *Silent Spring*, which raised awareness of the effect of human actions on the environment and changed the nature of environmental disputes from purely scientific debates to issues of science, economics, and politics.

The EPA released a comprehensive impact statement on the DDT ban in July, 1975. They concluded that the ban had very little effect on agriculture and public health. While the ban increased the cost of cotton farming by at least six dollars per acre, the EPA estimated that the cost passed on to consumers resulted in an overall price increase of 2.2 cents per person. The bulk of this cost went to the development and implementation of DDT alternatives, such as methyl parathion. Most important, the ban seemed to have little effect on crop yield in cotton and other agricultural sectors.

The most visible effect of the ban was the recovery of species brought to near extinction by its use, especially bird species harmed by decreases in eggshell thickness. DDT's chemical stability in the environment had made it possible for it to accumulate in the food chain, starting with an accumulation in plankton, chemical resistant insect strains, invertebrates such as mollusks, and fish. While certain levels of DDT were lethally toxic to these organisms, some of those that survived were eventually eaten by predatory fish and birds. The accumulated DDT was then either excreted by these animals or retained in fat tissues. The retained DDT adversely affected the reproductive success of several fish and bird species, including the bald eagle, peregrine falcon, and brown pelican. The study noted that evidence of increased eggshell thickness in threatened bird species had accompanied an increase in their populations which coincided with curtailed DDT use.

Finally, the DDT controversy coincided with the rise of American environmental activism. Beginning in 1967, groups including the Environmental Defense Fund, the National Audubon Society, and the National Wildlife Federation carried the cause of the environment to courtrooms and legislatures. For these groups, DDT served as a highly visible target for their overall concern that indiscriminate use of synthetic chemicals was damaging the ecological balance. Their victory in the DDT dispute set a precedent for the continued use of the judicial and legislative branches of government to regulate human control of the environment.

Bibliography

Beatty, Rita Gray. *The DDT Myth: Triumph of the Amateurs.* New York: John Day, 1973. Presents the pro-DDT perspective, arguing that its benefits in public health and agriculture far outweighed its potential risks. Journalistic, not scientific, in its presentation.

Carson, Rachel. *Silent Spring.* Boston: Houghton Mifflin, 1962. This is the book that catapulted environmental concerns into the public eye. Although it is lucid in its presentation, it lacks scientific data to back its assertions that indiscriminate chemical use will lead to long-term environmental and human catastrophe, a "silent spring." Argues that humans must live within the environment rather than dominating it from without.

Dunlap, Thomas R. *DDT: Scientists, Citizens, and Public Policy.* Princeton, N.J.: Princeton University Press, 1981. The author presents the history of DDT use in the United States from the pre-DDT history of insects, disease, and insecticides in the United States to the 1972 ban. The account is well balanced, and presents both sides favorably. Contains excellent appendices for those interested in the more technical aspects of DDT contamination, production, and metabolism; includes an ample bibliography for those interested in further study.

Whitten, Jamie L. *That We May Live*. Princeton, N.J.: D. Van Nostrand, 1966. Whitten was chairman of the U.S. House Appropriations Subcommittee for Agriculture during the years of DDT use and wrote this book as a response to Carson's *Silent Spring*. He argues that pesticides are necessary for the survival of humans on earth, and that DDT opponents distorted scientific claims to the contrary. Often lost in the attacks on the environmental movement, showing the difficulties both sides in the dispute had in dealing with inconclusive scientific data on the detriments associated with DDT.

Zimmerman, O. T., and Irvin Lavine. *DDT: Killer of Killers*. Dover, N.H.: Industrial Research Service, 1946. This small volume is one of the first comprehensive books published on the use, properties, and detriments of DDT. It is obviously pro-DDT in its stance, for studies which revealed even slight human health and environmental concerns had yet to be conducted. Although it details some environmental dangers, it asserts that such dangers are present only when the chemical is improperly used. Contains a section on the dosage to be used on various crops and other applications. A fine example of the enthusiasm with which the chemical was received.

William J. McKinney

Cross-References

Gorgas Develops Effective Methods for Controlling Mosquitoes (1904), p. 223; Steinmetz Warns of Pollution in *The Future of Electricity* (1908), p. 401; Insecticide Use Intensifies When Arsenic Proves Effective Against the Boll Weevil (1917), p. 640; Müller Discovers That DDT Is a Potent Insecticide (1939), p. 1146; Carson Publishes *Silent Spring* (1962), p. 1740; Manabe and Wetherald Warn of the Greenhouse Effect and Global Warming (1967), p. 1840; Rowland and Molina Theorize That Ozone Depletion Is Caused by Freon (1973), p. 2009; The British Antarctic Survey Confirms the First Known Hole in the Ozone Layer (1985), p. 2285.

COHEN AND BOYER DEVELOP
RECOMBINANT DNA TECHNOLOGY

Categories of event: Biology and chemistry
Time: 1973
Locale: Stanford, California

Cohen and Boyer pioneered techniques that now allow scientists to insert DNA from any source into bacteria and to detect the expression of the foreign genes in these simple cells

> *Principal personages:*
> STANLEY NORMAN COHEN (1935-   ), a physician and molecular geneticist who developed transformation in *E. coli* and helped pioneer recombinant DNA technology
> HERBERT WAYNE BOYER (1936-   ), a bacteriologist and molecular geneticist who first described the restriction enzyme EcoRI and helped pioneer recombinant DNA technology
> PAUL BERG (1926-   ), a biochemist and recipient of the 1980 Nobel Prize in Chemistry
> HUGH OLIVER SMITH (1929-   ), a molecular geneticist who described the first restriction endonuclease

Summary of Event

Recombinant DNA (deoxyribonucleic acid) technology, known also as genetic engineering or gene cloning, is a fascinating focus of scientific investigation that has, since its inception in 1973, revolutionized the field of molecular biology. This technology already has begun to have a profound effect on the quality of human life, and it allows scientists to address fundamental questions and problems in cell biology that were totally unapproachable using previous methods. Up until the advent of recombinant DNA techniques, it was virtually impossible to study individual genes or even sets of genes in detail, because of the complexity of higher organisms and the vast amount of DNA contained within each cell. If one were to stretch out all the DNA packaged in a typical human cell, for example, it would be nearly 2 meters long. Yet, the DNA that goes to make up the average human gene would be only about one-millionth of a meter long. This complexity of the organization of genes in higher organisms not surprisingly prevented scientists from obtaining large numbers of single genes in pure form. Now, with the aid of recombinant DNA technology, this is possible.

Recombinant DNA methods allow molecular biologists to add one or a small number of genes from essentially any organism to simple bacterial cells. More important, these foreign genes can be made to become an integral part of the bacterium. They will replicate along with the bacterial genetic material and thus be stably transmitted from one bacterial generation to the next. The foreign genes can be made also to be functional in their bacterial host—that is, they can be induced to make their normal gene products.

Bacteria are very simple single-celled organisms that are ubiquitous in nature. While some are capable of causing disease, the vast majority of bacteria are harmless to humans. Some, like the common intestinal bacterium, *Escherichia coli* (*E. coli*), are normal inhabitants of the human body and essential to human life. Each *E. coli* cell has a single circular DNA molecule, or chromosome, containing somewhere between two thousand and three thousand genes. In addition, some cells have one or more additional small circular DNA molecules called plasmids. A typical plasmid contains on the order of five to ten genes and is therefore much smaller than the *E. coli* chromosome. These plasmids are semiautonomous, meaning that while they are incapable of a cell-free existence, they generally remain separate from the larger chromosome and control and direct their own replication and transmission to each daughter cell at cell division. Sometimes, plasmids can confer useful properties on the host cell if they contain genes for resistance to certain antibiotics, viruses, and the like.

The "basic experiment" of recombinant DNA technology involves four essential elements: a method for generating pieces of DNA from different sources and splicing them back together, a "vector" molecule (often a plasmid) that can replicate both itself and any foreign DNA linked to it, a way to get this composite or recombinant DNA molecule back into a suitable bacterial host, and a means to select and separate out those bacterial cells that have picked up the desired recombinant plasmid from those cells that have not. The first step is accomplished by the use of two distinct classes of enzymes, known as restriction endonucleases and ligases. Restriction endonucleases have the unique property of cutting double-stranded DNA at specific recognition sites. Hundreds of different restriction enzymes are known, and each cuts DNA only at its unique recognition site—whenever a particular sequence of bases is encountered in the DNA molecule. DNA ligase, which can be purified from several bacterial or viral sources, is the enzyme that seals pieces of DNA together.

A suitable plasmid to use as a vector in recombinant DNA experiments is one that has a single recognition site for the particular restriction enzyme that will be used. Cleavage with the enzyme will thus result in the formation of linear DNAs from the circular plasmids, but none of the essential plasmid DNA will be lost. Mixing linearized plasmid DNA with the foreign DNA fragments of choice (usually cut with the same restriction enzyme) in the presence of DNA ligase results in joining of the various fragments together into reconstituted circles. Some of these will include both plasmid DNA and one or more foreign DNA fragments. These so-called reconstituted plasmids are then reintroduced back into *E. coli* host cells in a process called transformation. An essential feature of transformation is treatment of the host cells with calcium chloride, which weakens the cell walls and membranes, allowing the reconstituted plasmid DNA to be taken up inside the cells.

Selection of transformed cell clones that now contain plasmid DNA requires that these cells can be distinguished somehow from the much larger population of non-transformed bacteria. This is usually accomplished by choosing a plasmid vector that has one or more genes on it that confers upon the host cell resistance to a particular antibiotic. Plasmids commonly used in these experiments, for example, carry genes for either ampicillin resistance or tetracycline resistance or both. Growth of the transformation mixture in the presence of the antibiotic assures that only cells carrying plasmids will survive. These can be screened subsequently to determine which clones contain both plasmid DNA and the foreign DNA of interest. If all has gone well, these genetically engineered clones of bacterial cells will now replicate stably the foreign DNA, along with the rest of the chromosomal and plasmid DNA of each cell generation; the products of the foreign genes (RNA—ribonucleic acid— or protein) will be made as well.

By the early 1970's, the stage was set for the advent of recombinant DNA technology. DNA ligases had been discovered and purified independently in five separate laboratories in 1967. Hugh Oliver Smith described the first restriction endonuclease in 1970, and shortly thereafter Herbert Wayne Boyer described the isolation of EcoRI, a restriction endonuclease that has been of extraordinary importance in the development of cloning methods. Paul Berg and his group described construction of the first recombinant DNA molecules in a test tube, and at about the same time, researchers in Stanley Norman Cohen's laboratory reported on the first successful transformation experiments in *E. coli*.

In the fall of 1973, Cohen, at the Stanford University School of Medicine, and Boyer, at the University of California School of Medicine at San Francisco, were the first researchers to describe successfully a complete recombinant DNA experiment. Their report detailed the mixing and subsequent reconstitution of DNAs from two separate plasmids in *E. coli*. Shortly thereafter, they described experiments in which DNA from a plasmid found in an unrelated bacterium, *Staphylococcus aureus*, was successfully cloned in *E. coli*, and one year later they reported on the first successful cloning of animal genes in *E. coli*—genes encoding the precursor of the ribosomes (the structures on which cellular proteins are manufactured) from the South African clawed toad, *Xenopus laevis*. The recombinant DNA revolution had begun.

Impact of Event

Recombinant DNA technology is widely considered to be the most significant advance in molecular biology since the elucidation of the molecular structure of DNA in 1953 by James Watson and Francis Crick. It soon became apparent, however, that the technology opened a Pandora's box of social, ethical, and political issues that was unprecedented in scientific history. The research held the potential of addressing biological problems of fundamental theoretical and practical importance, yet it generated real concerns also, because some experiments might present new and unacceptable dangers. Even in the course of scholarly research with the best intentions, there was concern that a laboratory accident or an unanticipated experimental result might introduce dangerous genes into the environment, with *E. coli* carrying them. Not all the fears being expressed were realistic, but scientists at the time did not have a historical perspective on which they might be gauged.

Soon after the scientific concerns were first voiced, a conference was planned to allow many of the leading researchers in molecular biology to try to assess the potential dangers of recombinant DNA technology. The conference was held at the Asilomar Conference Center in February of 1975. Six months earlier, however, eleven world-respected authorities in molecular biology, including Cohen, Boyer, Berg, and others who helped develop recombinant DNA techniques, signed a letter that was simultaneously published in three English and American scientific journals. This letter called for a voluntary moratorium on recombinant DNA experiments until questions about potential hazards could be resolved. The development of a set of guidelines for recombinant DNA research, a modification of which was later adopted by the National Institutes of Health, was discussed at the Asilomar Conference. Levels of both biological and physical "containment" were defined, and each type of recombinant DNA experiment was assigned to an appropriate level. Some types of experiments were banned. In the years that followed the initial furor, guidelines have been modified accordingly, as many of the initial fears about possible dangers have proven groundless.

As predicted, recombinant DNA technology has proved to have extensive practical applications, particularly in the fields of medicine and agriculture. Virtually all insulin-dependent diabetics now take human insulin (made by genetically engineered bacteria) instead of porcine or bovine insulin. Human growth hormone, prolactin, interferon, and other rare human gene products with specific therapeutic uses in medicine are commercially available now only because they can be made in quantity by cloning methods. A proliferation of private gene-splicing companies has arisen to help meet the need for these new products. In agriculture, improved species of genetically engineered crop plants have been designed to help address problems in global food supplies. Of particular note is the effort to clone the bacterial genes for nitrogen fixation into crop plants, thus obviating the need for most fertilizers. Perhaps most significantly, recombinant DNA methods in laboratories around the world are helping to unravel some of the most basic questions in cell biology and to provide a better understanding of the processes controlling gene expression during development.

Bibliography

Cohen, Stanley N. "The Manipulation of Genes." *Scientific American* 233 (July, 1975): 24-33. Eminently readable and profusely illustrated, this article provides the general reader with an excellent historical perspective on the development of recombinant DNA technology and a sound description of the scientific processes involved. Discussion of the use of plasmids as vectors is particularly useful.

Grobstein, Clifford. "The Recombinant-DNA Debate." *Scientific American* 237 (July, 1977): 22-33. Written at the height of the public controversy over the risks and benefits of recombinant DNA research, this thoughtful article provides a unique perspective to the history of the phenomenon. Colorful, helpful illustrations complement Grobstein's balanced treatment of the sensitive issues.

Jackson, David A., and Stephen R. Stich, eds. *The Recombinant DNA Debate.* Englewood Cliffs, N.J.: Prentice-Hall, 1979. A collection of seventeen essays, written by recognized leaders in the fields of molecular biology, ethics, and philosophy, covering all aspects of the controversy. Of particular note are essays by Robert L. Sinsheimer and George Wald, two of the most vocal scientists who pushed for a halt to recombinant DNA research.

Knowles, Richard V. *Genetics, Society, and Decisions.* Columbus, Ohio: Charles E. Merrill, 1985. This broad-based college text was written for nonbiology majors with an interest in science and the social issues raised by the new advances in biology. Good treatment of basic principles of genetics, particularly as applied to humans. Chapter 19 provides a straightforward presentation of recombinant DNA, including the relevant science, history, applications, and controversial issues. Many useful illustrations.

Vigue, Charles L., and William G. Stanziale. "Recombinant DNA: History of the Controversy." *American Biology Teacher* 41 (November, 1979): 480-491. Written primarily for teachers of secondary school biology, this short article summarizes the history and controversy surrounding the recombinant DNA debate. Should be readily accessible to the average reader and a good first choice for further reading.

Jeffrey A. Knight

Cross-References

Avery, MacLeod, and McCarty Determine That DNA Carries Hereditary Information (1943), p. 1203; Watson and Crick Develop the Double-Helix Model for DNA (1951), p. 1406; Nirenberg Invents an Experimental Technique That Cracks the Genetic Code (1961), p. 1687; Kornberg and Coworkers Synthesize Biologically Active DNA (1967), p. 1857; Berg, Gilbert, and Sanger Develop Techniques for Genetic Engineering (1980), p. 2115; A Human Growth Hormone Gene Transferred to a Mouse Creates Giant Mice (1981), p. 2154.

ORGANIC MOLECULES ARE DISCOVERED
IN COMET KOHOUTEK

Category of event: Astronomy
Time: February, 1973-March, 1974
Locale: Hamburg, West Germany

Comet Kohoutek promised to be the "comet of the century"; although it fizzled, organic molecules were observed, which may be the basis for life on Earth

Principal personages:
> LUBOS KOHOUTEK (1935-   ), a Czechoslovakian astronomer who first observed Comet Kohoutek in 1969 and again in 1973 at the Hamburg Observatory in West Germany
> STEPHEN P. MARAN (1938-   ), the head of the National Aeronautics and Space Administration who was coordinator for comet observations
> CHET OPAL (1942-   ), an astronomer of the Naval Research Laboratory who helped produce a detailed map of hydrogen from ultraviolet photographic sequences
> PAUL D. FELDMAN (1939-   ), an astronomer who found large amounts of carbon in Comet Kohoutek from data gathered from a sounding rocket

Summary of Event

Lubos Kohoutek discovered the comet that would bear his name while looking for another comet—Biela's—not observed since the mid-nineteenth century. Kohoutek first observed this comet on July 26, 1969, on spectraplates made in 1968. His plates indicated a faint comet with a faint coma (the most prominent part of a comet) located between two novas. At this time, it was 644 million kilometers from Earth and more than 692 million kilometers from the sun. (This was an amazing find as comets that far out are rarely visible.) The measurements Kohoutek took showed that the comet had a magnitude of 14 degrees; condensed toward its center, it had a tail 1 arc minute long (an arc minute is an angular measure representing one-sixtieth of a degree). When Kohoutek made his discovery, the comet was still eight months from perihelion; gradually it brightened and became more distinct as it neared the sun until its magnitude measured 12.8 and its tail had quadrupled in length.

In early 1970, the magnitude had brightened to a magnitude of 10 with a tail of 8 arc minutes in length. Perigee came on March 13 with perihelion on the twentieth of March, which signaled a gradual fading of the comet. Its final observation was on May 24, 1970, when it made conjunction with the sun. Not until October was it "recovered"; observations continued until it faded finally from sight in April.

Kohoutek again discovered this comet while he was searching for a minor planet in February of 1973. At this appearance, its magnitude was 14.5 and it was tailless. In the next months, brightness varied because of an apparent two-stage coma. The inner coma was extremely condensed and measured 20 arc minutes across while the outer coma was faint and measured only 3 arc minutes across. The final observation of this appearance was made on October 22, 1973, although Comet Kohoutek was to be observed in perihelion on December 28, 1973, traveling at 161 kilometers per second and passing within 21 million kilometers of the sun. Most likely, Comet Kohoutek was making its first-ever approach to the sun and would not return for another encounter with the sun for perhaps a million years. After it had gone around the sun and had become visible again, its magnitude brightened steadily, from 10.5 in September to its brightest—third magnitude—in mid-December. The tail was its observed longest, too, in mid-December, when it was an estimated 18 degrees long. Predictions were that the comet would be as bright as the Moon—visible even in daylight—and the world prepared to view Comet Kohoutek in late December of 1973. Deemed "the comet of the century," it was to be seen easily through February.

Nevertheless, as time passed, it became obvious that Comet Kohoutek was not going to be the brilliant comet promised. The magnitude was not nearly the predicted value, and the comet certainly could not be observed in the daylit sky. Among the astronomers who observed and studied Comet Kohoutek were Stephen P. Maran, Chet Opal, and Paul D. Feldman. Although the general public was disappointed by the lack of a spectacular appearance and interest waned, astronauts followed its passage from the end of December. During January, 1974, it faded quickly, even though its tail developed to a magnificent 25 degrees in length. The comet dimmed and decreased in visible size until visual observations were impossible by April, 1974.

The colors described by the astronauts watching Comet Kohoutek—reds, blues, yellows, and golds—were early indications of the unique chemicals to be found in the comet. Unmanned space probes attempted to detect the rare "parent molecules" that break down into simpler molecules; these simple molecules form the bulk of the molecular portion of the comet, precursors of cometary atoms and radicals such as hydrogen, hydroxyl, and cyanogen. For example, water could be the parent molecule of hydrogen and hydroxyl, hydrogen cyanide that of cyanogen, and ethene that of carbon. The hydroxyl ion is one of these simple molecules; apparently, it originates in the proximity of 14,500 kilometers of the comet's nucleus. This is important because it is indicative of the size of the area in which the parent molecules are congregated.

Parent molecules may be broken down into simpler molecules by as many as fifteen different reactions that either break up or ionize neutral molecules or even form new ones. The only observations that can be made are after the reaction occurs, when end products are sent thousands of kilometers into space and where gas is so rare that collisions cannot occur with any regularity. As a possibility, consider that methane is freed from the comet surface. As it undergoes one of the possible reactions, it likely would be broken up into an organic carbon-hydrogen compound, carbon or hydrogen. Another scenario includes the chance that the elements carbon, hydrogen, oxygen, and nitrogen would combine and form molecules such as carbon monoxide, ammonia, or even water, cyanogen, or methane.

Parent molecules may exist in large numbers. One of the abundant molecules found has been atomic carbon (rather than molecular carbon), which probably is an offspring of carbon monoxide. Carbon monoxide was found to be evaporating in the comet as rapidly as the common water molecule. The implications are that carbon came from an outer layer laid down after the comet formed or that there is far more carbon monoxide than once thought.

Production rates of molecules had never been measured for any given comet; predictions of the numbers, kinds, and amounts of molecules are virtually impossible to make (with the exception of water found in such abundance as to explain the frequent occurrence of hydrogen, hydroxyl, and water radicals). Even if calculations were attempted, observations could present a biased sample—a sample not representative of the actual parent molecules that are vaporizing from the nucleus. Other than water, the only certain parent molecule is hydrogen cyanide, observed in Comet Kohoutek using a special antenna of the National Radio Astronomy Observatory. Methyl cyanide was detected in small amounts, making it an unlikely parent molecule with questionable production rates (in fact, methyl cyanide has been detected only in Comet Kohoutek). These discoveries lend support to the theory that comets may have been formed by aggregation of interstellar dust grains found far away from the sun, perhaps outside the solar system. Other possible parent molecules have even less chance of detection; for example, formaldehyde has never been observed, but it would dissociate most likely into hydrogen plus carbon monoxide immediately after leaving the nucleus.

It is the ionized portion of the atmosphere, called the cometary ionosphere, upon which interest has been focused. Quick reactions form the molecular ions carbon monoxide and nitrogen very near the nucleus; these may be precipitated by the photoionization caused by solar radiation. These rapid processes could occur only in the inner coma and would produce only limited numbers of ions near the nucleus. As Comet Kohoutek approached the sun, the heat vaporized the outer layers of ice into water vapor or steam. The solar radiation then ionized these water molecules so that they lost electrons and became positively charged. Once this charge existed, solar wind forced the molecules back into the comet's tail and away from the sun, producing the red light that was observed on Earth with telescopes and spectrographs.

Impact of Event

As the world awaited the giant Comet Kohoutek, astrologers and the religious alike began making predictions. *The Christmas Monster*, a tract released by a group entitled the Children of God, forecast that the comet would precede worldwide destruction.

Scientists believed it likely that the comet originated from a comet cloud orbiting the sun from 9,300 billion kilometers out (about one light-year). The cloud would be approaching the sun for the first time ever in its 2-million-year history. Coming from so far out, its material would have never been exposed to the evaporative effects of solar radiation, so it would be very "dusty." The discovery of methyl cyanide and hydrogen cyanide supports the theory that comets originate outside the solar system. These unusual molecules probably could not be formed in our system, as the heat of the sun would be likely to break them down before the comets themselves could form. These two molecules are found only infrequently; therefore, it is not surprising that one of the few places they have been observed is in interstellar nebulas.

Astronomers are searching for the significance that might be found in the discovery of organic molecules in comets. Some would suggest that life possibly originated from within comets. It is likely that natural radioactivity inside the core of the comet could produce enough heat to melt ice into pools of water more than 30 kilometers in diameter. These warm, central pools of water would be insulated by and protected by a thick shield of ice. Nevertheless, observations show the unlikelihood that these pools do exist. The ice layer is unstable and may split, but more commonly, known comets are too small to house these warm pools. Even if the pool does occur, there is the problem that there is no available energy source to create life. Laboratory experimentation has shown that molecules of hydrogen, methane, ammonia, and water can combine to form amino acids, but only when subjected to a lightninglike electrical charge. Perhaps amino acids have been formed in comets with the push of the small amount of radioactivity, which is, in reality, a contradiction of cause and effect.

Assumptions that life began in comets is dependent on three extremely improbable events: First, that the warm pools remained in cometary cores for millions and millions of years; second, that life began in these pools even though there was no energy source to boost the creation of larger molecules; and third, if life-forms were created, they were successfully transplanted to Earth.

Cometary passages are ripe for harvesting information that would increase astronomer's knowledge of comets. Comet Kohoutek was no exception; observed for months ahead of its perihelion, vast, well-coordinated observations were made employing Skylab astronauts, as well as radio observations, spectroscopy, and direct photography.

Bibliography

Brandt, John C., and Robert D. Chapman. *Introduction to Comets.* Cambridge, England: Cambridge University Press, 1981. The authors have tried to present a useful monograph focusing on cometary physics and its interrelationships. This somewhat complex book introduces comets—history, facts, and research—for students, scientists, and the general reader. Many charts and illustrations are included, as well as suggested readings.

Eberhart, Jonathan. "Eyes on the Comet." *Science News* 105 (May 4, 1974): 290-

291. A brief, easily understood article that points out that even though Comet Kohoutek was considered to be a flop, much usable information could be and has been gathered on its chemical nature. Primarily, Eberhart uses these data to point out that comets probably are formed outside the solar system.

Kronk, Gary W. *Comets: A Descriptive Catalog.* Hillside, N.J.: Enslow, 1984. The appearance of Comet Kohoutek in 1973 inspired Kronk to list and describe every observed comet from 371 to 1982. The comets, both long-period and nonperiodic, are listed chronologically with their respective brightest magnitudes, longest tail lengths, and largest coma diameters. A reference list is included.

Moore, Patrick. *Comets.* Rev. ed. New York: Charles Scribner's Sons, 1976. A well-illustrated, easy-to-understand volume touching on a wide variety of cometary topics. Many black-and-white photographs show clearly several famous comets, and many historic drawings and graphics are included. Comet Kohoutek's major physical facets are discussed, but little space is given to molecular studies. Glossary.

Whipple, Fred L. *The Mystery of Comets.* Washington, D.C.: Smithsonian Institution Press, 1985. This book presents the history of comets and is divided into two sections: the first discusses the current clues and theories from history and the second section provides information relative to the nature of comets, their origin, and their suspected relation to life on earth. Written for readers who are not science-oriented and scientists interested in comets.

Wilkening, Laurel L., ed. *Comets.* Tucson: University of Arizona Press, 1982. A collaboration of numerous authors that assembles most of the pertinent data regarding comets. Although intended for the novice as well as the specialist, it is heavy reading, requiring some technical knowledge of instrumentation and physics. Many different aspects of Comet Kohoutek are mentioned. Extensive references are given for each chapter. Glossary.

Iona C. Baldridge

Cross-References

Shapley Proves the Sun Is Distant from the Center of Our Galaxy (1918), p. 655; Lemaître Proposes the Big Bang Theory (1927), p. 825; Gamow and Associates Develop the Big Bang Theory (1948), p. 1309; Parker Predicts the Existence of the Solar Wind (1958), p. 1577; Skylab Inaugurates a New Era of Space Research (1973), p. 1997.

SKYLAB INAUGURATES A NEW ERA
OF SPACE RESEARCH

Category of event: Space and aviation
Time: May 14, 1973-February 8, 1974
Locale: Low Earth orbit (approximately 430 kilometers high)

*Skylab was the prototype for large orbital laboratories and demonstrated con-
clusively the ability of humans to function productively for prolonged periods in
microgravity*

 Principal personages:
  WILLIAM C. SCHNEIDER (1923-   ), the NASA project director with
   overall responsibility for Skylab's development and flight operations
  CHARLES ("PETE") CONRAD (1930-   ), a U.S. Navy captain and NASA
   astronaut, veteran of three previous spaceflights, and commander of
   Skylab 2, the first crew to live aboard the space station
  ALAN L. BEAN (1932-   ), a U.S. Navy captain and NASA astronaut
   who was a veteran of the Apollo 12 lunar mission and Skylab 3
  GERALD P. CARR (1932-   ), an aeronautical engineer and NASA astronaut who commanded the Skylab 4 crew
  JOSEPH P. KERWIN (1932-   ), a U.S. Navy commander and NASA
   astronaut who, as science pilot of Skylab 2, was the first physician to
   go into space
  OWEN K. GARRIOTT (1930-   ), a civilian electrical engineer and NASA
   astronaut who served as science pilot on Skylab 3
  EDWARD G. GIBSON (1936-   ), a solar physicist who served as science
   pilot in the Skylab 4 crew
  PAUL J. WEITZ (1932-   ), a U.S. Navy commander and NASA astronaut who served as pilot in the Skylab 2 crew
  JACK R. LOUSMA (1936-   ), a U.S. Marine Corps major and NASA
   astronaut who served as pilot in the Skylab 3 crew
  WILLIAM R. POGUE (1930-   ), a U.S. Air Force colonel and NASA
   astronaut who served as pilot in the Skylab 4 crew

Summary of Event

America's Skylab spacecraft sprang from the desire of the National Aeronautics
and Space Administration (NASA) for a program that could apply hardware developed for Apollo lunar missions to other manned spaceflight objectives. William C.
Schneider was appointed the NASA project director of Skylab. By late 1965, the
space agency had approved a manned earth orbiting laboratory as the program to
follow Apollo. The laboratory was to be created inside the third stage of a Saturn 5
rocket, with crews ferried to it by Apollo command modules. Two laboratories were

built, but before the first was launched on May 14, 1973, it was evident that funding constraints would prevent the second from ever going into space.

Skylab was seriously damaged when launch vibrations tore off a thermal shield, ripping away one of two large solar collectors and jamming the other closed. Mission planners considered abandoning Skylab immediately, since internal temperatures were too high and electrical power too low for it to carry out its tasks. A makeshift thermal shield was created by engineers at the Johnson Space Center in Houston, and the first crew of astronauts conducted a successful space walk to install it. A second space walk, more strenuous and daring than the first, freed the jammed solar panel and put Skylab in operating order.

Skylab was not the first space station. The Soviet Union launched Salyut 1 into orbit on April 19, 1971, and the crew of Soyuz 11 manned it for twenty-two days during June, 1971. Skylab, however, was huge in comparison to any previously manned space vehicle, and its ability to support human activity in orbit was both varied and extensive. Its pressurized interior volume of 370 cubic meters was almost four times greater than that of Salyut, and it was occupied by three separate three-man crews for periods ranging from twenty-eight to eighty-four days during its nine-month operational life, giving the United States an aggregate 12,351 man-hours of spaceflight experience. The difference between Skylab missions and all earlier manned spaceflights has been compared with the difference between taking a short trip in space in a vehicle the size of a car and living there in a three-bedroom house.

Skylab's sophisticated payload was geared to the Project's multiple research objectives. Highest priority was placed on gaining insight into an astronaut's ability to withstand prolonged spaceflight, particularly the physiological effects of exposure to microgravity. Skylab was also heavily instrumented for research in solar astronomy and Earth resources observation. Finally, it carried a varied assortment of corollary experiments and science demonstrations involving investigations in astronomy, biology, botany, and metals processing. The "corollaries" and "science demos," as the crews referred to them, were intended, at least in part, to discover technical and commercial applications for the microgravity environment, as NASA was already looking ahead to applications for future space stations.

As the crews went about their duties aboard Skylab, they were providing information also about two other significant sets of questions. At the root of one was a determination of whether human involvement in space missions made a sufficiently valuable contribution to justify the expense and danger. Specific experiments were included to evaluate the astronauts' ability to use tools but, in a sense, every task tested the arguments favoring their being there at all. A related agenda was to study the myriad astronaut/spacecraft interfaces and to evaluate the fixtures and amenities that affect the habitability of a spacecraft for missions of long duration. The astronauts were asked to comment on virtually every aspect of space living, from the decor of the station and the design of the eating utensils to the type and placement of various restraint systems at their work stations.

The interior space of Skylab was arranged with one eye to the needs of its varied research program and the other on what the comfort, convenience, and psychological preferences of the crew might dictate. The largest element of the spacecraft was the orbital workshop (OWS), which was divided into two decks. One deck was designed to offer something approaching conventional ships' quarters for the crew. It assumed the circular bulkhead closing off the bottom of the converted Saturn 5 third stage as the "floor" and provided a "ceiling" that gave the equivalent of normal headroom in many naval vessels. Between these two surfaces were "walls" separating the space into several compartments, which were entered through rectangular doors instead of round hatches. There were no engineering or structural requirements for this design, but it imposed an artificial sense of "up and down" on the occupants solely to evaluate how important this consideration might be. The second deck contrasted sharply with the first; it was a large open space more than 6 meters in diameter and approximately 12 meters long, containing no strong visual cues to help the crew orient themselves to a local vertical. The third manned volume, the Multiple Docking Adapter (MDA), was a tubular passageway whose circumference was lined with control panels. It presented the crew with a strong axial orientation and a very weak radial one.

About 270 different scientific and engineering experiments were dedicated to furnishing information about crew health and performance. Some thirty-five hundred hours of the astronauts' time aboard Skylab were spent in determining when the effects of weightlessness began to have a noticeable impact on health and whether the debilitation increased gradually and continuously throughout the mission or reached a plateau and stabilized. These experiments were incredibly thorough, causing the crew to believe that they were laboratory subjects. Their activities were so rigidly monitored that every mouthful of food, every urination, every solid waste elimination was reported. Never before had so much detailed information been collected for such a long period on a group of humans.

Skylab's vantage point beyond the atmosphere permitted observation of the sun in X-ray and ultraviolet wavelengths, opening up a wealth of information unobtainable on Earth. An unmanned module called the Apollo Telescope Mount (ATM), attached to the exterior of the laboratory, carried two X-ray telescopes, three ultraviolet telescopes, and a device for continuous study of the solar corona. These instruments could be operated in a semiautomatic mode according to a daily observing program, but were frequently controlled in a manual mode for ten to twelve hours per day in order to capture sudden developments, as the sun proved to be more active than was expected.

Complicated instrumentation to study Earth from space was a rather late addition to Skylab's payload. (Ironically, the value of Earth studies from orbit had not been generally appreciated until the Apollo flights began to underscore that Earth was a delicate oasis.) The Earth Resources Experiment Package (EREP) was included primarily to evaluate designs for a variety of equipment for remotely sensing the earth's surface and developing a data base to be used in interpreting images provided by spaceborne instrumentation. As the spacecraft viewed Earth from 450 kilometers high, researchers on the ground gathered simultaneous information from the fields, forests, lakes, and oceans that were being imaged. This data, called ground truth, was needed to evaluate the EREP images and to learn to recognize particular conditions and problems by the radiation "signatures" visible from space. In order to bring a large amount of the earth's surface under EREP's scrutiny, Skylab followed an orbit inclined 50 degrees to Earth's equator, allowing it to pass over every place between 50 degrees north and 50 degrees south latitude. The three manned missions to Skylab were spaced about ninety days apart so that the earth observing program could gather data spanning several seasons of the year. The crew selected for Skylab 2 were Charles ("Pete") Conrad, Joseph P. Kerwin, and Paul J. Weitz. Alan L. Bean commanded Skylab 3, with a crew of Owen K. Garriott and Jack R. Lousma. The astronauts chosen for Skylab 4 were Gerald P. Carr (the commander), Edward G. Gibson, and William R. Pogue. The information gathered by these astronauts was invaluable for future space exploration.

Impact of Event

Events provided the project with an immediate opportunity to demonstrate dramatically the contribution that astronauts could make to orbital operations. The first crew undoubtedly saved the $285 million spacecraft with their initial space walks, but they also repaired so many other problems with the spacecraft that they were nicknamed the "Fixit Crew." Subsequent crews also were frequently called upon to repair equipment, and astronauts were often able to obtain solar and Earth images that their automatic equipment would have missed, by reacting more quickly or by compensating for problems.

Although Skylab demonstrated humans' ability to make repairs and compensate for faulty equipment to an extent never thought possible when manned spaceflight began, regulating the workload proved more difficult than expected. Prior to occupancy, it had been intended that each astronaut would work eight hours a day, six days a week. It was soon evident that there was enough work to do to keep the crew busy sixteen hours a day, seven days a week. Eventually, the third crew balked and refused to accept such a grueling workload. It became apparent that the difficulties of even ordinary tasks took a heavy toll on endurance.

By the completion of the third crew's occupancy of Skylab, the project had clearly shown humans' ability to adjust physiologically to long-term space habitation. The degradation of the cardiovascular, skeletal, and muscular systems proved to stabilize after several weeks, and the crews that stayed in orbit longer even demonstrated an ability to recapture some of their preflight fitness through an intense regimen of appropriate exercise. This success notwithstanding, it was learned that the effects of prolonged spaceflight on the blood and vascular systems are extensive and in some respects severe. Red blood cell mass was found to decrease constantly while in orbit, with negative impact on the judgment and behavior patterns of the astronauts. This was compounded by the tendency for body fluids to migrate from the limbs to the upper body and head in microgravity conditions. A direct consequence of these observations was the recommendation that future space station crews be rotated every ninety days.

The astronauts' psychological reactions to their home in space helped shape the design of later space laboratories. Skylab's first deck, with its familiar up-down spatial references, was strongly preferred by eight of the nine astronauts. Accordingly, the interiors of Spacelab, first flown in 1983, and *Freedom*—to be the second generation space station of the United States—are arranged like hallways with one surface defined and decorated as a "floor." The Skylab crew were enthusiastic also about the porthole in their wardroom, provided for enjoyment rather than operational necessity. *Freedom* will have several cupola windows, which will serve both psychological and operational needs.

Skylab surprised scientists with the dramatic success of its solar observation program. Data obtained from more than 160,000 solar photographs revealed a star that was vastly more turbulent than expected and that responded rapidly in complex ways to disturbances in its magnetic field. These discoveries caused a complete rethinking of how the sun produces and distributes its energy.

The multispectral imaging technologies evaluated through the special EREP instrumentation led directly to a generation of powerful remote sensing satellites, which have proven extremely effective in studying an enormous range of Earth phenomena. Skylab's Earth imaging was credited also with discovering intriguing "hills and valleys" on the ocean surface, new fishing grounds, underground water for drought-stricken West Africa, new geothermal hot spots, and valuable deposits of oil and ores.

Bibliography

Canby, Thomas Y. "Skylab, Outpost on the Frontier of Space." *National Geographic* 146 (October, 1974): 441-493. *Geographic*'s combination of readable text and exceptional photography (selected from in-flight pictures taken by the crews) presents a good overview. Included is a photo-essay entitled "Skylab Looks at Earth," containing a representative selection of photographs illustrating some of the features visible to Skylab's Earth imaging systems.

Compton, W. David, and Charles D. Benson. *Living and Working in Space*. NASA SP-4208. Washington, D.C.: Government Printing Office, 1983. Provides a detailed official overview of the Skylab program from beginning to end, including the results of its research investigations. Intended for the general reader, the volume is well illustrated and more historical than technical.

Cooper, Henry S. F., Jr. *A House in Space*. New York: Holt, Rinehart and Winston, 1976. A highly readable discussion. Cooper has captured the essence of the experience of living for weeks aboard the spacecraft. Suitable for high school and adult readers, it illuminates both the news-making triumphs and many lesser moments of pleasure and frustration. Contains black-and-white photographs; however, no index is included.

Eddy, John A. *A New Sun: The Solar Results from Skylab*. NASA SP-402. Washington, D.C.: Scientific and Technical Information Office, 1979. A compendium of the photographs obtained with Skylab's ATM, together with background information and a description of the instruments. Text and photographs are ample and well organized in this 200-page book, suitable for laypersons with a good understanding of general science.

Gibson, Edward G. "The Sun as Never Seen Before." *National Geographic* 146 (October, 1974): 493-503. A short nontechnical article highlighting the discoveries made in Skylab's solar observing program. Illustrated with sixteen spectacular photographs of the sun and accompanied by text prepared by the science pilot of Skylab 4. The article conveys some of the reasons for scientific enthusiasm about Skylab's solar data. Recommended for a wide audience.

Robinson, George S. "The Biology of Skylab." In *Living in Outer Space*. Washington, D.C.: Public Affairs Press, 1975. Summarizes the major conclusions from the Skylab Life Sciences Experiments pertaining to human adaptation to spaceflight, and their implications for space colonization, including the legal and psychological ramifications of the impairments in human physiology. Robinson, a law scholar and NASA administrator, is writing for the advanced reader and specialist.

"Skylab." In *Life in Space*. Alexandria, Va.: Time-Life Books, 1983. A photo-essay containing several pages of text covering aspects of living aboard the spacecraft, from coping with weightlessness and space sickness to the equipment problems that plagued the first crew and the chaotic jumble of unstowed equipment with which the last crew struggled.

Richard S. Knapp

Cross-References

Glenn Is the First American to Orbit Earth (1962), p. 1723; The First Space Walk Is Conducted from Voskhod 2 (1965), p. 1787; The Orbital Rendezvous of Gemini 6 and 7 Succeeds (1965), p. 1803; The First Humans Land on the Moon (1969), p. 1907; *Columbia*'s Second Flight Proves the Practicality of the Space Shuttle (1981), p. 2180; Spacelab 1 Is Launched Aboard the Space Shuttle (1983), p. 2256; The First Permanently Manned Space Station Is Launched (1986), p. 2316.

MARINER 10 IS THE FIRST MISSION TO USE GRAVITATIONAL PULL OF ONE PLANET TO HELP IT REACH ANOTHER

Category of event: Space and aviation
Time: November 3, 1973-March 24, 1975
Locale: Earth to Venus and Mercury

The concept of gravity-propelled interplanetary space travel was first demonstrated by the United States space program in the mission of Mariner 10 to Venus and Mercury

Principal personages:

MICHAEL A. MINOVITCH (1935-), an American mathematician who discovered that interplanetary spacecraft could derive enormous propulsive energy from the gravitational fields of the planets they encountered

WALTER HOHMANN (1880-1945), a central figure in the German rocket school who published a classic study on rocket-propelled interplanetary voyages

WALKER E. (GENE) GIBERSON (1923-), the Mariner 10 project manager who directed the construction and flight of the spacecraft

Summary of Event

The potential for using a planet's gravitational attraction to obtain significant acceleration and direction changes on an interplanetary spacecraft was not widely recognized before the beginning of the space age. Virtually all astrodynamicists believed that the work of the German rocket scientist Walter Hohmann accurately assessed the minimum energy (and therefore the optimum) trajectories for all possible interplanetary missions. Missions to two or more planets were assumed to require very large chemical or nuclear rockets to provide the energy needed to exit from the encounter trajectory at the first planet, known as a Hohmann transfer orbit, and enter a new Hohmann transfer orbit to the next target.

In the early 1960's, Michael A. Minovitch, a young graduate student at the University of California at Los Angeles (UCLA), became the first mathematician to solve the vexing Restricted Three-Body Problem in celestial mechanics. His solution involved complex three-dimensional vector analysis and required the most powerful digital computers available to civilian research at that time. In the process, Minovitch discovered that a large amount of orbital energy could be exchanged between a space vehicle and a passing planet and that by carefully controlling such a vehicle's entry into a planet's gravitational sphere of influence, it would be possible to use that energy to propel the vehicle to virtually any other destination without any additional rocket propulsion. Using large amounts of computer time donated by UCLA and the

National Aeronautics and Space Administration's Jet Propulsion Laboratory (NASA/ JPL) at the California Institute of Technology, he then expanded his research to examine a great many possible multiple planet missions and identified numerous gravity-assisted trajectories and launch windows for missions of interest to planetary scientists.

In the late 1960's, NASA decided to give high priority to a 1973 Venus-Mercury mission using a gravity-assisted trajectory, since this represented the only practical opportunity to place a spacecraft in proximity to Mercury for the next decade. In 1970, the space agency obtained $3 million in start-up funding for what was then called the Mariner Venus/Mercury mission for 1973 (MVM73) mission, and a contract to manage the mission was awarded to JPL. Walker E. (Gene) Giberson was appointed project manager for the spacecraft by JPL.

From the outset, the MVM73 encounter trajectory was hotly contested between scientists who were most interested in seeing photographs of the planet's surface and those who wanted to evaluate many nonvisual factors. The objectives of the former would have been best served if the spacecraft passed both Mercury and Venus on their sunlit sides, but virtually all the other high-priority experiments required the spacecraft to pass on their night sides, and it was the latter trajectory that prevailed.

The spacecraft was built around an octagonal body with dual solar arrays. At right angles to the solar panels were a high-gain dish antenna and a 6-meter boom supporting a magnetometer. From the tip of the boom to the end of the dish antenna, the spacecraft measured 9.83 meters and weighed 503 kilograms in flight configuration. The science payload weighed 77 kilograms, including two television cameras, an X-band radio transmitter, a scanning electron spectrometer and scanning electrostatic analyzer, two magnetometers, an infrared radiometer, a charged particle telescope, and two ultraviolet spectrometers.

MVM73 lifted off from Cape Kennedy on November 3, 1973, atop an Atlas/ Centaur booster, which placed the spacecraft in a parking orbit 188 kilometers above Earth. The spacecraft traveled for only thirty minutes in this orbit before the Centaur second stage was refired, accelerating it to 40,900 kilometers per hour on a trajectory toward Venus. After successful trajectory insertion, the mission was officially designated as "Mariner 10."

The trajectory insertion burn was aimed so that the sun's gravity pulled Mariner 10 into a long-curving Hohmann transfer orbit to Venus. Hohmann had shown in 1925 that the optimum trajectory for an interplanetary probe was a segment of an elliptical orbit that barely touched the earth's orbit at one extreme and barely touched the target planet's orbit at the other. Such orbits require the spacecraft to travel halfway around the solar system from launch to encounter.

Upon arrival at Venus, that planet's gravity was expected to reduce the orbital energy of Mariner 10 just enough so that the spacecraft would follow a free-fall trajectory to an encounter with Mercury. Precise guidance was essential, as the window through which Mariner 10 had to pass to receive the desired amount of gravity assistance on its way to Mercury was only 400 kilometers in diameter, and any uncorrected error in aiming would be magnified one thousandfold by the time it reached Mercury.

Analysis of the trajectory achieved by the Centaur's second burn showed that if no corrections were made, the spacecraft would pass Venus on the sunward side at a distance of 48,300 kilometers. Mariner 10 relied on its own maneuvering engine to make a series of refinements in its course, called trajectory correction maneuvers (TCMs). A TCM was performed on November 13 to refine the flight path to within 1.5 percent of optimum, and a second TCM occurred in late January of 1974.

Mariner 10 missed hitting the exact center of its Venus encounter point by only 17 kilometers. It rounded the planet on February 5, 1974, with gravity bending the trajectory exactly as expected, and exited Venus' gravitational sphere of influence on the new course to Mercury. As it skimmed about 5,800 kilometers above the cloud tops, the cameras recorded more than four thousand pictures, while other instruments probed deeply into the atmosphere. From these data, scientists were able to gain clues to Venus' atmospheric composition and movement.

The gravity assist provided by Venus placed Mariner 10 into an elliptical orbit that had its perihelion (closest point to the sun) approximately at the point it encountered Mercury's orbit, and its aphelion (farthest point from the sun) at about the distance of Venus' orbit. The time it would take for Mariner 10 to circle this orbit would be almost twice as long as the eighty-eight days it takes Mercury to orbit the sun. Mission planners realized that this would cause the planet and the spacecraft to return to each other about every six months, allowing for additional "free encounters."

A TCM to refine the trajectory from Venus was executed on March 16, and Mariner 10 arrived at Mercury on March 29, 1974. After a twenty-one-week, 402-million kilometer journey from Earth, it was only 165 kilometers off dead center of its targeted encounter point. Its cameras showed a heavily cratered landscape, and the magnetometer detected a fairly strong magnetic field, implying that Mercury had a huge iron core. Virtually no atmosphere was detected, nor did the surface features show any evidence of erosion from water or wind.

Mariner 10's first encounter with Mercury achieved the accuracy needed for a second encounter in six months, but the planet and the spacecraft would have passed at a distance of 805,000 kilometers, precluding any useful science. Another TCM to reduce the miss distance and produce an opportunity for yet a third encounter in about twelve months was approved. A relatively large change in velocity was necessary, and it was decided to make two separate corrections twenty-four hours apart. These were executed on May 9 and 10, leaving Mariner 10 on a course that was within 1 percent of optimum.

Mariner 10 returned to Mercury on September 21, passing 48,000 kilometers above the Southern Hemisphere and returning two thousand high-resolution photographs. After this second flyby, the spacecraft was restored to cruise mode for the six month journey around the sun and back to Mercury again. Mariner 10's third visit to Mercury was its closest. It passed a scant 200 kilometers above the surface on March 16, 1975. Problems with Mariner's onboard video tape recorder resulted in receiving only 25 percent of each picture on Earth, but the imagery was spectacular. Most important, the third encounter fully substantiated the evidence that Mercury possesses an Earth-like magnetosphere. On March 24, the attitude control system ran out of fuel and communication with Mariner 10 was discontinued. The spacecraft had journeyed more than 1.6 billion kilometers in its 506-day mission, transformed an understanding of the inner solar system, and delivered the highest yield per cost of any interplanetary mission in NASA's history to that time.

Impact of Event

The gravity-assist technique first demonstrated in the mission of Mariner 10 has been used with brilliant success in several subsequent interplanetary missions and will be essential to most of the interplanetary missions planned for the remainder of the twentieth century. Gravity assists offer the possibility of much higher payoff per mission, since more targets can be reached with fewer spacecraft and smaller booster rockets. Also, travel time to the targets is greatly reduced. A spacecraft receiving a gravity slingshot from Jupiter and Saturn can reach Uranus in only nine years, whereas without the slingshot, it would take thirty years.

Pioneer 10 and Pioneer 11 were launched a year earlier than Mariner 10. Pioneer 10's encounter with Jupiter on December 3, 1973 accelerated the spacecraft to a speed that will cause it to escape the sun's gravity and exit the solar system. Pioneer 11 caught Jupiter from behind on December 3, 1974, and as it passed across the bow of the moving planet gravity pulled it into a tight 270 degree turn and accelerated it back toward a point on the opposite side of the sun, where it plunged past Saturn on September 1, 1979.

In the late 1970's and early 1980's, the planets Jupiter, Saturn, Uranus, Neptune, and Pluto were spaced about the solar system in a configuration that offered several opportunities for a single spacecraft to visit most of them using the gravity slingshot for retargeting itself at each successive encounter. NASA's plans to seize this opportunity involved two spacecraft—Voyager 1 and Voyager 2—launched in 1977. Voyager 1's trajectory took it within 277,000 kilometers of Jupiter on March 5, 1977. There it got a gravity assist to continue to Saturn, arriving on November 13, 1980. Voyager 2 was launched on a slightly different trajectory and passed through the Jovian moon system at a distance of about 650,000 kilometers from the planet on July 9, 1977. Jupiter's gravity redirected Voyager 2 on a course to Saturn in 1981 and then, with additional gravity slingshots, to Uranus in 1986 and Neptune in 1989.

The flight path for Galileo, the next generation of Jupiter probe, is like a complicated aerial trapeze act. It began its journey in late 1989 by going in the opposite direction, to Venus. Venus tossed it back to Earth, and as it streaked past Earth five months later, gravity accelerated it into a long orbit toward the asteroid belt, from which it will return in two years. Passing Earth again, it will get its third gravity kick, which will be enough to send it on a three-year trip to Jupiter.

Ulysses, also called the Solar-Polar orbiter, will attempt to go into a polar orbit around the sun. No previous interplanetary spacecraft has ever left the ecliptic plane, and no rocket in existence is powerful enough to provide the thrust to do this, even for a probe that weighs a mere 55 kilograms. To get such velocity, Ulysses will go first to Jupiter, where it will be whipped into a perpendicular orbit as it is tossed back at the sun.

Bibliography

Burgess, Eric. *Venus: An Errant Twin.* New York: Columbia University Press, 1985. This is a comprehensive and well-illustrated general discussion of what is known about Venus. Readers will find Mariner 10's discoveries chronicled, along with those of twenty other space probes. The discussion of Mariner 10 includes a good summary of the interplanetary flight to Venus.

Chapman, Clark R. *Planets of Rock and Ice.* Rev. ed. New York: Charles Scribner's Sons, 1982. Subtitled "From Mercury to the Moons of Saturn," Chapman's work puts the discoveries made by Mariner 10 and other planetary probes into the context of an emerging understanding of the solar system and the processes that have shaped the planets individually and collectively.

Cross, Charles A., and Patrick Moore. *The Atlas of Mercury.* New York: Crown, 1977. Based entirely on the images and data reported by Mariner 10, this is the best reference source for any study of the planet's surface features. Includes discussion of the history of observations of Mercury, background on the Mariner 10 spacecraft and its mission, and information on the magnetic field and atmosphere of the planet.

Gatland, Kenneth. *The Illustrated Encyclopedia of Space Technology.* New York: Harmony, 1981. Referenced for its discussion of interplanetary trajectories. Includes color diagrams of the Hohmann Transfer Orbit and the Voyager 1 and 2 encounter flight paths at Jupiter and Saturn. The discussion of Mariner 10 is superficial.

Murray, Bruce C., and Eric Burgess. *Flight to Mercury.* New York: Columbia University Press, 1976. The comprehensive account of the Mariner 10 mission. Murray, a leading planetologist, was in charge of the Mariner 10 imaging experiments. Burgess has written numerous articles and books on the space program, with emphasis on planetary exploration. Illustrated with more than one hundred photographs.

Strong, James. *Search the Solar System: The Role of Unmanned Interplanetary Probes.* New York: Crane, Russak, 1973. In addition to providing a general discussion of the role of unmanned spacecraft in exploring the planets, Strong provides a short explanation of the principle involved in the "gravity slingshot." His popular style is easy to read, and his pragmatic vision of appropriate planetary exploration for the remainder of the twentieth century has proved to be remarkably predictive in terms of the techniques employed.

Von Braun, Wernher, and Frederick I. Ordway III. *History of Rocketry and Space Travel.* 3d rev. ed. New York: Thomas Y. Crowell, 1975. Contains an excellent summary of the history of astronautics and the origin of ideas concerning how to accomplish interplanetary flight, by one of the men who participated in the development of rocketry. Features a comprehensive bibliography.

Washburn, Mark. *Distant Encounters*. New York: Harcourt Brace Jovanovich, 1983. A comprehensive discussion of the insights into the nature of Jupiter and Saturn made possible by the Voyager spacecraft, but with several valuable discussions of the trajectories chosen for the missions, including choices that were considered but not adopted. Illustrations are adequate and include drawings of the flight paths of the two spacecraft as they encountered the target planets.

Richard S. Knapp

Cross-References

Venera 3 Is the First Spacecraft to Impact on Another Planet (1965), p. 1797; Mariner 9 Is the First Known Spacecraft to Orbit Another Planet (1971), p. 1944; Pioneer 10 Is Launched (1972), p. 1956; Soviet Venera Spacecraft Transmit the First Pictures from the Surface of Venus (1975), p. 2042; Voyager 1 and 2 Explore the Planets (1977), p. 2082.

ROWLAND AND MOLINA THEORIZE THAT OZONE DEPLETION IS CAUSED BY FREON

Category of event: Earth science
Time: December, 1973-June, 1974
Locale: University of California, Irvine, California

Rowland and Molina warned that chlorofluorocarbon gases (Freons) may be destroying the ozone layer of the stratosphere

Principal personages:
F. SHERWOOD ROWLAND (1927-), a physical chemist whose specialty was radioactive isotopes and who was the codiscoverer of the chlorofluorocarbon/ozone effect
MARIO JOSÉ MOLINA (1943-), a physical chemist whose research was in chemical kinetics and who worked with Rowland
JAMES LOVELOCK, an English chemist whose measurements of chlorofluorocarbons were the stimulus for Rowland's research

Summary of Event

Ozone, a triatomic form of oxygen, is a bluish irritating gas of pungent odor, formed naturally in the upper atmosphere by a photochemical reaction with solar ultraviolet radiation. It is a rare gas that protects Earth from the most dangerous radiations of the sun. Ozone can absorb ultraviolet rays efficiently even when it is present in very small amounts; for example, if the ozone in the stratosphere were to fall on the earth's surface, it would create a dusting of only one-eighth inch in thickness. Altogether, the ozone layer removes one-twentieth of the sun's radiation, including the dangerous shortwave rays that can do great damage to living things. All biological systems have evolved under the protection of the ozone shield in the stratosphere. This shield blocks out ultraviolet-B (UV-B) radiation; however, human activity has for many years been depleting this ozone layer by the release of human-made, nearly inert, chlorofluorocarbon compounds. These compounds are commercially known as Freon.

Chlorofluorocarbons were discovered in 1928 by Thomas Midgley, Jr., a Du Pont Corporation chemist, who developed the chemical in response to General Motors Frigidaire Division's desire for a more efficient and safer refrigerant. In the 1950's, chlorofluorocarbons became widely used not only as household and commercial refrigerants and air conditioners but also as propellants in aerosol products and as solvents and intermediates in the synthesis of other fluorine compounds. They are nonflammable, have excellent chemical and thermal stability, and low toxicity. They are prepared by the reaction of hydrogen fluoride with carbon tetrachloride in the presence of a catalyst—antimony pentachloride and hydrogen fluoride.

Two researchers at the University of Michigan, Richard Stolarski, a physicist, and

Ralph Cicerone, an electrical engineer with a minor in physics, reported in 1973 on a study under contract for the United States National Aeronautical and Space Administration (NASA) Marshall Space Flight Center to determine the environmental effects of the proposed space shuttle. One of their principal areas of focus was hydrogen chloride emissions from solid-fuel rockets. They determined that the chlorine released from the shuttle's exhaust might deplete ozone. Their determinations were downplayed by NASA. Two Harvard University researchers, Michael McElroy and Steven Wofsy, had reached similar conclusions about chlorine from the shuttle.

In 1972, F. Sherwood Rowland, a physical chemist at the University of California at Irvine, had already made his name as a chemist who specialized in the study of radioactive isotopes. He first became interested in atmospheric chemistry as a result of a professional meeting he had attended in Fort Lauderdale, Florida, in 1972, which was organized by the United States Atomic Energy Commission (AEC). It was not the presentation but rather an item of gossip picked up during one of the breaks that sparked his interest. Lester Machta of the United States National Oceanic and Atmospheric Administration (NOAA) commented upon a conversation he had had with James Lovelock, an independent and creative English chemist, who had invented an electron capture gas chromatograph to detect atmospheric gases in minute amounts. With his invention, Lovelock had discovered a concentration of 230 parts per trillion of two commonly used chlorofluorocarbons in the atmosphere over western Ireland. Rowland later performed some calculations that showed that Lovelock's concentrations were very close to his rough estimate of the total amounts of chlorofluorocarbons being produced. He wondered what eventually became of the chemical.

Rowland reasoned that if all the chlorofluorocarbons ever released were still in the lowest part of the atmosphere, which is the troposphere, that meant nothing was destroying them in this layer. Yet, they had to go somewhere, and the only place was upward into the stratosphere, an upper portion of the atmosphere that is above approximately 11 kilometers. He believed that the chlorofluorocarbons would decompose with ultraviolet radiation. Eighteen months later, Rowland turned this casual commentary into an elaborate study.

In the summer of 1973, Lovelock published the results from his Irish study of human-made gas concentrations in the atmosphere. At about this time, Rowland was making research plans for the following academic year. He obtained permission from the AEC to conduct a study of chlorofluorocarbons. Ostensibly, this research was to set up ground rules on how to use the gases to trace the movement of air masses around Earth.

In the fall of 1973, Mario José Molina, born in Mexico, had completed his Ph.D. at the University of California, Berkeley, and had come to work with Rowland. He knew nothing about atmospheric chemistry; his research was in chemical lasers. Once he learned that all the chlorofluorocarbons ever released were still floating about the troposphere, however, he became curious. Rowland and Molina set out to determine what would happen to the chlorofluorocarbons in the atmosphere. By

November, 1973, Molina had established quickly that nothing happened to them in the troposphere. Chlorofluorocarbons do not react with living things, they do not dissolve in oceans, they do not get washed out of the air by rain, they do nothing except float around and gradually work their way upward into the stratosphere. It was a simple chemical deduction that they would be broken apart by ultraviolet radiation and that chlorine atoms would be released in the stratosphere. At the time, a few chlorine atoms seemed unworthy of concern; that is, not until Molina worked out the catalytic chains by which a single chlorine atom can scavenge and destroy many thousands of ozone molecules. The calculations were based on the 1972 assumption rate of 800,000 tons of chlorofluorocarbons released into the atmosphere per year.

Using detailed calculations for chemical reaction, Molina concluded that each chlorine atom from chlorofluorocarbons would collide with a molecule of the highly unstable ozone. The reaction did not end there. Once chlorine was freed from the chlorofluorocarbon, the by-product would be oxygen and a chemical fragment with an odd number of electrons called chlorine monoxide. The odd number of electrons, Molina knew, guaranteed that it would react with a free oxygen atom to achieve an even number of electrons. He calculated that, when the chlorine monoxide fragment met the free oxygen atom, the oxygen in chlorine monoxide would be attracted to the free oxygen atom and split off to form a new oxygen molecule. Chlorine would then be freed, which would collide with ozone, thus starting the cycle all over again. In short, the breakdown of chlorofluorocarbons by sunlight would set off a catalytic chain reaction in which one chlorine atom could gobble up 100,000 molecules of ozone, turning them into impotent molecules of oxygen.

Rowland and Molina had some doubt about their calculations, thinking that the numbers were too large to be correct. They asked for an opinion from Harold Johnston, a chemist at the University of California, Berkeley, an expert in atmospheric chemistry, and extremely knowledgeable about ozone and chemical reactions in the atmosphere. Johnston reassured them that their numbers were correct; however, he cautioned that their results would be attacked vigorously by those with vested interests in the chemicals. Nevertheless, they wrote up the study and sent it to *Nature*, where it was published in June, 1974.

Although the theory was originally scoffed at, such responses were short-lived, as subsequent laboratory work, atmospheric measurements, and computer modeling failed to find any significant errors in the theory.

Impact of Event

On September 26, 1974, the chlorofluorocarbon/ozone story made the front page of *The New York Times* in a story by Walter Sullivan describing calculations carried out by McElroy and his researchers at Harvard University. The work done by Rowland and Molina was also mentioned. Rowland and Molina's study, the University of Michigan study, and the Harvard findings set into motion a whole series of reactions. In October, 1974, a government committee recommended that the National

Academy of Sciences conduct a study on the validity of the chlorofluorocarbon/ ozone theory. In June of 1975, Johnson Wax, the nation's fifth largest manufacturer of aerosol sprays, announced that it would stop using chlorofluorocarbons in its products. In June, 1975, Oregon became the first state to ban chlorofluorocarbons in aerosol sprays. In October, 1976, the Food and Drug Administration (FDA) and the Environmental Protection Agency (EPA) proposed a phase-out of chlorofluorocarbons used in aerosols. In October, 1978, chlorofluorocarbons used in aerosols were banned in the United States. In August 1981, NASA scientist Donald Heath announced that satellite records showed that ozone had declined 1 percent. In October, 1984, an English research group led by Joe Farman detected 40 percent ozone loss over Antarctica during the austral spring, which was confirmed in August, 1985, by NASA satellite photographs showing the existence of an ozone hole over Antarctica. In August, 1987, the McDonald Corporation, which uses chlorofluorocarbons in the making of polyurethane foam containers for hamburgers, announced that it would stop using the chemical. In October, 1987, the Antarctica ozone expedition found chlorine chemicals to be the primary cause of ozone depletion. In March, 1988, the Du Pont Corporation announced it would cease manufacture of chlorofluorocarbons as substitutes become available. In May, 1988, preliminary findings of a hole in the ozone layer over the Arctic were discussed at a scientific conference in Colorado. In September, 1988, the EPA reported new evidence that showed it underestimated the degree of ozone depletion and announced that 85 percent cutbacks on chlorofluorocarbons were needed.

The debate on the total production ban of all chlorofluorocarbons will continue until an economical substitute for the chemical is developed. There is little doubt that chlorofluorocarbons are a threat to the ozone layer. Their impact has been far-reaching and significant for the survival of living things on Earth.

Bibliography

Cook, J. Gordon. *Our Astounding Atmosphere.* New York: Dial Press, 1957. Written in the popular style and very easy to read. Takes the mystery out of the atmosphere and presents a difficult subject in a most entertaining format. High school students should have no problem comprehending the subject matter. A few black and white photographs, but no graphics, mathematics, or tables.

Gribbin, John, ed. *The Breathing Planet.* New York: Basil Blackwell, 1986. A readable book; discusses climatic changes past and present, and looks at their effect on drought-stricken regions of the world. Two articles were written by Rowland: one on aerosol sprays and the other on chlorofluorocarbons in the stratosphere. Well illustrated; recommended as an excellent reference source for the upper-level high school student and the general public.

_____. *The Hole in the Sky: Man's Threat to the Ozone Layer.* New York: Bantam Books, 1988. An excellent paperback book that consists primarily of text, except for a few graphs and chemical equations. For the most part, should be used by the advanced high school and lower-level college student. Written by an acclaimed science writer and cosmologist. Focuses entirely on the ozone layer crisis and the frightening future the failure to curb Freons will have on all living things on Earth.

Lutgens, Frederick K., and Edward J. Tarbuck. *The Atmosphere: An Introduction to Meteorology.* 2d ed. Englewood Cliffs, N.J.: Prentice-Hall, 1982. An introductory-level college textbook; copiously illustrated. A review after each chapter and a review of vocabulary are included. Contains seven appendices and an extensive glossary. Well written; a good reference source on the atmosphere.

Roan, Sharon. *Ozone Crisis: The Fifteen-Year Evolution of a Sudden Global Emergency.* New York: John Wiley & Sons, 1989. An excellent book and highly recommended for readers on all levels. An extremely interesting commentary on the ozone crisis in language the nonscientist can understand. Contains a few graphs and photographs and an extensive bibliography. Includes a chronology of events relative to the chlorofluorocarbon/ozone problem. Written by a staff science writer for the *Orange County Register* in Santa Clara, California, who has written extensively about the ozone crisis, the greenhouse effect, beach erosion, wildlife preservation, and other environmental issues.

Earl G. Hoover

Cross-References

Teisserenc de Bort Discovers the Stratosphere and the Troposphere (1898), p. 26; Fabry Quantifies Ozone in the Upper Atmosphere (1913), p. 579; Midgley Introduces Dichlorodifluoromethane as a Refrigerant Gas (1930), p. 916; Piccard Travels to the Stratosphere by Balloon (1931), p. 963; Manabe and Wetherald Warn of the Greenhouse Effect and Global Warming (1967), p. 1840; The British Antarctic Survey Confirms the First Known Hole in the Ozone Layer (1985), p. 2285.

GEORGI AND GLASHOW DEVELOP
THE FIRST GRAND UNIFIED THEORY

Category of event: Physics
Time: February, 1974
Locale: Cambridge, Massachusetts

Georgi and Glashow developed the first of the unified field theories in subatomic physics, which, although tentative, could lead eventually to a single theory uniting all the laws of nature

Principal personages:

HOWARD GEORGI (1947-   ), an American physicist who worked with Sheldon L. Glashow to develop the first of the unified field theories in subatomic physics

SHELDON L. GLASHOW (1932-   ), an American physicist who, with Howard Georgi, developed the first unified field theory and won the 1979 Nobel Prize in Physics

ABDUS SALAM (1926-   ), a Pakistani physicist who shared the 1979 Nobel Prize in Physics with Glashow and Steven Weinberg and contributed to the Georgi/Glashow unified theory

JOGESH PATI, an Indian physicist who collaborated with Salam to develop a theory of proton decay, which led Georgi and Glashow to develop their unified theory

Summary of Event

Four known forces in the universe determine the behavior of everything, from the motion of planets and stars to the interaction and very form of matter itself. Those forces are: gravitational, electromagnetic, weak, and strong.

The gravitational force is generated by static mass (or mass in acceleration). It holds planets in their orbits and holds objects to the surface of planets. Although it is expressed over immense distances, it is the weakest of the four forces. The electromagnetic force exists between atoms and molecules. It is the force that drives machinery and electronic devices. The weak force exists at the atomic level and is responsible for the decay of radioisotopes; it is responsible for radioactivity. The strong force exists within the nucleus of the atom and holds together the elementary particles that make up protons and neutrons.

The goal of modern physics is the unification of all of these forces into a single, elegant theory that has become known as the "grand unified theory" (GUT). It is possible that two of the fundamental forces can be unified into subunified field theories (typically referred to as unified theories). It has actually been accomplished. There is a unified theory uniting the electromagnetic and weak forces called the electro-weak theory. The theory uniting the strong, weak, and electromagnetic forces has become known as the grand unified theory. Because it does not incorporate gravity, however, it is not the ultimate grand unified theory.

The term "unified field theory" has been well known since the time of German-American physicist Albert Einstein. Einstein, famous for his extraordinary formulation of the theory of relativity, which revolutionized science, tackled the task of uniting gravitational and electromagnetic forces in a single theory. Although he devoted much of his career to this aim, he failed utterly. Even before Einstein, unified field theories had been formulated already, although they were not recognized as GUTs at the time. Sir Isaac Newton formulated the classical quantification of gravity, uniting the localized effect of objects falling to the earth's surface to the astronomical effects of orbiting planets and stars. In the nineteenth century, James Clerk Maxwell devised a theory on electromagnetism, unifying the seemingly inconsonant notions of electricity, magnetism, and lightning into a single theory he called electromagnetism. Michael Faraday predated both Maxwell and Einstein when he made a premature attempt to unify electromagnetism and gravity in 1850.

Einstein rejected quantum physics, calling it a "stinking mess." His lack of acceptance of the only tool capable of delving into the atomic core was probably responsible for his failure to unify gravity and electromagnetism. Those physicists who came after Einstein and utilized this powerful tool of quantum mechanics, however, began to inch closer to the unification target.

Between 1961 and 1968, a series of papers were written by American physicists Sheldon L. Glashow and Steven Weinberg and Pakistani physicist Abdus Salam. In 1971, physicist Gerard 't Hooft brought the ideas of Glashow, Weinberg, and Salam together in a paper that demonstrated that they had predicted all the elements that would unite the quantum (subatomic) theories of the electromagnetic and weak forces. It became the first of the quantum unification theories, called the electro-weak theory. Glashow, Weinberg, and Salam won the 1979 Nobel Prize in Physics for their work.

Grand unification (as a hypothesis uniting three of the elementary forces) came about as a tentative supposition in 1973. Glashow teamed up with Howard Georgi, a Harvard colleague and physicist, and seized on the newly emergent theory of quantum chromodynamics (QCD), which lent a kind of order to the multifarious pieces uncovered from the heart of the atomic nucleus. They reasoned that although the differences in forces were very great (a difference of hundreds between the electromagnetic, weak, and strong), QCD allowed for enough definition that each of the three elementary particle forces could be defined by a single coupling constant (a term that defines the force's absolute strength).

In 1973, Salam published a paper he had written with physicist Jogesh Pati. In that paper, the first mention of an altogether radical notion appeared, the possibility that the proton—the most stable entity in the known universe—could, in fact, ultimately decay. This dramatic notion came about because they were able to class the building blocks of the nucleus with the tiny extranuclear components, electrons, known in subatomic jargon as the lepton family. If this general classification could be made, if one of the components of a proton could turn suddenly into an electron, then the whole of the atom itself would suddenly disappear, or decay. Hence, the concept of proton decay was related more or less directly to the decay of matter itself.

Glashow and Georgi's theory built on that notion and they called it SU(5), the particle physicist's method of naming complex mathematical groups or numerical definitions of theory. They rationalized through SU(5) that when a quark (the building block of a proton) decays during proton decay, it gives off a very heavy particle and the proton literally falls apart into leptons, which would finally translate into electrons and positrons. When they meet, forming photons, or light, the atom (matter) is gone forever.

The decaying of matter as a singular issue was a ponderous philosophical concept in itself. When the Salam-Pati paper was published in *Physical Review* in September, 1973, and mentioned proton decay, it was virtually ignored. Glashow and Georgi produced a paper entitled "Unity of All Elementary Particle Forces," which appeared in February, 1974. Like the Salam-Pati concept, it was met with considerable skepticism. Aggravating the debate was the fact that the Glashow-Georgi proton decay occurred in SU(5) for altogether different reasons than it did in the Salam-Pati theory. Finally, there was the absolute paucity of experimental evidence. Although SU(5) promised final unity, the deep thrust for the ultimate answers demanded ultimate experimental capacity.

Theoretical physics is tied tenuously to experimental physics in an uneasy symbiosis. The musings of theory must be borne out by experimental evidence, or the theory remains unproven. Meanwhile, the designs of experimental physicists are typically built on the stimulus of theorists. Quantum mechanics, as the study of the interior of the atom, depends on particle accelerators, which cause atoms to collide and split apart. Physicists collect the images of the atomic pieces (as traces on photographic plates) and examine them. As they delve deeper into the interior of the atom, however, larger and larger energies are required.

When physicists split the atom to verify elements of the electro-weak theory, they used the most powerful particle accelerators known, which demanded as much power as a large city for a single experiment. It is estimated that to verify the Glashow-Georgi theory, a particle accelerator many tens of millions of times more powerful than any in existence would have to be built. Far beyond any known science, such an accelerator built to match scientific theories of the 1990's would have to be 10 light-years in length. Thus, although the Glashow-Georgi theory seems well supported in concept, it is unknown when it will ever be supported by experimental evidence. For this reason, the theory can only be considered tentative.

Impact of Event

The Glashow-Georgi grand unified theory became the first physical theory to unify more than three of the elementary forces of nature: the electromagnetic, weak, and strong interactions. These forces are active in the subatomic world. Although the theory uses elements of physical theory supported by experimental evidence, it is not similarly supported as an independent theory by experimental data.

Einstein spent decades on his failed attempt at grand unification. Glashow and Georgi spent twenty-four hours. Their article in *Physical Review* in 1974 unified the work of physics for the previous half century. Yet, Einstein spent only two weeks hammering out relativity and laid out its formulations very briefly. The product of genius is not in effort spent but in its synthesis. Einstein synthesized the groundwork already laid by the century of physicists before him. Therefore, Glashow and Georgi, already very well acquainted with the ideas of quantum mechanics, realized that the final pieces to the puzzle were contained in the ideas of quantum chromodynamics, which allowed the final demands of a grand unified theory to be met. Nevertheless, as promising as it is, the Glashow/Georgi theory is still tentative.

When the ultimate grand unified theory is written—one that accounts for gravity as well as the electromagnetic, strong, and weak forces of nature—then the main task of physics will have been completed. Although there will be many adjunctives to the physical theory, the ultimate physical theory will have been achieved. Again, the energies required to test and to verify these physical GUTs are so large, it is unknown whether any of them will ever be supported by experimental evidence. Until that question is answered, physicists will face a philosophical impasse. Without experimental confirmation, the world of the theoretical physicist becomes separated from the experimental. When that happens, the definition of science as a unique philosophy will face a pivotal test. Such a separation will turn theoretical physics into what Georgi has called "recreational mathematical theology."

To make matters more confusing, since the mid-1970's, several unified field theories have sprung up, but none shows any promise or even capacity of being supported by experimental evidence. Many of them are elegant theories that have few theoretical flaws but are ultimately impossible to test. Many consist of unification through multiple dimensions beyond four—nearly impossible for the normal human mind even to grasp. They have been named "superstring," "supergravity," and "supersymmetry."

The grand unification theory that remains closer to the already proven individual elements of experimental physics is Glashow and Georgi's theory based on SU(5). Even if SU(5) is wrong, the main suppositions of proton decay and the ultimate linkage between the massive pieces of the nucleus and the leptons (electrons/positrons) are probably accurate. Only the intermediaries are still uncertain.

Bibliography

Crease, Robert P., and Charles C. Mann. *The Second Creation*. New York: Macmillan, 1986. In this magnificent book, Crease and Mann follow the making of twentieth century physics from its nineteenth century roots to the enigmatic mysteries of the late 1980's. Examines characters and personalities as well as the issues of physics. The detailed work of Glashow and Georgi is discussed, from the perspective of fundamental discoveries to the full implications of SU(5) as the grand unified theory.

Hawking, Stephen W. *A Brief History of Time*. New York: Bantam Books, 1988. In this eminently readable work, one of the most prominent physicists of modern times examines the universe from his view of creation to the present. Hawking examines the far-flung reaches of space and time, from black holes to the interior of the atom, and discusses the elementary particles of the atomic nucleus. Hawking also discusses the concept of grand unified theories and what they mean to physics. For a general audience. Illustrated.

Pagels, Heinz R. *The Cosmic Code*. New York: Simon & Schuster, 1982. This book describes quantum physics as "the language of nature." Pagels embarks on a literary quest to explain some of the most profoundly difficult topics in quantum physics clearly to the average reader. He succeeds and opens up the interior of the atom for a clear view of what is inside. Illustrated. One need not be a physicist to enjoy the material.

_____. *Perfect Symmetry*. New York: Simon & Schuster, 1985. In this follow-up to *The Cosmic Code* (cited above), Pagels delves deeper, providing a layperson's perspective. Covers the aspect of the atom's core and supports a lively discussion of the grand unified field theories. Touches on the frontiers of atomic physics and speculates on where the unified field theories will ultimately lead. Very readable and has become a classic of the genre.

Schwinger, Julian. *Einstein's Legacy*. New York: W. H. Freeman, 1986. In this beautifully illustrated book, Nobel Prize-winning physicist Julian Schwinger discusses the work of Einstein and his contributions to science. Although the last half of Einstein's life was devoted to his private struggles with grand unification, Schwinger devotes his single last paragraph to it. Nevertheless, the book is a must for any serious student of the master of physics.

Sutton, Christine. *The Particle Connection*. New York: Simon & Schuster, 1984. Sutton explains the details behind the tools of the physicist's trade: the particle accelerator. She describes how the machine is used and the nature of the particle chase at CERN, the European particle accelerator laboratory. Sutton also discusses SU(5) specifically as it relates to current particle accelerators and their practical limitations. Illustrated. Geared to the student with a reasonable grounding in science.

Dennis Chamberland

Cross-References

Thomson Wins the Nobel Prize for the Discovery of the Electron (1906), p. 356; Bohr Writes a Trilogy on Atomic and Molecular Structure (1912), p. 507; Rutherford Presents His Theory of the Atom (1912), p. 527; Rutherford Discovers the Proton (1914), p. 590; Einstein Completes His Theory of General Relativity (1915), p. 625; Lawrence Develops the Cyclotron (1931), p. 953; Cockcroft and Walton

OPTICAL PULSES SHORTER THAN
ONE TRILLIONTH OF A SECOND ARE PRODUCED

Category of event: Physics
Time: April, 1974
Locale: Bell Telephone Laboratories, Holmdel, New Jersey

Shank and Ippen shaped laser pulses to times less than one trillionth of a second in a composite dye laser system, opening new regimes of ultrafast electronics and ultrarapid physics and chemistry

Principal personages:
ALBERT EINSTEIN (1879-1955), a German-American physicist and 1921 Nobel Prize winner in physics who in 1916 developed the basic mathematical relation which allows the amplification of light
THEODORE H. MAIMAN (1927-   ), an American physicist who used the possibility of light amplification to produce the first laser in 1960
ERICH PETER IPPEN (1942-   ), an American electrical engineer who, together with Charles Vernon Shank, produced the first laser system
CHARLES VERNON SHANK (1943-   ), an American electrical engineer who produced the first laser system, which produced electronic pulses shorter than one trillionth of a second in duration

Summary of Event

Albert Einstein is known for his theories of relativity; however, his influence in physics was widespread and had surprising results. One of these results was the laser, discovered five years after his death. In 1916, Einstein noted that the known laws of radiation from hot bodies implied that light could be amplified. Einstein showed that radiation must be able to stimulate more of itself when interacting with matter from which the light had spontaneously radiated. Under the proper conditions, that stimulated radiation would amplify as it raced at the speed of light through the material. In addition, he gave an amazing mathematical relation from which to calculate the ratio of the stimulated radiation to the spontaneous radiation.

The light seen from the sun, a neon sign, or an incandescent light is spontaneously emitted. Light is only one part of the electromagnetic spectrum, which includes among its members radio, radar, and television radiation; these are stimulated radiations. Light is an electromagnetic wave with very short wavelength, while radio, radar, and television have long wavelengths. The mathematical relation that Einstein developed said that the ratio of stimulated to spontaneous emission, everything else being the same, depended on the third power of wavelength. Long waves (if they are to radiate in practical circumstances) must radiate by stimulation, while short waves radiate spontaneously. Decrease wavelength by ten, and the spontaneous radiation increases by one thousand. That very strong increase in spontaneous emission at short wavelength explains why a 100-watt light bulb costs less than one dollar, while a 100-watt radar transmitter costs thousands of dollars.

The laser is an extension to the short wavelengths of light of the electronics with which one is familiar at the long wavelengths of radio and television. This new laser electronics opens up the high frequencies and short time cycles characteristic of light to the benefits of conventional electronic radiations. Copious spontaneous emission characterizes common light sources and distinguishes them strongly from those electronic sources that can be modulated conveniently to carry sound, pictures, and other forms of information. Spontaneous light emission from common light sources goes in all directions and contains a wide range of jumbled frequencies, while the stimulated electronic emissions are directional and possess a narrow range of well-ordered, or coherent, frequencies, which are easily modulated. In order to tame the light to the stimulated emission of an electronic laser, the conditions for light amplification, which Einstein noted, were required.

It was more than fifty years after Einstein developed his formula for light amplification before scientists were able to fashion its instructions to produce a laser. Early in 1960, Theodore H. Maiman produced the first laser using a ruby crystal; later the same year, Ali Javan, an American physicist, constructed the first gas laser in neon. The discovery of the laser by Maiman opened up broad vistas in optical technology, which required additional discoveries and years to exploit. Most lasers operate as high-frequency oscillators, which generate a coherent beam of light by the process of emission stimulated from atoms or molecules by the beam while it is trapped within the laser material. The coherence of the laser causes the extremely high-frequency oscillations of the laser light to be very regular throughout the space they travel and the time they last, allowing properties of the laser beam, such as its time duration, to be altered radically.

Maiman had produced the red light of the ruby laser by illuminating a 5-centimeter-long, synthetic crystal of ruby with a powerful spiral flash lamp encircling the crystal. The red color of natural ruby is a weak fluorescence produced by a very small number of chromium impurities in an aluminium oxide crystal. The powerful light from the flash lamp pumped the chromium impurities of Maiman's ruby into the fluorescent state, producing an inverted population of that high-energy state. Silvered reflectors evaporated on the ends of the ruby crystal fed the red fluorescence back into the ruby, producing a series of strong, irregular laser pulses lasting millionths of a second. The chromium atoms, which are actually in the ion state inside the ruby, were the source of the stimulated emission, which was trapped and amplified by the simple optical "cavity" formed by the silvered reflectors at the ends of the ruby rod.

The duration of an electronic pulse is measured in a standardized set of units that decrease in duration by stages of one thousand. One thousandth of a second is a millisecond, and one millionth of a second—the typical duration of the ruby laser pulses observed by Maiman—is a microsecond. Normal electronics is quite capable of producing pulses which last one billionth of a second, or a nanosecond; but it was not until 1974 that pulses as short as one trillionth of a second, or one picosecond, were generated by Charles Vernon Shank and Erich Peter Ippen. The molecules used by Shank and Ippen in their ultrafast, subpicosecond laser were those of the strongly fluorescent dye, rhodamine 6G. The ultrashort pulses were shaped by use of an absorbing solution of DODCI dissolved in ethylene glycol, a common solvent inserted with the rhodamine 6G inside the laser cavity. Because the rhodamine dye fluoresces so rapidly, it is not possible to produce lasing in the rhodamine molecules by pumping with a normal flash lamp. Instead, the rhodamine was pumped with an extremely powerful argon laser beam in a complicated optical cavity containing five reflectors as well as the dye and absorber.

The short pulse experiments began in the early 1970's at the Holmdel laboratories of Bell Telephone in New Jersey. In 1972, Shank and Ippen began experiments in which both the rhodamine dye and the DODCI absorber were mixed together in one single solution of ethylene glycol, which was injected as a thin stream near the center of the laser cavity which, of necessity, had grown complex. The laser light was fed out of the cavity through a quartz acousto-optic coupler. They aimed at electronic light pulses shorter than any science had known.

There are no electronic photodetectors or oscilloscopes that can measure the very short pulses generated in apparatus such as that of Shank and Ippen. Instead of trying to catch the pulses in time, these short pulse durations are measured by the distance they move in space. The pulse is split into two parts. Each part is sent along a somewhat different path; then the two parts are recombined in a crystal of potassium dihydrogen phosphate (KDP), which responds strongly only when both beams are present inside the crystal. The beam paths are adjusted so that the crystal glows. Each laser beam consists of quick pulses spread over a short distance in space, and the glow occurs when both beams coincide within the KDP. The glow is picked up by a sensitive photomultiplier and then one path of the two beams is lengthened with respect to the other until the glow ceases. When the glow ceases, the pulses are arriving at the KDP at different times. A 1-picosecond pulse occupies only 0.3 millimeter, the path length difference that Shank and Ippen were seeking as a mark that they had generated picosecond pulses.

In April of 1974, Shank and Ippen reported in *Applied Physics Letters* that they had adjusted the argon laser pump source to produce about 2.5 watts of its blue laser light, added DODCI into the dye stream until stable pulse operation was observed, and measured the path differences over which the red dye laser pulses overlapped. The shortest distance was 0.15 millimeter. The laser pulses were more than a million times shorter than the original pulses of Maiman's ruby laser. Shank and Ippen had produced 0.5-picosecond pulses, pulses lasting only half a trillionth of a second.

Impact of Event

There were a variety of devices that generated short electronic pulses before Shank and Ippen's discovery. Photography with electronic spark discharges captures time events as short as one-tenth microsecond long, while high-speed flash lamps and electronics have resolutions of one-tenth nanosecond. Shorter time spans required lasers, and mode-locked, flash lamp-pumped, neodymium glass lasers had produced low repetition rate, irregular pulses of several picoseconds duration. Shank and Ippen had taken the principle of laser mode-locking and applied it to dye lasers to generate very reproducible, high-repetition rate, subpicosecond light pulses. In doing so, they opened up a new realm of fast electronics and a new measure of rapidity: femtosecond pulses. A femtosecond is the next unit in the progression, by one one-thousandth, of short time measures. It lasts only one thousandth of a picosecond and is over in one quadrillionth of a second, which is less than a cycle of visible light. Both Ippen and Shank were to play a strong role in giving technology and science the tool of femtosecond pulses.

In 1981, Shank and his coworkers collided two optical pulses together in a one-hundredth millimeter section of an absorber dye to produce 90-femtosecond-long optical pulses. The pulses had already been narrowed and sent traveling in opposite directions in a ring-shaped dye laser. They then took the 90-femtosecond pulses and sent them through a 15-centimeter section of single mode optical fiber and then onto a pair of optical gratings. Within the optical fiber, the pulses are temporally broadened, but then the blue end of the pulses catches up to the red end within the grating pair to produce compressed pulses only 30 femtoseconds long. These short pulses opened up a wide variety of previously hidden processes to scientists. Vibrational dynamics in molecules, photobiological mechanisms in living matter, and fast electronic properties of semiconductors are some examples of areas in which new science could now be explored. With even shorter pulses—in the range of 10 femtoseconds—entirely new classes of processes can be studied, including vibrations in liquids.

Improvement in techniques led researchers, including Ippen and his coworkers, to shorter pulse durations. In 1987, Shank and his associates at Bell Laboratories capped their assault on short optical pulses by using optical fibers and a sequence of grating and prisms to improve the final optical time compression. The final pulse width was only 6 femtoseconds long. A cycle of light in the middle of the visible spectrum is only 2 femtoseconds long, so it is quite clear that an ultimate limit is under assault in the production of ultrashort pulses visible to the eye. A common principle of physics and electronics dictates that short pulse duration demands a wide frequency (hence, color range) and that nature signifies the approach to the limit of short, visible pulses by transforming the normally red color of the dye light to white during the process that produces a 6-femtosecond pulse.

Bibliography

Alfano, R. R., and S. L. Shapiro. "Ultrafast Phenomena in Liquids and Solids." *Scientific American* 228 (June, 1973): 42-60. This excellent article presents the history of fast optical pulses, along with the state-of-the-art of fast laser pulses at a time just before the production of subpicosecond pulses. The article gives an excellent introduction to the background needed to understand the science and applications of ultrafast optical pulses.

Bova, Ben. *The Beauty of Light*. New York: John Wiley & Sons, 1988. This book gives an excellent summary of the fascinating properties of light; how these properties are used in art, science, industry, and technology; and how their beauty affects humans. Chapter 15 discusses lasers in an authoritative manner, without equations, and in understandable language. The book is a thoroughly enjoyable account of the science and art of light.

Heel, A. C. S. van, and C. H. F. Velzel. *What Is Light?* Translated by J. L. J. Rosenfeld. New York: McGraw-Hill, 1968. The book is a translation from the Dutch of a semitechnical description of light. The description of natural light is thorough, up to date, but not overbearing. The few equations are simple, and the text is supplemented with a large number of useful and attractive illustrations. Lasers are discussed briefly in the last chapter.

Ippen, Erich P., and Charles V. Shank. "Subpicosecond Spectroscopy." *Physics Today* 20 (May, 1978): 41-48. Ippen and Shank discuss the background, discovery, and applications of subpicosecond pulses in an understandable manner with illustrations, diagrams, and four equations. This is the authoritative article on the discovery of subpicosecond pulses for the general user.

Kaiser, W., ed. *Ultrashort Laser Pulses*. New York: Springer-Verlag, 1988. This technical monograph is the definitive reference on ultrashort optical laser pulses. The chapters are by various authorities, and while most chapters are readable, they require a science background. The introductory chapter is by Shank and contains illustrations.

Lasers and Light: Readings from "Scientific American." Introduction by Arthur L. Schalow. San Francisco: W. H. Freeman, 1969. These readings serve as an excellent introduction to the field of lasers and give a feeling for the sense of discovery pervading the early days following development of the laser. For more technical details on lasers and ultrafast laser pulses, see entries for O'Shea and Kaiser.

O'Shea, Donald C., W. Russell Callen, and William T. Rhodes. *Introduction to Lasers and Their Applications*. Reading, Mass.: Addison-Wesley, 1977. This undergraduate text covers the field of lasers in a detailed manner. Intended for science and engineering students, the text is not for browsing. Yet, the level is not too advanced; it can be used for selective reference.

Peter J. Walsh

Cross-References

Gabor Develops the Basic Concept of Holography (1947), p. 1288, The First Laser Is Developed in the United States (1960), p. 1672; Lasers Are Used in Eye Surgery for the First Time (1962), p. 1714; Tunable, Continuous Wave Visible Lasers Are Developed (1974), p. 2025; The First Commercial Test of Fiber-Optic Telecommunications Is Conducted (1977), p. 2078.

TUNABLE, CONTINUOUS WAVE VISIBLE LASERS ARE DEVELOPED

Category of event: Physics
Time: June, 1974
Locale: United States

Research culminated in the development of dye laser systems, which operated continuously and were tunable over a broad range of visible wavelengths, introducing a wide range of new applications for lasers

Principal personages:

JOSEPH A. GIRODMAINE (1933-), a Canadian-American physicist who produced a narrowly tunable, near infrared laser with Miller

ROBERT C. MILLER (1925-), an American physicist who produced, with Girodmaine, a narrowly tunable, near infrared laser, indicating the tuning possibilities of lasers

PETER P. SOROKIN (1931-), an American physicist who, with Lankard, developed the first dye laser

J. R. LANKARD, a physicist who developed the first dye laser, offering a practical broad-band laser source

OTTIS G. PETERSON (1936-), an American physicist who, with associates at Eastman Kodak Research Laboratory, produced the first continuously operating dye laser

HERWIG G. KOGELNIK (1932-), an Austrian-American physicist who designed the first astigmatically compensated, continuous wave, dye laser cavities

P. K. RUNGE, an American scientist who developed the first unconfined dye laser stream

R. ROSENBERG, a scientist who developed (with Runge) the first unconfined dye laser stream for use in continuous dye lasers, eliminating the need for dye cells

J. M. YARBOROUGH, an American scientist who produced continuous, tunable dye laser emission over the entire visible region

Summary of Event

The first laser was produced by the American physicist Theodore H. Maiman early in 1960 when he illuminated a 5-centimeter-long synthetic ruby crystal with the light from an intense flashlamp spiraling around the crystal. The ends of the ruby crystal had been coated with a thin layer of silver, forming a very short optical cavity in which the normally weak, red fluorescence of the ruby crystal could be trapped and amplified. The very intense red beam has become a hallmark of the laser.

The prescription for producing a laser had been implied in some relatively simple equations, first given by the great German-American physicist and Nobel Prize winner Albert Einstein in 1916, which related the rate of spontaneous emission to stimulated emission of light. Most of the light the eye sees is emitted spontaneously from atoms or molecules into all directions of space surrounding the emitter. Einstein's equation said that light of the same frequency as that spontaneously emitted could stimulate further emission from these atoms or molecules directed along the stimulating beam. Einstein found that light could be amplified in a directed beam under the proper conditions. More than four decades later, Maiman and Ali Javan discovered the proper conditions for ruby and for helium-neon.

These first lasers, and most of the lasers that soon followed, had fixed wavelengths that could not be easily changed. Visible light is often measured by millionths of a meter, or micrometers; ruby light, for example, has a wavelength of 0.694 micrometer. This wavelength may be changed only with difficulty, by changing the temperature of the ruby crystal or by applying pressure to the ruby crystal, and the wavelengths of gas lasers are even less amenable to change. For many applications, the discrete nature of the wavelength of the laser is of little consequence. In spectroscopic applications, however, there is a great need to change the wavelength of the light in a simple and rapid manner.

When compared with conventional, spontaneous light sources used in spectroscopy, lasers have intrinsic advantages of high power (energy per unit time), high intensity (power per unit area of illumination), high brightness (power per unit area and per unit angle), and high spectral brightness (brightness per unit wavelength). These advantages arise because the light waves in a laser are coherent and in unison, whereas those in a conventional light source are incoherent and scattered. If one focuses a laser beam with a good lens, the light can be squeezed down to the size of a wavelength, increasing the intensity of the laser beam enormously. Clever methods can be used to squeeze the laser into a short period or, alternatively, into a narrow band of wavelengths. If such methods could be developed to tune the laser frequency over a wide range, the intrinsic advantages of lasers could be brought to bear in the important scientific and technical areas where spectroscopy plays a role.

Spectroscopy employs the electromagnetic spectrum to measure properties of materials. The properties sought may be used to earmark individual atoms and molecules, in which case visible light is most useful, and the properties may give information about the bulk of a gas, liquid and solid. In this case, the infrared is often preferred. Whereas the applications of spectroscopy traditionally have been in measurements, the uses for the intense, tunable, narrow-band laser sources that have been developed have grown. These sources can be used to induce highly selective photochemical reactions, including the reactions used to separate nuclear isotopes for atomic reactors.

One of the first tunable laser sources was demonstrated in 1965 by Joseph A. Girodmaine and Robert C. Miller at Bell Laboratories. They employed the intense pulse from a neodymium laser to generate a pulsed output from a lithium niobate crystal that was tunable in the narrow infrared range from 0.97 micrometer to 1.15 micrometers. They wondered if the tuned emission could be produced in the visible, or if the tuning could be produced in a continuously operating laser.

A major advance toward broadly tunable lasers in the visible spectrum occurred with the development of the dye laser by Peter P. Sorokin and J. R. Lankard at the IBM Watson Research Center in 1966. They used the dye chloro-aluminum-phthalocyanine and illuminated it with pulses from a ruby laser. The dye absorbed the shorter-wavelength, higher-energy, red light from the ruby and emitted longer-wavelength, lower-energy light. With mirrors to return the emitted light to the dye, they produced the first dye laser. The ruby laser acted as a "pump" to produce lasing in the dye, just as the flash lamp had pumped the ruby in Maiman's first laser. The energy lost in the emission appeared as heat in the dye, requiring the dye to be cooled. Other dyes were introduced quickly by various researchers. Dyes have broad bands of emission wavelengths, and this is an obvious advantage in a laser that is to be tuned. Although a single dye still covers only a fraction of the visible spectrum, dyes were soon developed with other emission wavelengths that spanned much of the visible spectrum.

The disadvantage of dyes is that they emit so rapidly that they require extremely intense pumping, hence the ruby pump employed by Sorokin and Lankard. The rapid, spontaneous emission of the dyes, however, produces large stimulated emission amplification, so that only a very small portion of dye is needed to produce lasing. Thus, the intense pumping is needed in only a small volume of dye that was initially circulated in a transparent cell. Scientists were fortunate that intense lasers were available for focusing to pump the dye in the dye laser. Initially, these other laser sources were pulsed lasers, including both ruby and the neodymium laser used by Girodmaine and Miller. The need to introduce another laser beam between the mirrors of the dye laser cavity and to have the means to circulate the dye to remove the excess heat complicated the optical design of the dye laser cavity.

Soon scientists replaced one of the mirrors of the dye cavity by a diffraction grating. Optical gratings select only a narrow range of wavelengths for reflection along the cavity axis, depending on the orientation of the grating. By rotating the grating within the laser cavity, the output of the dye laser could be tuned across the broadband emission of the dye, and, with a single dye, pulsed emission could be tuned over about one-tenth the visible spectrum.

The intense and concentrated pumping required of the dye laser made continuous operation seem doubtful. In 1970, however, Ottis G. Peterson and his associates, S. A. Tuccio and B. B. Snavely at the Research Laboratories of Eastman Kodak, were able to design a dye laser that allowed continuous wave (CW) operation. They used a continuous argon ion laser to pump rhodamine 6G in a water solution, after adding some soap detergent. The water has a high heat capacity, affording excellent cooling for the dye, which was circulated at very high velocity through the pump focal region. The detergent acts to prevent agglomeration of the rhodamine molecules and to maintain the dye molecule in its lasing state.

The optical cavity used for the dye laser had grown complicated, and the optical quality of the laser beam suffered as a result. In 1971, Herwig W. Kogelnik, along with Erich Peter Ippen, A. Dienes, and Charles Vernon Shank at the Bell Laboratories, published a three-mirror design for a dye laser cavity that compensated for the astigmatism of the dye cavity. Astigmatism in the dye cavity distorts and enlarges both the focused pump spot and the entrapped laser spot. For optimum dye laser operation, these two spots should overlap completely and have a minimum size. In 1974, Ippen and Shank used this design in a dye laser to produce the first laser pulse shorter than one-trillionth of a second.

The improved performance of CW dye laser still suffered because of the cell containing the dye stream, typically 1 millimeter thin and moving at several meters per second. The cell walls became contaminated by particles burned at their surface by the intensely focused beam and became damaged by the high localized heating. In 1972, P. K. Runge and R. Rosenberg, also with Bell Laboratories, designed a nozzle that allowed unconfined flow of the dye stream as a thin sheet through the air and thus eliminated the need for a dye cell within the laser. Free-flowing dye streams became the standard practice for dye laser operation. The development of a fully tunable, visible dye laser was almost complete.

The apparatus was now in place for operation of a laser system that may be tuned precisely within the visible spectrum. When, in June of 1974, J. M. Yarborough produced continuous tuned emission from a set of overlapping dyes in a single laser system over the range from 0.415 to 0.790 micrometer, the spectrum was spanned. In all, Yarborough investigated eighteen dyes, some of which had not been previously lased. This dye set served as a reference set for researchers in the years to follow.

Impact of Event

Progress continued rapidly in improving the performance of CW tunable lasers. In 1977, H. W. Schroeder and his associates at the Institute for Applied Physics at the Technical University in East Germany used a ring-shaped laser to obtain tunable dye laser outputs, in a single frequency, at continuous powers near 1 watt. By 1982, T. F. Johnson and his coworkers at the research laboratory of Coherent, Inc., in Palo Alto, California, had produced a dye laser system that used a single argon ion gas laser to pump a set of eleven dyes, which gave a continuously tunable output that covered the range of 0.407 to 0.887 micrometer. They obtained more than 5 watts output from rhodamine 6G in the red region of the spectrum and more than 1 watt for the middle two-thirds of the visible spectrum. In addition, the rhodamine dye was capable of generating harmonic radiation in the near ultraviolet.

Continuous tunable laser systems were also developed with other types of lasers, often using techniques pioneered with dye lasers. The main requirement of any broadly tunable system is the presence of a broad, spontaneous emission curve in the active laser material. A number of solid-state and semiconductor lasers have been uncovered that meet the requirements of broad-band, continuous, tunable lasers.

Among the tunable solid-state lasers that have shown promise are those whose radiating atoms are the chromium and titanium ions. The secret to the broad-tuned laser operation is finding a good-quality crystal host in which these ions give broad emission. Chromium is the radiating atom in the ruby laser discovered by Maiman. Ruby is a crystal host of aluminum oxide that contains a small fraction of chromium oxide. Without the chromium oxide, the crystal is transparent sapphire, but the small percentage of chromium produces the red color of ruby and the intense red beam from a ruby laser. With sapphire as a host for the chromium in the ruby laser, the chromium in ruby does not give a very broad emission line, but in other hosts the output produced from chromium is quite broad. The first tunable laser using chromium was discovered in 1980 in the alexandrite host, formed from beryllium aluminum oxide, and there the chromium laser output spans the deep red region from 0.7 to 0.82 micrometer.

In contrast to chromium, the titanium ion in a host of sapphire does display very broad laser emission. First studied in 1982, titanium-sapphire produces a continuously tunable output over the range from 0.70 to 0.95 micrometer, from visible red to invisible near-infrared. The crystal may be pumped with an ion laser, or with the efficient semiconductor lasers, for CW operation. It should be noted that some semiconductor lasers form useful tunable sources whose output may be tuned by variation in current, temperature, and pressure.

In addition to the visible spectrum, the dye laser opened up a broad range of tunable laser applications in the ultraviolet and infrared by using unusual optical effects possible with lasers, often referred to as nonlinear effects. The electrical field within a laser beam depends on the intensity of the beam. By focusing even a rather weak laser beam in a material, the electrical field within the laser spot may be as large as the electrical field that holds the atoms and molecules of the material together. The material now responds by distorting the laser field so that, for example, multiples of the laser frequency may be produced. This effect allows the laser frequency to be doubled and the wavelength to be cut in half, converting dye red laser to near ultraviolet. When two different laser colors or frequencies are present in the spot simultaneously, both sum and difference frequencies may be generated. If the two laser frequencies are near each other in the visible band of one dye, the difference frequency generated is infrared light. Thus, a single visible dye laser under clever control can generate tunable ultraviolet and infrared beams in addition to visible. With a range of dyes, the complete region from ultraviolet through infrared may be scanned.

Bibliography

Bova, Ben. *The Beauty of Light*. New York: John Wiley & Sons, 1988. An excellent summary of light and how it is used in art, science, industry, and technology. Written simply and in nontechnical terms. Chapter 15 discusses lasers accurately without equations and in understandable language. The book is a thoroughly enjoyable account of the science and art of light.

Johnson, T. J. "Tunable Dye Lasers." In *Encyclopedia of Physical Science and Tech-*

nology, edited by Robert A. Meyers. New York: Academic Press, 1987. A comprehensive article on tunable dye lasers. The article is technical and contains equations and thus requires background. The excellent illustrations and fine writing allow judicious browsing for those with a nontechnical background.

O'Shea, Donald C., W. Russell Callen, and William T. Rhodes. *Introduction to Lasers and Their Applications.* Reading, Mass.: Addison-Wesley, 1977. This undergraduate text covers the field of lasers in a detailed manner. Intended for science and engineering students, the book is not for browsing. The level is not too advanced, however, so the book can be used for selective reference.

Schalow, Arthur L., ed. *Lasers and Light: Readings from "Scientific American."* San Francisco: W. H. Freeman, 1969. A fine introduction to the field of lasers that gives the sense of discovery pervading the early days following development of the laser.

Sorokin, Peter. "Organic Lasers." *Scientific American* 220 (February, 1969): 30-40. Organic lasers include dye lasers. The article by the discoverer of the dye laser is an understandable and authoritative article for the layperson. Summarizes the properties of dye lasers as they were known in the period immediately after their discovery.

Peter J. Walsh

Cross-References

Esaki Demonstrates Electron Tunneling in Semiconductors (1957), p. 1551; The First Laser Is Developed in the United States (1960), p. 1672; Lasers Are Used in Eye Surgery for the First Time (1962), p. 1714; Optical Pulses Shorter than One Trillionth of a Second Are Produced (1974), p. 2020; Optical Disks for the Storage of Computer Data Are Introduced (1984), p. 2262.

THE J/PSI SUBATOMIC PARTICLE IS DISCOVERED

Category of event: Physics
Time: August-September, 1974
Locale: Stanford, California; Brookhaven, New York

Ting and Richter discovered the J/psi particle, which established a new mass scale in particle physics and produced evidence for the existence of a charmed quark

> *Principal personages:*
> SAMUEL C. C. TING (1936-   ), the head of the Massachusetts Institute of Technology (MIT) research group using the Brookhaven Alternated Gradient Synchroton (AGS) who was the winner of 1976 Nobel Prize in Physics
> BURTON RICHTER (1931-   ), a professor of physics and the head of a research group using the Stanford Linear Accelerator; shared the 1976 Nobel Prize with Ting
> MARTIN PERL (1927-   ), the director of a research group at Stanford University and the University of California, Berkeley

Summary of Event

The most remarkable thing about the discovery of the J/psi is that it occurred almost simultaneously on both the East (Brookhaven, Long Island, New York) and West (Stanford, Palo Alto, California) coasts of the United States. Its discovery was not predicted by any theory, although later quark theorists were able to retrodict its discovery once the mass of the J/psi and of related particles were established.

The persistent weakness of quark theories is that there are no mass/energy predictions emergent. It always has been left to experimentalists to find the new particles in whatever energy regime they can be found.

In the summer of 1974, the two diverse groups who discovered the J/psi particles were at work on different accelerators employing totally different techniques.

A large research group formed around Samuel C. C. Ting of Massachusetts Institute of Technology (MIT) was using the Alternating Gradient Synchroton, located at Brookhaven National Laboratory in New York. At the time, this facility was one of the premier particle accelerators in operation in the United States because it accelerated protons to high energies in the tens of gigaelectronvolts (one gigaelectronvolt, or GeV, equals 1,000 megaelectronvolts, or MeV) region. The proton has a rest mass of 940 MeV/c^2; thus, these protons are highly relativistic in that most of their energy is in their sizable traveling speed and not in their rest mass. Ting and his group were having these accelerated protons hit extended beryllium targets. Beryllium is a relatively common light metal that is used as a target in many nuclear collision experiments because it presents particles with a sizable collision cross section. In August, 1974, it became evident that as the protons collided with the beryllium nuclei electrons, positrons and a new particle whose mass was later determined to be at $3,100$ GeV/c² emerged as collision products. This was a surprise, for at the time no theory clearly predicted such a massive particle, and Ting and his group called their new find a J-particle.

At the same time, another group of researchers at Stanford University and Lawrence Berkeley Laboratory of the University of California were using the Stanford Linear Accelerator (SLAC). This accelerator raises electrons and their antiparticles, positrons, to speeds almost the speed of light and then has them collide with each other. Electrons and positrons are leptons, very light particles. These electrons and positrons were colliding with each other in increasing steps of about 200 MeV, when suddenly the new particles appeared, which they called a psi-particle. The scheduled experimental runs at the time could have gone all the way up to 50 GeV.

Besides being a relatively massive particle, the most surprising thing about the J/psi was that it had such a long lifetime. Usually a meson of comparable mass would be expected to decay about ten thousand times faster than the observed lifetime of the J/psi. The experimental teams immediately realized that they had found a meson different from what was expected and perhaps an indicator of some new mass scale process or degree of freedom in particle physics.

A week after clearly recognizing the J/psi, the Stanford group found a new particle named the psi′ (psi-prime) at $3,684$ MeV/c², which decayed into the original J/psi plus two pions. It appeared that a whole family or subset of new mesons of comparable masses might be found, and soon thereafter reports of several such finds came from all over the world.

Fermilab, the highest-energy particle accelerator in operation at the time, identified several species of psi-particles by bombarding beryllium targets with 80,000 to 200,000 MeV photons.

It was not clear to theorists at the time what these new particles might be. Various suggestions mounted from an anomalous nuclear energy level to some sort of new lepton or even a magnetic monopole of some sort. It was not until two years later that the D-mesons were found again at Stanford's SLAC, and this led to substantial agreement as to what the J/psi might be.

The D-meson discovery hinted at the separation of quarks into various families, which, when combined with another antiquark of any family, yields mesons. The quarks (q), of which there are five families, are designated in terms of increasing mass by up (u), down (d), strange (s), charm (c), and bottom (b). A suspected heavier top (t) quark had not been found as of 1990, but appears necessary to complete the family picture. In the orthodox quark view, all mesons are made up of quark (q)-antiquark ($\bar{q}$) pairs.

In 1974, only the existence of u, d, and s quarks was suspected, and even then, most theorists viewed quarks as merely classifying schemes and not real constituents of elementary particles. The J/psi was later to be comfortably fitted as a charmed quark-antiquark ($c\bar{c}$) pair, and the D-meson confirmed that charm might indeed be a new family for quarks because it was composed of $c\bar{u}$ or $c\bar{d}$ structure. By early 1987, it was generally agreed by most orthodox particle theorists that the addition of charm quarks explained not only the J/psi and its related mesonic resonances but also the emergent family of D-mesons.

The J/psi was not the most primitive or lowest mass of the J/psi resonances. The original J/psi had a spin of 1. The more primitive eta sub c (for charm) particle, which allegedly should be the most primitive of the $c\bar{c}$ with a spin of 0, was discovered in 1984 and confirmed shortly thereafter with a mass of about 200 MeV/c^2 less than the J/psi. With the finding of the eta sub c as the lowest possible state of the charmed mesons, theoretical physicists became more certain that quarks might exist in nature and not be mere artifacts or clever classification rubrics.

J/psi decays into almost every other meson or mesonic resonance of lower mass. Eighty-six percent of the time it decays into mesonic hadrons, and about fifty different decay processes leaving stable hadrons have been observed in its decay. Also, more than sixty distinct mesonic resonances, which are relatively short-lived, have been observed as decay products of the J/psi.

These multifaceted decays make it a unique elementary particle and indicate that a new level of structure can explain such variety because lower-mass particles do not have such a rich set of decay products. Charm seems to fill this new structural format, although there are still serious questions as to why so many possible decays are evident. The J/psi is probed continually and is revealing even more possible decay products than ever envisioned in theoretical models of particle structure.

The discovery of the J/psi hinted at higher levels of structure being present in subatomic particles. Higher energies for accelerators would be demanded by particle physicists.

Impact of Event

In 1974, a new vista in particle physics was opened. The J/psi signaled a new family of charmed particles indicative of charmed quarks existing in nature. From 1974 to 1984, both charm and bottom (sometimes designated as beauty) were discovered experimentally, the latter being found via the Upsilon resonances at Fermilab, Batavia, Illinois. The Upsilon resonances came in at almost ten times the mass of the proton and three times the mass of the J/psi.

In the 1990's, ten charmed mesons consisting of charmed-anticharmed ($c\bar{c}$) quarks were discovered experimentally, ranging from the lightest to the heaviest. This very wide mass scale showed a tremendous richness in charmed quark structures, which intrigued theoretical physicists. The finding of these particles, explained by the charmed quarks, led to further subdividing of particle physics. The relatively light up (u) and down (d) quarks were classified as being members of one family and the strange (s) and charmed (c) quarks made up the second kinship group. With the experimental finds of the Upsilon resonances indicating the existence of a bottom (b) quark, physicists were led to postulate the existence of another top (t) quark to complete the heavy quark kinship grouping.

The search for the top quark has continued and is one of the main thrusts behind the building of the Superconducting Supercollider (SSC) particle accelerator, to be built in Texas in the 1990's. Because Fermilab was disappointed in not being able to find the top quark, it is now believed that energies of at least ten times this or more will be necessary to find the top quark.

An additional impetus to family classifications came from lepton, light particle, physics. Martin Perl, using the same SLAC used by Burton Richter, found the tau lepton in 1975. This made three families of leptons, the electron family, the muon family, and the tau family. By analogies borrowed from symmetry theories, there was mounting theoretical compulsion for postulating the existence of a similar three-family membership for quarks. Thus, when the Upsilon resonances were found at Fermilab, indicating the existence of a bottom quark, the top quark became the prime quest of experimental particle physics. Nevertheless, it could happen that no top quark exists at all; if this is the case, quark theories could fall by the wayside sometime in the future. The fundamental weakness of the quark models of particle structure is that, a priori, they are unable to predict the masses or energies at which various particles will be found. There are even doubts as to how well they are able to retrodict mass or energy levels once a new set of particles has been found experimentally. Retrodiction is much easier than prediction in any discipline, and so by fitting parameters, most theoretical physicists sense that they can explain the relatively high masses of the J/psi and the much heavier Upsilon resonances.

The best model of quark structure is quantum chromodynamics (QCD), in which at present there are thirty-six quarks. There are six quark flavors (up, down, strange, charm, bottom, and top), each of which can be of three distinct colors. This makes eighteen quarks when colors are counted, and when antiquarks with allowed anti-colors are counted the total is thirty-six. These quarks, or quark and antiquark combinations, are held together by gluons, massless spin 1 strong interaction carriers of which there are eight possible varieties. With these forty-four primitive inputs, all possible combinations of quarks and gluons with their appropriate antiparticles, almost all the hadrons can be explained.

The discovery of the J/psi gave quantum chromodynamics tremendous impetus which, when coupled with the discovery of the tau leptons, convinced the physics community that there was merit to considering all hadrons as being made up of quarks and gluons. As with nearly all new thrusts in theories, more problems were also presented to the theoretical community. One of the persistent questions is why quarks and leptons belong to families at all. If families exist, there might be a more primitive structure than quarks, which might explain the strange kinship relationships under which they interact. Another problem is that quarks have charges that are fractional electronic charges, one-third and two-thirds being the most likely. Leptons, at least those which are charged, have integral electronic charges. No quarks can be found outside the innards of particles, even though it can be explained through retrodiction and the introduction of new principles.

If the top quark is not found, the quarkist orthodoxy, buttressed in particle physics with the discovery of the J/psi in 1974, may be strangely silenced. The push to higher-energy accelerators such as the SSC, was greatly strengthened with the discovery of the J/psi, and this discovery is used as an exemplar in justifying the tremendous costs of such an effort.

The discovery of the J/psi caused particle physicists for the past two decades to promulgate the quarkist orthodoxy. Their successes in the experimental finds have been staggering, but there are still doubts as to what the quark model actually predicts, particularly in regard to the most intimate particle properties of mass, spin, angular momentum, and lifetimes.

The introduction of the charmed quark to explain the J/psi and the familial kinship notions to explain both quarks and leptons certainly explained the experimental finds, but not as completely as some might have wished. New finds are needed, and new physics must continue to pursue these finds. As with the actual finding of the J/psi in 1974, no one can predict comfortably when and where such finds might occur. That is what makes particle physics exciting: Nature does not have to succumb to classification schemes devised by physicists who are struggling to explain why so much diversity exists.

Bibliography

Bloom, Elliott D., and Gary J. Feldman. "Quarkonium." *Scientific American* 246 (May, 1982): 16, 66-77. An excellent review article on the status of quark theories in the early 1980's. Highly recommended for its balance between experimental discoveries and theoretical models and its depiction of how discoveries drive model makers and accelerator designers.

Glashow, Sheldon Lee. "Quarks with Color and Flavor." *Scientific American* 233 (October, 1975): 12, 38-50. Shows the condition of particle physics and the view of quark structures at the time the J/psi was discovered. Glashow presents a lucid and highly graphic picture of quark composites. Drawings are excellent.

Hitlin, D. G., and W. H. Toki. "Hadronic and Radiative Decays of the J/psi." *Annual Reviews of Nuclear and Particle Science* 38 (1988): 497-532. Reviews both the discovery of the J/psi and all the kindred resonances as well as their multifaceted decays. Represents the best up-to-date thinking and observations on the J/psi. Written largely for physicists; very thorough and wide ranging.

Quigg, Chris. "Elementary Particles and Forces." *Scientific American* 252 (April, 1985): 84-95. A general review of how quark structures shaped the development of both strong interaction and weak interaction theories. Outlines spectacular experimental particle discoveries from 1974 to 1985, discussing the role they played in the development of the total picture leading to unified field views. Well written with good graphics and tables outlining distinctions.

Wilczek, F. "QCD: The Modern Theory of the Strong Interactions." *Annual Reviews of Nuclear and Particle Science* 32 (1982): 177-209. Written by one of the theoretical proposers of charm; a complete discussion of how charm fits in the total particle picture. Also gives a good review of how the J/psi completed the second family of quarks in the QCD buildup.

John P. Kenny

Cross-References

Rutherford Discovers the Proton (1914), p. 590; Yukawa Proposes the Existence of Mesons (1934), p. 1030; Quarks Are Postulated by Gell-Mann and Zweig (1964), p. 1767; Friedman, Kendell, and Taylor Discover Quarks (1968), p. 1871; Gell-Mann Formulates the Theory of Quantum Chromodynamics (QCD) (1972), p. 1966; Georgi and Glashow Develop the First Grand Unified Theory (1974), p. 2014; The Tevatron Particle Accelerator Begins Operation at Fermilab (1985), p. 2301; The Superconducting Supercollider Is Under Construction in Texas (1988), p. 2372.

ANTHROPOLOGISTS DISCOVER "LUCY," AN EARLY HOMINID SKELETON

Category of event: Anthropology
Time: November, 1974
Locale: Hadar, Ethiopia

"Lucy," an early hominid skeleton dated at more than 3 million years old, precipitated a change in how human evolution is viewed

Principal personages:

DONALD C. JOHANSON (1943-　), the paleoanthropologist who co-organized the Hadar expeditions and discovered the skeletal remains of "Lucy"

MAURICE TAIEB (1935-　), a French geologist who recognized the significance of the Afar area in terms of fossil remains and approached Johanson about organizing an expedition to Hadar

TIM WHITE (1950-　), a physical anthropologist who worked with Johanson at Hadar and with Mary Leakey at Laetoli, and who recognized the similarities in the hominid fossils found at both sites

L. S. B. LEAKEY (1903-1972), the anthropologist who discovered the famous fossil sites which yielded australopithecine fossils and helped to make paleoanthropology fashionable after 1959

MARY LEAKEY (1913-　), an archaeologist and wife of L. S. B. Leakey whose work yielded valuable fossils and the famous fossilized hominid footprints

RICHARD E. LEAKEY (1944-　), an anthropologist and son of L. S. B. Leakey and Mary Leakey who has made many important fossil discoveries and has organized research at East Turkana in Kenya, a site that has produced numerous hominid fossils

Summary of Event

On November 30, 1974, Donald C. Johanson and another member of the International Afar Research Expedition (IARE) discovered small bones on the slope of a desert gully at Hadar in the Afar Triangle region of Ethiopia. These bones belonged to one individual, a unique hominid that did not resemble anything discovered previously. Named "Lucy" (after the Beatles' song "Lucy in the Sky with Diamonds"), the small skeleton was an amazing find and an important link in human evolution theories.

Major discoveries and theories in hominid evolution occurred during the twentieth century. From about 1900 to 1925, many were uninformed about hominid evolution. The term "hominid" has a very flexible definition, generally meaning an erect-walking primate that is an extinct ancestor to man. A hominid can be an ancestor of "true" man (modern man) or a relative, such as a modern primate. (The term "man" is used here to refer to both males and females of the genus *Homo*.) The few fossils that had been found before 1925 were from different geographic regions. They were also different from one another and no one knew exactly what they were, how they were related, or their age.

Early efforts in seeking the ancestors of man centered in Europe. Hominid fossils had been found in the Neanderthal Valley in Germany (Neanderthal man), Peking (Peking man), and Java (Java man), to name a few of the most famous. Other areas of the world were ignored as possible sites of hominid development. For example, although South Africa was recognized as having extremely old mammal and reptile fossils, no one looked for evidence of hominids there, since no hominid fossils had been found.

In 1924, a skull was found in South Africa that did not resemble man, but was not a baboon or chimpanzee. The skull was nicknamed the "Taung baby," since it was found at Taung and was estimated to be the skull of a six-year-old. The official name given was *Australopithecus africanus*. The discovery met with considerable skepticism and was basically ignored until 1936, when another skull was found. This skull appeared to be that of an adult australopithecine. Because of additional discoveries of fossils by the 1950's, most scientists accepted that two types of hominids had existed in South Africa: *Australopithecus africanus*, a slender type, and *australopithecus robustus*, a more primitive, robust type.

In 1959, L. S. B. Leakey and Mary Leakey discovered a skull of a large *Australopithecus robustus* at Olduvai Gorge in Tanzania. This discovery, known as "Zinj," was the first australopithecine found outside South Africa and the first to be reliably dated at 1.8 million years old. With the publicity surrounding the Leakeys' find, paleoanthropology (the study of ancient man) became fashionable to the general public and more funding was made available for further studies.

In 1972, Richard Leakey, the anthropologist son of L. S. B. Leakey and Mary Leakey, discovered a hominid skull at Koobi Fora in Kenya. He asserted that the skull was definitely *Homo*, or man, and that it was approximately 2.9 million years old. This skull was the oldest known fossil of man. If the more advanced homo existed at the same time as the obviously primitive australopithecines, then theories that homo evolved from australopithecines were wrong. Richard Leakey's discovery was a major find, one that generated excitement and controversy in the scientific world as theories were argued over and revised. (Later, more accurate dating placed the age of the skull at less than 2 million years.)

Also in 1972, Johanson and French geologist Maurice Taieb surveyed a potential fossil site in Ethiopia in the Afar Triangle region. After doing a preliminary survey, they decided to organize a formal expedition for the fall of 1973. This expedition was a cooperative international one, bringing American and French specialists together. During the late fall of the 1973 expedition, Johanson found leg bones and a knee joint of a 3-million-year-old hominid. These fossils were the oldest such fossils on record and indicated that hominids were probably walking upright 3 million years ago. In addition to the unique hominid fossils, Johanson's team discovered excellent mammal specimens. With such amazing finds in the first season, Johanson was able to raise more money for the second field season and entice other specialists to join the expedition.

During the second field season at Hadar in 1974, geologists generated detailed studies of the area, while other members of the team searched the area for fossils. In November, two of the oldest and finest hominid jaw fossils ever found were located. A few days later, a third jaw was found. Richard and Mary Leakey visited the site and confirmed Johanson's suspicion that the jaws could be homo with excessively primitive features. The jaws were dated at approximately 3 million years old, which made them the oldest known homo fossils.

On November 30, 1974, a few days after the Leakeys departed, Johanson found a nearly half-complete skeleton: "Lucy." For three weeks, everyone at the site collected several hundred pieces of bone, which made up approximately 40 percent of the skeleton. "Lucy" was a tiny-brained individual, approximately 1 meter tall. The sex of the skeleton was confirmed by the pelvic bones, which must be larger in females in order to permit the birth of large-brained babies. "Lucy" was an erect-walker, which gave certainty to theories that hominids walked erect at 3 million years B.C.

"Lucy's" bones were brought back to the United States for study in December, 1974. Through various techniques from 1975 to 1982, "Lucy's" age was determined to be close to 3.5 million years old, and the ages of the jaws found in 1974 and the knee joint found in 1973 were established at close to 4 million years old.

During the third and fourth seasons in 1975 and 1976 at Hadar, more hominid fossils were found. At site 333, the fragments of at least thirteen individuals of various ages and sexes were found scattered on a slope. These fossils were homo, and very different from "Lucy." The 1976 season also yielded stone tools, which strengthened the theory that the site 333 fossils were homo, since there is no evidence that australopithecines made or used tools. Because of the political conditions in Ethiopia, no expeditions were allowed until 1980, and that one was permitted only two weeks of field work.

Johanson and Tim White made extensive comparisons between the Hadar fossils and the fossils found at Laetoli, Tanzania, where Mary Leakey and White were working. These comparisons indicated that the Hadar and Laetoli hominids were similar and represented some developmental stage in-between apes and humans. This determination was a departure from Johanson's early belief that the fossils were homo. Johanson and White decided that, based on their fossil evidence, the Hadar and Laetoli hominids were an early, distinct australopithecine. They named these hominids *Australopithecus afarensis.* Their work has generated much interest from the scientific and general communities, but many questions regarding the human family tree remain to be answered.

Impact of Event

The discovery of "Lucy" was a significant development in the search for clues to understanding hominid evolution. "Lucy" was unique in that she was a very old, primitive, and small hominid that did not fit into the known hominid types. She was also the oldest and most complete hominid skeleton that had been found. Although only 40 percent of the skeleton was found, bones from both sides of the body were present; paleoanthropologists were able to reconstruct approximately 70 percent of her skeleton by using mirror imaging. With mirror imaging, existing bones are used to determine what the missing counterpart on the other side of the body looked like.

"Lucy" focused attention on the evidence that hominids were walking upright before brain size increased. Before 1974, the general assumption was that a large brain size was necessary before erect walking. "Lucy," however, was an upright walker with a primitive head and small braincase. The original standards used to define "human" were based on brain size since skulls, jaws, and teeth were the most common hominid fossils. Scientists recognize that brain size varies among modern individuals, and the ranges of brain size assigned to hominid types overlap with one another and with sizes assigned to other primates. Additionally, brain size is believed to have no significant relation to intelligence. Proponents of the theory that the brain developed before upright walking had to reject that theory based on the fossil evidence presented by "Lucy" and other hominid finds in the 1970's.

Because of the evidence of upright walking in a hominid estimated to be 3.5 million years old, and because of the small brain size, the question of why hominids began walking upright had to be reexamined. One previous theory was that manual dexterity, increased tool use, and brain development had forced some humans to stand erect in order to carry more with their hands. "Lucy's" hands were similar to modern man's, but no evidence has been found to suggest that australopithecines made or used tools. Various other theories explaining erect walking have been suggested and are being considered, but fossil evidence is scarce.

Johanson and White's proposal to assign a new name for the hominids represented by "Lucy," the site 333 fossils, and fossils from Laetoli caused considerable discussion. Johanson and White redrew the evolutionary family tree for humans, placing the *Australopithecus afarensis* as a common ancestor to both humans and the primates. People supporting the theory that "homo" had a separate evolutionary lineage from the primates are the most vocal opponents of the reclassification. White and Johanson acknowledge that their evolutionary tree is subject to change as new fossil evidence is acquired. Work continues in Africa to gain more information about the development of hominids, but the fossil record still has large gaps for which scientists can only speculate about the evolutionary changes in hominids.

Bibliography

Edey, Maitland A., and Donald C. Johanson. *Blueprints: Solving the Mystery of Evolution*. Boston: Little, Brown, 1989. This book traces the questions of evolution, beginning with groundwork laid by Carolus Linnaeus and Charles Darwin. Although interesting for background information, there is only brief mention of some major hominid fossil finds. The book is well written and is recommended for those interested in more detailed information on evolution at the molecular level. Illustrated, with a bibliography.

Herbert, Wray. "Lucy's Family Problems." *Science News* 124 (July 2, 1983): 8-11. Discussion of how *Australopithecus afarensis* fits into human evolution and the renewed controversy over whether "Lucy" represented something different from other hominid fossils found at Hadar. The article serves to illuminate the changing nature of theories in paleoanthropology. A relatively nontechnical, unbiased article for those interested in the further impact of "Lucy." Illustrated.

Johanson, Donald C. "Ethiopia Yields First 'Family' of Early Man." *National Geographic* 150 (December, 1976): 790-811. This article is a very brief summary of Johanson's work at Hadar, from the survey trip in 1972 through the 1975 field season. The text is minimal and easily understood. The photographs and illustrations accompanying the article are excellent.

Johanson, Donald C., and Maitland A. Edey. *Lucy: The Beginnings of Humankind.* New York: Simon & Schuster, 1981. This book is by far the most complete account of the discovery of "Lucy." Johanson and Edey write clearly, carefully explaining unfamiliar terms. The book relies on Johanson's journal notes and, in addition to information about the expeditions, tells of the preliminary work and the effects of "Lucy's" discovery on the scientific and general communities. Illustrated, with a bibliography.

Johanson, Donald C., and James Schreeve. *Lucy's Child: The Discovery of a Human Ancestor.* New York: William Morrow, 1989. Johanson continues the narrative he began in *Lucy.* He discusses the impact of "Lucy's" discovery and the controversy raised. Most of the book, however, focuses on field work undertaken in 1986 at Olduvai Gorge. This book offers additional insight into the difficulties and triumphs of field work and makes a nice companion piece for *Lucy.* Illustrated, with a bibliography.

Reader, John. *Missing Links: The Hunt for Earliest Man.* Boston: Little, Brown, 1981. Reader offers a well-written summary of the major people and discoveries in the history of paleoanthropology. The chapters are arranged so that the history flows from the Neanderthal discoveries in 1857 through the emergence of *Australopithecus afarensis* and related fossils by 1978. A good overall history. Illustrated, with a bibliography.

Virginia L. Salmon

Cross-References

Boule Reconstructs the First Neanderthal Skeleton (1908), p. 428; Zdansky Discovers Peking Man (1923), p. 761; Dart Discovers the First Recognized Australopithecine Fossil (1924), p. 780; Weidenreich Reconstructs the Face of Peking Man (1937), p. 1096; Leakey Finds a 1.75-Million-Year-Old Fossil Hominid (1959), p. 1603; Simons Identifies a 30-Million-Year-Old Primate Skull (1966), p. 1814; Hominid Fossils Are Gathered in the Same Place for Concentrated Study (1984), p. 2279; Scientists Date a *Homo sapiens* Fossil at Ninety-two Thousand Years (1987), p. 2341.

SOVIET VENERA SPACECRAFT TRANSMIT THE FIRST PICTURES FROM THE SURFACE OF VENUS

Category of event: Space and aviation
Time: October 22, 1975
Locale: Baikonur, Soviet Union

After a number of attempts, the Soviet space probes Venera 9 and 10 returned the first pictures from the surface of Venus

Principal personages:

ALEXANDER VINOGRADOV (1895-1975), a Soviet geochemist who specialized in studies of the chemical constitution of the earth and the solar system

KIRIL FLORENSKY (1915-1982), a Soviet geologist and head of the scientific team that interpreted photographs from the Venera space probes

ALEXANDER BASILEVSKY (1937-   ), a Soviet geologist who has played a principal role in interpreting the geology of Venus

ARNOLD SELIVANOV, a Soviet engineer who was responsible for much of the design of camera systems on the Venera spacecraft

MARGARITA NARAYEVA, a Soviet engineer who assisted in the design of camera systems on the Venera series of planetary probes

Summary of Event

By the late 1960's, the United States and the Soviet Union had each developed their own particular approaches to space exploration. The United States concentrated its planetary exploration efforts on the outer solar system, especially Mars. Soviet planetary exploration efforts were directed almost exclusively at Venus. Geologists and engineers who were involved in studies of space exploration and who played principal roles in the exploration of Venus include Alexander Vinogradov, Kiril Florensky, Alexander Basilevsky, Arnold Selivanov, and Margarita Narayeva.

The first probe to explore Venus was the United States probe Mariner 2, which was launched on August 27, 1962, and flew within 34,800 kilometers of Venus on December 14. Infrared observations made by Mariner revealed that Venus was far hotter than previously suspected, at least 350 degrees Celsius. Such a high temperature, far higher than an airless planet would have at Venus' distance from the sun, indicated that Venus must have a dense atmosphere of some heat-trapping gas, probably carbon dioxide.

The Soviet series of Venus missions began on April 2, 1964, with the launch of Zond (Probe) 1, but this probe ceased transmitting before reaching Venus. The next probes in the series, now renamed Venera (Venus), were launched on November 12 and 16, 1965. Venera 2 passed 24,000 kilometers from Venus on February 27, 1966, while Venera 3 impacted on Venus on March 1, the first human-made object to land on another planet. As with the first probe, contact was lost before the probes reached Venus.

Venera 4, launched from Tyuratam on June 12, 1967, was the first successful Soviet Venus probe. It reached Venus on October 18 and separated into two parts: a 380-kilogram lander that descended by parachute, and a "bus," the remainder of the probe, which continued past Venus and served as a communications relay. The lander transmitted data for ninety-eight minutes. Initially, Soviet space authorities claimed a successful landing, but later analysis suggested the probe failed about 25 kilometers above the surface. This misconception led to underestimates of Venus' atmospheric pressure. Veneras 5 and 6 were launched on January 5 and January 10, 1969, and arrived on May 16 and 17, respectively. They reported temperatures as high as 400 degrees Celsius, an atmosphere of at least 90 percent carbon dioxide, and atmospheric pressures up to sixty times that of Earth. Such pressures, equal to the pressure beneath 600 meters of water, crushed the probes before they reached the surface. Veneras 7 and 8 were modified on the basis of the earlier Venera findings and successfully landed on the surface and transmitted data. Venera 7 was launched August 17, 1970, and landed on December 15. Venera 8 was launched March 27, 1972, and landed on July 22. These probes found surface temperatures of 480 degrees Celsius and atmospheric pressure about ninety times that of Earth.

The Venera 9 and 10 spacecraft were launched on June 8 and 14, 1975. After a four-month voyage, they arrived at Venus on October 22 and 25. Instead of continuing past Venus, the spacecraft slowed and entered orbit around Venus, the first spacecraft to do so. Each lander then detached from its bus and descended, while the bus remained in orbit as a communications link.

Both Venera spacecraft landed on the eastern edge of Beta Regio, a large elevated region just north of Venus' equator. All that was known of Beta Regio at that time was that it was a highland that reflected radio signals strongly, indicating rough terrain. Later detailed radar mapping by American and Soviet Venus probes has shown that Beta Regio is a large volcanic plateau. Venera 9 landed on the northeast corner of Beta Regio; Venera 10 landed at the southeast corner, 2,000 kilometers to the south.

Unlike most landers of American design, the Venera spacecraft lacked legs. Instead, the base of the spacecraft was a circular skirt of crushable material. This design made the Veneras somewhat vulnerable to tipping. Venera 9 landed in very rough terrain, facing uphill on a slope of about 20 degrees. The spacecraft was tilted another 10 degrees by an object beneath it, probably a rock. The total tilt of the spacecraft was 30 degrees from the horizontal. Venera 10 landed in a smoother area. The surrounding terrain was nearly level, and the spacecraft was tilted only about 8 degrees by small rocks beneath it.

At the temperature of 480 degrees Celsius on the surface of Venus, the survival time of the spacecraft was brief. Because solid-state electronics are very vulnerable to heat, a long-lived probe would require a massive cooling system. All Venus landers have been designed for limited lifetimes, although they have thick insulation and are precooled to prolong their survival time. The insulation surrounding the Veneras was planned to provide about an hour of survival time, enough to transmit basic meteorological and surface chemistry data plus a panoramic photograph. The photographic apparatus consisted of a fixed television camera and a moving mirror that scanned the field of view. Both landers actually survived somewhat longer than expected and were transmitting a second panorama when their electronics finally failed. Viewing conditions on Venus were considerably better than expected. The Veneras had been equipped with artificial light sources in the expectation that the dense clouds of Venus would block most sunlight, but a surprisingly large amount of sunlight penetrated to the surface, which was about as bright as on an overcast day on Earth. The dense atmosphere was clear near the surface, with visibility of at least several hundred meters.

Venera 9 photographed a scene of cobbles and small boulders averaging a few tens of centimeters in size. One conspicuous set of boulders appeared to have once been a single slab, broken into three pieces, all of which showed an apparent layered structure. The Venera 10 scene showed broad expanses of slablike or layered rock outcrops with patches of pebbly material. Similar slabby rock surfaces have been photographed elsewhere on Venus by later landers in the Venera series, indicating that processes that form layered rocks are widespread on the planet. The Venera 9 and 10 chemistry probes could detect only a limited number of chemical elements. The low potassium contents of rocks at the landing sites suggested that the rocks probably were equivalent to terrestrial basalt, a dark, silica-poor volcanic rock. This finding has been supported by data from later probes.

Impact of Event

As similar as Venus is to Earth in size and location in the solar system, the two planets are radically different in many ways. The dense clouds that cover Venus once led to free speculation about the possibility of abundant water and life on Venus. Even though the United States probe Mariner 2 had sent data in 1962 indicating that the surface of Venus was very hot, there was a widespread reluctance on the part of many scientists, and the public alike, to believe that Venus, seemingly so Earth-like, was actually as hostile as it is.

Because of its global cloud cover, physical conditions on Venus were poorly known when the Soviet Union began its Venus explorations. The clouds hid the true surface conditions from view, conditions so hostile that a number of early spacecraft did not survive to reach the surface. It was only by learning from early failures that spacecraft robust enough to survive briefly on Venus were built eventually. Landing on another planet is a great technical accomplishment; landing on a planet with the surface conditions of Venus and successfully returning data and photographs is an achievement of the highest order.

On a planet where spacecraft survive only an hour after landing, every second of telemetered data is a significant increase in knowledge. As important as the meteorological data returned by Venera 9 and 10 were, the most significant achievements of these landers were the first pictures of the surface of Venus, indeed, of any other planet. It had been supposed that the surface of Venus might be very dark; it turned out to be much lighter than expected. With no water, only wind was likely to cause erosion on Venus. Many scientists expected a bland surface of wind-blown dust and smoothly worn rocks. The pictures returned by the Venera spacecraft superficially look rather mundane, but actually show that there is an unexpected level and variety of rock-forming processes at work on the surface of Venus. The photographs show that some processes break down and transport rock fragments. Most rock-forming processes on the surface of Earth are related in various ways to water. What might break down rocks and form layers on a waterless world like Venus is still a subject of debate. The layered rocks could be lava flows, layers of volcanic ash, or possibly layers of wind-blown sediment. In place of the water-deposited cementing minerals that bind such rocks on Earth, rocks on Venus might be cemented by chemical reactions with Venus' dense atmosphere. Also, it may be hot enough on Venus for partial melting of some rocks to occur.

Bibliography

Basilevsky, Aleksandr. "The World Next Door." *Sky and Telescope* 77 (April, 1989): 360-368. A well-illustrated survey of current knowledge of Venus, with many striking radar images from Venera orbiters. The article compares the evidence for crustal movements on Venus with plate tectonics on Earth.

Burgess, Eric. *Venus: An Errant Twin*. New York: Columbia University Press, 1985. Perhaps the most comprehensive book on Venus available for general audiences. The book surveys the history of Venus exploration by telescope and spacecraft, as well as the findings from Soviet and American missions.

Canby, Thomas Y. "Are the Soviets Ahead in Space?" *National Geographic* 170 (October, 1986): 420-459. A well-illustrated examination of the Soviet space program, with numerous diagrams and photographs of Soviet launch vehicles and spacecraft.

Carr, Michael H., et al. *The Geology of the Terrestrial Planets*. NASA SP-469. Washington, D.C.: Government Printing Office, 1984. A comparison of the planets from Mercury to Mars, with many photographs from planetary spacecraft. The chapter on Venus contains a detailed history of missions to that planet.

Oberg, James E. *Red Star in Orbit*. New York: Random House, 1981. An account of Soviet space exploration that attempts to penetrate the secrecy that has surrounded the Soviet space program. The primary emphasis is on manned space exploration.

Saunders, R. Stephen. "Venus: The Hellish Place Next Door." *Astronomy* 18 (March, 1990): 18-28. A summary of existing knowledge of Venus, written in nontechnical language. The principal emphasis is on the different evolutionary histories of the climates of Venus and Earth.

Von Braun, Wernher, Frederick I. Ordway III, and David Dooling. *Space Travel: A History*. Rev. ed. New York: Harper & Row, 1985. A comprehensive history of space exploration, with lengthy coverage of the early history of rocketry. The emphasis is on manned spaceflight, but there is a good summary of planetary missions.

Steven I. Dutch

Cross-References

The First Rocket with More than One Stage Is Created (1949), p. 1342; Sputnik 1, the First Artificial Satellite, Is Launched (1957), p. 1545; Luna 3 Provides the First Views of the Far Side of the Moon (1959), p. 1619; Venera 3 Is the First Spacecraft to Impact on Another Planet (1965), p. 1797; Mariner 9 Is the First Known Spacecraft to Orbit Another Planet (1971), p. 1944; Viking Spacecraft Send Photographs to Earth from the Surface of Mars (1976), p. 2052.

KIBBLE PROPOSES THE THEORY OF COSMIC STRINGS

Category of event: Astronomy
Time: 1976
Locale: Imperial College, London, England

Kibble proposed the theory of cosmic strings, which provided a workable explanation of how matter formed into stars, galaxies, and clusters

> *Principal personages:*
> THOMAS KIBBLE, the physicist who worked out the initial theory of cosmic strings
> NEIL TUROK, a protégé of Kibble who worked out a computer simulation on the evolution of strings
> ANDREAS ALBRECHT, a scientist who worked on a computer simulation on the evolution of cosmic strings
> EDWARD WITTEN, the scientist who proposed that cosmic strings could conduct electricity

Summary of Event

According to the big bang theory, the universe began with a colossal explosion some 15 to 20 billion years ago. Modern advances in particle physics and other branches of the physical sciences have allowed scientists to formulate theories that describe the events immediately following the big bang. According to theory, from the moment of the creation of the universe until a point in time some 10^{-43} seconds later, all four of the forces of nature consisted of one superforce. The universe consisted of energy; there were no elementary particles as yet. Physicists refer to this state as one of symmetry. In other words, the universe would have appeared to have the same properties in all directions. As minute increments of time passed, the symmetry was broken as individual forces began to appear. First came gravity, and then the strong nuclear force. At about 10^{-12} seconds, the weak nuclear force and electromagnetism began to exist as independent forces. The appearance of these forces allowed the formation of elementary particles and then, within minutes, the first atomic nuclei. After much expansion and cooling of the universe, the first atoms were formed. Physicists estimate that this latter event took place about 700,000 years after the initial explosion that created the universe.

Prior to the formation of the first atoms, the vast number of free electrons in the universe interacted with light emitted at the instant of the big bang. After most of the electrons had become involved in the formation of atoms, matter and light were decoupled and the universe became transparent to radiation. At this juncture, reduced light pressure allowed bits of matter to begin to form larger masses.

According to the first theories of galactic formation, gravitational forces, acting in the early universe, caused matter to form lumps. These lumps, in turn, attracted great clouds of gas and dust, and from these huge rotating masses, individual stars were born. Stars that were formed close to one another remained gravitationally bound and formed huge multibillion-star assemblages called galaxies. Individual galaxies were attracted by gravity to form clusters, and clusters were bound to superclusters.

More modern theories have begun to cast some doubt on this scenario. One of the major problems that cosmologists are now trying to solve is the question of how or why bits of matter lumped together in the early universe. At the same time, any successful model of the universe must also explain why vast areas are totally devoid of matter. It has been argued that there may have been some irregularities in the initial explosion that caused some unevenness in the distribution of matter. Because the cosmic background radiation—the remnant of the big bang fireball—is the same in intensity from all parts of the sky, however, this idea is difficult to accept. One point on which most cosmologists do agree is that the universe has not existed long enough to form galaxies by means of the gravitational model as described. The reasoning for this conclusion is that by the time atoms had begun to bond together and form lumps of matter, these lumps would have drifted so far apart from one another in the expanding universe that gravitational forces would have been too weak to draw them together to form stars, galaxies, and clusters.

One possible solution for this problem of why matter was able to collect rapidly enough, and in great enough quantities, to form bodies such as stars and galaxies lies in the possible existence of a bizarre substance known to scientists as "dark matter." There are many forms of matter which, in theory, could be classified as dark matter. Scientists, however, have reduced this list of possibilities to two. First, dark matter may be in the form of tiny particles called neutrinos. Once thought to be massless, neutrinos are now believed to possess a small mass. If this is true, then most of the mass in the universe might well be in the form of neutrinos. The second possibility is a particle known as a WIMP, or weakly interacting massive particle. These particles, which are about ten times as massive as protons, have been predicted by some theoretical models of the universe.

There is some direct evidence for the existence of matter that is not detectable optically. Studies of rotation curves for various spiral galaxies indicate that these galaxies, including the Milky Way, are not rotating at the rate that theory indicates they should. It is believed that this unexplained rotation is a result of the presence of matter—matter that cannot be observed. Current observations have suggested that as much as 90 percent of the matter in the universe has not yet been detected.

In 1976, physicist Thomas Kibble, working at the Imperial College in London, was considering the possible effects of modern theories of unified fields on the universe. He was particularly concerned with that fraction of a second after the big bang when the forces began to assume their separate identities. His mathematical model suggested that shortly after the big bang, the rapidly cooling universe developed flaws that appeared to be stringlike in nature. This rapid cooling of the universe would produce what is called a phase transition and is analogous to the cracks and other flaws formed when water is frozen into ice. Kibble's strings were described as slender strands of highly concentrated mass-energy. These remnants of the original fireball, according to theory, are much thinner in diameter than a proton and as long as the known universe. A segment of cosmic string 1.6 kilometers long would weigh more than the entire earth. This large mass suggests that strings must have been formed early in the history of the universe when there was an excess of energy.

Computer simulations, conducted by Neil Turok and Andreas Albrecht, indicate that as the universe expanded and rapidly cooled immediately after the big bang, defects in space-time formed long, continuous chains. Within these chains or strings, symmetry still exists. The forces of nature exist as one force, and as a result, there are no atomic particles. As the universe expands, strings evolve. Rapid vibrations within any one string may cause portions of that string to overlap. When this occurs, the loop that has been formed breaks off from the string. These loops may be of any size, from microscopic to several light-years across.

According to the theory of cosmic strings, the loops undergo rapid oscillations. These oscillations, the speed of which may approach the speed of light, cause the emission of gravitational waves. These waves, which had been predicted by Albert Einstein's theory of general relativity, are ripples in the fabric of space-time. As a string radiates this energy, it will shrink eventually and disappear. It has been estimated that a loop of cosmic string 1,000 light-years in circumference would radiate away in 10 to 100 million years.

It has been determined that strings have a finite lifetime. Therefore, scientists ponder if there are any left in the universe today. Researchers working on string theory have determined that the smallest loop that could have been formed in the primeval universe and still exist must have had an initial diameter of at least 1 million light-years. It is also theorized that if there are currently any strings in existence, they are widely dispersed, perhaps as much as 1 billion light-years from Earth.

A modification of cosmic string theory by Edward Witten suggests that strings might be superconductors of electricity. It has been calculated that currents as great as 100 quintillion amperes could be induced. The flow of electrical current produces a magnetic field, so strings should be surrounded by intense fields. Particles, trapped and accelerated within these fields, would glow. Perhaps the observation of radiation from these particles might one day provide the first observational evidence for the existence of cosmic strings.

Impact of Event

The theory of cosmic strings may offer a solution to the problem of how stars, galaxies, clusters, and superclusters were formed from what was believed to be a smooth, evenly expanding plasma. At the same time, strings may explain why galaxies are distributed as they are and why huge areas of space are virtually empty.

There is a distinct possibility that cosmic strings may have been the form of dark matter that acted as a nucleus and provided the gravitational attraction that caused bits of matter to adhere in the early universe. If so, strings would also have some influence on the distribution of galaxies and groups of galaxies. In the 1980's, astronomers discovered a huge group of galaxies. These thousands of galaxies were distributed in a sheetlike structure. This formation, which is more than 500 million light-years in length, was dubbed "The Great Wall" by astronomers.

During the 1980's, astronomers were studying the motion of galaxies near the Milky Way. They concluded that hundreds of galaxies, including the Milky Way, are being gravitationally drawn toward a massive celestial body. This body, which is being called "The Great Attractor," is believed to be centered about 150 million light-years from Earth. Because cosmic strings theoretically would project vast, incredibly powerful gravitational fields, they must be taken into consideration when examining possible causes for The Great Wall and The Great Attractor.

A superconducting string might explain why there are vast voids in the universe. As current passed through the string, a strong magnetic field would be set up around the string and electromagnetic waves would propagate away from it. The gases near the radiating string would heat up, expand, and form a bubble of matter around the string. According to a modern theory of the structure of the universe, galaxies are formed where bubbles intersect. In this scenario, matter is blown away from strings instead of being attracted to them, so an explanation for cosmic voids is given.

Although there is no observational evidence that can prove conclusively that cosmic strings exist, scientists have considered the effects strings would have on the space around them. In 1986, radio telescopes, probing the center of the Milky Way, observed an image of threadlike structures. Some astronomers believe that these structures might be glowing strings.

The phenomenon known as gravitational lensing might provide evidence for the existence of cosmic strings. In the formation of a gravitational lens, a massive body located between Earth and a distant galaxy causes light from the galaxy to split around it. When the light is observed from Earth, there appears to be two different galaxies. It is possible, according to theory, that a cosmic string could cause such a phenomenon. The first gravitational lens was discovered in 1978, and since that time, about six have been confirmed. All, however, have been caused by a luminous body. A recent discovery might indicate the presence of a string. Two University of Hawaii astronomers have reported what appears to be a chain of galactic pairs. This could be evidence of a string causing gravitational lensing, or, perhaps, a completely unrelated phenomenon.

Other investigations may confirm the existence of cosmic strings. One is the search for some sign of unevenness in the cosmic background radiation. A string would reveal itself by slightly heating the microwave radiation in front of it and by slightly cooling the radiation behind it as it oscillates through space. Sensitive instruments aboard satellites are looking for such a pattern of heating and cooling.

According to astrophysicists, the effects of strings should be detected by gravity waves. These distortions in space-time produced by the rapid oscillations of loops of string have not yet been observed. Future research will have to wait for the development of more sophisticated equipment.

Bibliography

Abell, George O., David Morrison, and Sidney C. Wolff. *Realm of the Universe.* New York: Saunders College Publishing, 1987. A well-illustrated volume covering topics in astronomy from the beginning of the universe to the evolution of galactic clusters. Several chapters are devoted to stellar evolution and modern theories of cosmology. A portion of the volume also discusses the solar system. Intended for a college course in general astronomy, the book would be appropriate for the informed layperson.

Bartusiak, Marcia. "If You Like Black Holes, You'll Love Cosmic Strings." *Discover* 9 (April, 1988): 60-68. A very readable article covering the evolution of the concept of cosmic strings from the first string theory in 1976 to modern modifications of that theory.

Hartman, William K. *The Cosmic Voyage: Through Time and Space.* Belmont, Calif.: Wadsworth, 1990. A well-written, well-illustrated volume covering a host of topics in both the solar system and stellar astronomy. This volume was intended for use as a textbook for a general astronomy course at the college freshman level. It would be accessible to the informed layperson.

Press, William H., and David N. Spergel. "Cosmic Strings: Topological Fossils of the Hot Big Bang." *Physics Today* 42 (March, 1989): 29-35. A very technical paper covering the structure, dynamics, and evolution of cosmic strings. The theory of superconducting strings is introduced. The reader should have a working knowledge of differential equations and advanced physics.

Trefil, James S. *The Dark Side of the Universe: Searching for the Outer Limits of the Cosmos.* New York: Charles Scribner's Sons, 1988. This volume discusses the origin, structure, and fate of the universe. Topics in modern cosmology such as cosmic bubbles, dark matter, and cosmic strings are well presented. The reader should have some background in modern physics and astronomy.

Vilenkin, Alexander. "Cosmic Strings." *Scientific American* 257 (December, 1987): 94-98. A technical article on the nature of cosmic strings and evidence for their existence. The reader should have a background in physics and astronomy.

David W. Maguire

Cross-References

Rutherford Discovers the Proton (1914), p. 590; Einstein Completes His Theory of General Relativity (1915), p. 625; Lemaître Proposes the Big Bang Theory (1927), p. 825; Hubble Confirms the Expanding Universe (1929), p. 878; Gamow and Associates Develop the Big Bang Theory (1948), p. 1309; De Vaucouleurs Identifies the Local Supercluster of Galaxies (1953), p. 1454; Tully Discovers the Pisces-Cetus Supercluster Complex (1986), p. 2306; Supernova 1987A Corroborates the Theories of Star Formation (1987), p. 2351.

VIKING SPACECRAFT SEND PHOTOGRAPHS TO EARTH FROM THE SURFACE OF MARS

Category of event: Space and aviation
Time: July 20, 1976, and September 3, 1976
Locale: Mars

The Viking landing missions were the first successful spacecraft to return photographs from the surface of Mars

> *Principal personages:*
> JAMES S. MARTIN, JR., the project manager for the NASA Langley Research Center
> ISRAEL TABACK, the deputy project manager (Technical) for the NASA Langley Research Center
> GERALD A. SOFFEN (1926-   ), a project scientist at the NASA Langley Research Center
> A. THOMAS YOUNG (1938-   ), the mission director at the NASA Langley Research Center
> G. CALVIN BROOME, the Lander science instruments manager at the NASA Langley Research Center

Summary of Event

For centuries, the planet Mars has fired the human imagination to speculate about the wonders that might exist on its surface. Earth-based telescopes revealed numerous features suggestive of seasonal change, running water, growing vegetation, and even canals. The possibility that Mars has some form of intelligent life was not ruled out. With this in mind, it seems only natural that one of the first objectives of the space age would be to go to Mars and see what the planet truly holds.

The first close-up look at Mars came in 1965, when the Mariner 4 spacecraft flew past the planet. In the short time that this spacecraft was near Mars, it took twenty-one detailed photographs of the planet's surface. These photographs showed no evidence of canals or life, only a cold, cratered world much like Earth's moon. The 1969 flights of Mariners 6 and 7 confirmed the earlier data. Mars was not a "living planet," but a cold, dead one. Some scientists believed that Mars no longer qualified as a priority in the planning of future space missions.

It was very fortunate for the exploration of Mars that the 1971 Mariner 9 mission was not canceled. Mariner 9, unlike the three previous missions, was to be an orbiter. It would spend nearly a full year in orbit around Mars, photographing almost the entire Martian surface. The planet Mariner 9 revealed was far different from what the earlier Mariner photographs had depicted. Mariner 9 photographs showed huge volcanoes, dried river channels, glacial deposits at the polar caps, and an enormous rift valley that spanned nearly 4,800 kilometers. Mars was no longer seen as a dead planet but as one that was geologically alive. The question of life being found on Mars was, therefore, once again considered.

The results of the Mariner 9 mission set the stage for the very sophisticated Viking mission. This mission would utilize a two-component spacecraft that consisted of an Orbiter and a Lander. The Orbiter would provide detailed photographs of the surface of the planet, as well as collect data on its atmospheric composition, wind speeds, and temperature variations. The Lander would photograph the surface first-hand, sample and analyze soil specimens, record the Martian weather conditions, and conduct experiments to search for the presence of life. It was hoped that the life-search experiments would prove positive and therefore shed light on the question concerning the origin of life. Viking would prove to be the most ambitious robotic mission flown; its results were eagerly awaited by scientists around the world.

The Viking mission to Mars was first defined in 1968 by the National Aeronautics and Space Administration (NASA). Viking was not, however, the original plan for the exploration of Mars. Its predecessor, Voyager, was discussed from 1965 until 1967, and then dropped. The original Voyager proposal called for a flyby, orbiter, and later lander missions. Successive missions were to have been even more ambitious and costly. The giant Saturn 5 moon rocket was to have been the launch vehicle. Voyager never got beyond the discussion stage as a result of budget cuts, but a less costly mission was sought to replace it.

The Viking spacecraft that eventually flew to Mars was a combination of the original Voyager design and that of the lunar Surveyor spacecraft. The techniques of soft-landing upon another planetary body had been well developed in the Moon missions and required a slight modification only to accommodate the thin Martian atmosphere. Viking would include two Orbiter-Lander pairs that were launched separately and placed into slightly different orbits around Mars. The Orbiters first would examine proposed landing sites and then certify them according to criteria required for a safe landing. Only then would the Landers be released and directed toward their targets. The actual landing would be accomplished by a gradual slowing, resulting first from aerodynamic drag and then by parachute. Retro-rocket firing would complete the soft-landing, with no greater impact velocity than a minor jolt.

Viking 1 was launched from Kennedy Space Center at Cape Canaveral on August 20, 1975, with Viking 2 being launched on September 9, 1975. Key personnel from the NASA Langley Research Center, which managed the Viking landing missions, were James S. Martin, Jr., A. Thomas Young, Israel Taback, G. Calvin Broome, and Gerald A. Soffen.

The Viking missions were not the first to attempt a soft-landing upon Mars. Two Soviet spacecraft (Mars 2 in 1971 and Mars 6 in 1973) apparently crashed into the Martian surface. A third Soviet spacecraft (Mars 3 in 1971) did soft-land but stopped operating after only twenty seconds. These missions taught NASA scientists that landing upon Mars would be no easy task. It was hoped that the two-component concept of the Viking spacecraft would avoid the problems encountered by the Soviets.

The Viking 1 spacecraft achieved Martian orbit on June 19, 1976. It was placed into a highly elliptical orbit that brought it as close as 1,513 kilometers and as far as 33,000 kilometers. This orbital configuration provided the combination of close-up and wide-angle photography methods necessary to evaulate properly the Martian surface.

Although the computers on board the Viking spacecraft contained preselected landing sites, the Orbiter first had to examine these sites and then had to certify their suitability for a safe landing. This became a very frustrating aspect of the mission. Photographs taken by the Mariner 9 spacecraft showed smooth, flat landing sites, which proved to be rough and unsuitable for landing. When choice after choice of sites failed to meet the landing criteria, the actual landing had to be pushed back from its scheduled date. Safety factors had priority in site selection.

In the early morning hours of July 20, 1976, the computers on board Viking 1 initiated the landing sequence to commence. The anxious scientists waited almost twenty minutes for the telemetry to confirm the events taking place on Mars, because Viking 1 was more than 321 million kilometers from Earth at that time. As each confirmed stage came into Mission Control at the Jet Propulsion Laboratory (JPL) in Pasadena, California, scientists breathed sighs of relief. Confirmation of the safe landing of Viking 1 came at 5:12 A.M. Pacific daylight time.

Shortly after Viking 1 landed, its twin-camera system began taking pictures and transmitting them back to Earth. The cameras on board could operate individually for panoramic views or together for stereoscopic pictures. Also, they had the capability to photograph in black and white, or, through the use of filters, which produce color images. Because their method of photography was line-scan, it required almost five minutes to complete a single image. As the imaging progressed, the spacecraft relayed to Earth each line-scan as it was taken. At Mission Control, the scientists viewed tantalizing narrow strips of images as they grew into full pictures. It was a historic moment for human exploration.

The first complete image from the surface of Mars showed one of the Lander's footpads resting firmly on a surface consisting of fine soil and scattered rocks. Evidence of volcanic rocks and wind erosion was clearly visible. Red was truly the color of Mars.

While the Lander was busy photographing on the surface, the Viking 1 Orbiter began a systematic photographic survey of its own that would eventually produce high-resolution photographs of nearly the entire surface. Programmed changes in the spacecraft's orbit would bring it to within 30 kilometers of the Martian moon, Phobos. The resulting photographs showed a cratered world that may represent one of the most primitive objects in the solar system.

Viking 1 was joined in orbit by Viking 2 on August 7, 1976. Viking 2 began its site selection process as Viking 1 had done. Because of the success of Viking 1, the landing site for Viking 2 could be more ambitious and go to a more exciting location, where the chances of finding life might be greater. Together, this pair of spacecraft would rewrite the knowledge of Mars and present science with a picture of a dynamic planet that was not unlike Earth in many respects.

Impact of Event

The primary objectives of the Viking missions to Mars were to provide data in an attempt to determine where Mars fits in when compared to the evolution of the earth and Moon and to determine if life existed elsewhere in the solar system. For centuries, Mars has been a leading candidate in the search for extraterrestrial life because it shares certain similarities with Earth. Most important of these was the possibility of finding water in liquid form on its surface. It seems apparent that wherever water is present, the potential for some form of life is very good. The ever-changing appearance of the Martian polar caps seemed to confirm the presence of water.

The Viking Landers had three instruments that could detect various forms of life. The first was its cameras. Provided that any life-form was at least as large as lichen, it could be photographed directly. The next instrument was the gas chromatograph/mass spectrometer, which could detect organic molecules in the Martian soil. The finding of these molecules would present a serious argument for the presence of life (past or present). Perhaps the most ambitious of the three instruments were the life-detection experiments. These were designed to detect any unusual activity in the Martian soil that might be construed as a life function. Each of the three experiments searched for signs of metabolic processes such as those produced by bacteria, green plants, and animals on Earth. It is important to remember that these experiments were designed to detect life-forms as they occur on Earth. Perhaps some purely Martian organism was not detected because of its unique characteristics, but that is unlikely.

Although the biological experiments performed as designed, they could not confirm the presence of any life-form. They discovered that Martian soil mimics life when it is exposed to water vapor. These chemical reactions, which were attributed at first to a life process, were the result of oxidants present in the soil and atmosphere of Mars. This was a particularly disturbing discovery, because oxidants such as peroxides and superoxides tend to break down organic matter and living tissue. This suggests that if life did exist on Mars, it would be destroyed quickly by these compounds.

The Viking missions to Mars did not prove or disprove conclusively the existence of life there, but the missions did provide a wealth of other information about Mars. Geological observations made from the Orbiters portray a planet that in many ways is as dynamic as Earth. Huge volcanoes tower over the flat plains of the Martian northern hemisphere, while the southern half is seen as a cratered, frozen world. Dried-up river valleys with their associated features were clearly in evidence in the north, and glacial features were obvious in the polar regions. Valles Marineris (Valley of the Mariners), a huge rift valley that has no comparison on Earth, spans more than 4,800 kilometers. Prior to the Viking missions, no one had guessed how complex the Martian geology would be.

The Viking missions not only returned valuable information about Mars but also demonstrated what a robot spacecraft can do far from Earth. The two Viking Landers proved that spacecraft could work independently and successfully for extended periods of time. The Landers pushed around rocks, recorded "Marsquakes," observed wind storms and snowfalls, and even repaired a troublesome mechanical arm. Viking rates as one of the most successful missions ever flown.

Bibliography

Bane, Don. *Viking: The Exploration of Mars.* NASA EP-208. Washington, D.C.: Government Printing Office, 1984. This booklet presents a wealth of descriptive information on the Viking missions from the point of lift-off to the very last data sent back from Mars. It is beautifully illustrated with the best of the Viking photographs.

Carr, Michael H., and Nancy Evans. *Images of Mars: The Viking Extended Mission.* NASA SP-444. Washington, D.C.: Government Printing Office, 1980. This booklet serves as a good continuation of the work by Bevan M. French cited below. Although it presents good background information on the Viking missions, its best feature can be found in the explanations of the photographs presented.

French, Bevan M. *Mars: The Viking Discoveries.* NASA EP-146. Washington, D.C.: Government Printing Office, 1977. This booklet describes the initial results obtained from the Viking 1 and 2 missions. It presents a clear and concise report that is readable at most levels. An excellently reviewed bibliography complements the text and illustrations.

Greeley, Ronald. *Planetary Landscapes.* London: Allen & Unwin, 1985. This book offers the reader both the fundamentals of planetary geology and an in-depth look at specific topics. The chapter that discusses Mars is comprehensive and utilizes many of the Viking images. It is an extremely well-illustrated and readable book. A handy reference for those interested in the planets.

Moore, Patrick, and Garry Hunt. *Atlas of the Solar System.* Chicago: Rand McNally, 1983. This work provides one of the most comprehensive reviews of the solar system available. It combines knowledge gained from spacecraft investigations with that of ground-based studies. Its content and illustrations are extremely well presented. The chapter on Mars is extensive and up-to-date, with numerous photographs from the various Mars missions. Excellent reference for further study.

Mutch, Thomas A., et al., comps. *The Geology of Mars.* Princeton, N.J.: Princeton University Press, 1976. It is unfortunate that this work was published just after the Viking 1 landing; therefore, it contains only minimal reference to that mission. This text, nevertheless, represents perhaps the most comprehensive discussion of Martian geology available up to the Viking mission.

National Aeronautics and Space Administration, Viking Lander Imaging Team. *The Martian Landscape.* NASA SP-425. Washington, D.C.: Government Printing Office, 1978. This book complements the work by Spitzer (cited below) by presenting a detailed view of Mars as seen from the planet's surface. Details of the missions are explained in chapter 1, with spectacular photographs following in subsequent chapters. Explanations of the features illustrated are extensive.

Spitzer, Cary R., ed. *Viking Orbiter Views of Mars.* NASA SP-441. Washington,

D.C.: Government Printing Office, 1980. This book represents a detailed look at the Martian surface as seen from orbit. Very good descriptions of the features are represented, along with numerous photographs. Accompanying the text is a thorough glossary and suggestions for further reading.

Paul P. Sipiera

Cross-References

Venera 3 Is the First Spacecraft to Impact on Another Planet (1965), p. 1797; The Lunar Orbiter 1 Sends Back Photographs of the Moon's Surface (1966), p. 1825; Mariner 9 Is the First Known Spacecraft to Orbit Another Planet (1971), p. 1944; Soviet Venera Spacecraft Transmit the First Pictures from the Surface of Venus (1975), p. 2042; Voyager 1 and 2 Explore the Planets (1977), p. 2082.

DEEP-SEA HYDROTHERMAL VENTS AND NEW LIFE-FORMS ARE DISCOVERED

Category of event: Earth science
Time: 1977
Locale: Galápagos Rift, Pacific Ocean

Corliss and Ballard discovered deep-sea hot springs and collected previously un-known life-forms uniquely adapted to exploit these submarine oases

Principal personages:
ROBERT D. BALLARD (1942-   ), a marine geologist who has popu-larized the use of deep submersibles in scientific research
JOHN B. CORLISS (1936-   ), an oceanographer who collected and iden-tified some of the first vent community organisms
JOHN M. EDMOND (1943-   ), a marine geochemist who was involved in the collection and analysis of hydrothermal vent waters

Summary of Event

Serendipity and science are not strange bedfellows. Some of the truly great scien-tific discoveries have been made when researchers stumbled upon a new phenome-non while pursuing another, often unrelated, goal. That is serendipity—a pleasant surprise discovered in the course of another study. The discovery of previously un-known life-forms and life-styles in the deep sea falls into that category.

The conception by earth scientists of how the earth functioned underwent a dra-matic revolution in the early to mid-1960's, when the theories of seafloor spreading and plate tectonics were combined and formalized. The earth's surface is composed of a number of rigid plates that move in somewhat predictable ways. The plates move away from each other along spreading centers that tend to be located in the major ocean basins (for example, one major spreading center extends down the length of the Atlantic Ocean). Spreading centers in the oceans are associated with exten-sive submarine mountain chains, or ridges, that are roughly symmetrical on either side of a central valley that runs down the length of the chain. In this way, they differ from most continental mountain chains. The plates collide with each other in other places, called convergent boundaries. One of the most obvious indicators of a colli-sion boundary is the presence of geologically young mountains and active volcanoes (for example, the Andes mountains of South America).

Earth scientists understood long ago that the force necessary to push or pull these plates across the surface of the earth must be phenomenal. The theories suggest the circulation cells of molten rock deep in the earth rise near the surface directly be-neath the central valley in spreading-center mountain chains. The molten rock from below creates and forces its way into cracks in the overlying plate. Injecting new molten rock into these cracks forces the older, rigid rocks on either side to move away from the central valley. In this way, new rock is formed in the central valleys, older rocks are pushed aside, and everything on the plates on either side of the spreading center moves away from the ridge. Obviously, this is an extremely slow process which can cause substantial movement and change over geologic times (typically millions of years).

As earth scientists began to examine the implications of the theory of plate tectonics, it became obvious that the concentration of molten rock input in the central valleys should heat the water in the cracks between the rocks that compose the valley floor. Drawing analogies with hot spring and geyser fields on continents (for example, those in Yellowstone National Park), some scientists suggested that there may be equivalent hot springs in the deep ocean.

In 1977, a group of oceanographers had selected an area of the Pacific Ocean near the Galápagos Islands for a detailed search for hydrothermal ("hot water") springs. The goals of the team included locating and making on-site observations of hydrothermal activity with the deep submersible *Alvin*. In order to focus their search and use *Alvin* most efficiently, the team towed a remotely operated camera and instrument-laden sled behind their ship and slightly above the ocean bottom. Perhaps more important than any pictures, the temperature sensors on the sled would identify any slight increase in bottom water temperature—a good indicator of hydrothermal activity. The oceanographic team included Robert D. Ballard from the Woods Hole Oceanographic Institute near Boston. Ballard had made extensive use of the deep-diving submersible *Alvin* and the remotely towed instrument sled in other geologic investigations of ridge systems. Other members of the team included John B. Corliss of Oregon State University and John M. Edmond of the Massachusetts Institute of Technology. The remaining team members were marine geologists and other oceanographers who were interested in measuring aspects of the hydrothermal activity when, and if, a distinct system was found.

Several likely areas for investigation were identified based on temperature variations detected by the remote instruments. In addition, photographs indicated the presence of large clam shells in areas where temperature anomalies were observed. Using the presence of the clam shells as the best indicators of some unusual phenomenon, members of the team dispatched *Alvin* for close-up observations.

When *Alvin* and the scientists arrived at the bottom (more than 2,500 meters below the surface), they found unequivocal evidence for hydrothermal springs. In their target areas, the water near the rocky bottom was shimmering because of the difference in temperature between the normal bottom water (about 3 degrees Celsius) and that coming from cracks in the bottom. The scientists noticed that many of the rock surfaces in the areas of activity were covered with a thin dusting of white material. They hypothesized that the hot water coming out of the rocks was transporting dissolved minerals that stay dissolved in hot water but were "frozen out" when the hot water mixed with the very cold bottom water. In a few places, the location of the hot water spring was concentrated to such an extent that when the minerals solidified, they formed a hollow chimney. As a result, some of the hydrothermal areas resembled a factory with billowing smokestacks.

The geological discoveries paled by comparison with the discovery of densely packed biological communities surrounding the vents. The inhabitants of these communities included gigantic clams, tube worms, and crabs, as well as a few organisms new to science. Even those that were not familiar looking, upon closer examination, proved to be variations of known species. In addition to finding these biological oases surrounding active vents, the researchers found the remains of communities surrounding inactive vent areas. The obvious association between the communities and the vents—and especially the death of the communities around dormant vents— suggested that the source of nourishment for the organisms was based on the vent. Most of the knowledge of biological communities comes from near-surface environments wherein the sun is the ultimate source of energy and green plant photosynthesis provides the base for most food chains. In the hydrothermal vent communities, scientists were faced with a thriving ecosystem whose basic food chain was a mystery.

Water samples taken near the vents were analyzed and found to contain high concentrations of a type of dissolved sulfur called sulfide. Such concentrations of sulfide were not unexpected by the chemists who collected the samples, but now, in a search to find the base of this food chain, they had to consider the biochemistry of the waters. Large volumes of the vent water were filtered, and the filters were examined for evidence of bacteria. Bacteria that utilize dissolved sulfide as the basis of their biochemistry were known from isolated near-surface environments (for example, hydrothermal springs). Not surprisingly, the filters revealed massive quantities of the bacteria. Subsequent examination of the larger animals, like the filter-feeding clams and worms, revealed remains of similar bacteria that had been filtered out of the water.

The conclusion reached by the researchers, and subsequently refined and verified by others, was that in this world without sunlight, there are thriving communities that use the internal heat of the earth as their basic energy source. The base of the food chain in these communities rests on chemicals leached from the heated rocks by water and carried to the ocean floor's surface where bacterial colonies thrive. It is apparent also that the vents are not permanent (they may last only a few decades), and when they cease to supply the sulfide-laden water, the communities collapse.

Impact of Event

The discovery of hydrothermal vents and their associated communities near the Galápagos Islands set off a worldwide search for other deep-sea vent systems. Meanwhile, the analysis of the data from the original finds increases the knowledge of that system.

The original team of researchers had built their hypotheses upon existing theories. Therefore, they had focused their search on hydrothermal vents in order to answer some fundamental geological questions regarding the means of formation of many of the world's major mineral deposits. Most mining geologists had recognized that some of the minerals they mined were associated with rocks derived from the deep ocean. How those rocks got exposed on continents was answered by the theory of plate tectonics. How those minerals were associated with the rocks was not clear until the submarine hydrothermal vents could be studied in detail.

The vent communities launched an intensive research effort on the part of marine biologists. It is interesting to note that the original discovery team—focused on answering geological questions—did not include a biologist. The geologists from the research team found themselves showing pictures to, and trying to answer the questions of, marine biologists around the world. Many questions about the detailed functioning of the ecosystem remain to be answered even now. Of particular interest is the question of how new vents are colonized.

In the final analysis, the discovery of the vents was a classic example of the scientific method in practice wherein researchers built upon the accumulated body of knowledge and hypothesized the existence and location of a natural phenomenon. The scientific method could not have helped the scientists predict the association of the vents with their unique biota. Such serendipitous discoveries serve to remind scientists and the general public alike that there are still major portions of the earth's surface, especially the oceans, that remain unexplored.

Bibliography

Ballard, Robert D. *Exploring Our Living Planet*. Washington, D.C.: National Geographic Society, 1983. This is a well-written and well-illustrated book devoted to exploring aspects of the revolution in earth science spawned by plate tectonics. Ballard provides an in-person account of discoveries made during dives in *Alvin*, including the dives in the Galápagos Rift that led to discovery of the hydrothermal vents. There is an excellent index and an adequate bibliography.

Corliss, John B., et al. "Submarine Thermal Springs on the Galápagos Rift." *Science* 203 (March 16, 1979): 1073-1083. This is a feature summary article that was coauthored by the members of the research team. Although the article is written in a scholarly style, the sense of excitement over the discovery of the new communities comes through, especially through the use of photographs. The article contains some technical data and an authoritative bibliography that college-level readers should find useful.

Francheteau, Jean. "The Oceanic Crust." *Scientific American* 249 (September, 1983): 114-129. This article is part of a special issue of *Scientific American* devoted to summarizing the knowledge of the earth in the light of plate tectonics. It provides the reader with a source of information on the oceanic crust, which underlies and is the foundation for all submarine hydrothermal vents. The author briefly touches upon the phenomenon of hydrothermal vents in this accessible and well-illustrated article. It provides an easy way to put the hydrothermal vents into the context of plate tectonics.

Jones, Meredith L., ed. *Hydrothermal Vents of the Eastern Pacific: An Overview*. Vienna, Va.: INFAX, 1985. This book contains the published papers resulting from a conference held in Philadelphia to summarize the biological data to that point. The articles tend to be highly technical but they contain a wealth of information. The articles are profusely, although technically, illustrated. Each article contains an extensive bibliography. The introduction provides a quick history of major research events leading up to the conference.

Macdonald, Ken C., and Bruce P. Luyendyk. "The Crest of the East Pacific Rise." *Scientific American* 244 (May, 1981): 100-116. An excellent article by members of the multinational research team that explored the hydrothermal vents of the northern East Pacific Rise near the Gulf of California. As is usual for a *Scientific American* article, the text is not too technical and is complemented by excellent photographs and illustrations. A bibliography is included but it is too brief to be much help.

Richard W. Arnseth

Cross-References

Wegener Proposes the Theory of Continental Drift (1912), p. 522; The German *Meteor* Expedition Discovers the Midatlantic Ridge (1925), p. 805; Heezen and Ewing Discover the Midoceanic Ridge (1956), p. 1508; Hess Concludes the Debate on Continental Drift (1960), p. 1650; The Sealab 2 Expedition Concludes (1965), p. 1792; The *Glomar Challenger* Obtains Thousands of Ocean Floor Samples (1968), p. 1876.

HEEGER AND MacDIARMID DISCOVER THAT IODINE-DOPED POLYACETYLENE CONDUCTS ELECTRICITY

Categories of event: Chemistry and physics
Time: 1977
Locale: University of Pennsylvania, Philadelphia

Heeger and MacDiarmid treated polyacetylene with iodine vapor to form an electrically conducting polymer

Principal personages:
ALAN J. HEEGER (1936-   ), an American physicist, head of the Institute for Polymers and Organic Solids, editor of *Journal of Synthetic Metals,* and pioneer in the physics of conducting polymers
ALAN G. MACDIARMID (1927-   ), a New Zealand inorganic chemist and a pioneer in the chemistry of conducting polymers
HIDEKI SHIRAKAWA, a Japanese chemist who discovered a useful form of polyacetylene while working at Tokyo Institute of Technology, Japan

Summary of Event

While polymers, or plastics, are known to possess many useful properties, electrical conductivity was not usually counted as one of them. This changed, however, when three scientists came together in 1977 at the University of Pennsylvania in Philadelphia. The discovery they made not only changed and broadened the understanding of materials but also made possible new and unusual applications.

As background, four possible types of conductivity must be clarified. Insulators conduct electricity very poorly and usually are made from ceramics or common plastics. Semiconductors are mediocre electrical conductors usually containing silicon or germanium. They can be treated (doped) with very small amounts of other elements to increase their electrical conductivity ten- to one-hundredfold. A conductor conducts electricity and heat easily. Conductors usually are metals—silver and copper being the best conductors known. Some metals, when cooled to very low temperature, offer no resistance to electricity and become superconductors. Until 1977, very few known nonmetallic materials were conductors. One type of compound known to conduct electricity was known as a linear-chain material, since their structure was made up of long chains of atoms. The impetus for the study of these materials was provided in 1964 by W. A. Little of Stanford University. He predicted that if a linear-chain material could be designed to the right specification, it might exhibit superconductivity, not only at low temperatures but also at room temperature. Since a room-temperature superconductor would be of great technological importance, Little's idea inspired a worldwide effort to synthesize and study such compounds. One such compound was poly(sulfur nitride) otherwise known as poly-thiazyl. First discovered in 1910, this compound consists of a zigzag linear chain of alternating sulfur and nitrogen atoms. In the early 1970's, Alan J. Heeger, Alan G. MacDiarmid, and others found polythiazyl to have good electrical conductivity, par-ticularly along the direction of the chain. The conductivity in one direction is many times greater than other directions. This led these compounds to be termed one-dimensional metals and further increased interest in linear-chain compounds. Poly-thiazyl was also found to be a superconductor in 1975, but only at the extremely low temperature of 0.3 Kelvin. In spite of these promising results, there are several prob-lems with polythiazyl; for example, it is difficult to make and is so unstable it can explode without warning.

The stable carbon-containing compound closest in form and structure to poly-thiazyl is polyacetylene, a polymer of acetylene gas. A molecule of polyacetylene has a formula of $(CH)_x$ in the form of a linear zigzag chain of carbon atoms, each carbon with a single hydrogen atom attached. Polyacetylene had been known since 1955 as a useless black insulating powder. In the early 1970's, however, a graduate student in Hideki Shirakawa's laboratory at the Tokyo Institute of Technology was trying to make polyacetylene the usual way from acetylene gas but made a mistake in the quantities he used. Instead of a dark powder, Shirakawa's student found he had a lustrous silver film which could be stretched like plastic food wrap. This new synthesis of polyacetylene, as developed and refined by Shirakawa, involves coating the inside of a glass vessel with a chemical catalyst that encourages polymerization, the linking together of small molecules into long polymer chains. When acetylene gas, a small molecule of four atoms, is released into the vessel, a silvery film begins to grow on the glass. Within five minutes, a layer of pure polyacetylene as thick as a piece of paper coats the vessel. After washing off impurities, the film can be peeled from the sides of the glass and stored under vacuum or an inert gas. Since poly-acetylene decomposes easily in air, preparing it demands skills in glassblowing and in vacuum line techniques. These skills had been developed and highly refined by MacDiarmid, Heeger, and their students in their work on silicon compounds and polythiazyl in the early 1970's. When MacDiarmid visited Shirakawa's laboratory in 1976, he learned about the new form of polyacetylene. He invited Shirakawa to spend a year with them at the University of Pennsylvania. With Shirakawa's aid, Heeger and MacDiarmid were able to prepare the new form of polyacetylene and began to study its chemical and physical properties. One of the first things they did was to treat it with iodine vapor to see if polyacetylene would react with iodine in a manner similar to polythiazyl. As the iodine vapor swirled about the silvery poly-acetylene film, they noticed a rapid change in color to deep gold, resembling the color of polythiazyl. Testing the film, they found that its conductivity had increased some twelve orders of magnitude until it was behaving like a metal. They had made the first of a new class of compounds known as conducting polymers. Later tests showed that the iodine had removed electrons from the carbon atoms making up the polymer chain and this was what led to increased conductivity.

Many other chemicals were added to polyacetylene to see if its conductivity could be changed. Those that worked fall into two classes: chemicals that remove electrons and chemicals that add electrons. Removing electrons is termed *p*-doping, while adding electrons is termed *n*-doping. Either technique turns polyacetylene into a golden material with electrical conductivity similar to a metal. In addition to iodine, a number of other *p*-dopants were found: bromine, arsenic pentafluoride, and perchloric acid. Adding electrons was technically more difficult, but eventually *n*-dopants like sodium or sodium naphthalide were discovered. Unlike semiconductor dopants, much larger amounts of dopants are needed to change a polymer from an insulator to a conductor. Up to several percent of the polymer chain units must be doped, instead of a few parts per million found in semiconductors.

A major problem with chemical doping is being able to control the process. In order to solve this problem, Heeger and MacDiarmid developed electrochemical doping of polyacetylene, a process useful for all conducting polymers. If polyacetylene is immersed in certain conducting solutions and a voltage applied to the polymer, electric current will flow and controllable doping will take place. This is a very important and useful technique, which led directly to polymer batteries.

While many conducting polymers now exist, polyacetylene is still considered the prototype for all conducting polymers and is the most extensively studied. It has also shown the highest conductivity of any polymer found, almost as high as the metal copper.

Impact of Event

The primary importance of the discovery of conducting polyacetylene is that it gives scientists both a new class of materials and the ability to modify an important property of these materials: namely, their electrical conductivity. This discovery allows one to combine certain useful properties of polymers with those of a metal. Thus, materials can be prepared that are lighter than metals, moldable, strong, and will conduct electricity or heat well. These polymers can be used in certain applications where the use of metals leads to problems. For example, one of the first commercial uses of conducting polymers was for electrodes in lightweight, rechargeable batteries. This avoided both material problems such as corrosion of the electrodes, the environmental problems of mining of metals, and disposal of used batteries containing toxic metals.

Other places where conducting polymers can be used in place of metals include plastic solar cells, which convert sunlight to electricity and plastic electrodes in fuel cells. A fairly simple application being developed depends on the ability of conducting polymers to absorb static electricity and act as electromagnetic shields to stop radiation leakage from electronic devices such as computers. Currently, metal- or carbon-filled plastics serve this purpose, but the processability and higher conductivity of doped polymers gives them special advantages.

Conducting polymers like polyacetylene have some new and unusual properties all their own. The color of these materials depends on their level of *p*- or *n*-doping. They are electrochromic, meaning that through electrochemistry their color can be changed. Some of these polymers can exist in up to four different colors, depending upon to what percentage they have been doped or whether they are *n*- or *p*-doped. Here, there are a host of uses, some very simple, others more esoteric. A thin film of conducting polymer can be used in an electronic shade, in a shutter or in a visual display device. Conducting polymers will make possible optical data storage and optical switches and transistors, allowing their use in computers which are powered by light instead of electricity.

Since the doping level of a conducting polymer is sensitive to its chemical environment, sensors that can measure the concentration of something such as a drug have been prepared. Other medical uses are emerging. For example, study of conducting polymers has led to a greater understanding of how biological membranes work. Drug delivery systems using soluble conducting polymers as carriers for drugs are also being studied.

Since 1977, several conducting polymers have been made which are both soluble and stable, something not found with polyacetylene. These new polymers can be processed into a desired shape or blended with commercially available polymers. Some of these soluble and stable polymers can even be prepared on existing cloth or yarn giving conducting cloth or fibers. The entire field of conducting polymers has expanded rapidly, as is shown by the number of worldwide patents for their use— from 5 in 1988 to 154 by June, 1990.

Bibliography

Alper, Joseph. "Conductive Polymers Recharged." *Science* 246 (October 13, 1989): 208-210. A nontechnical article on conducting polymers and their applications. Contains a special section on how research workers overcame early problems.

Chien, James C. W. *Polyacetylene: Chemistry, Physics, and Material Science.* New York: Academic Press, 1984. A scientific monograph on polyacetylene up to 1984. Technical and demanding to read.

Epstein, Arthur J., and Joel S. Miller. "Linear-Chain Conductors." *Scientific American* 241 (October, 1979): 52-61. An early article which gives a good sense of the place of polyacetylene among linear chain molecules. Slightly difficult to read but with good diagrams and figures.

Freundlich, Naomi. "Putting a Real Charge into Plastic." *Business Week*, December 11, 1989, 114. A nontechnical explanation of conducting polymers.

Kaner, Richard B., and Alan G. MacDiarmid. "Plastics That Conduct Electricity." *Scientific American* 258 (February, 1988): 106-111. This is the best and easiest-to-read general article on this subject. Written by MacDiarmid and one of his students, it contains accurate information, useful figures, and explanations of how conducting polymers work.

Skotheim, Terje A. *Handbook of Conducting Polymers.* New York: Marcel Dekker, 1986. The only comprehensive book in this field. Fairly technical and demanding, it is a valuable source of information on the history of polyacetylene, the many varieties of conducting polymers, and their applications.

David F. MacInnes, Jr.

Cross-References

Baekeland Invents Bakelite (1905), p. 280; Carothers Patents Nylon (1935), p. 1055; Shockley, Bardeen, and Brattain Discover the Transistor (1947), p. 1304; Bell Telephone Scientists Develop the Photovoltaic Cell (1954), p. 1487; The Microprocessor "Computer on a Chip" Is Introduced (1971), p. 1938; Bell Laboratories Scientists Announce a Liquid-Junction Solar Cell of 11.5 Percent Efficiency (1981), p. 2159; Optical Disks for the Storage of Computer Data Are Introduced (1984), p. 2262.

ASTRONOMERS DISCOVER THE RINGS
OF THE PLANET URANUS

Category of event: Astronomy
Time: March 10-11, 1977
Locale: Indian Ocean

Elliot's astronomy team accidentally made the first discovery of a set of planetary rings in more than three hundred years, fueling new speculation on ring characteristics

Principal personages:

JAMES ELLIOT (1943-   ), an astronomer who was the head of the Uranian occultation mission

EDWARD (TED) DUNHAM, a Cornell University doctoral candidate in astronomy who operated data-recording equipment on the occultation mission

DOUGLAS MINK, a computer programmer who operated data-recording equipment on the occultation mission

ROBERT MILLIS (1941-   ), an astronomer who was the head of the groundbased observatory team in Perth, Australia

CARL GILLESPIE, the mission director for the National Aeronautics and Space Administration (NASA)

JIM MCCLENAHAN, the mission director for NASA

Summary of Event

In 1973, Gordon Taylor of the Royal Greenwich Observatory predicted a stellar occultation by the planet Uranus. An occultation, which is similar to an eclipse, occurs when a planet moves to block the view of a star. Occultations are useful to scientists; by closely timing the disappearance and reappearance of the star, the astronomers can measure precisely the diameter of the planet. Furthermore, by watching how the starlight behaves as it disappears, scientists can learn about the temperature of the planet's atmosphere. Previous occultations had been observed of Jupiter, Neptune, and Mars.

This particular occultation, of a star known as SAO 158687, was the first ever predicted for Uranus. Because of its importance, Cornell University astronomer James Elliot persuaded the National Aeronautics and Space Administration (NASA) to let his team observe the event in the Gerald P. Kuiper Airborne Observatory (KAO).

The KAO is a C-141 cargo jet converted for use in astronomical missions. Named for the pioneer astronomer Gerald P. Kuiper, the KAO carries a 0.9-meter-diameter telescope and its controlling computer. Because the KAO can reach altitudes of 12,300 meters, above the bulk of the atmosphere, it provides much clearer views than ground-based telescopes. In the case of the Uranian occultation, the KAO would provide another advantage: It could fly to the Indian Ocean, where the event would be visible for the longest time.

Rather than rely on human eyesight or cameras to view the event, the astronomers used a photometer, a very sensitive light detector having ten thousand times the resolution of photographic equipment. Affixed to the telescope, it would divide the faint light from the star and the planet into three wavelength ranges and record their intensity onto both magnetic tape and a paper strip. As the occultation progressed and the star vanished behind Uranus, the continuous photometer record would provide the most precise "image" of the event.

The KAO left Australia's Perth International Airport, heading southwest, at 10:37 P.M. local time, on March 10, 1977. The flight would take more than ten hours. On board were the flight crew, the telescope equipment operators led by NASA mission directors Carl Gillespie and Jim McClenahan, and Elliot and his assistants Edward (Ted) Dunham and Douglas Mink.

Several months prior to the flight, an error had been discovered in Taylor's original prediction of the occultation. There was a chance now that the occultation would not occur at all. On board the KAO, Elliot was wary that the error would cause them to lose this opportunity, so he started the continuous data recording forty-one minutes before the forecast occultation time. Only six minutes after the recording began, the photometer recorded a sudden drop in the starlight which lasted for several seconds. This "dip" was so faint that it went unnoticed until Dunham saw it on the data recording almost a minute later. Elliot suspected that the dimming may have been caused by clouds in the upper atmosphere or a momentary failure in the tracking system, but no one on the mission had a definite explanation.

Four minutes after the first dip, a second dip occurred. This time, meteorologist Pete Kuhn, who monitored the water vapor in the air, verified that no clouds could have caused the dip (or "secondary occultation," as it was properly called). The team laughed when Elliot joked that it may have been caused by the "D-ring" of Uranus (that is, the fourth ring from the planet), which, in fact, it was. This was not the first time that this ironic joke had been made. When the prediction error had first been discovered, the astronomers realized that the event they intended to view might not occur; yet, they were committed to the flight and continued their preparations. Joseph Veverka, Elliot's colleague at Cornell, joked that if the event did not happen, Elliot could still use the data to determine the region in which a Uranian ring system could be found. This became a regular joke among Elliot's team, because it was widely "known" that only Saturn had rings. Clearly, if the occultation did not occur, the scientists could not justify the mission to NASA by providing information on hypothetical rings.

On the KAO, a third dip happened a minute after the second. Because of the brief duration of the dips, Elliot raised the possibility that "small bodies like thin rings" were occluding the star. It is important to remember that Saturn's rings are tens of thousands of kilometers wide (measured radially). Despite the team's jesting, any Uranian rings "should" have been very wide as well. Now, as they saw the starlight dimmed by some sort of structure much narrower than Saturn's rings, Elliot pictured a zone, or "belt" containing many full-size moons, as opposed to a set of true rings, composed of small particles. As Elliot wrote later, the team sometimes spoke of "thin rings" when talking about this belt, but, in truth, a ring only about 10 kilometers wide seemed "farfetched" to them.

To confirm the theory of the "belt of satellites," they continued observing the star after it passed from behind Uranus itself, to see if the star would be occluded by more moons on the other side of Uranus. The confirmation came when the five new secondary occultations occurred as expected. This made a total of ten of the mysterious occultations: five to the "left" of the planet and five to the "right."

When the KAO landed back in Australia at 8:43 A.M., Elliot spoke with his colleague Robert Millis, who had also observed the occultation from an observatory in Perth. From this vantage point, Uranus did, indeed, miss the star and no planetary occultation was recorded; but because Millis, like Elliot, had begun recording early, his equipment had registered five secondary occultations on one side of the planet. (Worldwide, eight astronomical teams had observed the occultation of Uranus, but only the KAO team had observed both sets of secondary occultations.) Elliot's and Millis' teams discussed the findings. The dips in the starlight must have been caused by either a set of five rings or the "belt of satellites." Once again, they dismissed the ring hypothesis because the star was blocked for such a short time—only a second or two in each case—and it seemed impossible that a planet could have rings only a few kilometers wide.

Nevertheless, one fact might contradict the belt-of-satellites hypothesis. When the star was occluded by Uranus itself, its light was naturally blocked completely. Yet, during the mysterious "secondary occultations," the starlight was only dimmed, not blocked. A moon of Uranus would have cut off the light, unless it only partially blocked the star's image. Elliot considered the possibility of ten such "grazes" to be "highly improbable, [but] not impossible. In any case, it seemed more likely than narrow rings." Elliot realized that there was a definite way to determine the answer.

From the telescope's point of view, Uranus' known satellites, as well as any new satellites or rings, would orbit the planet in what appeared to be a wide circle with the planet in their center, like a bullseye. If this new discovery were a belt containing ten or more new satellites, each would appear at a different distance from Uranus, and each of the dips would indicate that. On the other hand, if these *were* rings, then the spacing between the dips to one side of the planet would correspond exactly to that on the other side.

Elliot examined the chart recording made by the photometer, intending to disprove finally the ring hypothesis, but, instead, found matching spacings that could be caused only by rings. Later, Mink and Dunham reexamined the record by computer, and found the spacings to be even closer than Elliot had first believed.

Elliot and his colleagues had found something completely unexpected: a new set of rings which, unlike those of Saturn, were narrow, sharp-edged, and so dark that they could be seen only with twentieth century equipment.

Impact of Event

In later years, further occultation studies of Uranus revealed four more rings, and in 1986 the Voyager 2 space probe discovered two more, for a known total of eleven. Prior to their discovery, most theories of planetary rings had tried to explain why only one planet, Saturn, has any. The surprising nature of the Uranian rings has changed this focus: Scientists now want to know what causes different types of rings. Also, they want to know why these rings are so narrow. The particles that make up a ring orbit the planet like tiny moons. Particles collide with others nearby; this tends to make the orbits spread out until the particles circle, without colliding, in a broad ring. This is the case at Saturn. Yet, most of the Uranian rings are less than 13 kilometers wide. Some scientists believe that these rings are relatively young, and have not had time to spread out this way. Peter Goldreich and Scott Tremaine of the California Institute of Technology (Caltech) theorized that a ring could be kept from spreading by the gravitational influence of two tiny "shepherd moons," one which orbits just inside the ring, and the other outside, thus "shepherding" the particles into a tight ring. Uranus' furthest-out "epsilon" ring is now known to have two shepherd moons; others have been discovered at Saturn.

Astronomers are curious about why these rings are so dark. Whereas Saturn's rings are reflective water ice, these new rings are too dark to be seen in any telescope. The particles are blacker than coal dust. It is still not known what they are made of.

Scientists no longer consider rings to be uncommon around Jovian (large) planets such as Uranus and Neptune. The Voyager space probes, as expected, found a ring at Jupiter—but this ring is unlike those found previously, for it is made of gas, not particles.

Finally, the Uranian discovery proved that stellar occultations can be used to find planetary rings as well as its other uses. Occultation studies of Neptune have found evidence of rings there.

Bibliography

Corliss, William R., ed. *The Moon and the Planets: A Catalog of Astronomical Anomalies.* Glen Arm, Md.: The Sourcebook Project, 1985. One in a projected twenty-five-volume series whose purpose is to "collect and categorize all phenomena that cannot be explained readily by prevailing scientific theories." This edition calls attention to the unusual darkness and narrowness of the rings of Uranus, the "disarray" among Neptune's moons, some anomalous wet areas on Mars, and so on. Each entry is referenced scrupulously to established journals and suggests possible explanations to some of its anomalies.

Elliot, James L. *The Ring Tape.* Cambridge, Mass.: MIT Press, 1985. This is the audio tape made aboard the Kuiper Airborne Observatory during the Uranian occultation mission. Excerpted in both *Rings* and *Sky and Telescope*, it provides a unique opportunity to listen to discovery in progress.

Elliot, James L., Edward Dunham, and Robert L. Millis. "Discovering the Rings of

Uranus." *Sky and Telescope* 53 (June, 1977): 412-416, 430. Elliot's first-person account of the Uranian discovery, including excerpts from the audio tape made in the KAO cabin during the mission. *Sky and Telescope* is aimed at amateur astronomers; therefore, the piece is accessible to most readers with a high school science background.

Elliot, James L., and Richard Kerr. *Rings: Discoveries from Galileo to Voyager.* Cambridge, Mass.: MIT Press, 1984. Discusses all planetary rings known and hypothetical, as of 1984. Includes a more in-depth retelling by Elliot of the KAO mission. Some chapters written by Kerr, who has reported on Jovian and Saturnian discoveries for *Science* magazine. Clear and entertaining for general readership.

Hunt, Garry, and Patrick Moore. *Atlas of Uranus.* New York: Cambridge University Press, 1988. One of the few books devoted to the planet, it was published late enough to include information from the Voyager 2 flyby. Contains close-up maps of all moons and information on all eleven rings. This is a "coffee-table" book: large-page format, easy to follow, and with many attractive pictures.

Miller, Ron, and William K. Hartmann. *The Grand Tour: A Traveler's Guide to the Solar System.* New York: Workman, 1981. A general-readership book with an unusual focus: It examines all bodies in the solar system at least 1,600 kilometers in diameter, in size order beginning with Jupiter and going down to the largest asteroids. The paintings are stunningly effective (if one conveniently forgets the fact that the real-life scenes would not be as well lit). This book contains rare views of the Uranian and Saturnian rings from underneath.

Shawn Vincent Wilson

Cross-References

Voyager 1 and 2 Explore the Planets (1977), p. 2082; The First Ring Around Jupiter Is Discovered (1979), p. 2104; Astronomers Discover an Unusual Ring System of Planet Neptune (1982), p. 2211.

APPLE II BECOMES THE FIRST SUCCESSFUL PREASSEMBLED PERSONAL COMPUTER

Category of event: Applied science
Time: April, 1977
Locale: Cupertino, California

The introduction of the preassembled Apple II personal computer by Apple Computer created the market for home, education, and small business computing

Principal personages:

STEPHEN WOZNIAK (1950-   ), the cofounder of Apple Computer and designer of the Apple II computer

STEVEN JOBS (1955-   ), the cofounder of Apple Computer

MIKE MARKKULA (1942-   ), the former Intel marketing manager who became a founder and the initial financial backer of Apple Computer

REGIS MCKENNA (1939-   ), the owner and namesake of the Silicon Valley public relations and advertising company who handled the Apple account

MICHAEL SCOTT (1943-   ), the first president of Apple Computer

FREDERICK RODNEY HOLT, a former analog engineer at Atari who designed the power supply for the Apple II

CHRIS ESPINOSA (1961-   ), the high school student who wrote the BASIC program shipped with the Apple II

RANDY WIGGINTON (1960-   ), a high school student and Apple software programmer

Summary of Event

The development of the computer during the 1940's expanded the capacity to do intensive mathematical calculations, doing them far more rapidly than teams of scientists could. Initially, it was thought that computers would be effective only for scientific work; in the 1950's, it was shown that business could make great use of them. Into the 1960's, however, the idea that a small computer could be of use to the average person was an idea held by only a few radicals within the industry. Rather than any industry radicals, however, it was a pair of counterculture individuals from the Silicon Valley—the high technology area between San Francisco and San Jose—that eventually launched the personal computer revolution.

Both Steven Jobs and Stephen Wozniak had attended Homestead High School in Los Altos, California, although at different times, and both developed early interests in technology, especially computers. In 1971, Wozniak built his first computer from spare parts. Shortly after this, he was introduced to Jobs. Jobs had already developed an interest in electronics (he once telephoned William Hewlett, cofounder of Hewlett-Packard, to ask for parts), and he and Wozniak became friends. Their first business venture together was the construction and sale of "blue boxes," illegal devices that allowed the user to make long-distance telephone calls for free. The business expanded in 1972 when Wozniak began attending the University of California at Berkeley and Jobs went to Reed College in Oregon. The following year, Wozniak began working at Hewlett-Packard, where he studied calculator design; Jobs took a job at Atari, the video company. The friendship paid off again when Wozniak, at Jobs's request, designed the game "Breakout" for Atari, and the pair was paid seven hundred dollars.

In 1975, the Altair computer, a personal computer in kit form based on the Intel 8080 microprocessor, was introduced by Micro Instrumentation and Telemetry Systems (MITS). Shortly thereafter, the first personal computer club, the Homebrew Computer Club, began meeting in Menlo Park, near Stanford University. Wozniak and Jobs began attending the meetings regularly. Wozniak eagerly examined the Altairs that others brought. He thought that the design could be improved. When MOS Technology offered its new 6502 microprocessor chip for twenty dollars, he decided to test his ideas. Before beginning work on the new design, Wozniak wrote a version of BASIC (the programming language) for the new chip. After that task was accomplished, he turned to the computer design. In only a few more weeks, he had produced a circuit board utilizing the MOS chip, along with interfaces that connected it to a keyboard and a video monitor. A hobbyist at heart, Wozniak showed the machine at a Homebrew meeting and distributed photocopies of the design.

Jobs, however, saw the potential opportunity that this machine, which he named an "Apple," presented. He talked Wozniak into forming a partnership to develop personal computers. Jobs sold his car and Wozniak sold his two Hewlett-Packard calculators; with the money, they ordered printed circuit boards made, speeding up the production process. Their break came when Paul Terrell, a retailer, was so impressed that he ordered fifty fully assembled Apples. Within thirty days, the computers were completed, and they sold for the ominous sounding fee of $666.66.

During the summer of 1976, Wozniak turned his attention to developing an improved version of the Apple. The new computer would come with a keyboard, an internal power supply, BASIC built into the on-board memory, slots for adding peripheral cards to link with printers and other devices, and color graphics, all enclosed in a case. The output would be seen on a television. Jobs and Wozniak calculated the price for the complete machine to be twelve hundred dollars.

Jobs then began to seek the help they would need to implement their design successfully. After consulting with Nolan Bushnell, his former boss and founder of Atari, Jobs was led to Mike Markkula, a former marketing manager at both Fairchild Semi-conductor and the Intel Corporation. Wealthy through stock options, Markkula had retired at an early age. After visiting the young entrepreneurs in Jobs's garage, however, Markkula decided to help them with a business plan. In a few months, in return for a $91,000 personal investment and a guarantee for a $250,000 line of credit at Bank of America, Markkula received one-third share of the now incorporated Apple Computer. At about the same time that Jobs was introduced to Markkula, two other key figures were brought aboard. Frederick Rodney Holt, an analog engineer at Atari, was brought aboard to design the power supply. Wozniak, familiar only with digital (binary) and not analog (wave) electronics, was unable to design the power supply. Holt not only designed the light, fanless power supply but also convinced Jobs and Wozniak not to challenge the Federal Communications Commission over the television interface. Instead, the interface was given over to third-party developers so that users, not Apple, would be in violation. Another key figure was Regis McKenna, the head of the Regis McKenna Public Relations agency. His was the best of the public relations firms that served the high-technology industries of the valley, and Jobs wanted it to handle the Apple account. At first, McKenna rejected the offer, but Jobs's constant pleadings finally convinced him. The agency's first contributions to Apple were the design of the colorful striped Apple logo and a color ad in *Playboy* magazine, designed to broaden the appeal for the machine beyond the small group of hobbyists. Both proved to be wise moves.

In February, 1977, the first Apple Computer corporate office was opened in Cupertino, California. By this time, two of Wozniak's friends from Homebrew, Randy Wigginton and Chris Espinosa—both high school students—had joined the company. Their specialty was in writing software. Espinosa worked through his Christmas vacation so that BASIC could ship with the computer. It was also at this time, at Markkula's suggestion, that Michael Scott was selected as company president. Markkula did not want to manage the company and believed that Jobs was not ready to do it. (Scott had worked with Markkula at Fairchild Semiconductor.)

The team pushed ahead to complete the new Apple in time to display it at the First West Coast Computer Faire in April. Jerry Mannock, formerly with Hewlett-Packard, was hired to create the design for the case. The first cases, made by another company, were uneven in quality, but the group worked out the worst flaws. At this time, the name Apple II was chosen for the new model. The Apple II computer debuted at the West Coast Computer Faire and included a combination of innovative components. The motherboard was far simpler and more elegantly designed than any previous computer, and the power supply was cooler and lighter. The ease of hooking the Apple II up to a television monitor made it much more attractive to potential buyers. Within and without, the Apple II was sleek and powerful.

The company continued to grow in 1977. Demand for the Apple II increased, and production doubled every few months. They earned $2.7 million in 1977, and the personal computer was well on its way to establishing itself with consumers.

Impact of Event

The introduction of the Apple II computer launched what was to be a wave of new computers aimed at the home and small business markets. Within a few months of the Apple II's introduction, Commodore introduced its PET computer and Tandy/Radio Shack brought out its TRS-80. Apple continued to work on producing the peripheral cards that would help the demand to increase. Prospects were so good for the company that two large venture capitalist groups, Rockefeller-backed Venrock

Associates and Arthur Rock & Company (who had launched Intel), invested large sums in Apple. At the end of 1977, however, there was no way to know that Apple would become the dominant player in the personal computer field. Both Commodore and Tandy/Radio Shack had presumed strengths in their well-established distribution lines. Apple had worked out a distribution deal with the newly established ComputerLand franchise chain, however, and Commodore's marketing tactics soured their potential relationships with retailers. Nevertheless, the future at the close of 1977 was still unclear. Following up on a suggestion made by Markkula at a December strategy meeting, Wozniak began work on creating a floppy disk system for the Apple II. The cassette tape storage that all personal computers were relying on, unfortunately, was not very reliable, and it was very time-consuming. Floppy disks, which had been introduced for larger computers by International Business Machines (IBM) in 1970, were reliable and fast. Markkula wanted one to show at the Consumer Electronics Show in January.

As with everything that interested him, Wozniak spent almost all of his time learning about and designing a floppy disk drive. Once he understood the principles involved, he designed a simpler control circuit, then eliminated the need for synchronization circuitry altogether. Finally, with Wigginton's assistance, Wozniak wrote the software used to read from and write to the disk. When the final drive shipped in June, 1978, it made possible real development of software for the computer, opening up the market to nonhobbyists. It also positioned Apple ahead of both Commodore and Tandy/Radio Shack.

The final element that cemented Apple's place in the personal computer market occurred in 1979 when Personal Software, Daniel Fylstra's software marketing company, introduced VisiCalc. VisiCalc was the first "spreadsheet" program—a program that featured financial data stored in table form, along with the ability to make some data dependent upon other data. This meant that "what if" financial forecasting could be quickly and easily done. For the first time, there was now a compelling reason for the purchase of an Apple II, and businesses purchased both.

By 1980, Apple had sold 130,000 Apple II's. That year, the stock went public, and Jobs and Wozniak, among others, became wealthy. Three years later, Apple Computer became the youngest company to make the Fortune 500 list of the largest industrial companies. By then, IBM had entered the field, and begun to dominate, but the Apple II's earlier success ensured that personal computers would not be a market fad: By the end of the 1980's, 35 million personal computers were in use.

Bibliography

Freiberger, Paul, and Michael Swaine. *Fire in the Valley: The Making of the Personal Computer*. Berkeley, Calif.: Osborne/McGraw-Hill, 1984. Freiberger and Swaine, both longtime reporters on the personal computer world, tell the story of the growth of the industry. Based largely on extensive personal interviews. Photographs, index.

Garr, Doug. *Woz: The Prodigal Son of Silicon Valley*. New York: Avon, 1984. A slim paperback focusing on Wozniak. Unfortunately, it fails to explore his creative insights. Illustrations.

Levering, Robert, Michael Katz, and Milton Moskowitz. *The Computer Entrepreneurs: Who's Making It Big and How in America's Upstart Industry.* New York: New American Library, 1984. These writers on American business provide more than a reference book, less than a complete story. A cross section of personal computing in the early 1980's. Provides profiles on major figures in the personal computer revolution. Elementary, but informative. Illustrations, index.

Levy, Steven. *Hackers: Heroes of the Computer Revolution.* Garden City, N.Y.: Doubleday, 1984. Levy, a well-known writer on the personal computer, tells the story of the people, often behind-the-scenes, who inspired the personal computing revolution. Based on extensive interviews and archives. Notes, index.

Moritz, Michael. *The Little Kingdom: The Private Story of Apple Computer.* New York: William Morrow, 1984. The *Time* reporter tells the early story of Apple Computer from the corporate perspective. Based on interviews and largely anecdotal. Index.

Rose, Frank. *West of Eden: The End of Innocence at Apple Computer.* New York: Viking Press, 1989. Free-lance journalist Rose focuses primarily on the post-Jobs era at Apple, although discussion is devoted to the company's formation. Based on interviews. Good bibliographic note. Index.

Stine, G. Harry. *The Untold Story of the Computer Revolution: Bits, Bytes, Bauds, and Brains.* New York: Arbor House, 1985. Focuses on evolution of the computer from its beginnings. Some coverage of Apple. Photographs.

Young, Jeffrey S. *Steve Jobs: The Journey Is the Reward.* Glenview, Ill.: Scott, Foresman, 1988. Young, a founding editor of *Macworld* magazine, provides a cogent picture of Jobs. Based on extensive interviews and firsthand knowledge. Photographs, bibliography, notes.

George R. Ehrhardt

Cross-References

Eckert and Mauchly Develop the ENIAC (1943), p. 1213; The First Electronic Stored-Program Computer (BINAC) Is Completed (1949), p. 1347; UNIVAC I Becomes the First Commercial Electronic Computer and the First to Use Magnetic Tape (1951), p. 1396; Backus' IBM Team Develops the FORTRAN Computer Language (1954), p. 1475; Hopper Invents the Computer Language COBOL (1959), p. 1593; Kemeny and Kurtz Develp the BASIC Computer Language (1964), p. 1772; The Microprocessor "Computer on a Chip" Is Introduced (1971), p. 1938; The IBM Personal Computer, Using DOS, Is Introduced (1981), p. 2169; IBM Introduces a Personal Computer with a Standard Hard Disk Drive (1983), p. 2240.

THE FIRST COMMERCIAL TEST OF FIBER-OPTIC TELECOMMUNICATIONS IS CONDUCTED

Category of event: Applied science
Time: May, 1977
Locale: Chicago, Illinois

The invention of fiber-optics technology revolutionized the telecommunications industry and advanced medical procedures

Principal personages:

THEODORE H. MAIMAN (1927-   ), an American physicist and inventor of the solid-state laser, which made fiber-optics telecommunications practical

NARINDER S. KAPANY (1927-   ), the Indian-born coinventor of the fiberscope who developed the first practical glass-coated glass fiber and coined the term "fiber optics"

CHARLES K. KAO (1933-   ), a Chinese-born electrical engineer who first suggested the use of optical fibers for the transmission of telephone signals

ALEXANDER GRAHAM BELL (1847-1922), the Scottish American who invented the telephone and the photophone, a device that transmitted messages via light waves

SAMUEL F. B. MORSE (1791-1872), an American artist and inventor who was responsible for the development of the electromagnetic telegraph system

Summary of Event

Ever since Samuel F. B. Morse, inventor of the telegraph, sent his famous message "What hath God wrought?" by means of electrical impulses traveling at the speed of light over a 66-kilometer-long telegraph line strung between Washington, D.C., and Baltimore in 1844, scientists have sought to develop faster, less expensive, and more efficient means of conveying information over great distances. The impact of the telegraph, or "lightning wire," as it came to be known, was immediate and far-reaching. It was used at first to report stock market prices and the results of political elections. Telegraphy played a large role in the American Civil War; in fact, the first transcontinental telegraph sent was a message from Stephen J. Field, Chief Justice of the California Supreme Court, to President Abraham Lincoln on October 24, 1861, declaring that state's loyalty to the Union. By 1866, telegraph lines reached all across the continent, and a telegraph cable had been laid beneath the Atlantic Ocean, linking the Old World with the New World.

Another American inventor, a Scottish emigrant to the United States, made the leap from the telegraph to the telephone. Alexander Graham Bell, as a teacher of the deaf, was interested in the physiology and physics of speech. In 1875, he began experimenting with ways to transmit sound vibrations electrically. He realized that an electrical current could be modulated to resemble the vibrations of speech. Bell patented his invention on March 7, 1876; on July 9, 1877, he founded the Bell Telephone Company.

In 1880, five years after inventing the telephone, Bell invented a device called the "photophone" with which he demonstrated that speech could be transmitted on a beam of light. Light is a form of electromagnetic energy. It travels in a vibrating wave. By modulating the amplitude, or height, of the wave, a light beam can be made to carry messages. Bell's invention employed a mirrored diaphragm that converted sound waves directly into a beam of light. At the receiving end, a selenium resistor connected to a headphone reconverted the light into sound. "I have heard a ray of sun laugh and cough and sing," Bell wrote of his invention.

Although Bell demonstrated that he could transmit speech over distances of several hundred meters with the photophone, the device was cumbersome and unreliable and never caught on like the telephone. One hundred years would pass before large-scale commercial application of Bell's dream of talking on a beam of light would become a reality.

Two key technological advances were needed first: development of the laser and development of high-purity glass. In 1960, Theodore H. Maiman, an American physicist and electrical engineer working at Hughes Research Laboratories in Malibu, California, built the first laser. The laser produces an intense, narrowly focused beam of light that can be modulated to carry huge amounts of information. It soon became apparent, however, that even bright laser light could be broken up and absorbed by smog, fog, rain, and snow. In 1966, Charles K. Kao, an electrical engineer working for the Standard Telecommunications Laboratories in England, proposed that glass fibers be used to transmit the message-bearing beams of laser light without disruption.

Optical glass fiber is made from common materials consisting basically of silica, soda, and lime. To make the optical fiber, the inside of a thin-walled silica glass tube is coated with successive layers of extremely thin glass. Typically, a tube is lined with a hundred or more such layers. The tube is then heated to 2,000 degrees Celsius and collapsed into a thin glass rod, called a preform. The preform is then pulled into thin strands of fiber. The fibers are coated with plastic to protect them from being scored or scratched and then sheathed in flexible cable.

The first glass fibers produced contained many impurities and imperfections, which resulted in significant light losses. Signal repeaters were needed every few meters to boost the fading light pulses. In 1970, however, researchers at the Corning Glass Works, in New York, developed a fiber pure enough to carry light at least 1 kilometer without amplification.

The new fiber-optics technology was seized upon quickly by the telephone industry. It was anticipated that a bundle of optical fibers having the diameter of a pencil could carry several hundred telephone calls at the same time. Optical fibers were first tested by telephone companies in big cities, where ever-increasing, high-density phone traffic often overloaded the capacity of underground conduits. On May 11, 1977, American Telephone & Telegraph Co. (AT&T), in a cooperative venture with Illinois Bell Telephone, Western Electric, and Bell Telephone Laboratories, began the first commercial test of fiber-optics telecommunications in downtown Chicago. In the keynote speech at the opening ceremonies for the Chicago system, AT&T's vice-president of engineering and network services, Morris Tanenbaum, noted that "there is a long, often torturous path that must be traveled between a research discovery and its application in a practical cost-effective system." The experimental system in Chicago represented the culmination of nearly twenty years of research and development, much of it conducted at a furious and highly competitive pace.

The Chicago test was intended to evaluate the potential of fiber-optics telecommunications under actual operating conditions and to acquire experience in the installation, operation, and maintenance of such a system in a congested, urban area. The system consisted of a 2.4-kilometer cable laid beneath the city's streets in existing telephone ducts. The cable, 1.3 centimeters in diameter, linked an office building in Chicago's business district with two telephone exchange centers. Voice, data, and video signals were coded into pulses of laser light and transmitted through the hair-breadth glass fibers. The tests demonstrated successfully that it was possible for a single pair of fibers to carry nearly six hundred telephone conversations with a high degree of reliability and at a reasonable cost.

Six years later, in October, 1983, the Bell Laboratories succeeded in transmitting the equivalent of six thousand telephone signals through an optical fiber cable 161 kilometers long. Since that time, countries all over the world, from England to Indonesia, have developed optical communications systems.

Impact of Event

Hailed as a revolutionary new technology, the greatest impact of fiber optics has been in the telecommunications industry. Optical fibers provide greater information-carrying capacity—a single fiber can now carry thousands of conversations with complete freedom from electrical interference. Made of common materials—glass and plastic—they are less expensive than copper wire, weigh less, and take up considerably less space. Because the laser-light signals do not degrade over distances, they do not require the regular and periodic amplification that electrical signals passing through a copper wire require. In terms of both economics and logistics, fiber optical technologies have revolutionized the telecommunications industry.

One of the first uses of fiber optics and perhaps its best-known application is the fiberscope, a medical instrument that permits internal examination of the human body without surgery or X-ray techniques. The fiberscope, or endoscope, developed in the late 1950's by Narinder S. Kapany, a physicist and president of Optical Technology, Inc., consists of two fiber bundles. One of the fiber bundles transmits bright light into the patient, while the other conveys a color image back to the eye of the physician. The fiberscope has been used to look for ulcers, cancer, and polyps in the stomach, intestine, and esophagus of humans. Medical instruments, such as forceps, can be attached to the fiberscope, allowing the physician to perform a range of medical procedures, such as clearing a blocked windpipe or cutting precancerous polyps from the colon.

Bibliography

Boraiko, Allen A. "Fiber Optics: Harnessing Light by a Thread." *National Geographic* 156 (October, 1979): 516-535. A fascinating and lively account of the development of fiber optics intended for the average reader. Provides a historical context for the breakthrough discoveries of the 1960's and 1970's in the areas of laser and glass fiber technologies. Based in part upon interviews with scientists and researchers working in the field of fiber optics. Includes twenty-eight stunning color photographs.

Boyle, W. S. "Light-Wave Communications." *Scientific American* 237 (August, 1977): 40-48. Describes the experimental fiber-optics system set up in Chicago in May of 1977, the first commercial test of "light-wave" telephone service. The article provides detailed, technical information on the laser technology involved as well as the operating principles of optical fiber transmission. Includes diagrams showing cross sections of lasers, optical fibers, and cables.

Bruce, Robert V. *Bell: Alexander Graham Bell and the Conquest of Solitude.* Ithaca, N.Y.: Cornell University Press, 1990. Reprint of the 1973 edition. Considered by many to be the definitive biography of Bell. Bruce describes Bell's invention within the context of the history of American science and technology. Heavily illustrated. Includes a bibliography and index.

Inventors and Discoverers: Changing Our World. Washington, D.C.: National Geographic Society, 1988. Beautifully illustrated and well-written tribute to inventors and their inventions. Contains fascinating details of the lives of the men and women who made discoveries that dramatically changed the world in which they lived. The essay by science writer Stephen Hall, "The Age of Electricity," covers the contributions made by Morse and Bell. Includes an introductory essay by Daniel Boorstin, American historian and former Librarian of Congress.

Kapany, N. S. *Fiber Optics: Principles and Applications.* New York: Academic Press, 1967. Assumes an understanding of physics at the undergraduate level. The first part of the book provides a theoretical background, while the second part covers the applications of fiber optics in various fields, including medicine, photo-electronics, and high-speed photography.

Nancy Schiller

Cross-References

The First Transcontinental Telephone Call Is Made (1915), p. 595; Transatlantic Radiotelephony Is First Demonstrated (1915), p. 615; The First Transatlantic Telephone Cable Is Put Into Operation (1956), p. 1502; The First Laser Is Developed in the United States (1960), p. 1672.

VOYAGER 1 AND 2 EXPLORE THE PLANETS

Categories of event: Space and aviation, and astronomy
Time: September, 1977-1989
Locale: Cape Canaveral, Florida

The Voyager space vehicles began the most ambitious expedition of the twentieth century; their imagery and data revolutionized the concept of the solar system

Principal personages:
DONALD M. GRAY (1929-   ), the navigation team chief of Voyager, National Aeronautics and Space Administration, Jet Propulsion Laboratory

RAYMOND L. HEACOCK (1928-   ), the Voyager 1 project manager who was awarded the prestigious James Watt International Medal for engineering achievement

CHARLES E. KOHLHASE (1935-   ), the manager of the Voyager Mission Planning Office

ELLIS D. MINER (1937-   ), a space scientist at NASA

BRADFORD ADELBERT SMITH (1931-   ), the head of the Voyager imaging team

LAURENCE SODERBLOM (1944-   ), an expert on Galilean satellite morphology

STEPHEN PATRICK SYNOTT (1946-   ), a leader of the Voyager Optical Navigation Team and associate manager member of the Imaging Science Team

Summary of Event

Often referred to as the "grand tour," two highly sophisticated spacecraft, Voyager 1 and 2, made a journey past the large outer planets of the solar system and transmitted to Earth detailed scientific information of immense value. The mission spanned twelve years from 1977 to 1989.

As the other planets orbit the sun, each at its own rate, they are, in rare cases, aligned in a pattern that presents an opportunity for a spacecraft launched from Earth to fly past each of them in turn. Such an alignment occurs only once every 175 years, and it also presents the most economical approach in both time and money. It was with these considerations in mind that the National Aeronautics and Space Administration (NASA) carried out the Voyager program, clearly one of its most successful. Key personnel included Donald M. Gray, Raymond L. Heacock, Ellis D. Miner, Bradford Adelbert Smith, and Laurence Soderblom.

Rarely has the scientific community been so lavish in its praise of a scientific endeavor as it has for Voyager. Science writer Arthur E. Smith thought highly of the two Voyagers.

Although the planet alignment phenomenon had been long known to scientists, the actual flight trajectory required complex calculations and considerable work by mission analysts before an optimum flight trajectory was adopted. Historian Craig B. Waff credits the important orbital tactic called "gravity assist," in which a planet's gravity is used to alter the direction and increase the velocity of a spacecraft, to Michael Minovitch, a graduate student working at NASA's Jet Propulsion Laboratory (JPL). Waff credits Gary Flandro, a graduate student in 1965 at JPL with first demonstrating that "Jupiter, Saturn, Uranus, and Neptune would soon be positioned in such a way as to allow a single spacecraft, launched during any one of five annual launch windows, in the 1976-1980 period to encounter all four planets."

The fact that the last such favorable planet alignments had occurred during the term of President Thomas Jefferson was used widely by proponents of such a mission.

As remarkably successful as Voyagers 1 and 2 were, it should be noted that they were preceded by two simpler, yet highly successful, spacecraft known as Pioneer 1 and Pioneer 2. They provided information about Jupiter and Saturn, which contributed heavily to the success of Voyager.

Pioneer 10 was launched on March 3, 1972, weighing only 258 kilograms, tiny for a spacecraft, yet containing much miniaturized equipment. Its sole target was the planet Jupiter. Arriving in the vicinity of the gas giant in December, 1973, it experienced huge clouds of radiation one hundred times that which is lethal for humans, yet it survived with minor damage. Passing Jupiter, it continued returning information for four years until its modest 8-watt transmitter could no longer be heard. Jupiter's gravitational pull provided the "gravity assist," or slingshot effect, which has taken it deep into interstellar space in a direction toward the star Aldebaran.

Pioneer 11, launched on April 6, 1973, was even more successful. Its different trajectory took it closer to Jupiter, braving the radiation belts and allowing it to be propelled toward Saturn. The pictures it returned of both planets were far better than any ever obtained by telescopes. Both Pioneers 10 and 11 have passed out of the solar system, having paved the way for the Voyagers.

The minigrand tour of the Voyager, so called because Pluto was eliminated from the program, required a spacecraft design far more sophisticated than the Pioneer's, since it would be required not only to survive intense radiation but also to be capable of detecting and reacting to various internal problems that might arise. Earth commands would require far too much time for corrective action because of the much greater distance to the outer planets. In short, the Voyager had to be strong, intelligent, and long-lived.

Launched on September 5, 1977, Voyager 1 made its closest approach to Jupiter on March 5, 1979, six months after launch and four months ahead of Voyager 2. Passing the planet, it encountered the moons Amalthea, Io, Ganymede, and Callisto and passed to within 1 million kilometers of Europa. As a result of the startling flood of information transmitted to earth by Voyager 1, scientists and engineers at JPL sent new commands to Voyager 2's computers, changing some of its programmed instructions, which resulted in a new trajectory for maximum sensing and photography.

During March and April, 1979, Voyager 1 took more than sixteen thousand photographs before continuing its journey toward Saturn. One of its earlier photographs taken on September 18, 1977, shows for the first time in a single frame the relationship of the earth and the Moon: more as a dual planet than as a planet with an accompanying satellite. Voyager 1 encountered Saturn and its moons in early November, 1980. The closest approach to the outer moon, Titan, occurred on November 11. It swung beneath Saturn's ring system and behind the planet, as viewed from Earth, making its closest approach to the moon Iapetus on November 14, 1980. Voyager 1's last images of Saturn were taken on December 19, 1980, and it now speeds deeper into space in the direction of the constellation Ophiuchus.

Voyager 2 arrived at Jupiter in July, 1979, with a different trajectory from that of Voyager 1 and at different angles, which permitted photographs of the opposite hemispheres of Callisto and Ganymede, high-resolution images of Europa, and Jupiter's ring.

Just as the eighty-hour gravitational tug of Jupiter had propelled the tiny Voyager 2 toward Saturn, so now giant Saturn pulled it into a new direction: directly toward Uranus, a journey of fifty-three months. As this outward journey from Saturn began, Voyager 2 now became the true trailblazer, going where no spacecraft had ever gone before. At Uranus, which orbits the sun at a distance nineteen times that of Earth, Voyager 2's on-board systems continued to perform well after the four and one-half year trip from Saturn. Precisely as planned, the gravitational pull of Uranus bent Voyager's trajectory toward its last assignment, planet Neptune. The scientific instruments continued to perform another three years and seven months before reaching Neptune on August 24, 1989.

At Neptune, Voyager 2 had traveled about 7 billion kilometers, a distance so great that its radio signals, which travel through space at the speed of light, required about four hours to reach Earth. At this distance, Neptune receives relatively little sunlight and thus appeared dim to Voyager 2. Longer exposures were therefore necessary, but longer exposures meant smearing of the images, because of Voyager's speed of 27 kilometers per second. This (and other problems) was overcome by JPL scientists by pivoting the spacecraft and/or the scan platform slowly during the exposures.

The flyby of Neptune and its moons was a complete success. Approaching the planet, Voyager 2 swung over its north pole, and during this time it was hidden from Earth by Neptune for forty-nine minutes. It then encountered Triton, Neptune's moon, which has an atmosphere, the last planetary objective of the mission. It now heads into deep space, transmitting data as long as its systems remain active.

Impact of Event

Both the technological success and the scientific success of the Voyager missions were remarkable and can hardly be overstated. The mechanical and electronic components of the spacecraft functioned exceedingly well over the twelve-year span.

Voyager 1 created an explosion of excitement when, passing through the equatorial plane of Jupiter in March, 1979, it detected a ring around the planet. The discovery was unexpected, since no evidence had ever been presented supporting its possible existence. Mission planners agreed to devote a single photograph to a "one shot" search for a ring, and luckily that was enough.

As for Jupiter's atmosphere, Voyager 1 confirmed and greatly refined the findings of Pioneer 10 and 11, which saw it as a caldron of boiling, multicolored gases moving with incredible energy from some internal source. High-resolution images of the Great Red Spot, which has been observed in Jupiter's upper atmosphere since the early 1800's, showed exquisite detail. In addition to confirming that the spot is cooler than its surroundings and therefore at a higher attitude, infrared scans detected methane, ethane, and ammonia.

Callisto, the second largest of Jupiter's moons and the most heavily cratered, was determined to have a diameter about one and one-half times that of Earth, although its density is only about one-third as great. This suggested the satellite is about one-half water and ice. The fact that no deep craters were found supports this model, since water-ice walls are not strong enough to stand very high and tend to flow down into the crater floor. Though the scientists expected to see craters on Io, none was seen. The extremely active volcanoes on Io were the most startling discovery of the mission, and it is now believed to be the most geologically active object in the solar system. In appearance, Europa exhibits extensive cracks and faults quite unlike any seen before. They are very long, crisscrossing like blood vessels, some extending partway around its surface.

In February, 1981, Voyager 1 made its closest approach to Saturn, its final planet target. While still 48 million kilometers out, it made the first of many shocking discoveries. Spreading out radially across the B ring were spokelike features, which were later presumed to be electrically charged particles that rotate with the planet's magnetic field.

At Jupiter in 1979, Voyager 2's first success was a huge one: It provided improved photographs of the ring and it produced the first high-resolution image of Europa. Both Voyagers returned great quantities of new information about the planet Saturn, its rings, and its moons, information that has resulted subsequently in answered questions, in the formation of new theories, and in raising yet more questions.

All five major moons of Uranus were imaged (led by Stephen Patrick Synott of the Imaging Science Team) by Voyager 2: Miranda, Ariel, Umbriel, Titania, and Oberon. It was discovered that these moons, except Umbriel, have undergone tectonic activity. This is in sharp contrast to the previously examined cratered, ice-covered moons of Saturn. At Neptune, Voyager 2's final planet, new discoveries came quickly: two new moons and the shadow of a cloud, something never seen before. The presence of the shadow made possible a determination of the cloud height. Also discovered was a ring system of three rings and a dust "sheet" in the equatorial plane. A large dark spot, a smaller dark spot, and a small white spot stand out as Neptune's most noticeable features. Triton exhibited complex surface features resembling a cantaloupe, signs of past volcanic activity, and a large polar ice cap. The icy planet is made of rock and frozen nitrogen.

Charles E. Kohlhase, mission design manager for the Voyager project, stated: "There is little doubt that Voyager is the greatest mission of planetary discovery ever undertaken in the 20 year history of the United States Space Program." Now well beyond the solar system, Voyager 2 is expected to continue sending useful information about the environment of space for at least twenty more years.

Bibliography

Beatty, J. Kelly. "Rendezvous with a Ringed Giant." *Sky and Telescope* 61 (January, 1981): 7-18. This article has some of the first remarkable photographs that began the revolution in planetary studies. These provided more questions than answers and increased the anticipation and expectation of the next target, Uranus.

Berry, Richard. "Voyager: Discovery at Uranus." *Astronomy* 14 (May, 1986): 6-22. New and startling discoveries at the seventh planet. In-depth analysis of the ring system and good discussion of tectonic activity on Uranian satellites.

Berry, Richard, and Robert Burnham. "Voyager 2 at Saturn." *Astronomy* 9 (November, 1981): 6-30. Beautifully illustrated, an analysis of the Voyager 2 encounter in nontechnical, yet complete, coverage.

Elliot, James, and Richard Kerr. "How Jupiter's Ring Was Discovered." *Mercury* 14 (November/December, 1985): 162-171. The discovery of Jupiter's most hidden structure demonstrates science and scientists at their best.

Kohlhase, Charles E. "Voyager's Path of Discovery." *Astronomy* 14 (February, 1986): 14-22. Kohlhase, the mission manager, gives insight in this overview of the Voyager mission, one that began as a substitute for a more ambitious project.

Littmann, Mark. "The Triumphant Grand Tour of Voyager 2." *Astronomy* 16 (December, 1988): 34-40. This inspiring account illuminates the technological achievements scientists and engineers performed, ensuring the success of the mission.

Miner, Ellis D. "Voyager's Last Encounter." *Sky and Telescope* 78 (July, 1989): 26-29. This very readable account is a summation of the most cost-effective mission ever produced.

Smith, Arthur. *Planetary Exploration*. Wellingborough, England: Patrick Stephens, 1989. This account of both American and Soviet interplanetary expeditions is an overview of the history of this technology and the revolutionary scientific data it generated.

Sohus, Anita, and Ellis Miner. "Voyager Mission to Neptune." *Mercury* 17 (September/October, 1988): 130. Written about a year before the close encounter. Presents an excellent discussion of the mission for the general reader.

Waff, Craig B. "The Struggle for the Outer Planets." *Astronomy* 17 (September, 1989): 44. A fascinating presentation of the history and evolution of the space program for exploring the planets.

Richard C. Jones

Cross-References

Franklin and Burke Discover Radio Emissions from Jupiter (1955), p. 1492; Pioneer 10 Is Launched (1972), p. 1956; Astronomers Discover the Rings of the Planet Uranus (1977), p. 2068; The First Ring Around Jupiter Is Discovered (1979), p. 2104; Astronomers Discover an Unusual Ring System of Planet Neptune (1982), p. 2211.

GRUENTZIG USES PERCUTANEOUS TRANSLUMINAL ANGIOPLASTY, VIA A BALLOON CATHETER, TO UNCLOG DISEASED ARTERIES

Category of event: Medicine
Time: September 16, 1977
Locale: University of Zurich Hospital, Zurich, Switzerland

Gruentzig developed the balloon catheter method to carry out percutaneous transluminal angioplasty, a technique that clears narrowing of coronary arteries, successfully countering the disease

Principal personages:
> ANDREAS GRUENTZIG (1939-1985), a cardiologist who developed balloon angioplasty and perfected its use
> DAVID A. KUMPE, a radiologist who collaborated with Gruentzig
> CHARLES T. DOTTER (1920-   ), a cardiac radiologist who pioneered percutaneous transluminal angioplasty in 1964 with Melvin P. Judkins

Summary of Event

Arteriosclerosis—often called hardening of the arteries—is a disease in which arteries become hard and inelastic. The disease may affect a few arteries or it may be spread widely throughout the body. In some cases, arteriosclerosis causes hemorrhages because the diseased arterial walls become brittle and rupture. Arteriosclerotic hemorrhage in brain arteries results in apoplexy (stroke) in many people afflicted with the disease. In other cases, arteriosclerosis may cause loss of circulation in the extremities, leading to gangrene and the need for amputation of a limb. Complications of severe arteriosclerosis are a major cause of death resulting from cardiovascular disease.

Although severe arteriosclerosis is seen most often in middle-aged and aged people, some hardening of the arteries is present in people of younger ages. For example, large-scale studies of American soldiers in their late teens and early twenties showed evidence of the disease in about 80 percent of the test population. Furthermore, arteriosclerosis is much more common in men than in premenopausal women, and its incidence is virtually epidemic in the industrialized nations of the world. The prevalence and severity of the disease are particularly high in Northern Europe, in the United States, and in Australia.

The factors associated with the development of arteriosclerosis include overeating, high fat intake, sedentary life-style, emotional stress, and smoking. High blood cholesterol levels are viewed as being especially valid indicators of the future development of the disease, if it is not present already. It is viewed that diminished cholesterol intake is particularly valuable as a preventive of the disease.

Despite the fact that millions of people afflicted with arteriosclerosis endure painful, disabling, or even lethal consequences, the methodology for successful treatment of the disease was slow in coming. In the 1960's, nonsurgical methods reportedly provided the afflicted person little more than a chance to live with the disease. Furthermore, surgical intervention by bypass techniques was deemed to be confined to a few highly specialized vascular surgeons to aid the vast number of arteriosclerotic patients in industrialized countries.

The first relief of the problem occurred when Charles T. Dotter and Melvin P. Judkins, at the University of Oregon, introduced transluminal angioplasty. (See "Transluminal Treatment of Arteriosclerotic Obstruction," 1964.) These pioneers devised a method to overcome the arteriosclerotic narrowing and blockage of leg arteries, which previously had doomed victims of the disease to amputation of afflicted legs when gangrene developed. The Dotter-Judkins method is carried out in the following way: The identification of the diseased portion of the artery is made by fluoroscopy. Then, a very long, 0.13-centimeter-diameter, coiled spring catheter guide is inserted gently into the artery through a surgical incision. The catheter guide is then directed with a fluoroscope until its tip passes through the arteriosclerotic region of the blocked artery. Next, a 0.25-centimeter hollow catheter is slipped over the guide and advanced along, until it, too, passes through the arteriosclerotic blockage. This enlarges the initial opening (lumen) generated in the diseased artery. Wherever possible, a second 0.51-centimeter catheter is used to enlarge the arterial lumen further. Finally, the catheters and the catheter guide are withdrawn carefully. Successful transluminal angioplasty results in normal blood flow through the artery. Utilization of this pioneering methodology reversed the disease in many patients treated by Dotter, Judkins, and their associates.

In the early 1970's, Andreas Gruentzig, of the University of Zurich, developed a new type of catheter for use in transluminal angioplasty. This catheter resulted in a small balloon that could be inflated after it was placed into the diseased portion of an artery. Inflation of the balloon catheter opened up the artery. The placement technique used was very similar to that of Dotter and Judkins. Yet, Gruentzig's balloon catheter had the advantages of simplifying the operation and speeding it up.

Gruentzig began to modify the methodology to enable use of balloon angioplasty to clear arteriosclerotic blockage in the much smaller coronary arteries that feed the heart. He expected the modified method to allow appropriate patients to avoid long, complicated, and more dangerous surgical bypass methods that were in general practice at the time. Gruentzig's efforts to develop the new technique included animal experiments and postmortem human studies.

By 1977, Gruentzig and his coworkers were ready to test coronary balloon angioplasty on an appropriate human subject. They carried out the first operation (see J. Willis Hurst, "The First Coronary Angioplasty as Described by Andreas Gruentzig," 1986) on a thirty-seven-year-old insurance salesman who had exhibited severe exercise-induced angina pectoris (cardiac-induced chest pain) because of arteriosclerotic blockage of one coronary artery. Gruentzig reportedly stated that the patient was enthusiastic about being able to avoid coronary bypass surgery "in spite of the fact that he was to be the first person ever treated with the new technique." On September 16, 1977, the operation was carried out in the cardiac catheterization laboratory at the University Hospital in Zurich, Switzerland. Gruentzig reported that "the catheter was advanced to the arterial block with no difficulty." He also noted that "the dilation of the balloon, twice, to relieve the blockage caused no chest pain and normalized the coronary blood pressure" in the patient. After completion of the operation, Gruentzig jubilantly stated that his colleagues were surprised that the procedure was so easy.

Sensibly wishing to avoid premature announcement of the success of the method, Gruentzig—with the cooperation of the press—avoided media exposure of the technique until 1978. At that time, the success of his first five balloon angioplasty cases were published in *The Lancet*. Balloon catheterization, named percutaneous transluminal coronary angioplasty (PTCA), had become a reality.

A 1979 report by Gruentzig and David A. Kumpe ("Technique of Percutaneous Transluminal Angioplasty with the Gruentzig Balloon Catheter") describes the methodology used. The advantages of the Gruentzig method over other procedures available were described as being the diminished incidence of dangerous blood clot formation resulting from balloon catheter construction; the additional safety feature—that balloon expansion cannot occur beyond a predetermined diameter— even if the desired pressure of the PTCA system is exceeded accidentally, and the ease and speed of the operation. Gruentzig cautioned that the technique should be performed by experienced practitioners who have been trained in an institution at which it is practiced often. In addition, he noted that optimum PTCA requires close cooperation between the radiologist, the surgeon, and other members of a complex operating team.

Gruentzig moved to Emory University in Atlanta, Georgia, amid much acclaim from his peers. He went on to perfect the technique and inculcate its wide use.

Impact of Event

Andreas Gruentzig carried out the first balloon angioplasty in 1977; without Gruentzig's efforts most physicians believe that successful coronary angioplasty would have taken much longer to develop. Yet, the evolution of the methodology that enabled him to design the technique began forty-eight years earlier. In 1929, the first cardiac catheterization was effected by Werner Forssmann, who wished to find a means for injecting therapeutic drugs into the heart, to treat serious medical conditions. The efforts of Forssmann, including his initial clandestine adventure—using himself as the experimental subject—are discussed in J. Willis Hurst's chapter in *Coronary Arteriography and Angioplasty* (1985), edited by Spencer B. King III and John S. Douglas, Jr.

Thirteen years later, another essential event occurred. This was the development of selective coronary arteriography, by Mason Sones. Sones's technique made it possible for others to carry out modern diagnostic and therapeutic endeavors in the area. Without it, the efforts of Dotter and Judkins, and Gruentzig's endeavors, would not have been possible.

The results of Gruentzig's work are particularly important because now it is generally agreed that very few atherosclerotic arteries cannot be reclaimed from the disease by prudent utilization of balloon catheters. Their use is very widespread, and the consequences have made life longer, safer, and better for millions of atherosclerosis sufferers.

Much of Gruentzig's work was limited to dilation of one diseased artery, and many patients exhibit multivessel disease. Fortunately, as sequels to Gruentzig's effort, many advances have been made in the years since the first balloon angioplasties were performed. The procedures have changed, and handling of coronary artery disease has been facilitated to make the technique much more useful in such multivessel disease.

Furthermore, many patients once viewed as being too high risk for PTCA can be accommodated now. Part of the expansion of its application results from the development of new safety measures (for example, addition of oxygen carriers to the arteries and to new surgical devices). Also, lasers now are used to facilitate removal of atherosclerotic blockage from arteries, and many physicians believe that the foreseeable future will do away with the need for catheters.

Bibliography

Dotter, Charles T., and Melvin P. Jones. "Transluminal Treatment of Arteriosclerotic Obstruction." *Circulation* 30 (1964): 654-670. This pioneering article explains the Dotter-Judkins technique of percutaneous transluminal angioplasty in treatment of arteriosclerotic leg arteries. The methodology used, the clinical cases studied, and the expectations of better future methods are presented. Ten references recount development of the work and present useful aspects of work on coronary artery disease.

Friedman, Steven G. *A History of Vascular Surgery.* Mount Kisco, N.Y.: Futura, 1989. This brief, but interesting, book highlights many of the most important developments in vascular surgery from antiquity to the late 1980's. Chapter 13, "Recent Advances," is of particular interest because it puts the work of Dotter, Gruentzig, and other contributors to percutaneous transluminal angioplasty into useful perspective.

Gruentzig, Andreas. "Results from Coronary Angioplasty and Implications for the Future." *American Heart Journal* 103 (1982): 779-783. The article describes the Gruentzig methodology for coronary balloon catheterization and its results on hundreds of patients. The high success rate of the procedure and its advantages over coronary bypass operations are discussed. Criteria for patient choice are mentioned as well as future implications of the procedure.

Gruentzig, Andreas, and David A. Kumpe. "Technique of Percutaneous Transluminal Angioplasty with the Gruentzig Balloon Catheter." *American Journal of Roentgenology* 132 (1979): 547-552. This article describes the Gruentzig (original spelling Gruntzig) method in detail. Particularly interesting are pictures of the catheters and their guidance system, methodology of balloon inflation, and the case reports. The authors describe the overall results of three hundred operations performed and followed up since 1971. Eighteen useful references provide more detail where desired.

Hurst, J. Willis. "The First Coronary Angioplasty as Described by Andreas Gruentzig." *American Journal of Cardiology* 57 (1986): 185-186. This brief paper recounts the balloon angioplasty carried out in Zurich, in September, 1977. The basis of the technique, Gruentzig's jubilance at its success, and his good sense in holding back premature publication of the results of the operation are described. Aspects of Gruentzig's personality and scientific ability are revealed.

King, Spencer B., III, and John S. Douglas, Jr., eds. *Coronary Arteriography and Angioplasty.* New York: McGraw-Hill, 1985. The book details many important developments in the area. For example, chapter 1 describes development of cardiac catheterization from a clandestine self-test by Forssmann to Gruentzig's work on PTCA. Other features include chapter 15, "Selection for Surgery or PTCA" and chapter 17, on results of PTCA after Gruentzig and others had carried it out many times.

Tommaso, C. L. "Management of High Risk Coronary Angioplasty." *American Journal of Cardiology* 64 (1989): 33E-37E. This paper describes advances in high-risk coronary angioplasty made in the period since Gruentzig's original work. These advances include specialized balloon catheters, infusion of oxygen-carrying substances to the artery treated, and other valuable new methodology. Gruentzig's belief that better techniques would develop is validated in this article.

Sanford S. Singer

Cross-References

Carrel Develops a Technique for Rejoining Severed Blood Vessels (1902), p. 134; McLean Discovers the Natural Anticoagulant Heparin (1915), p. 610; Gibbon Develops the Heart-Lung Machine (1934), p. 1024; Blalock Performs the First "Blue Baby" Operation (1944), p. 1250; Favaloro Develops the Coronary Artery Bypass Operation (1967), p. 1835; Barnard Performs the First Human Heart Transplant (1967), p. 1866; DeVries Implants the First Jarvik-7 Artificial Heart (1982), p. 2195.

ROHRER AND BINNIG INVENT
THE SCANNING TUNNELING MICROSCOPE

Category of event: Physics
Time: 1978-1981
Locale: Zurich, Switzerland

By developing the scanning tunneling microscope, Binnig and Rohrer began a trend that is leading to the visualization and the control of matter at the atomic level

Principal personages:

GERD BINNIG (1947-   ), a West German physicist who was the co-winner of the 1986 Nobel Prize in Physics

HEINRICH ROHRER (1933-   ), a Swiss physicist who was the cowinner of the 1986 Nobel Prize in Physics

ERNST RUSKA (1906-1988), a West German engineer who was the former director of the Institute of Electron Microscopy of the Fritz Haber Institute in West Berlin and cowinner of the 1986 Nobel Prize in Physics

ANTONI VAN LEEUWENHOEK (1632-1723), a Dutch naturalist who was the inventor of the first optical microscope

Summary of Event

The field of microscopy began at the end of the seventeenth century when Antoni van Leeuwenhoek developed the first optical microscope. In this type of microscope, a magnified image of the sample is obtained by directing light onto it and then taking the light through a lens system. Van Leeuwenhoek's microscope allowed him to observe the existence of life in a scale invisible to the naked eye. Since then, development in the optical microscope has revealed the existence of single cells, pathogenic agents, and bacteria. During the nineteenth century, however, the German physicist and lensmaker Ernst Abbe showed that objects smaller than half the wavelength of the light used for viewing cannot be resolved by an optical microscope. This limitation, called "Abbe's barrier," results from the wave nature of light. Consequently, optical microscopes cannot resolve objects that are smaller than about 400 nanometers, or one thousand-millionth of a meter. That is about two thousand times greater than the diameter of a typical atom. Hence, to observe objects smaller than 400 nanometers, the wavelength of the radiation used in the observation must be smaller than the wavelength of light.

The next breakthrough came during the early 1920's, when advances in theoretical physics predicted a wave behavior as well as a particle behavior of electrons. The wave properties were predicted by Louis de Broglie in 1925 and confirmed by Clinton J. Davisson and Lester H. Germer of the Bell Telephone Laboratories in 1927. It was found that high-energy electrons have shorter wavelengths than low-energy electrons and that electrons with sufficient energies exhibit wavelengths comparable to the diameter of the atom. In 1927, Hans Busch showed in a mathematical analysis that current-carrying coils behave as electron lenses and that they obey the same lens equation as optical lenses. Using these findings, Ernst Ruska developed the electron microscope in the early 1930's.

The basic principle behind the operation of the electron microscope (often referred to as "transmission electron microscope," or TEM) is the same as that of the optical microscope except for the viewing method. The image of the sample is focused on a fluorescent screen or photographic plate instead of the eye. By 1944, the German corporation of Siemens and Halske manufactured electron microscopes with a resolution of 7 nanometers; modern instruments are capable of resolving objects as small as 0.5 nanometer. This allowed the viewing of structures as small as a few atoms across as well as the viewing of large atoms and large molecules. Yet, the electron beam used in this type of microscope imposed some limitations on its applicability. First, to avoid the scattering of the electrons, the samples must be put in vacuum, which limits the applicability of the microscope to samples that can sustain such an environment. Most important, some fragile samples, like organic molecules, are inevitably destroyed by the high-energy beams required for high resolutions. On the other hand, in 1956, J. A. O'Keefe of the United States Army Mapping Service suggested a microscope that overcomes Abbe's barrier. In such a microscope, the sample must be scanned back and forth across the light (or electron) beam; the resolution is then limited only by the size of the beam. Eric Ash of University College, London, built such a microscope in 1972; he was able to resolve an object with a size equal to one two-hundredth of the wavelength of the microwave radiation he used for viewing. The same principle was then used in developing the scanning electron microscope (SEM). This technique, widely used during the 1970's and 1980's, allowed some three-dimensional surface relief impressions of the specimens. The size of the beam was confined to about 10 nanometers, the resolution of the microscope.

From 1936 to 1955, Erwin Wilhelm Müller developed the field ion microscope (FIM), the first microscope to enable a direct viewing of the atomic structure of surfaces. In this technique, the sample is placed at the tip of an extremely sharp needle, which is placed in an evacuated chamber across from a fluorescent screen (or photographic plate). Then, a small amount of helium gas is entered into the chamber and a high positive voltage is applied to the needle. Helium atoms are then attracted to the needle and stripped out of their electrons; the resulting ions are repelled toward the fluorescent screen. The pattern obtained at the screen is an image of the tip. The use of this technique is limited to samples capable of sustaining the high electric fields necessary for its operation. In the early 1970's, Russel D. Young and Clayton Teague from the National Bureau of Standards (NBS) developed the "Topografiner," a new kind of FIM with the sample placed at a large distance from the tip. The tip is scanned across the surface of the sample with a precision of about a nanometer. The precision in the three-dimensional motion of the tip was obtained by using three legs made of piezoelectric crystals. These materials change shape in a reproducible manner when subjected to a voltage. The extent of expansion or contraction of the crystal depends on the applied voltage. Thus, the operator can control the motion of the probe by varying the voltage acting on the three legs. The resolution of the topografiner is limited by the size of the probe.

The idea for the scanning tunneling microscope (STM) arose when Heinrich Rohrer from the International Business Machine Corporation (IBM) at the Zurich research laboratory met Gerd Binnig in Frankfurt in 1978. By 1980, Binning and Rohrer built their first instrument. At 2:00 A.M. on March 18, 1981, they observed the first evidence of electron tunneling through a vacuum barrier. Images with a resolution at the atomic level were observed some weeks later. This type of microscope is very similar to the topografiner. In both spectrometers, a fine probe is scanned over the surface of the sample; the differences are in the distance separating the tip from the sample and in the voltage applied at the tip. In the STM, the tip is kept at a height of less than a nanometer away from the surface, and the voltage applied between the specimen and the probe is low. Under these conditions, the electron cloud of atoms at the end of the tip overlaps with the electron cloud of atoms at the surface of the specimen. For a thin tip, the overlapping occurs between clouds of only two atoms, one from the tip, the other from the sample. This overlapping results in a measurable electrical current flowing through the vacuum or insulating material existing between the tip and the sample. Since there is no actual contact between sample and probe, the current is called tunneling current; it depends on the position of the sample relative to the tip, the composition of tip and sample, and the voltage applied between the two. When the probe is moved laterally across the surface with the voltage between the probe and sample kept constant, the change in the distance between the probe and the surface (caused by surface irregularities) results in a change of the tunneling current. Two methods are used to translate these changes into an image of the surface. The first method involves changing the height of the probe to keep the tunneling current constant; the voltage used to change the height is translated by a computer into an image of the surface. The second method scans the probe at a constant height away from the sample; the voltage across the probe and sample is changed to keep the tunneling current constant. These changes in voltage are translated into the image of the surface. The constant height mode allows faster scan rates with a reduced depth resolution. On the other hand, the dependence of the tunneling current on the applied voltage at every position of the probe is used to distinguish the various atoms composing the surface. The main limitation of the technique is that it is applicable only to conducting samples or to samples with some surface treatment.

Impact of Event

The STM has had a great impact on scientific research as well as on technological advances. The scanning tunneling microscope has several advantages over the previous microscope techniques. For example, the voltages used for the tunneling are low and nondestructive; the spectrometer does not require an evacuated medium for its operation as long as an insulator is separating the probe from the sample. The preciseness of the manipulation of the probe allows an unprecedented three dimensional high resolution of surfaces. Another advantage of the microscope is its capability of distinguishing between the different atomic species on the surface of the specimen. Observations were made on metals, semiconductors, superconductors, and on biological matter.

The STM found many industrial applications involving both the design and quality control stages. In the manufacturing of diffraction grating masters, for example, the STM was used to guide the ruling machine cutting the grooves and to examine the quality of the final product. The combination of its lateral and vertical high resolutions makes the STM the only device capable of visualizing the grooves in those devices. On the other hand, the STM was used in the design of recording heads as well as in the manufacturing of the stampers used to dig holes through compact disks. Further applications are envisioned in the electronic component manufacturing industry.

Most important, the STM was successfully used in the manipulation of matter at the atomic level. In October, 1989, by letting the probe sink into the surface of a metal-oxide crystal, researchers at Rutgers University were able to dig a square hole of about 250 atoms across and 10 atoms deep. A more impressive feat appeared in a report in the April 5, 1990, issue of *Nature*; M. Eigler and Erhard K. Schweiser of IBM's Almaden Research Center spelled out their employer's three-letter acronym using thirty-five atoms of xenon. This ability to move and place individual atoms precisely raises several possibilities, which include custom-made molecules, atomic scale data storage, and ultrasmall electrical logic circuits.

The success of the STM has led to several new microscopes developed for the study of other features of sample surfaces. While they share the same scanning probe technique in their measurements, these techniques use different processes for the actual detection. The most popular among these new devices is the atomic force microscope (AFM). This device measures the tiny electric forces that exist between the electrons of the probe and the electrons of the sample without the need for electron flow, which makes the technique particularly useful in imaging nonconducting surfaces. Other scanned probe microscopes use physical properties such as temperature and magnetism to probe the surfaces.

Bibliography

Binnig, Gerd, and Heinrich Rohrer. "The Scanning Tunneling Microscope." *Scientific American* 257 (August, 1985): 50-56. This article, written by the developers of STM, is intended for the general reader. Very instructive.

——————. "Scanning Tunneling Microscopy: From Birth to Adolescence." *Review of Modern Physics* 59 (July, 1987): 615-625. This is the Nobel Prize acceptance lecture delivered jointly by the authors on December 8, 1986. It covers the historic developments of the scanning tunneling microscope. Extensive bibliography.

Golovchenko, J. A. "The Tunneling Microscope: A New Look at the Atomic World." *Science* 232 (April, 1986): 48-53. This article gives a precise explanation of the concepts involved without going into too much detail. Intended for the general reader.

Hansma, Paul K., and Jerry Tersoff. "Scanning Tunneling Microscopy." *Journal of Applied Physics* 61 (January, 1987): R1-R23. Useful for one with a technical background. It is rich with technical information, and it includes an extensive bibliography.

Hansma, P. K., V. B. Elings, O. Marti, and C. E. Bracker. "Scanning Tunneling Microscopy and Atomic Force Microscopy: Application to Biology and Technology." *Science* 242 (October, 1988): 209-216. This article provides a good, though not easy, treatment of the biological applications of the technique and a description of the operation of the atomic force microscope (AFM).

Jones, R. Edwin, and Richard L. Childers. "Electron Microscopes" and "The Field Ion and Scanning Tunneling Microscopes." In *Contemporary College Physics.* Reading, Mass.: Addison-Wesley, 1990. These two essays are perfect for someone with a high school background in physics who wants a brief simple description of the operating principle of some of the various devices used in microscopy. The essays are part of a feature of this undergraduate physics textbook called *Practical Physics*, which deals with various applications of physics.

Quate, F. Calvin. "Vacuum Tunneling: A New Technique for Microscopy." *Physics Today* 39 (August, 1986): 26-33. The details and accuracy of the historic developments of the technique are impressive; it also includes a copy of one of the pages of Binnig's laboratory notebook. While some of the details require some physics background, the general reader can bypass them and enjoy the rest of the information.

Schwarzschild, Bertram. "Physics Nobel Prize Awarded for Microscopies Old and New." *Physics Today* 40 (January, 1987): 17-21. A good description of the historical development of both the electron microscopy technique and the scanning tunneling microscopy technique.

Trefil, James. "Seeing Atoms: With the New Scanning-Probe Microscopes, It's Become Almost Routine." *Discover* 11 (June, 1990): 54-60. Contains the simplest treatment of the principles involved. Discusses the advantages and limitations of the microscope. Trefil successfully uses examples from everyday life to explain the concepts involved.

Wickramasinghe, H. Kumar. "Scanned-Probe Microscopes." *Scientific American* 261 (October, 1989): 98-105. For one interested in the impact of the development of the scanning tunneling microscope. Contains a list of the various probe microscopy techniques developed.

Taha Mzoughi

Cross-References

Zsigmondy Invents the Ultramicroscope (1902), p. 159; Ruska Creates the First Electron Microscope (1931), p. 958; Müller Invents the Field Emission Microscope (1936), p. 1070; Müller Develops the Field Ion Microscope (1952), p. 1434; Esaki Demonstrates Electron Tunneling in Semiconductors (1957), p. 1551.

BROWN GIVES BIRTH TO THE FIRST "TEST-TUBE" BABY

Category of event: Medicine
Time: July 25, 1978
Locale: Oldham General Hospital, Oldham, Lancashire, England

Brown gave birth to the first "test-tube" baby aided by the pioneering efforts of Edwards and Steptoe

Principal personages:
> LESLEY BROWN (1947-   ), the mother of the first "test-tube" baby and an exceedingly brave and persistent woman
> GILBERT JOHN BROWN (1941-   ), the father of the first "test-tube" baby and a very happy and proud husband and parent
> PATRICK CHRISTOPHER STEPTOE (1913-1988), the obstetrician who perfected the laparoscope used to obtain oocytes from the ovary and who delivered Louise Joy Brown
> ROBERT GEOFFREY EDWARDS (1925-   ), a biologist and professor of human reproduction at the University of Cambridge who designed the methods for in vitro fertilization and early embryo development
> JEAN M. PURDY, a laboratory technician and former nurse who was vital to the monitoring of the eggs, sperm, and embryos in culture

Summary of Event

One cannot describe adequately in a short space all the drama, excitement, and controversy over the birth of Louise Joy Brown on July 25, 1978. An interested person could read books about the pain, intensity, and happiness surrounding the momentous occasion involving a couple desperately wanting a baby, a dedicated doctor who loved his patients, a biologist who was able to obtain fertilization and embryo growth in vitro, a technician who was extraordinarily dedicated, the wives and families of the caregivers, other hospital workers, and the incessant press.

As a medical student in the 1930's, Patrick Christopher Steptoe was confronted by the extreme sadness of women coming to the clinic with blocked tubes that prevented fertilization and travel of an embryo to the uterus for development. After practicing in London and learning as much as possible about problems of infertility, he moved to the position at Oldham General Hospital in 1951. After some time, he developed the technique of laparoscopy to study the internal abdominal area. He used a flexible source of cool light and fiber optics to visualize, move, and even take samples from the ovaries and other organs through a small incision in the naval.

In the 1950's, Robert Geoffrey Edwards was a student in zoology and then genetics working primarily with the development of mouse embryos. He also worked on the immunology of reproduction and on embryos. In time, he was able to obtain human ovaries to use to allow the maturing of the oocytes (in humans and many other mammals the female germ cell is an oocyte and does not completely mature until after fertilization) in a culture dish in the laboratory. In 1963, at the University of Cambridge, he spent many months trying to fertilize ripe human oocytes with spermatozoa in a culture dish.

Edwards and Steptoe met at a scientific meeting in 1968 in London. They decided to collaborate, even though it would be a sacrifice driving 266 kilometers for Edwards and Jean M. Purdy, his technician. They spent the next several years working on perfecting the techniques of oocyte ripening and fertilization in the culture dish. In 1969, they were confronted by critics and supporters after they published a paper in the journal *Nature* about the ripening of twenty-four of fifty-six human oocytes and subsequent fertilization of some of them in culture. At this point, they began to involve patients. Steptoe obtained by laparoscopy already ripe oocytes from the ovary follicles, areas of the ovary where they matured. Edwards and Purdy, who designed the apparatus to retrieve the oocytes, would at least attempt to do the fertilization using the husband's spermatozoa and would follow any subsequent embryonic development, called cleavage. Within a short time, they had routine fertilization and some development in the culture dish, called in vitro. At first, the embryos quit developing after the eight-cell stage, but then with a different culture medium and some serum from the patient, they actually got blastocysts, the first stage of differentiation of cells and the one that implants in the uterus. Even though they had great success, their application for research funds to the Medical Research Council was denied.

In 1971, they moved their research to Kershaw's Cottage Hospital in Royton, near Oldham, where they had sterile facilities, including an operating area, a small laboratory, a culture room, and an anesthetic room. Edwards attended a conference on the ethics of the research in Washington and to the cheers of the audience answered the critics. In England, he had the backing of an ethics committee that considered many medical issues. In 1972, they started replacing embryos into the mothers. They were not successful and during an interval in 1973 they stopped working on the project. Finally, in the summer of 1975, they had success, a pregnancy.

The elation of those early weeks was ended when the ectopic pregnancy (the embryo had implanted in the part of the tube adjacent to the uterus) had to be terminated at seven weeks. The report, however, made medical history in the journal *The Lancet* on April 24, 1976. A few more pregnancies occurred and ended prematurely on their own before Edwards and Steptoe decided to use the natural reproductive cycles instead of hormone treatment. They obtained real success in late 1977. Lesley Brown was the second patient to go through the regime of urine collection every three hours to check natural hormones in order to decide when the oocyte would be ready to be ovulated. A ripe oocyte was collected on the first try and it was fertilized without problems. The eight-cell embryo was placed into her uterus just after 12 A.M. on November 13, 1977. In three weeks, Lesley Brown was informed that the hormone level screening from urine and blood tests was positive for pregnancy. In the next few months, there were three more pregnancies and the realization that the hormone fluctuations in humans made evening replacement of the embryos more efficient. The peace and quiet of that time might also have helped.

There was incredible tension because of the press; it may have had some effect on the pregnancy, as Lesley Brown developed toxemia. She had an elevated blood pressure, and her baby was not growing as fast as usually. The principals had tried to protect her and the other patients, but someone at the hospital actually sold privileged information about the patients. She had to stay with Steptoe's or her relatives, as reporters hounded both her home area and the work of her husband, Gilbert John. She spent the last eight weeks of her pregnancy at Oldham General Hospital under an assumed name. Besides the press problems, there was the continuing worry about the fetus' growth and possible placental problems. When the time came for the birth, the fluid around the fetus was analyzed and found to be favorable. The decision was made that the weight and head diameter had progressed past a premature state. A cesarean section was planned and carried out with official Department of Health and Social Security filming and the world almost literally watching. At 11:47 P.M. on July 25, 1978, Louise Joy Brown at five pounds and twelve ounces—the world's first "test tube" baby—was born.

Two of the other pregnancies spontaneously aborted; the fourth one resulted in a normal birth to a woman in a Glasgow hospital under the care of her own gynecologist in January, 1979, one month prematurely. Steptoe and Edwards were greeted by a standing ovation that winter at the Royal College of Obstetricians and Gynecologists as they presented their research.

Impact of Event

The controversies and acclaims over in vitro fertilization, early development in the laboratory, and implantation into the mother's uterus began in 1966, with Edward's report of in vitro fertilization. It has not ended. Subsequently, there have been thousands of babies born from Australia to England to the United States and Japan; the various centers have had widely varying success rates, but the average reported births is 14 percent of fertilizations. Actually, the success rate for births from natural fertilization is about 35 percent. One might assume, however, that with the spermatozoa checked for viability and normality and only the perfect oocytes used, the success rate would be higher than when there is no choosing.

There are many references that describe the views of all sides of all the arguments, in favor of and opposed to more research on human embryos and on the implantation of the embryo in the uterus. Research is allowed in England on embryos or "preembryos" up to fourteen days of development. The research must be aimed at improvement of approved techniques, including those for "test tube" babies. These cells are treated with respect because they are potential humans.

Arguments in favor of more research and applications of this technique include the projected increase in the level of success of "test tube" babies, the establishment in culture of cells of the early stages that might be used to replace tissues in adults or children that have defective organs, and the understanding of normal human development so that developmental abnormalities might be prevented. In most clinics, any extra embryos not transplanted are still the property of the parents and are frozen, but may be given up as donations to other couples or for research. Decisions about parentage and responsibilities are still being decided, but most donations are anonymous so that neither donor nor recipient knows the origin.

Many who speak against the technique fear abnormal offspring, but this has been rare and even the mandatory amniocentesis does not reveal an increased incidence of abnormalities. Others rightfully worry about what is termed "informed consent." Edwards and Steptoe always state how they explained that the methods "might" result in a pregnancy. Gena Corea, in her book *The Mother Machine* (1985), gives evidence that many of the women in the early experiments were research subjects, not really patients, and she doubts that Edwards and Steptoe understood this. The feeling from both Corea and Edwards/Steptoe is that no matter what they said, these women would have subjected themselves to anything to have a child. Corea cautions that this is not really a consent as much as a coercion caused by societal and family pressures to have a child to be a complete woman. There are also those who argue that this process is unnatural and could lead to genetic screening for certain perfect offspring and destroying the others for increasingly trivial reasons including the "wrong" sex. The various ethics committees connected with hospitals and clinics, if properly constituted and informed, can keep watchful and helpful eyes on the processes.

Bibliography

Brown, Lesley, and John Brown, with Sue Freeman. *Our Miracle Called Louise: A Parent's Story.* New York: Paddington Press, 1979. This personal account is a reflection of the sorrows, drama, and joy surrounding the birth of their desperately wanted child. Moreover, it is revealing as to literal pain and fears that were endured by this strong couple. Early life stories also included. Lovely homey style; short with pictures.

Corea, Gena. *The Mother Machine.* New York: Harper & Row, 1985. A feminist perspective on many current reproductive technologies that take control away from the woman. The other side of the story emphasizing some of the difficulties and realities. Some very thoughtful and provocative opinions that must be voiced before some techniques go too far. Each chapter contains detailed notes, explanations of processes, well referenced and indexed. Part 3 discusses in vitro research.

Edwards, Robert G. *Test-Tube Babies.* Burlington, N.C.: Scientific Publications Division, Carolina Biological Supply, 1981. A fifteen-page booklet that describes the early in vitro research, the actual processes involved, the pregnancies, and some of the ethical problems. Clearly written for the general reader and liberally illustrated with pictures of equipment and embryos. Graphs help to explain the hormone measurements. Further reading suggestions are given.

Edwards, Robert G., and Ruth E. Fowler. "Human Embryos in the Laboratory." *Scientific American* 223 (December, 1970): 44-54. A presentation of early in vitro work with oocyte maturation, fertilization, and embryo development. Pictures and drawings of fertilization, embryos, and their chromosomes. Some discussion and graphics of abnormal embryos. Explanations are basic and clear.

Edwards, Robert, and Patrick Steptoe. *A Matter of Life.* New York: William Morrow, 1980. This short book is written by both of the men who were involved with the research and the clinical aspects of the first "test tube" baby. Well-explained, personal styles that give some biographical information on both, a detailed account of their research and clinical work, and the controversies surrounding it. The reader will share their passion for their work and their patients.

Gold, Michael. "The Baby Makers." *Science 85* 6 (April, 1985): 26-38. An account of the work of the first American clinic in Norfolk, Virginia, with Howard and Georgeanna Jones; written for nonexperts. Diagrams of the processes and pictures of several of the children. Some discussion of infertility in general. A very personal account of the participants, including some of the controversies, especially in the beginning.

Grobstein, Clifford. "External Human Fertilization." *Scientific American* 240 (June, 1979): 57-67. An account by a well-known biologist about human reproduction in vitro and in vivo, including a large section on public policy and arguments surrounding the technical advances in the field. Many illustrations. Three references are given.

Halpern, Sue. "Infertility: Playing the Odds." *Ms.* 17 (January/February, 1989): 146-156. A discussion of infertility in the United States and some personal experiences with many attempted remedies from technological to adoption. Straightforward, factual, and nonjudgmental. Addresses given for infertility clinics and also for a support group, Resolve.

Moore, Keith L. *The Developing Human.* Philadelphia: W. B. Saunders, 1988. A medical textbook that is technical in some aspects, but reads very well and has many diagrams and pictures that are helpful for basic understanding of the processes of human reproduction. Pictures of abnormalities. Excellent discussion of fetal membranes. Timetables of development on inside covers. Well referenced and indexed.

Judith E. Heady

Cross-References

Ivanov Develops Artificial Insemination (1901), p. 113; Harrison Observes the Development of Nerve Fibers in the Laboratory (1907), p. 380; Bevis Describes Amniocentesis as a Method for Disclosing Fetal Genetic Traits (1952), p. 1439; Donald Is the First to Use Ultrasound to Examine Unborn Children (1958), p. 1562; Clewell Corrects Hydrocephalus by Surgery on a Fetus (1981), p. 2174; Daffos Uses Blood Taken Through the Umbilical Cord to Diagnose Fetal Disease (1982), p. 2205; The First Successful Human Embryo Transfer Is Performed (1983), p. 2235.

THE FIRST RING AROUND JUPITER IS DISCOVERED

Category of event: Astronomy
Time: March 4-7, 1979
Locale: Jet Propulsion Laboratory, Pasadena, California

Jupiter became the third planet known to possess a ring; it was photographed by the Voyager 1 probe and raised new questions about the physical properties of the Jovian system

Principal personages:
RAYMOND L. HEACOCK (1928-   ), the Voyager project manager
EDWARD C. STONE (1936-   ), the Voyager project scientist
BRADFORD ADELBERT SMITH (1931-   ), the imaging science team leader

Summary of Event

The exciting discovery in 1977 of a system of rings around the planet Uranus raised the possibility that other large planets in the outer solar system might possess such features. Proposals to search for rings around Jupiter, however, initially did not attract much enthusiasm. Prior to the Uranian discovery, several prominent astrophysicists had developed elaborate models to explain why Saturn—apparently alone among the outer planets—had a ring system. Since conditions for distant Uranus and Neptune were not well known—the Uranian ring discovery came inadvertently during sensitive measurements of the planets' occultation (blocking) of star light from the Kuiper Airborne Observatory at an altitude of more than 12,000 meters— many of these models depended upon contrasting circumstances at Saturn with those in the Jovian (Jupiter) system: atmospheric characteristics, positions of satellites, radiation emissions, and the like. Before 1977, there was a fairly broad consensus in the scientific community that Jovian conditions would not allow for a ring system. In any case, such a system, if significant, surely would have been observed long ago around Jupiter, a planet substantially larger than Saturn and only about half as far from Earth.

There remained, however, the possibility of a tenuous ring system (possibly one with unusual properties that would pose further challenges to standard models) that may have escaped telescopic detection or gone unnoticed by observers not expecting to find it. Since the 1950's, there had been occasional reports of what looked to some observers like an unexplained shadow on the surface of Jupiter in certain photographs. Moreover, in December, 1974, Pioneer 11 recorded anomalous radiation levels near the Jovian equator. Although not equipped with photographic equipment, Pioneer 11 carried instruments to locate and measure radiation drops in the "shadows" of Jovian satellites, the result of charged particles being absorbed by the satellites and thus reducing background radiation. Pioneer 10, which passed Jupiter in December, 1973, identified these areas, thus providing parameters for general radiation levels near Jupiter (which turned out to be exceedingly high).

Pioneer 11 replicated these experiments, but in the process also recorded an unexpected drop in radiation levels on both the inbound and outbound legs of the flyby, at about 1.8-1.9 Jovian radii, suggesting absorption of charged particles by solid matter. Pioneer 11 data also showed a drop in micrometeoroid impacts with the spacecraft as it approached the Jovian equator. One explanation would be that some mechanism reduced the general levels of dust in low Jovian latitudes and concentrated this material near the equator.

On the basis of these rather enigmatic readings, members of the Voyager Imaging Team (Raymond L. Heacock, Edward C. Stone, and Bradford Adelbert Smith), preparing flyby routines for two spacecraft in 1979, proposed a search for a ring system. Team managers, though dubious, finally consented to a one-exposure photographic search to be programmed for Voyager 1. On March 4, 1979, Voyager 1 found the ring in the predicted location. After careful analysis, the team announced the discovery three days later. Within days, the University of Hawaii observatory on Mauna Kea confirmed the ring by ground sighting.

The Jupiter ring system is some 9,000 kilometers in breadth and only 30-35 kilometers thick. Its outer edge, which Voyager showed to be sharp and well defined, was 58,000 kilometers from the planet. Its close proximity to Jupiter (less than two planetary radii) suggested it must be composed of relatively large chunks of material. At this distance, forces of radiation and Jupiter's particle ionization effects would reduce quickly the angular momentum of dust and small particles and send them into the atmosphere. The sharp outer edge of the ring, the team speculated, may be caused by the gravitational resonance with the small Jovian satellite Amalthea. (Such resonances disturb stability of small objects and tend to limit them to certain orbital paths.)

Voyager 2, now hastily reprogrammed to study the ring, presented new problems with its photographs taken July 9-11, 1979. These showed a complex ring system, much brighter and dustier than had been expected. Clearly, much of the material in the rings consisted of dust and small particles only a few microns in diameter. Particles of this size could survive only a few centuries, possibly less, amid the maelstrom of radiation and ionization in the vicinity of Jupiter. Unless the ring system was a very recent phenomenon—a proposition counter to the thinking of most planetary scientists—it had to be receiving new material continuously from somewhere else in the Jovian system.

An early candidate as a material source for the rings was astonishing volcanic activity on Io, one of the major Jovian satellites. Photographs of Io showed huge volcanic eruptions in progress, even as the Voyager spacecraft passed through the system. Some scientists proposed that material entered the rings not only from Io but also from a great many sources, including satellites and micrometeoroid bodies. In the complexity of resonances and forces emanating from the large number of Jovian satellites, the ring system was a kind of broom sweeping up particles from throughout the system.

Further observation established that, although some material might be entering the rings from a variety of exotic sources, replenishment, for the most part, came from larger bodies inside the ring system. Examination of Voyager data revealed the presence of a small satellite, christened Adrastea, less than 50 kilometers in diameter and orbiting very near the outer edge of the ring system, and still another, smaller satellite (Metis) in a very similar orbit. Scientists generally have accepted the theory that the rings contain many objects perhaps hundreds of meters in diameter—Voyager's cameras could resolve individual bodies only down to about 1 kilometer—slowly being pulverized by radiation and micrometeoroid bombardment. Adrastea and Metis were important discoveries for another reason: They were the first to be observed of a classification of moons or satellites generally called "shepherds." Their presence prevents the spread of ring particles and confers upon many ring systems remarkably clean and sharp boundary zones. Some researchers already had theorized the presence (eventually confirmed) of shepherd moons to explain characteristics of the ring system around Uranus, and similar bodies are now known to exist in the Saturn system. Discovery of the Jovian shepherds was particularly fortunate at the time since location of the more plentiful shepherds of rings around Saturn proved a more difficult task than expected.

Impact of Event

Detection of the Jovian rings provides both an object lesson in the nature of thought processes at work in scientific communities and important new information about the character of the outer solar system. The enormous quantities of data returned by the Voyagers also raised entirely new questions concerning planetary and solar system origins.

Clearly, some scientists became so entrapped in the theoretical models required to explain the long-known rings of Saturn that they failed to interpret correctly several kinds of pre-Voyager data suggesting the presence of Jovian rings. One wrote off the drop in radiation levels near the equator of Jupiter (first recorded by Pioneer 10 six years before Voyager 1 photographed the rings) as a false reading, possibly caused by equipment malfunction. The Hawaii team that obtained ground confirmation of the rings—using only a moderately large reflector with a mirror diameter of 2.2 meters—once alerted to look for the rings, accepted photographic evidence that most likely would not have led to the same conclusion before the Voyager 1 mission. The ring discovery showed that theories and models, valuable as they are for maintaining the elements of rigor and verification in research, occasionally lead scientists to look for only what they are predisposed to find and to overlook that which they previously have ruled out.

The Jupiter ring system, in itself, turned out to be a relatively minor chapter in a burst of discovery about the outer planets. The Uranus observations in 1977 and Voyager's flyby of Jupiter in 1979 shattered ring theories based on Saturn models and collectively implied that ring systems are typical of the giant planets beyond Mars. The subsequent arrival of Voyager 2 at Saturn released a torrent of information about ring systems and underscored the relative simplicity of the Jovian ring. Voyager went on to explore the Uranus system in 1984, as astronomical evidence of a ring around Neptune appeared and confirmed the Neptune ring by direct observation in 1989.

The ring system around Jupiter has been incorporated into a far more complex set of models than most scientists could have perceived prior to the Voyager missions. Ring systems are not the smooth and uniform objects they were once thought to be. Instead, they contain myriad structures, many of which still elude explanation. The rings of Jupiter, although hardly as visually spectacular as the Saturn system, nevertheless, appear to be more typical of ring formation and history than the major rings of Saturn—ironically now viewed once again as exceptional, if not unique, objects. The Jupiter ring system, those of Uranus and Neptune, as well as some of the minor rings of Saturn, are all made up of extremely small particles and some, astonishingly close to their primaries, raise a host of unsolved questions. By far the most unsettling aspect of these rings is that many of them appear to be much younger than the estimated 4.5-billion-year age of the solar system. So, too, do most of the tiny shepherd moons, particularly those well inside the Roche limit, wherein solid bodies should not be able to survive the gravitational stresses caused by their primaries.

Computer models suggest that satellites with diameters about 20-50 kilometers, lodged amid the debris of the rings and subject to fierce radiation, could not survive much more than 1 billion years. Smaller bodies might be even younger, perhaps only a few thousand years old. (One scientist speculated that a tenuous ring of water ice particles discovered around Saturn might have been formed around the time of the American Civil War.)

Jupiter's ring system, like most, is ever-changing. Material accretes (attaches) to the shepherd moons, while other material is stripped away by radiation and other effects. Shepherd moons may sweep clean for a time wide swaths in the particle belts. Most of the rings appear doomed to destruction by the forces around them, but others will replace them. The rings of Jupiter belong to a planetary system almost as complex as the solar system itself, a dynamic and incredibly varied system the study of which has breathed new life into planetary sciences.

Bibliography

Belton, M. J. S., Robert A. West, Jurgen Rahe, and Margarita Pereyda. *Workshop on Time-Variable Phenomena in the Jovian System*. Washington, D.C.: Government Printing Office, 1989. Technical conference on various phenomena including the dynamics of the Jovian ring. Bibliography included.

Burns, Joseph A., Mark R. Showalter, Jeffrey N. Cuzzi, and James B. Pollack. "Physical Processes in Jupiter's Ring: Clues to Its Origin by Jove!" *Icarus* 44 (1980): 339-360. Discussion of the physical forces involved in creating and maintaining Jupiter's tenuous ring system. Recounts early theories proposed before discovery of the shepherd moons and the mechanics of the shepherd system.

Burns, Joseph A., Mark R. Showalter, and E. G. Morfill. "The Ethereal Rings of Jupiter and Saturn." In *Planetary Rings*, edited by Richard Greenberg and Andre Brahic. Tucson: University of Arizona Press, 1984. Technical discussion of the properties of the ring around Jupiter and similar structures around Saturn. These constitute a ring class with extremely small, short-lived particles requiring continual replenishment and "herding" by small moons. The ring system discovered around Jupiter clarified some of the problems in understanding several phenomena in the rings of Saturn revealed by the Voyager 2 flyby in 1981.

Cuzzi, Jeffrey N., and Larry W. Esposito. "The Rings of Uranus." *Scientific American* 257 (July, 1987): 52-66. The best discussion in popular scientific literature of the dynamics of narrow, dusty rings such as those around Uranus and Jupiter. Particularly useful on the nature of shepherd moons.

Elliot, James, and Richard Kerr. *Rings: Discoveries from Galileo to Voyager.* 2d ed. Cambridge, Mass.: MIT Press, 1987. Highly readable and nontechnical account of the series of ring discoveries around Jupiter, Saturn, and Uranus from 1977 to 1986. Authors describe these as one of the major scientific events of the late twentieth century and discuss problems and theories derived from them. Useful account of technology involved in the discoveries and of the human element of excitement in the research teams. Extensive bibliography.

Esposito, Larry W. "The Changing Shape of Planetary Rings." *Astronomy* 15 (September, 1987): 6-15. Excellent summary, by the chair of the Voyager Rings Science Working Group, of a decade of discovery, showing that the trend of research supports theories that the Jovian rings—and probably those of Uranus as well as some Saturnian rings—may be much younger than their primary planets.

Hunt, Garry E., and Patrick Moore. *Jupiter.* New York: Rand McNally, 1981. Atlas and description of the Jovian system for lay readers which incorporates large amounts of Pioneer and Voyager data. Features excellent diagrams and composite photographs of the ring system.

Jewitt, D. C., and G. E. Danielson. "The Jovian Ring." *Journal of Geophysical Research* 86 (1981): 8691-8697. Early summary of apparent physical properties of the rings and questions raised of origin and maintenance under measured conditions.

Owen, Tobias, et al. "Jupiter's Rings." *Nature* 281 (1979): 442-446. The initial report on existence of the rings from the research team, summarizing properties and problems presented by the discovery.

Smith, Bradford A., et al. "The Jupiter System Through the Eyes of Voyager." *Science* 204 (1979): 927-950. Summary of observations of the Voyager spacecraft, which places the rings in context with other knowledge about the Jovian system.

Ronald W. Davis

Cross-References

Franklin and Burke Discover Radio Emissions from Jupiter (1955), p. 1492; Pio-

AN ANCIENT SANCTUARY IS DISCOVERED
IN EL JUYO CAVE, SPAIN

Categories of event: Archaeology and anthropology
Time: August, 1979
Locale: Five miles west of Santander, on the northern coast of Spain

Freeman and Echegaray excavated the remains of a fourteen-thousand-year-old sanctuary cave in Spain, providing the first material evidence of communal religious rituals

Principal personages:
> LESLIE G. FREEMAN (1935-   ), the president of the Institute for Prehistoric Investigations and a professor of anthropology at the University of Chicago
> JOAQUIN GONZALEZ ECHEGARAY, the director of the Altamira Museum and Research Center, Santander, Spain

Summary of Event

Leslie G. Freeman and Joaquin Gonzalez Echegaray have cooperated in a number of excavations in Spain. In 1968 to 1969, their discovery of the thirty-thousand-year-old mold of an Aurignacian burial in Cueva Morin, captured in fuzzy outline by the dripping water and fine sediments of the cave, made archaeological history.

In the 1970's, Freeman and Gonzalez Echegaray again teamed up to explore the subsistence practices of the last industrial phase of the Paleolithic, the fourteen-thousand-year-old Magdalenian culture. The abundance and variety of plant and animal life in Spain's north coastal Cantabrian region was noteworthy, and even when glaciation made other regions uninhabitable, the area supported a dense human population. The research team decided to begin their search in three Cantabrian caves: Altamira (the famous site whose breathtaking painted bisons were the first examples of Paleolithic cave art recognized by the scientific world), Rascano, and El Juyo.

Freeman and Gonzalez Echegaray dug in some of the unexplored sections of Altamira and the subalpine site of Rascano, but they recovered little from these already well-explored caves. In 1978, they shifted their activities to El Juyo, where earlier testing had revealed a deep deposit of Magdalenian materials rich in subsistence-related data. Sealed by a rockfall and subsequent stalagmite formation, El Juyo's living floors had remained undisturbed for at least fourteen thousand years.

The initial phase of the El Juyo excavations revealed the much hoped for evidence of economic activities. Gonzalez Echegaray and Freeman discovered that El Juyo's Magdalenian inhabitants were a nomadic group experienced in hunting red deer and gathering shellfish, whose seasonal migratory pattern may well have taken them some 16 kilometers distant to the site of Rascano to hunt ibex.

One mid-August morning, however, the excavators uncovered evidence of a very different kind: a stone slab, 2 meters by 1 meter by 15 centimeters and weighing nearly a ton. This huge piece of stalagmite had been carried at least 11 meters and placed in position near the back of the cave by the prehistoric inhabitants. Beneath it were several smaller stones, set upright to support the big stone, and a row of twenty-six unbroken bone spearpoints.

The archaeological team suspected the presence of an elaborate grave, but beneath the stone they instead discovered a complex earth feature. For two months, they meticulously scraped away layer after layer of this feature, guided at times by only the faintest changes in soil color and texture. What emerged was a picture of an elaborate group effort on the part of the Magdalenian builders of the place and possible evidence for the earliest religious sanctuary ever discovered in a Paleolithic deposit.

During the first phase of Magdalenian construction, the stone debris on the cave floor had been cleared to make a triangular space. A trench was then excavated in this area. The bottom of the trench was covered with clean sand, on which a lump of white clay, some sea shells, and deer ribs and feet were carefully arranged and covered with more clean sand and a layer of rose-colored ocher (red iron pigment mixed with animal fat) An antler tine, point down, was thrust into the ocher, and loose dirt from the surrounding cave floor was packed into the trench to a height level with the top of the antler. Finally, the whole area was covered with a layer of clean sand and more ocher.

In the next construction phase, a mound about 0.8 meter thick was built over the filled trench. This mound consisted of alternating levels that Freeman and Gonzalez Echegaray labeled "offering" layers (thin layers of burnt animal remains, vegetal materials, and ocher) and "fill" layers (thick levels composed of carefully arranged circular lots of earth). To create the fill layers, Magdalenian builders employed the same technique children use to make sand castles: They filled straight-sided, cylindrical containers (each 10 centimeters long and 10 centimeters in diameter) with packed clay. The clay cylinders were carefully turned out of the containers and arranged in groups of seven to form "rosettes," with one central lot surrounded by a circle of six others. The empty spaces between the cylinders were filled with clean clay, and the surface of each cylinder was covered with a layer of colored clay—red, yellow, green, or black—to form flowerlike patterns. When completed, the mound was topped with a few bone spearpoints, some lumps of red ocher, and sealed with a yellow clay layer ringed by twenty-six unbroken spearpoints. The hole was finally capped with the enormous limestone slab.

Gonzalez Echegaray and Freeman also uncovered a smaller structure oriented at a right angle to the first and joined to it by a tube lined with some black, greasy organic residue. The building pattern was similar, with the substitution of a lump of red ocher and a hollow filled with black earth (perhaps the remains of an object more perishable than the antler tine used in the other mound). Here, the clay cylinders were used to make a double line pattern instead of rosettes. Stones outlined the mouth of the joining tube, and the entire small mound was encircled with a stone and clay wall that completely separated the small structure from the large one. Finally, another big rectangular stone slab had been propped upright at an 80-degree angle to the horizontal and set parallel to the length of the smaller structure. This unusual complex of structures was accompanied by some small circular pits containing sea shells, bits of ocher, and eye bone needles. Oddly, many needles had been placed in a vertical position in the pits. Most intriguing of all, however, was the stone sculpture that stood guard over the cave complex.

The stone for the sculpture had been carried from outside the cave mouth and set into place at the back, facing the entrance and directly overlooking the small structure. The rock, which stands 36 centimeters high and 33 centimeters wide, was then shaped by a Magdalenian artist into the representation of a human head. Natural flaws in the stone which suggested a mouth and an eye were extended with chisel marks to outline lips and teeth. Further engraving added a second eye, a nose, a moustache, hairline, and beard. Close inspection revealed additional surprises. A natural fissure divides the stone vertically into two distinct halves, and the ancient artist had added some peculiar features to the left side: a deeply chiseled tear duct, a triangular engraving suggesting an animal nose, a muzzle with a single long fang, and what proved, upon laboratory examination, to be dots of paint added to suggest whisker roots. In short, the face in the cave was a dual entity, half human and half feline.

As a whole, the complex of El Juyo must have been the product of group activity, for the movement of the stone slabs and the elaborate fill construction would have required many people to accomplish. The complex is completely unlike an ordinary Paleolithic living cave and suggests a much more esoteric symbolic meaning. Freeman and Gonzalez Echegaray had discovered the remains of the earliest known evidence of complex religious belief and group ritual.

Impact of Event

Although the recovery of prehistoric artifacts suggesting religious meaning is not unusual, the richness of El Juyo's remains is unique. The very complexity of its construction makes interpretation difficult, and numerous explanations have been offered. Freeman and Gonzalez Echegaray point out that religious ritual (symbolic behavior employed to influence the supernatural) is largely arbitrary and obscure. It, therefore, cannot be explained in terms of its usefulness in solving everyday problems. Such behavior often involves a belief in spiritual entities and the cooperation of a number of people in activities aimed at influencing such beings. Ideally, an archaeologist searching for evidence of prehistoric ritual behavior hopes for a material representation of the spiritual entity and remains resulting from group participation in symbolic interactions with this entity.

El Juyo's excavators did not find the kind of cultural material—food debris and broken tools—usually associated with everyday living activities. Only selected animal bones, unbroken spearpoints and needles, and ocher were found in the sanctuary area. Furthermore, the complex structure of the mounds and the size of their covering slabs suggests the cooperative efforts of many people.

Most intriguing of all is the stone face, whose dual nature is apparent only when examined closely, from a peculiar angle, and with special lighting. Prehistoric depictions of animals and human beings are not unique to El Juyo, but the combining of both into one face is totally unknown elsewhere. Freeman and Gonzalez Echegaray believe this dualism may have been intended to give the statue a "public" meaning to all onlookers and an "occult, esoteric" meaning to a chosen few. They argue that it could represent an example of "complementary oppositions" (red/white, projecting antler tine/hollow depression, spearpoints in mounds/needles in pits, human/ feline, human/savage), symbolisms common among modern populations.

The complex might also have been utilized in shamanistic activities. Anthropologists speculate that shamans—primitive religious practitioners found throughout the world today—were possibly the earliest type of religious specialists. Interestingly, modern shamans are often believed to be able to change from human to animal (wolf, jaguar, and the like), thus giving rise to tales of werewolves and shape shifters. Perhaps the stone face was used in rituals in which a select few were initiated into the mysteries of the shaman.

A symbol system as complex as El Juyo provides much opportunity for speculation. It would seem that the complex probably was a sanctuary where some kind of group ritual was carried out. Others have argued that it may also represent the loss of social equality and the beginning of a status system in which some people had access to a restricted knowledge unavailable to others. The debates will undoubtedly continue for many years. The work of Freeman and Gonzalez Echegaray has provided the materials with which to attempt the reconstructions of humankind's most fascinating cultural manifestation—the beginnings of formalized religious rituals and group belief.

Bibliography

Freeman, Leslie G., and Joaquin Gonzalez Echegaray. "El Juyo: A 14,000-Year-Old Sanctuary from Northern Spain." *History of Religions* 21 (August, 1981): 1-19. This article provides a detailed description of the El Juyo excavations. It also contains an extensive discussion of the importance of the finds to the knowledge of the development of prehistoric religion. Its theoretical orientation reflects that of the excavators; yet, not all anthropologists might agree with their intepretations of the finds. Illustrated, with useful references.

Freeman, Leslie G., Richard G. Klein, and Joaquin Gonzalez Echegaray. "A Stone Age Sanctuary: Archaeological Finds in Spain Reveal the Existence of Religious Behavior 14,000 Years Ago." *Natural History* 92 (August, 1983): 46-53. This article includes a brief description of the El Juyo excavations, a very good general outline of the cultures of the Upper Paleolithic with particular focus on the Cantabrian region of Spain, and a brief discussion of the evidence for Paleolithic religion. Most useful for someone with limited knowlege of Paleolithic archaeology. Illustrated.

Pfeiffer, John E. *The Creative Explosion: An Inquiry into the Origins of Art and Religion*. New York: Harper & Row, 1982. This book is a good summary of evidence for Paleolithic art and religion. The importance of the El Juyo discovery is considered in the context of the development of Paleolithic belief systems and world view. Speculative, with conclusions not all anthropologists would accept, but well written and thought provoking.

_____. *The Emergence of Humankind*. 4th ed. New York: Harper & Row, 1981. This popular book on paleoanthropology includes an interesting chapter on the development of religion among Paleolithic peoples and puts the finds of El Juyo in the more general context of human evolution as a whole. A bit outdated in its descriptions of fossil finds, but useful for someone with limited knowledge of the field.

_____. "Inner Sanctum." *Science 82* 3 (January/February, 1982): 66-68. This is a brief, popular article on the El Juyo excavations. It makes the suggestion that the finds reflect the rise of institutionalized status and privilege. Intriguing speculations and lively descriptions. The artist's reconstruction of the El Juyo cave is very interesting.

Suzanne Knudson Engler

Cross-References

Evans Discovers the Minoan Civilization on Crete (1900), p. 67; The French Expedition at Susa Discovers the Hammurabi Code (1902), p. 169; Bingham Discovers an Inca City in the Peruvian Jungle (1911), p. 491; Carter Discovers the Tomb of Tutankhamen (1922), p. 730; Seventeen-Thousand-Year-Old Paintings Are Discovered in Lascaux Cave (1940), p. 1176; Scientists Date a *Homo sapiens* Fossil at Ninety-Two Thousand Years (1987), p. 2341.

BERG, GILBERT, AND SANGER DEVELOP
TECHNIQUES FOR GENETIC ENGINEERING

Categories of event: Chemistry and biology
Time: 1980
Locale: United States and England

Berg, Gilbert, and Sanger's development of techniques for sequencing and manipulating DNA initiated a new field of research

> *Principal personages:*
> PAUL BERG (1926-), the Willson professor of biochemistry at Stanford Medical Center who was a cowinner of the 1980 Nobel Prize in Chemistry
> WALTER GILBERT (1932-), the Carl M. Loeb professor at Harvard University who was a cowinner of the 1980 Nobel Prize in Chemistry
> FREDERICK SANGER (1918-), a professor of biochemistry at the University of Cambridge who was a cowinner of the 1980 Nobel Prize in Chemistry

Summary of Event

Deoxyribonucleic acid (DNA) is often called an information-containing molecule. Genes, which are made of DNA, reside in the nucleus of a cell and directly control the kinds of proteins a cell can make. These proteins give a cell its specific metabolism. Thus, to understand how a cell functions and in order to manipulate these genes, it is necessary to understand how the information is contained in the DNA and how it is used to produce specific kinds of proteins. One critical limitation in gaining such an understanding has been the enormous size and complexity of the DNA molecules found in cells. Fortunately for researchers, every DNA molecule is built up by linking, in end-to-end fashion, a large number of smaller molecules into two chains. Only four different subunits, called nucleotides, are used to construct any DNA. These nucleotides are called adenine (A), thymine (T), guanine (G), and cytosine (C). Every A in one chain is matched with a T in the other. Likewise, C's are always matched with G's. Thus, the enormous complexity of a human chromosome, which consists of a single long DNA molecule containing hundreds of millions of nucleotides, can be simplified by knowing that it is nothing more than two chains that consist of a linear sequence of A's, C's, T's, and G's.

In the early 1960's and early 1970's, techniques were developed that allowed the laborious determination of the nucleotide sequence of a very small piece of ribonucleic acid (RNA, found in cells and similar to DNA). Given that the DNA in a single human cell contains more than 5 billion nucleotides in forty-six long strands called chromosomes, however, it soon became clear that more powerful techniques for the analysis of DNA molecules would be needed if biologists were to make sense of this incredible complexity and gain the ability to manipulate genes directly.

Walter Gilbert and his colleague Allan Maxam had been using the techniques of bacterial genetics to try to understand how the bacterium *Escherichia coli* was able to turn on its ability to use milk sugar as a source of energy. Gilbert realized that he would have to know the DNA sequence, that is, the linear sequence of nucleotides, of the genes that controlled this process in order to understand fully how the bacterium was able to produce this switch in metabolism. In early 1975, Gilbert, Maxam, and a visiting scientist named Andrei Mirzabekov began to experiment with various chemical treatments that could cut DNA molecules after specific nucleotides as a way to sequence the molecule. DNA to be sequenced was first labeled by adding a radioactive molecule to one end of the DNA. The sample was then divided into four portions. Each portion was treated with a different set of chemical reagents such that in one tube, the DNA was cut after adenine, while in other tubes the DNA was cut after thymine, cytosine, or guanine, respectively. The various fragments produced in each reaction were separated by size using the technique of gel electrophoresis, a technique so precise at separating DNA molecules that it can be used to distinguish two molecules that differ in size by only one nucleotide out of three hundred.

The separated fragments of DNA were detected by autoradiography. In this technique, X-ray film was exposed by the radioactivity of the DNA fragments which produced dark bands on the developed film. Following autoradiography, an investigator can know which chemical treatment can be produced in a fragment of a given size. The first nucleotide in a DNA molecule produces a fragment one nucleotide long, which appears in only one of the four chemical treatments and thus is sensitive to that particular chemical cleavage. The second nucleotide in the sequence can be determined by observing which chemical treatment made a radioactive fragment two nucleotides long. This analysis continues until the full sequence is determined. Approximately 250-300 nucleotides can be determined from a single set of reactions, and very long stretches of DNA sequence can be obtained by linking the sequence of overlapping fragments.

Frederick Sanger had already made major contributions to an understanding of the mechanisms by which genes control the functions of a cell. He had won the 1958 Nobel Prize in Chemistry for his work leading to a practical method for determining the amino acid sequence of proteins. Sanger, like Gilbert, however, realized that a full understanding of the function of a gene would require a technique for the easy sequencing of DNA. While Gilbert had used a chemical cleavage technique for sequencing DNA, Sanger chose to use a biological technique. Sanger took advantage of a discovery by Joachim Messing and coworkers that DNA from any source could be linked end-to-end with the DNA of a small virus called M13. This technique was called "cloning." The viral DNA and the DNA cloned (linked) to it were easily prepared in large quantities by growing bacteria. The cloned DNA was split into four equal portions for sequencing. Sanger's technique was called the dideoxy method, as the cloned DNA was mixed with a mixture of radioactive nucleotides and special dideoxy nucleotides. When DNA polymerase, the enzyme responsible for assembling nucleotides into long strands, was added, new DNA was made. DNA polymerase always pairs A's in the template with T's in the newly made DNA and vice versa. Likewise, DNA polymerase always paired C's with G's. The newly made DNA was "complementary" to the cloned template DNA, and the nucleotide sequence of this new DNA could be used to determine the nucleotide sequence of the cloned DNA. If DNA polymerase used a special dideoxy nucleotide, instead of a normal nucleotide in lengthening the growing DNA, the lengthening reaction would stop, because no further nucleotides can be added to a DNA ending with a dideoxy residue. The length of a DNA fragment produced in a mixture containing dideoxy A then determined the position of T's in the cloned template DNA. Similarly, the length of fragments ending with dideoxy C would determine the position of G's in the template, and so on.

As with Gilbert's technique, the nucleotide sequence was actually determined by separating all the newly formed radioactive DNA fragments by gel electrophoresis. These DNA fragments were detected by autoradiography and the sequence determined by first finding the fragment, which was one nucleotide long, and seeing in which dideoxy reaction it was found. If the dideoxy A reaction produced a one-nucleotide-long DNA fragment, then the first nucleotide of the cloned DNA was a T. Completing the DNA sequence then involved observing which dideoxy reaction caused the production of a DNA fragment two nucleotides long, then three, and so on. Again, nearly three hundred bases could be sequenced at one time, but Sanger's technique proved to be both simpler and faster than Gilbert's.

While the techniques of Sanger and Gilbert were designed to describe the nucleotide sequence of any piece of DNA, other techniques for manipulating DNA were required for genetic engineering to be possible. Paul Berg was one of the founders of the technique of cloning genes from two different organisms. These hybrid DNA molecules could then be produced in sufficient amounts to sequence easily. Or, the genes could be mutated and put back into the cells from which they were obtained to determine the effects these specific changes had on gene function. Genes from one organism could easily be introduced into the cells of another by the same techniques, thus adding new functions to organisms. These techniques were called genetic engineering. Berg became interested in studying the DNA of a small virus, called SV40, in the late 1960's. Although it infected monkey cells normally to produce new viruses, SV40 could cause mouse cells to become cancerous. SV40 was therefore a tumor virus and had been studied extensively as a model for virally induced cancer.

In 1971 to 1972, Berg, along with David A. Jackson and Robert H. Symons, developed a technique for cloning genes called dA:dT tailing. In this technique, any two DNAs could be joined by adding adenine residues to one DNA, while thymine residues are added to the other DNA. Since adenine formed pairs with thymines, these two pieces of DNA linked together and a large, hybrid DNA was formed. In his first experiments, Berg cloned genes from the bacteria *E. coli* into SV40 virus. This hybrid DNA was then used to infect monkey cells and the added genes were a part of the SV40 virus chromosome. The infected cells were a rich source for the cloned DNA which could be isolated, sequenced, and studied. Later modifications of this basic technique allowed the cloned genes to function in the monkey cells. Manipulating genes from any organism by use of these techniques and their more recent modifications became routine.

Impact of Event

The information obtained from the techniques of cloning and DNA sequencing has revolutionized the understanding of how genes, cells, and organisms function. Incredibly complex processes such as the function of the nervous system and brain, the development of embryos, the functioning of the immune system, and the genetic contribution to cancer were now understandable at a molecular level. The science of genetic engineering (changing the genes of bacteria, plants, and/or animals directly rather than by classical breeding schemes) was now routine. As a result of the ability to clone and manipulate genes, bacteria can make human proteins such as insulin or growth hormone and plants can be produced that are resistant to herbicides or viral infections. These same techniques have allowed researchers to diagnose and cure genetic diseases in animals by replacing a defective gene in a cell with a normal one. These techniques will ultimately prove useful in the diagnosis and cure of common human genetic diseases such as cystic fibrosis and muscular dystrophy.

The DNA sequencing techniques of Gilbert and Sanger are still used routinely in laboratories across the world. The complete DNA sequence of the chromosomes of certain large viruses has been obtained and the DNA sequence of the bacterium, *E. coli*, the common brewer's yeast, and the fruit fly *Drosophila* should be completed by the mid-1990's. An ultimate goal of sequencing is to obtain the DNA sequence of the 5 billion nucleotides of a single human cell. This information will be a bonanza for the diagnosis and treatment of other genetic diseases and will provide the ultimate blueprint for the genetic capability of a human being.

Berg's technique for cloning DNA has been largely superseded by new methods that use restriction enzymes. These enzymes act as molecular scissors and can cut a DNA molecule producing "sticky ends" which allow any two cut DNA molecules to be relinked. The advantage of this technique over Berg's is that it is much faster and easier, but more important, the two DNA fragments can be readily separated from each other again by using the same restriction enzyme. The understanding of gene function allows these cloned DNAs to function as normal genes in the organisms in which they have been introduced. It is now possible to take essentially any gene from any organism and have it function in the cells of another organism, no matter how unrelated they may be.

Clearly, the explosion in information resulting from the exploitation of Berg, Gilbert, and Sanger's techniques is revolutionizing understanding of the biological world. These techniques have given biochemists tools for altering the genes of organisms in ways never before imagined.

Bibliography

Alberts, Bruce, et al. *Molecular Biology of the Cell.* 2d ed. New York: Garland, 1989. A complete and up-to-date exposition of the major areas of knowledge that directly resulted from recombinant DNA technology. College-level text.

Lewin, Benjamin. *Genes IV.* New York: Oxford University Press, 1990. A college-level text describing the techniques and discoveries in recombinant DNA technology and DNA sequencing. Includes a glossary and references.

Suzuki, David T., and Peter Knudtson. *Genethics.* Cambridge, Mass.: Harvard University Press, 1989. A discussion of the inevitable clash between the capability provided by the new DNA technology, genetic engineering, and human ethics.

Watson, James D., et al. *Molecular Biology of the Gene.* 4th ed. Menlo Park, Calif.: Benjamin/Cummings, 1987. An excellent, readable, college-level account of the discoveries that came from the basic science of recombinant DNA technology. Amply illustrated, with good references.

Watson, James D., and John Tooze. *The DNA Story.* San Francisco: W. H. Freeman, 1981. A highly readable historical account of the major discoveries in genetic engineering.

Joseph G. Pelliccia

Cross-References

Avery, Macleod, and McCarty Determine That DNA Carries Hereditary Information (1943), p. 1203; Watson and Crick Develop the Double-Helix Model for DNA (1951), p. 1406; Sanger Wins the Nobel Prize for the Discovery of the Structure of Insulin (1958), p. 1567; Nirenberg Invents an Experimental Technique That Cracks the Genetic Code (1961), p. 1687; Kornberg and Coworkers Synthesize Biologically Active DNA (1967), p. 1857; Cohen and Boyer Develop Recombinant DNA Technology (1973), p. 1987; The First Commercial Genetic Engineering Product, Humulin, Is Marketed by Eli Lilly (1982), p. 2221; A Genetically Engineered Vaccine for Hepatitis B Is Approved for Use (1986), p. 2326.

EVIDENCE IS FOUND OF
A WORLDWIDE CATASTROPHE AT
THE END OF THE CRETACEOUS PERIOD

Category of event: Earth science
Time: 1980
Locale: Gubbio, Italy

Luis and Walter Alvarez discovered evidence that the earth was hit by a large asteroid 65 million years ago

Principal personages:

LUIS W. ALVAREZ (1911-1988), a professor emeritus of physics at the University of California, Berkeley, and recipient of the 1968 Nobel Prize in Physics

WALTER ALVAREZ (1940-), a professor of geology at the University of California, Berkeley, who studied the Cretaceous clay layer at Gubbio, Italy

FRANK ASARO, a nuclear chemist who analyzed the chemistry of the clay layer

HELEN MICHEL, a nuclear chemist who helped to analyze the elements in the clay layer

EUGENE MERLE SHOEMAKER (1928-), a geologist with the U.S. Geological Survey who is the foremost authority on asteroid and meteorite impact

DAVID MALCOLM RAUP (1933-), a professor of geology at the University of Chicago and proponent of the idea of cyclic mass extinction

CHARLES OFFICER, a professor of geology at Dartmouth College who has challenged the asteroid impact hypothesis

CHARLES LUM DRAKE (1924-), a professor of geology who suggested that the late Cretaceous extinctions were caused by massive volcanic eruptions

Summary of Event

The geologic history of the earth is punctuated by a number of mass extinctions in which large numbers of different species disappeared at nearly the same time. These mass extinctions have been recognized for more than one hundred years and are used to mark the major time boundaries of the last 600 million years of geologic history. The most interesting and intensively studied mass extinction was the one that occurred at the end of the Cretaceous period about 65 million years ago. This event has held the fascination of the general public and generations of professional geologists because the great monsters of the past—the dinosaurs—made their last stand at this time. Why did they and many other animals completely disappear from the earth at this time? Many hypotheses have been suggested (supernova explosions, climate change, receding seas, disease, among many others), but a central unifying theory has eluded geologists. Until the 1980's, the general consensus among professional geologists was that the mass extinction was the result of significant, but gradual, environmental changes in the oceans and on the continents that were too severe for many animals to survive. In 1980, a hypothesis was put forth by four scientists at the University of California, Berkeley, that revolutionized and revitalized the study of the Cretaceous extinction and reshaped modern thinking about the basic nature of geologic change. Luis W. Alvarez, Walter Alvarez, Frank Asaro, and Helen Michel published a landmark paper in the June 6, 1980, issue of *Science* that presented data that suggested that the extinction was caused by the impact of a large asteroid.

As so often happens in science, this discovery resulted from research that initially was conducted on an entirely different problem. Walter Alvarez, a geologist with very wide interests, was working on the paleomagnetism of the Cretaceous and Tertiary rocks of north central Italy. He sought to correlate the magnetic changes preserved in the rocks near Gubbio, Italy, with similar changes that had been found in the ocean floors. Alvarez had chosen to study the marine limestones that are very well exposed in Battacione Gorge near Gubbio. These limestones contained fossils that clearly showed the terminal Cretaceous extinction and were overlain by a peculiar 1-2-centimeter-thick clay layer that was formed at the same time as the extinction. Alvarez was naturally curious about the clay layer and asked how long it took for the deposition of the clay layer. Geologists and geophysicists have developed techniques to determine the general ages of rocks and fossils, but none of these methods is precise enough to recognize the boundaries of an event that represents an interval of only a few thousands of years' duration that occurred 65 million years ago. Another new technique was needed, and Walter Alvarez approached his father, Luis, with the problem. Luis Alvarez was a brilliant scientist of enormous curiosity and drive who had won the 1968 Nobel Prize in Physics for his work on subatomic particles. Captivated by the problem, the elder Alvarez suggested to his son that they have the clay analyzed for meteoritic (extraterrestrial) debris. Luis Alvarez knew that the influx of meteoritic dust was nearly constant and he reasoned that in slowly accumulating sediments such as a marine clay, the amount of dust would be both measurable and a direct indication of the total time for deposition of the layer. Unfortunately, meteoritic dust is not easily distinguished from normal clays, and Luis Alvarez suggested that they analyze the clay chemically for elements that might indicate the meteoritic contribution to the clay. They enlisted the aid of two nuclear chemists at Berkeley, Asaro and Michel. In order to solve the problem, Asaro and Michel determined the abundances of twenty-eight elements within the clay and the surrounding limestones. The results showed that all but one of the elements had similar abundances among all the samples. The lone exception was iridium, a platinum-like element that was up to 160 times higher in the clay than in the limestones. This iridium anomaly meant that the dating technique could not be accurate (it was far too high for normal influx), but it was the key to the latest Cretaceous event.

Geologists believe that most of the iridium of the earth was concentrated in the core of the earth when the iron and other chemically similar elements (including iridium) migrated to the center of the earth early in its history. One result of this core formation is that iridium is exceedingly rare in the crust of the earth. To find the relatively high levels of iridium in the clay required an extraordinary explanation. Walter Alvarez initially thought the iridium was produced by the explosion of a nearby supernova. Detailed analysis of the elements and their isotopes, however, showed that this was highly improbable. The Berkeley scientists then proposed that the iridium was imparted directly to Earth with the impact of a large asteroid, approximately 10 kilometers in diameter.

Most asteroids orbit the sun in paths between Mars and Jupiter, but some (Apollo objects) do have Earth-crossing orbits. Eugene Merle Shoemaker—the foremost authority on asteroid and meteorite impacts—calculated that the rate of impacting had been fairly constant over the last 2 billion years and that the probabilities suggest that an impact of a 10-kilometer asteroid would occur on the average of every 100 million years. Shoemaker's calculations showed that an impact envisioned by the Berkeley group was possible (even highly likely), but the crowning piece of evidence for impact—the crater—was missing. The scientists had calculated that the crater would have been approximately 150-200 kilometers in diameter. Only three craters of this size were then known, and none was of the right age. The missing crater, however, was not a critical flaw in the hypothesis. There was a 70 percent chance that the asteroid hit in the oceans and a 50 percent chance that any crater formed in the Cretaceous seafloor would have been destroyed over the last 65 million years by tectonic processes. Still, a crater of the estimated size and the correct age would have sealed the case for asteroid impact.

Although the Berkeley group had reported only three sites of the iridium anomaly (at Italy, Denmark, and New Zealand), they were confident of their thesis and anticipated that other sites would soon be found. Later, many more sites were located, and their idea of impact was more generally accepted. The main emphasis of the original article, however, was much broader than simple impact. They suggested that the impact was the central cause for the late Cretaceous mass extinction and they proposed a general scenario to describe the event. The asteroid would have blasted both terrestrial and meteoritic debris into the atmosphere and the smaller dust particles would have quickly encircled the earth. The dust would later settle to the earth to form the clay layer. While in the atmosphere, the dust would have darkened the day skies for up to three years (later, it was shortened to a few months), thereby essentially shutting down all photosynthesis in the oceans and on the continents. The loss of photosynthesis would have rapidly collapsed the food chains of the world, and the catastrophic extinction of animals would have followed. Although many other effects of the impact were later recognized, the Berkeley group had made the critical point: The extinction was not gradual but catastrophic.

Impact of Event

In linking asteroid impact with mass extinction, the Berkeley group ignited a firestorm of controversy and the consequent explosion of creative research. Not only were geologists drawn into the controversy but also biologists, chemists, and astronomers became deeply involved in the questions surrounding the impact hypothesis.

The immediate response from the geologic community was to search for evidence to support or undermine the hypothesis. Several groups and laboratories around the world began to search for evidence for iridium anomalies in other Cretaceous-Tertiary boundary clays, and they were soon rewarded by finding more than fifty sites within three years. Studies of known impact craters had shown that the shock was caused by impact-generated fractures throughout many of the minerals (particularly quartz). These sets of fractures are characteristic only of impacts, and similarly "shocked" quartz minerals were found in the boundary clays. Also found in the clays were microtektites, small glasslike beads of material that had been melted by impact and cooled quickly when ejected into the atmosphere.

All these data seemed to indicate definitely that an impact had occurred, but a number of reputable scientists argued that the data were not conclusive and the known evidence was better explained by massive volcanism. Charles Officer and Charles Lum Drake argued that since iridium had been detected in the volcanic eruptions in Hawaii, the iridium-rich layer could have been produced by volcanism on a grand scale. They pointed out that the huge volcanic flows of the Deccan Traps of India are of the right age (about 65 million years old) to have created the anomaly. Nevertheless, shocked quartz has not been found in association with the volcanic flows.

Research into the processes of mass extinction was stimulated as well. David Malcolm Raup and John Sepkoski, both paleontologists at the University of Chicago, analyzed statistically the occurrences of extinction among families of organisms and recognized that mass extinctions appear to have occurred about every 26 million years. These cycles are of such long duration as to eliminate most, if not all, terrestrial processes and suggest extraterrestrial causes. Raup and others think that impacts could cause cyclic mass extinctions, and the report of iridium anomalies at boundaries other than at the latest Cretaceous strengthened their argument. If impacts are cyclic, then there must be some astronomical process that periodically spins asteroids (or comets) across the earth's orbit. Astronomers proposed several ideas, including the presence of a dim, distant companion star of the sun. The search for this star, called Nemesis, has begun at Berkeley, but no good candidates have yet been identified.

Evolutionary theory has also been profoundly affected by the impact hypothesis. Most modern models of evolution state that species gradually change through time by genetic mutations acted upon by natural selection. Extinction caused by asteroid impacts adds another dimension to this view. Impacts cannot be anticipated, they dramatically affect the environment for very short periods of time, and organisms cannot adapt quickly enough to survive. Impacts can, therefore, change the development of life by exterminating otherwise dominant life-forms (dinosaurs, for example) and open up ecological niches for other organisms (such as mammals in the Tertiary).

The basic concepts of geologic change have been altered by the impact hypothesis. Generations of geologists have been strong adherents to uniformitarianism, the idea that geologic change is caused only by processes still at work today and that the changes are generally gradual. Impacts are not gradual, and they are now seen as major factors in the history of Earth. Catastrophic events, therefore, are now thought of as possible and plausible causes of many of the major events of the geologic past.

Bibliography

Alvarez, Luis, Walter Alvarez, Frank Asaro, and Helen V. Michel. "Extraterrestrial Cause for the Cretaceous-Tertiary Extinction." *Science* 208 (June 6, 1980): 1095-1108. The landmark paper that proposed the asteroid impact hypothesis based upon the iridium anomaly found at Gubbio, Italy. A technical report that is useful for the interested student of science or the history of science.
Goldsmith, Donald. *Nemesis: The Death Star and Other Theories of Mass Extinction*. New York: Berkley Publishing Group, 1985. An exceptionally well-written account of the asteroid impact theory and Nemesis written by one of the major investigators in the search for Nemesis.
Grieve, Richard A. F. "Impact Cratering on the Earth." *Scientific American* 262 (April, 1990): 66-73. An excellent article on the number, nature, and mechanics of impacts on Earth. This well-illustrated article gives geologic evidence of impact at known craters and how it applies to the asteroid impact theory.
Hsü, Kenneth J. *The Great Dying*. San Diego: Harcourt Brace Jovanovich, 1986. An excellent, but somewhat technical, work by one of the most influential earth scientists today. He gives a complete summary of the impact theory and the resulting extinction. Hsü also relates many interesting insights into the lives and thinking of the people researching the Cretaceous-Tertiary extinction. Highly recommended.
Raup, David M. *The Nemesis Affair: A Story of the Death of Dinosaurs and the Ways of Science*. New York: W. W. Norton, 1986. A well-written account of the asteroid impact theory by the major advocate of periodic extinction. This wonderful book not only gives a clear, historical account of the development of the impact theory but also gives the layperson insights into the process of science itself.
Wilford, John Noble. *The Riddle of the Dinosaur*. New York: Alfred A. Knopf, 1986. A good popular summary of the modern interpretations of dinosaurs. Several chapters address the extinction of these animals and treat the asteroid impact theory clearly and fairly.

Jay R. Yett

Cross-References

Wegener Proposes the Theory of Continental Drift (1912), p. 522; Bjerknes Discovers Fronts in Atmospheric Circulation (1919), p. 675; Hess Concludes the Debate on Continental Drift (1960), p. 1650.

THE INFLATIONARY THEORY SOLVES LONG-STANDING PROBLEMS WITH THE BIG BANG

Categories of event: Physics and astronomy
Time: 1980
Locale: Stanford University, Stanford, California

Guth proposed a new theory of cosmology that says that the current slow linear expansion of the universe was rapid and exponential for a very brief period near the beginning of time

Principal personages:

ALAN GUTH (1947-   ), a physicist who, with Henry Tye, developed the first theory of inflation

HENRY TYE (1946-   ), a physicist who persuaded Guth that cosmological considerations were relevant to particle physics; codeveloper of the inflation idea

A. D. LINDE (1926-   ), a physicist who improved Guth's theories by making the "bubble universe" of inflation large enough to agree with present observations

PAUL J. STEINHARDT (1952-   ), a physicist who contributed to the improvements necessary to make inflation a viable theory

ANDREAS ALBRECHT (1927-   ), a physicist who worked with Steinhardt on the developing inflationary theory

Summary of Event

The inflationary cosmological model was proposed by Alan Guth in 1980 in an attempt to remove some serious difficulties from the standard big bang model of cosmology as well as to explain some curious phenomena associated with particle physics. (Cosmology is the large-scale study of the universe, including its origin.)

The big bang is the currently accepted cosmological theory. It explains many of the important observed features of the universe, such as the universal 3-degree background radiation discovered by Arno A. Penzias and Robert W. Wilson in 1964, the recession of the galaxies discovered by Edwin Powell Hubble in 1927, and the relative distribution of the lighter elements.

According to the big bang theory of cosmology, the present universe emerged out of an incredibly small region of space—essentially a point—that expanded in a fiery explosion some 10 to 20 billion years ago. Very little is known about the actual moment of creation itself, but other theories have had some success in describing the evolution of the universe after 10^{-43} seconds, which is called the Planck time. Albert Einstein's theory of gravity can describe the universe back to that instant, but the theory fails beyond that point because the density and small size of the universe are such as to require a theory of quantum gravity, which has not yet been developed.

In the original primordial universe, the temperature was so high that there was no distinction between matter and energy—it was too hot for material particles to form. As the universe cooled, elementary particles were able to crystallize out of the high-energy sea; further cooling allowed these particles to form atoms, then molecules, then clouds, then stars and planets, and finally people. Emerging out of a region of space smaller than a proton with an initial temperature in excess of 10 billion Kelvins, the universe has expanded to one that is more than 10 billion light-years in diameter with a temperature of only 3 Kelvins.

There are three serious problems with the standard big bang theory that inflation attempts to solve: the horizon problem, the smoothness problem, and the flatness problem. The horizon problem exists because the standard model cannot explain why the universe is so uniform over such great distances. If the different parts of the universe are so separated that they cannot communicate with one another, even at the speed of light, then why do they look exactly the same? Just as animals on widely separated continents evolve differently, so different parts of the universe should evolve differently. What was the mechanism that kept the various parts of the universe evolving along exactly the same path? Normally, homogeneity is maintained by "mixing," but the universe is too large to mix. The smoothness problem has to do with the formation of galaxies, which represent inhomogeneities in an otherwise very uniform universe. These deviations from perfect uniformity must originate in tiny inhomogeneities in the very early universe. The problem lies in the fact that even tiny deviations from perfect smoothness in the early universe will result in enormous nonuniformity 10 billion years later. The universe today is almost perfectly uniform, except for the occasional galaxy, so the early universe must have been extraordinarily uniform. Yet, it is not known how it could have been so smooth. The flatness problem relates to the energy density or, equivalently, the mass density of the universe. If the energy density is only slightly higher than the "critical value," the big bang will reverse eventually and be followed by a big contraction. If the density is slightly lower than this critical value, the current expansion will continue forever. If the density is exactly equal to the critical value, corresponding to a "flat" universe, the expansion will continue, but at a gradually slowing rate. Measured values of the present energy density yield results that are very close to the critical value. The flatness problem exists because any initial deviation from perfect flatness in the early universe is magnified tremendously by the subsequent expansion of the universe. To account for the current measured flatness of the universe, the early universe must have been flat to within one part in a thousand trillion (10^{15}). This is an unbelievable constraint on the allowed values for the initial flatness. It still needs to be determined where it came from.

The horizon, smoothness, and flatness problems are dealt with in the standard big bang model by simply assuming the specific initial conditions necessary to account for the present observed features of the universe. The fact that these conditions are assumed without any theoretical basis makes them arbitrary, which is considered to be a weakness in a physical theory. A good physical theory explains things; it does not assume them.

Guth's inflationary cosmology solves all three of these problems by modifying the big bang slightly. Guth suggested that the very early universe, which he called a "bubble," underwent an initial very rapid exponential expansion that was much faster than the linear expansion of the standard big bang model. This bubble inflated to become the present "observable" universe, which is thus embedded in a much larger "unobservable" universe. This exponential expansion increased the size of the universe by a factor of 10^{50}, from smaller than a proton to larger than a softball. Inflation began at 10^{-35} seconds after the big bang and ended at 10^{-33} seconds.

According to the inflationary model, the observable universe grew out of a much smaller region of space—the inflationary bubble—than was formerly thought. The material in this small, primordial bubble was, thus, very densely packed and able to interact mutually for a longer period, homogenizing itself so that when it began to separate, it would all be evolving in the same way, thus solving the horizon problem. When the inflation occurred, the universe expanded dramatically, maintaining its newly established homogeneity, thus solving the smoothness problem. The rapid period of expansion inflated the expanding "bubble universe" far more rapidly than predicted by the standard model. This means that the observable universe is only a small part of the entire universe; any portion of the larger universe beyond a distance of 10 to 20 billion light-years is invisible to those on Earth because the universe is not old enough for a signal to have traveled from them. The flatness of the universe increased with the size of the universe, just as the flatness of a square drawn on a balloon increases as the balloon is blown up. By making the entire universe much larger, inflation explains successfully why the visible portion is so flat.

Furthermore, the inflationary model is able to explain the approximate distribution and size of the galactic clusters that populate the universe. In the original primordial bubble, the homogeneity would be limited by the laws of quantum mechanics, which state that there will be small fluctuations even in a perfectly uniform region of space. These small fluctuations were magnified dramatically by inflation until they became the large structures that are seen as galaxies.

There were a few problems with Guth's original formulation. The most serious was the length of the inflationary epoch, which was too short to produce a universe of adequate size. In 1984, A. D. Linde, Paul J. Steinhardt, and Andreas Albrecht solved this problem and improved the inflationary theory until it is now widely accepted as the most likely explanation for the observed features of the present universe.

Impact of Event

The theory of the inflationary universe falls into the category of what is called speculative science. It is frequently impossible to verify such theories, and they sometimes attract followers on the basis of their symmetry, or simplicity, or what physicists call "beauty."

As of 1990, it is impossible to develop any theory that describes accurately the evolution of the very early universe because a theory of quantum gravity does not yet exist. Albert Einstein's theory of general relativity describes the large-scale behavior of the gravitational bodies in the universe. The theory of quantum mechanics describes the small-scale behavior of the atomic and subatomic world. To describe the very early universe requires both theories; there is so much mass involved and the density is so high that gravitational effects are significant, and the size of the early universe is so small that it requires quantum theory for its explanation. At present, quantum theory and gravity are not compatible, so a complete description of the early universe requires additional progress in theoretical physics.

In the meantime, the inflation theory has made the current theory of the big bang much more attractive by explaining many things that the big bang simply assumed. The inflation theory has also helped to explain the actual origin of the universe itself, one of the deepest mysteries in science and actually a question highly relevant to philosophy and religion. In a remarkable application of quantum theory, Guth has calculated that the tiny fluctuations present even in a vacuum—an "empty" region of space—might be adequate to initiate the process of inflation. According to quantum theory, which is very well established, a vacuum is not completely inactive. There must be a small energy field present that is fluctuating about zero. There is a probability that one of these fluctuations could erupt and produce a new universe. The laws of physics permit this because a universe such as Earth's, has almost no net energy in it. The positive energy associated with all the matter—Einstein's $E = mc^2$—is balanced by the negative energy associated with the gravitational force. If the universe is indeed flat, which both measurements and inflationary theory seem to suggest, then it has no net energy, indicating that it could have erupted from an "empty" vacuum, without violating the law of conservation of energy.

The inflationary theory has also shed some light on theories of particle physics. According to particle physics, the electromagnetic force, the strong nuclear force, and the weak nuclear force were all united under one of the grand unified theories (GUTs) in the early universe. This union was possible because of the high temperatures present at the time. As the universe cooled, the individual forces were "frozen out" of the energetic background and separated from one another in the way they are currently seen. (This freezing-out process is analogous to the kind of phase change that takes place when a liquid freezes or a gas condenses.) The inflationary expansion of the early universe was caused by the release of energy that is always associated with a cooling phase change.

In shedding light on fundamental questions of both cosmology and particle physics, the inflationary theory has made a significant contribution to science. Further progress in theoretical physics will have to occur, however, before this theory can be embraced with confidence.

Bibliography

Carrigan, Richard A., and W. Peter Trower, eds. *Particle Physics in the Cosmos.* New York: W. H. Freeman, 1989. This book is a collection of articles reprinted from *Scientific American*, some of which discuss the important role that inflationary theories play in theories of cosmology. In particular "The Inflationary Universe" by Alan Guth and Paul J. Steinhardt (May, 1984), "The Structure of the Early Universe" by John Barrow and Joseph Silk (April, 1980), and "The Large Scale Structure of the Universe" by Joseph Silk, Alexander Szalay, and Yakov Zel'dovich (October, 1983) are excellent. This book is actually better than the original articles because some of the authors have added postscripts updating some of the material.

Gribbin, John. *The Omega Point: The Search for the Missing Mass and the Ultimate Fate of the Universe.* New York: Bantam Books, 1988. This book, by a popular science writer with professional training in cosmology, is one of the most accessible to the nonscientist.

Harrison, Edward R. *Cosmology: The Science of the Universe.* Cambridge, England: Cambridge University Press, 1981. Discusses the big bang and continuous creation cosmologies. A good overview. For a wide audience. Bibliography.

Pagels, Heinz R. *Perfect Symmetry: The Search for the Beginning of Time.* New York: Simon & Schuster, 1985. This book is written by another top scientist in the field of cosmology. It contains much historical material as well as a discussion of the current status of cosmological theories. There are several sections that discuss the various aspects of the inflationary theory.

Silk, Joseph. *The Big Bang.* Rev. ed. New York: W. H. Freeman, 1989. This excellent book contains a complete discussion of the big bang, including its historical development. The author is a prominent scientist working in the field of cosmology. The book contains an excellent introduction to inflationary theories, with an unusual diagram showing the relative size of the inflationary universe and the visible universe.

Karl Giberson

Cross-References

Schwarzschild Develops a Solution to the Equations of General Relativity (1916), p. 630; Lemaître Proposes the Big Bang Theory (1927), p. 825; Gamow and Associates Develop the Big Bang Theory (1948), p. 1309; Penzias and Wilson Discover Cosmic Microwave Background Radiation (1963), p. 1762; Georgi and Glashow Develop the First Grand Unified Theory (1974), p. 2014; Kibble Proposes the Theory of Cosmic Strings (1976), p. 2047.

GRIESS CONSTRUCTS "THE MONSTER," THE LAST SPORADIC GROUP

Category of event: Mathematics
Time: January 14, 1980
Locale: Institute for Advanced Study, Princeton, New Jersey

The classification of all finite simple groups, started in 1830, was completed when Griess constructed "the Monster," a group with 8×10^{53} elements, by hand

Principal personages:

JOHN GRIGGS THOMSPON (1932-), an American mathematician and winner of the Fields Medal in 1970 for his contributions to simple group theory

RICHARD BRAUER (1901-1977), a German-born mathematician and a pioneer of fundamental approaches used in simple group theory

CLAUDE CHEVALLEY (1909-), a French mathematician who clarified the connection between finite and infinite simple groups

DANIEL GORENSTEIN (1923-), an American mathematician who played an important role in proving the classification theorem and in its subsequent revision

ROBERT LOUIS GRIESS, JR. (1945-), an American mathematician who constructed the largest sporadic simple group, "the Monster"

Summary of Event

There are few more general and powerful concepts in mathematics than that of a group. Groups describe the properties of objects as varied as the real numbers and the integers, card shuffling, codes, crystals, elementary particles, and the Rubik's cube puzzle—indeed, any object with some symmetry. The fundamental building blocks of finite groups are called the simple finite groups, much as atoms are the basic units used in constructing molecules. So it was with some excitement when mathematicians completed the proof of the classification of all the finite simple groups: The theorem states that any finite simple group either belongs to one of eighteen infinite families or is one of twenty-six exceptions, called "sporadic" groups. The excitement reached a climax when Robert Louis Griess, Jr., constructed the final finite simple group, "the Monster," so named for its size.

In proving the classification, more than ten thousand journal pages have been filled by more than one hundred mathematicians, mainly from the United States, England, and Germany, but also from Japan, Australia, and Canada. The majority of the proof was published in some five hundred articles between the late 1940's and the early 1980's, though efforts to solve the problem span the entire 150-year history of group theory. Simple groups having been recognized as fundamental from the start, the problem of classifying them gradually evolved into a field complete with its own specialized techniques.

Group theory was originally invented by the French mathematical prodigy Évariste Galois in the 1830's to answer an age-old question about the solutions to polynomial equations. He defined a group to be a set of elements that satisfy four properties based on ordinary arithmetic: closure, identity, inverse, and associativity. The group of numbers under addition is an infinite group; an example of a finite group is the set of six ways to rotate or flip an equilateral triangle onto itself. The inverse of one of these moves would be the move that reversed the original motion, and the identity is doing nothing at all.

Any finite group can be decomposed uniquely or factored into its component simple groups, just as any whole number can be uniquely decomposed into its prime factors. One can get an intuitive feel for what it means to factor a group by considering the above symmetry group of the equilateral triangle. The motions on the triangle can be naturally considered as a combination of flips and rotations. In this way, the symmetry group is factored into the two-member flipping group and the three-member rotation group. Because these factor groups cannot be factored any further, they are called simple. As it turns out, any group with a prime number of elements must be simple.

Though Galois had laid the foundation, it was not until the 1860's that his work was understood and built upon by other mathematicians. In 1870, Camille Jordan established the existence of five additional infinite families of simple groups. One family consists of what are called the alternating groups, and the remaining four families are known as the classical simple groups. In 1889, Otto Hölder's proof that any finite group has a unique decomposition spurred a systematic search for finite simple groups. By 1900, the list of simple groups with less than two thousand elements was completed. In addition, there were five "Mathieu" groups, discovered by Émile Mathieu in 1861, that apparently did not fit into any of the established families and so were the first sporadic groups.

Meanwhile, between 1888 and 1894, Sophus Lie, Elie Cartan, and others had succeeded in classifying all of a certain type of infinite group called Lie groups. It became apparent that there was a connection between some of the simple Lie groups and the classical finite simple groups. In addition to those corresponding to the classical simple groups, there were five "exceptional Lie groups." In 1901 and 1905, the American mathematician Leonard E. Dickson was able to derive another infinite family of finite simple groups, the finite analog of one of the exceptional Lie groups. It was not until 1955 that the French mathematician Claude Chevalley constructed another finite group by clarifying the relationship between the Lie groups and their finite counterparts. Over the following six years, other mathematicians extended Chevalley's analysis and uncovered the remainder of the sixteen families of "Lie type," including ones with no strict analogy among the Lie groups. They are the only infinite families besides those of prime order and the family of alternating groups.

The approach toward a proof of a complete classification proceeded on two flanks. First, there were attempts at restricted classification theorems, where all simple groups with a given property were enumerated. By making an exhaustive list of properties, the full theorem could be completed. Second, there were attempts to construct new sporadic groups from scratch. Though largely unfruitful, these efforts produced a few successes that provided insight for the categorical approach. An example of a restricted classification theorem is the celebrated "odd order theorem," which states that all nonprime simple groups have an even number of elements. Originally conjectured by the English mathematician William Burnside in 1901, this theorem was not proven until 1963, by Walter Feit and John Griggs Thompson. Their proof is significant for the tools developed as well as the result itself and so provided a great impetus for the classification. In particular, groups of even order must have an element, called an involution, which when combined with itself forms the identity. That nonprime simple groups must have an involution meant that the pioneering work of the German mathematician Richard Brauer on involutions would be central in characterizing simple groups.

In 1966, Zvonimir Janko found the first sporadic group in more than a century as a counterexample while trying to prove a certain restricted classification theorem. Janko's example provided a theme for constructing other sporadic groups as exceptions to restricted classsification theorems. With the recent classification of the infinite families, a new simple group came as a surprise to the mathematical community. As more sporadics were discovered—at a rate of one or two a year—it seemed possible that an effort at a full classification would be futile if there were an infinite number of sporadic groups. By 1972, however, a complete classification seemed conceivable, and the American mathematician Daniel Gorenstein presented a sixteen-step program in a series of lectures at the University of Chicago. The projected thirty-year time period to accomplish the program was shortened to less than ten years by the work of Michael Aschbacher, among many others. By the end of the decade, it was clear that no significant theoretical obstacles remained.

On January 14, 1980, Griess announced that he had succeeded in constructing "the Monster," the largest and final sporadic group. Calculated by hand, "the Monster" has 808,017,424,794,512,875,886,459,904,961,710,757,005,754,368,000,000,000 elements and contains most of the other sporadic groups embedded within it. A prior proposal for constructing "the Monster" by computer was estimated at a cost of $3 million for one year of computing power. The clinching stroke in the proof of the classification came, however, in January of 1981, when Simon Norton demonstrated the uniqueness of "the Monster" group.

Impact of Event

The classification was the culmination of the effort of a large part of modern finite group theory and represented a powerful new tool. After completion there was some feeling that the field had "worked itself out of a job." Yet, as consequences of the theorem continued to develop, mathematicians realized that there was still much work ahead. Aside from implications for fields external to finite group theory, a fair portion of internal work remained because of the extraordinary length and complexity of the theorem. An organized effort toward simplification and "revisionism" was started among the team.

Applications outside the theory of finite groups include several theorems in related branches of mathematics, such as number theory, model theory, the theory of algorithms, and coding theory. The Mathieu groups were known previously to be related intimately to the theory of codes, and some of the other sporadic groups were first derived from their connection with self-correcting codes and the packing of spheres in very high dimensions. In general, questions in related fields that can be reduced to questions concerning all finite simple groups can be settled simply by checking through them, case by case. In comparison, the techniques developed were of limited use in other fields, which is perhaps surprising considering the fundamental nature of the theorem. Because of the structural similarity between finite simple groups and other mathematical structures, it is hoped that the general approach will provide guidance to the classification in these other cases. In addition, some of the intermediate results in proving the classification have been of external interest. The highly symmetrical "Monster" group, for example, appears unexpectedly in the coefficients of certain functions, though the precise connection has yet to be clearly elucidated.

A revision was necessary for several reasons. Questions arose concerning human verifiability of a proof that is more than ten thousand pages long and depended upon long machine calculations. Some doubted that the proof was done correctly or optimally because it was so lengthy. It was surprising that so complex a taxonomy of simple groups could grow out of so simple a question. Careful examination of the proof, however, suggests that the complexity is inherent in the problem. As time passes and as no new simple groups are discovered, confidence in the validity of the proof has grown. Moreover, the theorem evolved historically as a patchwork of many individual, sometimes redundant, sections. The program of "revisionism" involved culling together the essential parts of roughly a hundred articles and either tidying them up or rederiving them. Alternative methods of proof provided a check, especially if they depended less or not at all on prior results from the classification theorem. Even before the full theorem was complete, in 1970, Helmut Bender had introduced methods for revision, some of which were used in the later stages of the proof. The ongoing revision is completed in stages, and the increased unity and succinctness of approach is hoped to compress the proof to roughly three thousand pages. This "second generation proof" leaves a more structurally sound legacy for future mathematicians to benefit from the essential insights of the classification.

Bibliography

Cartwright, Mark. "Ten Thousand Pages to Prove Simplicity: Two Hundred Years of Research Was Needed to Prove a Fundamental Theorem in Group Theory. The Miracle Is That It Was Proved at All." *New Scientist* 105 (May 30, 1985): 26-30. This brief popular account traces the historical development of the classification as the triumph of the combined efforts of many key personages. Mathematical concepts are kept to a minimum.

Gallian, Joseph A. "Finite Simple Groups." In *Contemporary Abstract Algebra*. 2d ed. Lexington, Mass.: D. C. Heath, 1990. Prefaced by a few pages of historical background, this chapter in a textbook for upperclass undergraduates surveys finite simple groups, using classification according to size, or the "range problem," as a starting point. Includes exercises, programming exercises, and a diverse introductory bibliography.

_____. "The Search for Finite Simple Groups: The Eighty Year Quest for the Building Blocks of Group Theory Reflects Sporadic Growth Spurts Whenever New Basic Techniques Were Discovered." *Mathematics Magazine* 49 (September, 1976): 163-180. Gallian has researched the history of group theory and has written several articles intended for college students. In this article, the history of finite simple groups and the range problem is chronicled as would be appropriate for a student familiar with basic group theoretic concepts. Includes bibliography.

Gardner, Martin. "The Capture of the Monster: A Mathematical Group with a Ridiculous Number of Elements." *Scientific American* 242 (June, 1980): 20-32. Complex text, which contains some detail concerning "the Monster" and other sporadic groups.

Gorenstein, Daniel. *The Classification of Finite Simple Groups*. New York: Plenum Press, 1983. One of the driving forces behind the classification, Gorenstein has taken a central role in its documentation and revision. Here lies the lengthy record for the nonspecialist professional mathematician. Some of the philosophy involved in the revision is included here and in its sequel, *Classifying the Simple Groups*. A personal account can be found also in *A Century of American Mathematics*.

_____. "The Enormous Theorem: The Classification of the Finite Simple Groups Is Unprecedented in the History of Mathematics, for Its Proof Is 15,000 Pages Long. The Exotic Solution Has Stimulated Interest Far Beyond the Field." *Scientific American* 253 (June, 1985): 104-115. Writing for the layperson, Gorenstein succeeds in explaining the nature of groups and simple groups in intuitive terms using plain examples. He also manages to allow glimpses into some of the processes that went into the classification and why it was such an important and complex endeavor. Includes a short, more technical, bibliography.

Hammond, Allen L. "Sporadic Groups: Exceptions, or Part of a Pattern?" *Science* 181 (July 13, 1973): 146-148. A brief exposition for the layperson about sporadic groups, their connection with error-correcting codes, and some of their uses.

Hurley, James F., and Arunas Rudvalis. "Finite Simple Groups." *The American Mathematical Monthly* 84 (November, 1977): 693-714. A mathematical sketch of the development of the classification theorem, intended for college mathematics students. Includes an extensive mathematical bibliography.

David Wu

Cross-References

The Study of Mathematical Fields by Steinitz Inaugurates Modern Abstract Algebra (1909), p. 438; Noether Publishes the Theory of Ideals in Rings (1921), p. 716.

VON KLITZING DISCOVERS
THE QUANTIZED HALL EFFECT

Category of event: Physics
Time: February 5, 1980
Locale: High Field Magnetic Laboratory of the Max Planck Institute, Grenoble, France

Von Klitzing discovered the quantized Hall effect, which provides the most accurate measurements of certain fundamental constants of nature

Principal personages:
> EDWIN HERBERT HALL (1855-1938), an American physicist who discovered the Hall effect in 1879
> KLAUS VON KLITZING (1943-   ), a German physicist and winner of the 1985 Nobel Prize in Physics
> ROBERT BETTS LAUGHLIN (1950-   ), an American physicist who, among other scientists, provided explanations for the integral and fractional quantized Hall effect

Summary of Event

With the explosive growth in microelectronics technology, semiconductors have been among the most well-studied of materials. The understanding of the basic microscopic phenomena in these crystals has advanced to the point where semiconductors are manufactured and manipulated on a molecular level. The properties of silicon, for example, are perhaps the best understood among solids. The measurement of these properties often reflects the complicated nature of the microscopic structure of these materials as well as the specifications of the particular sample.

Given such detailed knowledge, the discovery of a novel fundamental phenomenon in semiconductors was startling. The measurement of the quantized Hall effect depends only upon fundamental constants of nature and not on sample irregularities or impurities. The properties of a solid are especially dependent on a host of internal and environmental parameters, such as the geometry, temperature, purity of the sample, and the history of its preparation. It was surprising to find a manifestation of quantum mechanical behavior in macroscopic samples that is so distinct and precise.

The Hall effect is a class of phenomena that occur when a material carrying current is subject to a magnetic field perpendicular to the direction of the current. As first observed in 1879 by the American physicist Edwin Herbert Hall, an electric voltage results in a direction perpendicular to both the current and the magnetic field. The ratio of this voltage to the current is the Hall resistance. In comparison, the normal electrical resistance is the ratio of the voltage in the direction of the current to the current. In the "classical" Hall effect, which occurs for a wide range of temperatures, the Hall resistance increases linearly with the strength of the magnetic field. The constant of proportionality depends on the individual characteristics of the sample and is a measure of the density of electrons that carry current. The classical Hall effect is well described by what is called an "electron gas," where the motion of the conducting electrons of the solid is considered to be independent from each other and can freely wander within the crystal matrix. The case of the quantized Hall effect, however, requires a two-dimensional electron gas, where the electrons are confined to a plane of conduction. This is realized at the interface between a semiconductor and an insulator, where an electric field draws the semiconductor's electrons toward the two-dimensional interface. Temperatures of a few Kelvins are needed to keep the electrons stuck to the surface.

It was demonstrated in 1966 that electrons confined to motion in such a plane, typically 10 nanometers thick, result in new quantum mechanical effects. The motion of the electrons is quantized, that is, their energies assume one of several evenly spaced discrete values. The number of electrons that can assume a particular energy, called a "Landau level," is proportional to the strength of the magnetic field. Increasing the field strength also increases the spacing between the Landau levels. At low temperatures, the electrons try to minimize their energy, and therefore, the Landau levels are filled sequentially by energy. The highest filled energy is called the "Fermi level." Raising the magnetic field effectively lowers the Fermi level, since more electrons can then be accommodated per Landau level. Alternatively, the Fermi level can be altered simply by changing the number of electrons.

Certain general aspects of the quantized Hall effect were, in fact, predicted in 1975 (five years prior to Klaus von Klitzing's experiments) by the Japanese theorists Tsuneya Ando, Yukio Matsumoto, and Yasutada Uemura of the University of Tokyo. They recognized that when every Landau level is either completely filled or completely empty, the electrical resistance should vanish. Under these conditions of "integral filling," the Hall resistance would be a certain ratio of fundamental constants divided by the number of filled levels and would be independent of the geometry. Unfortunately, the theory was only approximate and would not have been considered reliable for the actual experimental situation. The crucially important aspects of the extreme precision and the robustness of the effect under varying conditions were unforeseen. Also, in experiments as early as 1977 performed by von Klitzing's co-worker Thomas Englert, slight plateaus in the Hall resistance were visible in some samples. These anomalous plateaus were considered unexplained by any published theories.

Von Klitzing's research through the 1970's included studies of silicon devices in high magnetic fields and under conditions of mechanical stress. In 1980, von Klitzing decided to investigate the anomalies in the Hall resistance. The high-quality samples he used were "metal oxide semiconductor field-effect transistors," or MOSFETs, constructed by his collaborators Gerhardt Dorda of the Siemens Research Laboratory in Munich and Michael Pepper of the University of Cambridge. A layer of insulating oxide is sandwiched between a metal strip, which provides a voltage potential, and the silicon, which supports the two-dimensional electron gas at its surface. The samples were typically about 0.4 millimeter long and .05 millimeter wide. By increasing the voltage on the metal electrode, more electrons could be drawn to the surface of the semiconductor, thereby raising the Fermi level.

Von Klitzing took his experiment to the High Field Magnetic Laboratory of the Max Planck Institute in Grenoble, France, to make measurements using their 20-tesla magnet, the magnetic field strength of which is roughly one million times stronger than the earth's at ground level. Von Klitzing found for practically every sample that the Hall resistances were equal to the same fundamental ratio divided by integers to within a few percent, extending over well-developed plateaus in the variation of the Fermi level. The subsequent high-precision results published were measured using the more stable 15-tesla magnet at the University of Würzburg. The accuracy improved to five parts per million, with the primary source of inaccuracy being the instability of the resistance standard. The ratio of the resistance at different plateaus, for example, was the ratio of integers to one part in thirty million. During the plateau regions, the electrical resistance fell very nearly to zero, ten times lower than any nonsuperconducting metal. Moreover, the resistivity continues to decrease as the temperature approaches absolute zero.

The surprising features of the quantized Hall effect sent theoreticians into a flurry of activity. Impurities had been conventionally thought of as either trapping or deflecting electrons off their paths, giving rise to electrical resistance and causing variation in measurements from sample to sample. The seeming lack of involvement of impurities or defects was particularly enigmatic. In 1981, preliminary calculations by University of Maryland theorist Richard E. Prange suggested that although an electron can be trapped in a "localized state" around a defect in the crystal, under the condition that the Landau levels are integrally filled, the current lost to the trapped electron is exactly compensated by an increase in the velocity of electrons near the defect. The electrons move like a fluid, where flow speeds up around a barrier so that the total transported volume remains the same.

Theoreticians came to realize more generally that not only do impurities not cause resistance, but ironically they also are responsible for the plateaus in the Hall resistance as the magnetic field or the Fermi level is varied. The localized states act as a reservoir between Landau levels. As the Fermi level rises past complete filling of the conducting states of a given Landau level, only localized states are left to be filled up. The conducting electrons are effectively unaffected, giving rise to the constancy of the current as the Fermi level is varied.

Impact of Event

The measurement accuracy of the fundamental ratio found in the quantized Hall resistance subsequently improved to one part in 10^8, and resulted in several immediate benefits. After a series of tests in independent laboratories was completed by the end of 1986, the quantized Hall effect was adopted as the international standard for resistance. The fine-structure constant, which is related to the fundamental Hall ratio by the speed of light, is a measure of the coupling of elementary particles to the electromagnetic field. Complementing high-energy accelerator experiments, the improved determination of the fine-structure constant provides a stringent test for theories of the fundamental electromagnetic interactions.

Soon after the integral quantized Hall effect was explained, a new "fractional" quantized Hall effect was discovered in 1982 by Dan C. Tsui, Horst L. Störmer, and Arthur Charles Gossard of Bell Laboratories. The type of sample they used for creating the two-dimensional electron gas, called a heterojunction, was made by a process called molecular beam epitaxy, where a layer of gallium arsenide positively doped with aluminum is grown onto a substrate layer of pure gallium arsenide. The gallium arsenide electrons are attracted toward the positively doped semiconductor and thus build up into a layer at the interface. The new device was a more perfect crystal and had better conduction properties, which were crucial for a successful observation of the fractional quantized Hall effect. In the fall of 1981, they brought their sample to the Francis Bitter Magnet Laboratory at MIT, where they used the 28-tesla magnet. They were searching at high fields and temperatures less than 1 Kelvin for an "electron crystal," where the electronic orbitals become arranged into a lattice. Instead, they found the same kind of plateaus and drops in the resistance observed in the integral quantized Hall effect, but occurring when only one-third or two-thirds of a Landau level is filled. Since then, many other fractions have appeared.

Theoretical investigations indicated that the observations could not be explained by an electron solid and demanded a radical description of the electronic behavior. In 1983, Robert Betts Laughlin gave a remarkable explanation in terms of a "quantum electronic liquid," in which the motions of the electrons are strongly affected by each other. The electronic liquid is incompressible: Rather than causing the density to increase, squeezing on the liquid causes a condensation of exotic fractional charges. These fractional charges play the role that electrons do in the integral Hall effect and so cause the plateaus at fractional values.

The impact on the field of physics reaches far beyond the accuracy of the measurement of the Hall resistance. Although the effect itself is not expected to be commercially significant, the MOSFET is essentially identical to components that may be important in the following generation of computers. Additionally, similarities have emerged between the physical mechanisms of the fractional quantized Hall effect and that of high-temperature superconductors. Common features include a two-dimensional structure, low resistivity, and the collective motion of a macroscopic number of particles. The primary significance of the quantized Hall effects lies in revolutionizing and deepening an understanding of electronic properties of solids in high magnetic fields.

Bibliography

Halperin, Bertrand I. "The 1985 Nobel Prize in Physics." *Science* 231 (February 21, 1986): 820-822. Halperin, professor of physics at Harvard, has contributed theoretical insights to the understanding of the quantized Hall effect. This article introduces the concepts of the quantized Hall effect at a basic level and provides some historical background. (Note that the comment concerning even-denominator fraction plateaus was written before they were discovered in 1987.) Includes a photograph and brief biographical notes.

_____. "The Quantized Hall Effect: This Variation on a Classical Phenomenon Makes It Possible, Even in an Irregular Sample, to Measure Fundamental Constants with an Accuracy Rivaling That of the Most Precise Measurements Yet Made." *Scientific American* 254 (April, 1986): 52-60. Most of this article is devoted to explaining the quantized Hall effect in terms of simple classical ideas, without resorting to quantum mechanics. Includes diagrams, an illustration of the device, and a four-entry bibliography.

Klitzing, Klaus von. "The Quantized Hall Effect." *Reviews of Modern Physics* 58 (July, 1986): 519-631. This lecture, presented at the Nobel ceremonies, is intended for the nonspecialist scientist. Though some of the material is advanced, the historical remarks and discussion about establishing resistance standards are still useful. Includes an extensive section with data characteristic of the quantized Hall effect and a technical bibliography.

MacDonald, Allan H., ed. *Quantum Hall Effect: A Perspective*. Boston: Kluwer Academic Publishers, 1989. Mostly a compilation of significant research articles in the history of the Hall effect; it does, however, include the readable article "The Discovery of the Quantum Hall Effect" first published in *Metrologia* 22 (1986): 118-127 by von Klitzing's graduate adviser, G. Landwehr. Landwehr gives a personal, anecdotal account of von Klitzing's work, giving the insider's details of the history, such as why von Klitzing's results were originally rejected for publication. The first section includes a primer on the subject.

Schwarzschild, Bertram. "Von Klitzing Wins Nobel Physics Prize for Quantum Hall Effect." *Physics Today* 38 (December, 1985): 17-20. The journalistic article gives a chronological account of the developments surrounding the discovery of the quantized Hall effect. Includes a bibliography.

David Wu

Cross-References

Shockley, Bardeen, and Brattain Discover the Transistor (1947), p. 1304; Bardeen, Cooper, and Schrieffer Explain Superconductivity (1957), p. 1533; Esaki Demonstrates Electron Tunneling in Semiconductors (1957), p. 1551; The Microprocessor "Computer on a Chip" Is Introduced (1971), p. 1938.

PLUTO IS FOUND TO POSSESS A THIN ATMOSPHERE

Category of event: Astronomy
Time: May, 1980
Locale: Tucson, Arizona

Fink investigated specific infrared absorption bands of Pluto and detected significant amounts of methane gas believed to be unusual for such a small planet

Principal personage:
UWE FINK (1939-   ), an American astronomer eminent in planetary spectroscopy who first detected methane in Pluto's atmosphere, water-ice in Saturn's rings, and water-ice on Jupiter's and Saturn's satellites

Summary of Event

Although Pluto was discovered in 1930, not much was known about this planet until photometry made possible detailed spectroscopic observations in the 1970's and 1980's. Its orbit, for example, is the most eccentric in the solar system, varying in distance from 4.4 to 7.4 billion kilometers. The very elliptical orbit results in significant temperature variations during its orbital year.

In 1976, methane ice was discovered on its surface. Previously, astronomers assumed Pluto's surface was dark in order to explain its dimness; but since methane ice reflects a high percentage of sunlight, researchers now explain that Pluto is not faint because it is dark but because it is small. Photoelectric observations made of Pluto have revealed considerable brightness variations. Analysis of these variations indicates that they are caused by surface reflection irregularities.

Previous investigations using absorption spectra had given suspicion of a methane atmosphere, but instrument image tubes at that time were not sensitive enough in the stronger absorption bands to be conclusive. Response was generally weak in wavelengths below 900 nanometers (a nanometer is one billionth of a meter) where absorption bands may be present but are difficult to detect.

The detection of methane in the atmosphere of Pluto was accomplished in May, 1980, by a team led by Uwe Fink of the University of Arizona. The observations were completed at the 155-centimeter Catalina telescope. The team constructed a low-resolution spectrograph to which was attached a silicon charge-coupled device (CCD). A CCD typically contains a quarter million microscopic, light-sensitive diodes in a space the size of a postage stamp. CCDs are capable of detecting both bright and faint objects and are much more sensitive than photographic plates. The CCD was manufactured by Texas Instruments and arranged in an array containing twenty-five hundred elements in a 500-by-500 matrix only 15.2 microns square. To enhance the detection capability, the array was cooled to −120 degrees Celsius. The electronics as well as the detector were built at the California Institute of Technology for use by the investigation group of the Space Telescope Wide Field Planetary Camera.

A spectrograph was assembled at the Lunar and Planetary Laboratory of the University of Arizona and designed for use with the CCD array. The spectrograph utilized a transmission diffraction grating, with a total spectral range from 570-1,100 nanometers and a resolution capability of 1.1 nanometers for first-order spectra. The longer wavelength threshold of 1,100 nanometers was determined by the CCD sensitivity, and the shorter wavelength limit was set by the response of a special filter built to eliminate overlapping spectra orders.

Pluto was observed, and its light was allowed to fall upon the spectrograph entrance slit which was imaged directly on the CCD array. The width of the slit, although very small at 18 millimeters, allowed adequate coverage of the background sky. When contrasted to Pluto, the sky background was distinct but only one-tenth as strong as that of Pluto.

In the data reduction, the signals from Pluto were added at each wavelength from eleven tiny detector units called pixels. Next, the background sky counts and direct current offsets caused by the CCD array were subtracted from the five pixels on either side of Pluto. Data were also taken from standard stars and reduced using the same method. Dividing Pluto's average brightness by the star's average brightness effectively removed sensitivity variations that occurred across the detector array. This technique of ratio comparison allowed the astronomers to locate regions of no absorption in the atmosphere. When the data were compared to the sky-only spectrum, the researchers noted no perceptible data fluctuations in the position of Pluto's spectrum.

The ratio spectra indicated definite methane bands at 620, 720, 790, 840, 860, 890, and 1,000 nanometers; the strongest of these bands was located at 890 nanometers. The observed band saturation was definitive evidence for gaseous methane absorption, as opposed to liquid or solid-state absorption. Methane, in both the liquid and solid states, exhibits abundant linear absorption with no visible saturation. One would not expect a contribution of more than a few percent from either the liquid or solid states of methane. The absorption spectra of Pluto, therefore, clearly showed the presence of a methane atmosphere.

The researchers made an estimate for the total atmospheric pressure on Pluto at 0.05 of that of Earth. This value was based on calculations of the partial pressure of methane alone in Pluto's atmosphere, along with variables such as the mass of the methane molecule, estimates of Pluto's gravitational acceleration, as well as the abundance of methane molecules in a given column of atmosphere. Fink's research team measured significant amounts of gaseous methane and believed that, although methane surface frost still may be present, it would not be an important factor in explaining the absorption spectra. They also did not rule out the possibility of a heavier gas being mixed in with the methane, which might be necessary for the planet to hold onto its atmosphere if its mass is quite small.

Astronomers, in general, believe water—in addition to methane—is a major part of the planet because of its low density as well as the general abundance of water elsewhere in the solar system. The presence of water on Pluto has not been confirmed as the spectrum of methane completely overcomes the spectral bands of water.

The atmosphere of Pluto is believed to be caused by the sublimation of methane from ice to gas on its surface. The effect will be more pronounced when it is closer to the sun at perihelion, resulting in more intense surface heating. Some of the methane that is vaporizing will drift toward the poles and freeze at the surface. This effect will be enhanced as Pluto travels farther out from the sun, causing less methane to subliminate. It is believed that Pluto's poles remain covered with methane throughout its 248-year period of revolution.

Impact of Event

Investigations were undertaken as early as 1944 to search for an atmosphere on Pluto. All results proved negative until 1980, when spectral measurements of high resolution were obtained revealing methane gas. What was particularly surprising to the astronomers was that such a small planet as Pluto even had an atmosphere. The significance of methane in Pluto's atmosphere can present a problem depending upon its mass and radius. Assuming a smaller than currently accepted mass, calculations will place the radius at a maximum of 950 kilometers to have a stable atmosphere. This would lead to a surface reflectivity (albedo) of greater than 100 percent, which is unrealistic. On the other hand, for a slightly larger radius of 1,100 kilometers, then the albedo drops to a reasonable 67 percent.

A mass increase of 50 percent over the latter model leads to a radius of 1,400 kilometers, which falls within the limits determined by the technique of speckle-interferometry, a relatively new photographic method designed to "freeze-out" interfering motions of the earth's atmosphere. Also, if the mass of Pluto used in the determination is too small, then the addition of a heavier gas is required to maintain atmospheric stability over a long period of time. As of 1990, the calculations of the total atmospheric pressure are not refined enough either to confirm or refute the additional heavier gas.

The amount of methane gas, as originally measured by Fink, was possibly overestimated in spite of the small values that were measured. In fact, one study showed that the atmospheric column density on Pluto is no more than $\frac{1}{900}$ that of Earth.

Occultations of stars by Pluto in later years have also verified the presence of a thin atmosphere. An occultation occurs when a planetary body passes in front of a star, obscuring its light. The atmosphere would produce a pronounced dimming of the star before it was occulted. For Pluto to maintain a methane atmosphere, the surface temperature must be in a state of equilibrium. If the temperature is too cold, methane will freeze out. A temperature too high will result in an upward pressure exceeding the surface pressure causing a rapid loss of atmosphere.

Investigations of Pluto's spectrum demonstrate that the degree of absorption varies with its rotational period of 6.4 days. An atmosphere that is uniform fails to explain this variation. Methane frost on the surface will explain this variation in rotational period, however, with or without an atmosphere.

Fink's discovery team believes that Pluto's small size, high inclination of orbit, and large distance from the sun have traditionally made Pluto a very unusual type of planet. The discovery of its moon, however, and detection of an atmosphere will enhance its image and establish Pluto as a more typical member of the solar system.

Bibliography

Brown, Robert H., and Dale P. Cruikshank. "The Moons of Uranus, Neptune and Pluto." *Scientific American* 253 (July, 1985): 38-47. A nontechnical discussion of ground-based studies of the outer solar system prior to the Voyager 2 flyby. Charts displayed indicate that Pluto's absorption bands are characteristic of methane frost. A diagram is presented demonstrating why eclipses between Pluto and its moon Charon take place during only two brief periods.

Croswell, Ken. "Pluto: Enigma on the Edge of the Solar System." *Astronomy* 14 (July, 1986): 6-22. The observational history of Pluto is covered from Percival Lowell and Clyde Tombaugh to the 1978 discovery of Pluto's moon Charon. The eclipse cycle of Pluto and its moon is discussed as well as spectroscopic studies of Pluto's surface and atmosphere. A table of physical and orbital characteristics is included.

Eberhart, Jonathan. "Pluto: Limits on Its Atmosphere, Ice on Its Moon." *Science News* 133 (September, 1987): 207. Important facts are compiled concerning Pluto and its moon, such as the discovery of the moon, the eclipse relationship, infrared spectrometry studies, and speculations about Pluto's thin atmosphere.

Fink, Uwe, et al. "Detection of a Methane Atmosphere on Pluto." *Icarus* 44 (October, 1980): 62-71. Although somewhat technical in this key article, Fink describes the actual observations, instrumentation used, and data reduction that led to the discovery of methane gas on Pluto. Prior photometric and spectroscopic research is thoroughly discussed and a comprehensive reference list is included.

Littmann, Mark. "The Smallest Planet." In *Planets Beyond*. New York: John Wiley & Sons, 1988. A comprehensive observational history of Pluto is accompanied by detailed photographs and diagrams. Pluto's atmosphere and surface is discussed in the light of current knowledge. An account of the discovery of Pluto's moon is presented as originally interpreted from almost imperceptible image distortions on photographs.

Time-Life Books, eds. "Outposts." In *The Far Planets*. Alexandria, Va.: Time-Life Books, 1988. A fascinating and superbly illustrated chapter on Pluto and its moon summarizes the present knowledge. Multicolored diagrams are a unique feature. Tables depict the planets' size and orientation in space and orbit. A glossary and detailed bibliography are included.

Michael L. Broyles

Cross-References

Tombaugh Discovers Pluto (1930), p. 944; Astronomers Discover the Rings of the Planet Uranus (1977), p. 2068; Voyager 1 and 2 Explore the Planets (1977), p. 2082; The First Ring Around Jupiter Is Discovered (1979), p. 2104; Astronomers Discover an Unusual Ring System of Planet Neptune (1982), p. 2211.

RADAR OBSERVATIONS SHOW THAT MAYAN AGRICULTURAL CENTERS ARE SURROUNDED BY CANALS

Category of event: Archaeology
Time: June, 1980
Locale: Guatemala

Radar images indicated the presence of a network of canals in areas surrounding Mayan population centers

Principal personages:

JOHN LLOYD STEPHENS (1805-1852), an American attorney, explorer, and minister to Central America who wrote extensively about his explorations of the Mayan ruins

FREDERICK CATHERWOOD (1799-1854), an English illustrator and explorer

DIEGO DE LANDA (1524-1579), a Spanish Franciscan priest who was appointed bishop over the Yucatán

Summary of Event

Like the Aztec and Inca civilizations that had been discovered and documented by Central and South American explorers Hernán Cortés and Francisco Fernández de Córdoba in the 1500's, the Maya were contacted as early as 1502. After the plundering of the Aztec capital Tenochtitlán, near present-day Mexico City, in 1521, the Spanish conquistadors turned their attention south. Cortés sent one of his captains, Pedro de Alvarado, with a small force of Spanish soldiers and a contingent of twenty thousand native troops, and shortly thereafter, Guatemala and San Salvador were under Spanish rule. In addition to the diseases that were introduced to the continent, the conquest also involved the destruction of the Indian identity. This reached its zenith on July 12, 1562, with the destruction of an entire Mayan library of pictographic books and codices by Friar Diego de Landa. The loss to the understanding of this pre-Columbian civilization is rivaled only by the razing of the library at Alexandria, Egypt, in A.D. 415. To date, only four of these volumes are known to remain.

Clashes between the Spanish and various native tribes occurred throughout the entire Central American region and lasted almost twenty years. What was revealed to the Europeans was a vast landscape of empty cities. It was apparent that the conquered Mayan nation was a shadow of its former self.

In 1840, these empty ruins were brought to light again by the explorations of John Lloyd Stephens and Frederick Catherwood, who documented the remains of eight ruined cities. These cities included Quirigua, Copan, Palenque, and Uxmal. The region was lightly populated, with large areas of rain forest being systematically cut down and burned to provide growing space and fertilizer for several years of farming. The practice, commonly called "slash and burn" farming, was being carried out by the indigenous population. This practice is characteristic of present-day subsistence-level farming in tropical rain forests. This type of agriculture, however, could not have supported the estimated 14 million people that had occupied those now dead cities. The Aztec in central Mexico used more advanced techniques to include *chinampas* (artificial islands) and irrigation, while in South America, the Inca practiced terracing and fertilization. At that time, however, there was little evidence found in Central America of these advanced horticultural techniques. What Stephens and Catherwood did discover was a complex system of *cenotes* (reservoirs) as well as artificial *aguadas* (wells). Farmers used these wells, not knowing that they were constructed earlier. Dry season excavations of these wells indicated that they were man-made. A survey of one locale revealed more than forty such constructions. It was not known who had built these structures and why they were abandoned.

Part of the mystery is that the Mayan civilization began its decline about A.D. 900 leaving behind its cities and petroglyphs. Several theories say that the culture declined either as a result of conquest by neighboring nations or as a result of famine causing the Mayas to migrate and merge with one of the other cultures. Until the late 1970's, the prevailing thought was that a climatic change caused a drought, with the slash and burn type of farming unable to support a large population under those extreme conditions. It became apparent that the lack of any large-scale land surveys, combined with the limited field work done on nonurbanized areas, would continue this debate over how a large Mayan population had supported itself.

From October, 1977, until August, 1980, the National Aeronautics and Space Administration (NASA) made available an airborne, side-looking radar to begin detailed mapping of northern Belize and northeastern Guatemala. The radar was designed originally for the radar mapping of the surface of the planet Venus, but a more suitable application appeared to be the mapping of the swamps and jungles that make up two-thirds of the Mayan territory. This would allow researchers to see through the tree cover and marshland, and for the first time allowed a large region to be investigated for the presence of man-made structures. The focus of this mapping was on the site of Tikal, the largest site identified with Mayan culture.

The radar provided an image that varied from 10 to 21 kilometers in width, with a resolution of 20 meters. Many of the large buildings cast shadows and the large, flat surfaces reflected the radar, showing up as spots of light. What surprised the researchers was the appearance of intricate patterns of lines in areas known to be swamps. Because these lines were confined to low-lying land, the immediate label given to them was canals, as used for either transportation or drainage. If further examination proved this correct, then it meant that instead of subsistence-level agricultural practices, the Mayas had developed a more sustainable form of farming.

The next step was confirmation of this data with on-site excavations. Ground examination indicated a system of raised fields with the canals acting as drainage conduits, therefore correlating with the information gathered by aerial imaging. In addition, an excavation found canals 1 meter wide separating fields that were 3 meters square. Because of the radar's resolution, however, these features remained for the field archaeologist to interpret. It was found that many of these fields were artificially raised to make them more suitable for agriculture. The Mayas may have utilized these raised fields to grow a variety of food and fiber plants. Analysis of pollen and other plant debris indicates the presence of corn, cotton, and amaranth. Unfortunately, the presence of these plants can be explained by either human cultivation or the self-reproduction of wild species. Water lilies are also associated with these raised areas. Studies have shown that these plants provide a good mulch for the crops that could have been grown on these elevated islands.

The radar could not discriminate between ancient and modern structures, as such; railroads, highways, airstrips, and natural geologic structures were picked up also. Of additional interest, however, is what appears to be the presence of some paved areas, possibly roads built by the Mayas.

Impact of Event

This land-mapping technique has altered the way archaeology is viewed and has added much to existing knowledge about the Mayan civilization. Traditionally, archaeology has been thought of as a science requiring laborious methods and manpower to accomplish the work of literally uncovering past civilizations. With this new tool of aerial radar-mapping, however, it has become significantly easier to screen large, unexplored areas for the presence of man-made structures. The visual image of hundreds of laborers removing debris has been replaced by that of technicians poring over airborne images.

Knowledge of the Mayas has also been changed. It was never fully understood why the Mayas would build large cities in what appeared to be vast swamps. Because these lands, when drained, could provide a stable agricultural base, the Mayas might have considered this land the most valuable. The extent of the network of canals indicates a high degree of centralization; an estimated 2,500 square kilometers (or 250,000 hectares) of land in Belize and Peten region alone were subjected to this type of hydraulic engineering.

The work was not easily accomplished, as the indigenous culture did not have the wheel, or any draft animals, or iron tools prior to the arrival of Western explorers. This means that all the work was done manually, using stone tools and baskets to move the debris. A research team tried to replicate the construction of a raised field, using tools and techniques that the Mayas would have employed. From this work, an estimate was made on the amount of time that it would have taken Indian laborers to build the discovered fields: For the task to have been completed in fifty years, it would have taken a minimum of 500,000 people. Another possibility is that this system of canals may have evolved during the life span of this culture. Like all Mesoamerican cultures, the Mayan culture had at its base the farmer. It is reasonable to assume that during the height of their society, there were as many as 10 million people working the land.

This revelation also helps to reevaluate how the agricultural capabilities of the tropical rain forest are viewed. The practice of slash and burn farming bankrupts the fragile soil, while the use of silt and aquatic plants dredged from those canals can act to build the topsoil, add additional nutrients, and lead to more productive farming practices. It is hoped that the people of Central America will take advantage of the new knowledge gained in the study of the Mayas. A shift appears to be starting, with the training of native farmers in a more regenerative style of agriculture. This is the most lasting impact of the discovery and study of the canal systems of the Mayan civilization.

Bibliography

Adams, R. E. W., W. E. Brown, Jr., and T. Patrick Culbert. "Radar Mapping, Archaeology, and Ancient Maya Land Use." *Science* 213 (September 25, 1980): 1457-1463. This article outlines the discovery of waterways through the use of aerial radar. An excellent bibliography at the end of this article covers the work done by previous investigators studying Mayan land use.

Hunter, C. Bruce. *A Guide to Ancient Maya Ruins*. Norman: University of Oklahoma Press, 1974. An interesting book on Mayan cities. Careful reading of the text reveals the agricultural misconceptions that were prevalent. Good, basic book on Mayan culture.

Ivanoff, Pierre. *Maya*. New York: Madison Square Press, 1973. This book is a photographic record of fifteen different Mayan cities. It is one of the best books for a student interested in major cultural landmarks of the Mayas.

Stephens, John L. *Incidents of Travel in Yucatán*. 2 vols. Reprint. Mineola, N.Y.: Dover, 1963. These two volumes are an excellent firsthand account of the rediscovered ruins on the Yucatán peninsula. The drawings done by Catherwood enhance Stephens' narration.

Turner, B. L., II, and Peter D. Harrison. "Prehistoric Raised-Field Agriculture in the Maya Lowlands." *Science* 213 (July 24, 1981): 399-404. This article discusses at length the excavation of a Mayan settlement in the Belize lowlands.

Charles A. Bartocci

Cross-References

Bingham Discovers an Inca City in the Peruvian Jungle (1911), p. 491; Anthropologists Claim That Ecuadorian Pottery Shows Transpacific Contact in 3000 B.C. (1960's), p. 1624.

THE U.S. CENTERS FOR DISEASE CONTROL RECOGNIZES AIDS FOR THE FIRST TIME

Category of event: Medicine
Time: 1981
Locale: United States

The U.S. Centers for Disease Control was the first to report the AIDS epidemic by citing cases of pneumocystis pneumonia in Los Angeles

Principal personages:

JAMES CURRAN, an epidemiologist and director of AIDS research efforts at the Centers for Disease Control in Atlanta

JOEL WEISMAN, a physician practicing in Los Angeles, who with Dr. Michael Gottlieb, an immunologist at the University of California, Los Angeles, identified the first cases of AIDS

GRETE RASK, a Danish surgeon practicing in Zaire who became the first documented European to be infected with the AIDS virus

GAETAN DUGAS, an Air Canada flight attendant who developed Kaposi's sarcoma and spread the infection in the homosexual communities; given the name "patient zero"

Summary of Event

In 1976, people in a village along the Ebola river on the Zaire-Sudan border experienced a virulent and horrifying disease that came suddenly. A trader from the nearby village of Enzara suffering from fevers and profuse and uncontrollable bleeding reported to the teaching hospital in Moridi. Within days of his admission, 40 percent of the nurses and several doctors were stricken. By the time World Health Organization officials and Centers for Disease Control (CDC) staff arrived, thirty-nine nurses and two physicians had died from Ebola fever. Later that year, another insidious disease, manifested by undue malaise, unrelenting pneumonia, skin lesions, and cachectic weight loss, was making its round in the village of Abumombazi, Zaire, close to the Ebola River.

Notable among the first affected in Africa was a Danish surgeon, Dr. Grete Rask, who had devoted much of her professional life in medical service to the people of the former Belgian Congo. Having studied stomach surgery and tropical diseases, she settled in the remote village of Abumombazi as a physician in a small ill-equipped hospital. Sterile rubber gloves, disposable needles and syringes, and adequate blood banking systems were almost nonexistent. Needles were used over and over again until they were too dull. Surgical gloves were washed, rewashed, and turned inside out until worn through. Surgeons risked contracting various blood-borne diseases while providing essential services to their patients. Unfortunately, patients also were at risk of being inoculated with or receiving some virulent infections by the application of these unsanitary practices. As the only surgeon in this Zairian village hospital, Rask operated on her patients often with her bare hands, using poorly sterilized equipment.

In 1976, Rask developed grotesquely swollen lymph glands, severe fatigue, continuous weight loss, and was suffering from intractable diarrhea. Later, she labored for each breath and finally decided to return to her native Denmark to die. For months, doctors tested and examined the surgeon but were unable to make a diagnosis. Doctors could not understand why several health problems attacked the frail woman. Her mouth was covered with yeast infections, staphylococcus bacteria spread in her bloodstream, and her lungs were infected with unknown organisms. She was dying, and they could not help. Serum tests showed her immune system as being almost nonfunctional. Her body lacked T cells, the body's main defense. Biopsies showed no lymphatic cancer. The doctors were stumped. She died on December 12, 1977.

The autopsy revealed that millions of *Pneumocystis carinii* organisms caused the rare pneumonia that slowly ravaged and suffocated Rask. This protozoan became the landmark organism in the identification of the new disease. Questions were raised as to where and how she contracted this infection, but answers were not available. It was not, however, Ebola fever.

Pneumocystis carinii was found first in guinea pigs in 1910 by a Brazilian scientist named Dr. Carini. Subsequent research determined that it is a common resident of every terrain and can be found dormant even in an individual with a normally functioning immune system. This strange new disease depleted the patient's immune system, leaving the patient's body vulnerable to unusual and rare infections such as cryptococcal, fungal, cytomegalovirus, toxoplasmosis, and papovavirus infections. Meningitis, tuberculosis, and *Pneumocystis carinii* with the Kaposi's sarcoma were present also.

The 1970's was an important decade: It was the Bicentennial decade, the post-hippie decade, and the decade that saw the end of the Vietnam War; it was the decade of self-expression and the decade that heralded the beginning of the insidious plague acquired immunodeficiency syndrome (AIDS).

Clinical epidemics of cryptococcal meningitis, progressive Kaposi's sarcoma, and esophageal candidiasis were recognized in Zaire, Zambia, Uganda, Rwanda, and Tanzania. This syndrome was termed "slim disease" in these countries because of the sudden unintentional weight loss of the affected individuals resulting in the severely emaciated appearance. Kaposi's sarcoma—a multifocal malignant neoplasm manifested primarily by purplish vascular nodules in the skin and other organs— had become a common finding in the affected patients. During the same period, similar clinical manifestations were noted, primarily in homosexual males in New York City and San Francisco who developed Kaposi's sarcoma of the skin, oral candidiasis, weight loss, fever, and pneumonia.

One of the very first identified cases in North America was a Canadian flight attendant, Gaetan Dugas, who would later be known as "patient zero." Dugas' job afforded him the travel outlet to experience life to the fullest. In 1978, he developed purplish skin lesions and was informed that he had Kaposi's sarcoma, a nonmalignant skin cancer. He went about his regular routines with no concerns. When news of more Kaposi's sarcoma in the homosexual population was publicized, he contacted Drs. Alvin Friedman-Kien and Linda Laubenstein at New York University. His affliction was diagnosed as cancer. In desperation, he went to bath houses and in dimmed lighting had anonymous sex. In Europe, homosexual men with symptoms were those who had either visited the United States or those whose sexual partners had visited the United States. Some were immigrant Africans.

The CDC embarked on a major investigation to track patients and their contacts in an effort to determine the primary causative factor, origin, mode of its transmission, and why its nidus was focused on homosexual men. European and African health care providers, with assistance from major international agencies, were involved, likewise, in the search for answers and to determine why women in Africa were affected at the same rate as men. Hepatitis B was prevalent. Parasitic organisms were found lodged in the gastrointestinal tract of homosexual men, producing devastating enteric diseases. These unusual infections were attributed to several sexual practices of homosexual men.

By mid-1980, fifty-five men had been diagnosed with one or more of the opportunistic infections associated with this virus in the United States. Ten were diagnosed in Europe and many unknown numbers in Africa. CDC officials, such as James Curran, decided to make public the information on AIDS. This report was edited to minimize fear; the "Morbidity and Mortality Weekly Reports" were used as the vehicle to announce to physicians, health departments, and health workers the presence of this epidemic under the headline, "Pneumocystitis Pneumonia—Los Angeles" on June 5, 1981. It appeared as a trite case report (from patients' reports by Joel Weisman and Michael Gottlieb) that noted the links between pneumocystis pneumonia, cytomegalovirus, and oral candidiasis.

Impact of Event

This report opened the door for this deadly killer to be evaluated. The virus was to be found in the blood and was transmitted through blood transfusions. The virus also would be found in the umbilical cord and passed from mother to fetus. The virus would be passed through needles and would become a menace for intravenous drug abusers. The virus would be found in semen and become a threat to sexual partners of affected individuals, both men and women. The virus with the opportunistic infections producing AIDS would be the most feared and dreaded epidemic of the twentieth century. AIDS came at a time when the priority of the government was to curtail spending on domestic affairs.

Since the first public report in 1981, the number of affected individuals has multiplied. Global estimates of persons diagnosed with AIDS are 254,169, with an unknown number dead. The estimate of human immunodeficiency virus positive cases is in the millions. The U.S. statistical number for AIDS infected persons, as of

June 1, 1990, was 139,765, with a 60 percent fatality rate.

Cases have been categorized according to the following risk behavior/risk groups: sexually active homosexual men, 59 percent; intravenous drug abusers, 21 percent; homosexual/intravenous drug users, 7 percent; blood transfusions, 2 percent; heterosexual sexual activity, 5 percent; blood disorders, 1 percent; perinatal infants of affected mothers, 2 percent; undetermined causes, 3 percent. AIDS is an epidemic in evolution. Medical science has yet to determine the history of this syndrome. No one really knows when it started or where it will end.

In 1986, the unusual case of a fifteen-year-old black male from St. Louis, Missouri, who died in May, 1969, was made public. His symptoms of bronchial pneumonia, swollen lymph glands, resulting in a rapid fifteen-month decline, with multiple unrelenting infections so baffled his physicians that they saved some tissue and blood samples for future testing. Tests performed by Robert Gary at Tulane University in 1987 confirmed the presence of HIV (human immunodeficiency virus) in the serum and tissues of this patient. Tissue autopsy revealed he had Kaposi's sarcoma throughout the soft tissue of his body. He was diagnosed ten years posthumously with AIDS as the cause of death. An area of special impact is that of sexual affiliations and life-style. "Safe sex" was a term heard frequently; monogamy and protection from sexually transmitted diseases by the use of condoms were recommended.

This report also led to heated discussions regarding the origin of the virus and how it was transmitted to humans. In 1980, Robert Gallo had hypothesized that this virus probably originated in Africa, was carried by Portuguese to the Orient and the West Indies, infecting Haitians and finally reaching the Americans in its current form. This theory was revitalized during the spread of Ebola fever. A blood-borne virus, it quickly spread through sexual intercourse and through sharing contaminated needles in local hospitals. Similarly, it was being spread through sexual contacts— heterosexuals in Africa and Haiti, homosexuals in Europe and the United States. This theory has had little validation for its support. At the present time, research is continuing toward management modalities that not only treat the opportunistic infections that occur but also can prevent the transmission of the virus to uninfected persons.

Bibliography

Check, William A. *AIDS*. New York: Chelsea House, 1988. For a wide audience. C. Everett Koop discusses AIDS from a historical perspective.

Drotman, D. Peter, and James W. Curran. "Epidemiology and Prevention of Acquired Immunodeficiency Syndrome (AIDS)." In *Public Health and Preventive Medicine*, edited by Kenneth Fuller Maxey. 12th ed. East Norwalk, Conn.: Appleton-Century-Crofts, 1985. The textbook of public health issues. This article discusses the major aspects of AIDS, including the international patterns.

Gottlieb, Michael S., et al. *CDC Mortality and Morbidity Weekly Review—June 5, 1981*. Atlanta: The Atlanta HHS Publication, 1981. The landmark publication that the CDC used to announce the presence of AIDS in the United States. Discusses doses of pneumocystis pneumonia infections in five homosexual men in the Los Angeles area. The editorial after the article discusses homosexual life-style, immuno-suppression, the presence of cytomegalovirus, and *Pneumocystis carinii* infections.

Ma, Pearl, and Donald Armstrong, eds. *AIDS and Infections of Homosexual Men.* 2d ed. Stoneham, Mass.: Butterworth, 1989. Discusses sexually transmitted diseases, in addition to AIDS. For a general audience. Informative. References.

Medical World News—The News Magazine of Medicine—November 23, 1987. San Francisco, Calif.: Miller Freeman, 1987. Discusses the fifteen-year-old teenager whose death in St. Louis in 1969 may have been caused by AIDS. Reports that tests performed on stored blood and tissue samples confirmed the presence of the human immune virus. Documents the first known infected individual.

Shilts, Randy. *And the Band Played On: Politics, People, and the AIDS Epidemic.* New York: St. Martin's Press, 1987. A histopolitical account of the AIDS epidemic. Gives almost a day-by-day recounting of various events that surround this syndrome from the first time it was identified. Shilts goes further to chronicle some of the earlier cases and their eventual outcome.

Margaret I. Aguwa

Cross-References

Rous Discovers That Some Cancers Are Caused by Viruses (1910), p. 459; Horsfall Announces That Cancer Results from Alterations in the DNA of Cells (1961), p. 1682; Daffos Uses Blood Taken Through the Umbilical Cord to Diagnose Fetal Disease (1982), p. 2205; A Genetically Engineered Vaccine for Hepatitis B Is Approved for Use (1986), p. 2326; Medical Researchers Develop and Test Promising Drugs for the Treatment of AIDS Patients (1991), p. 2382.

A HUMAN GROWTH HORMONE GENE TRANSFERRED TO A MOUSE CREATES GIANT MICE

Category of event: Biology
Time: 1981-1982
Locale: Philadelphia, Pennsylvania

Brinster and associates demonstrated that control genes and DNA sequences could be used from species that were not closely related to express transferred foreign genes

Principal personages:
RALPH BRINSTER (1932-   ), a developmental biologist and veterinarian who developed culture techniques for mammalian embryos and conducted the "giant mouse" experiment
RICHARD PALMITER (1942-   ), an American molecular biologist who developed the fused mouse metallothionein-thymidine kinase plasmid used to transfer and express the human growth hormone gene
ROBERT HAMMER (1952-   ), a postdoctoral fellow in Brinster's laboratory who conducted many of the experiments

Summary of Event

To manipulate the genome of animals directly has been the dream of biologists and geneticists for more than one hundred years. The ability to control the development and expression of phenotypic characteristics is a genetic holy grail. It is somewhat unfair to point to one group or one event; what is apparent today is the culmination of many previous discoveries. A generation of hard work by hundreds of molecular biologists has brought humankind to the present summit in genetics with its breath-taking view.

Experiments by several investigators were especially significant and dramatic. Based on techniques for gene splicing worked out in bacteria, and those for maintaining and manipulating embryos, gene transfers have been accomplished in higher animals. Some credit goes to John Gurdon and coworkers at the Medical Research Council in England for their unexpected success with a direct microinjection of spliced DNA (deoxyribonucleic acid) into frog oocytes; in addition, coworkers demonstrated in 1971 that oocytes can express foreign genetic material. They showed that oocytes have the capacity to incorporate a small amount of DNA into their nuclei, though only a small fraction of it is expressed. This was a refinement of the microinjection that Gurdon used for transferring single cells from one embryo to another in the late 1960's. In 1977, a similar experiment was done on mammalian culture cells by Michael Wigler's group at Cold Spring Harbor and in 1980 on fertilized eggs by Francis Ruddle's group at Yale.

For twenty-five years, Ralph Brinster at the University of Pennsylvania had been working on manipulating the mouse genome and made significant contributions to the methodology in the field. Seeing the potential of Richard Palmiter's mouse metallothionein gene, Brinster initiated a collaboration to produce transgenic mice. He first had to overcome the ability to grow mammalian embryos in vitro (in glass) up to the blastocyst stage. The second obstacle included the uptake of at least one copy of the gene by the embryo. The third included the ability to insert specific regulator DNA flanking sequences and to control their expression. The first obstacle was overcome by Brinster in 1963 using mouse embryos. His culture techniques have become a standard for mammalian embryologists. His technique, in which the embryo is grown under an oil layer in his medium containing pyruvate, allowed for consistent and reproducible blastocysts that could then be transplanted into pseudopregnant mice. In addition, Brinster found that testicular teratocarcinoma cells could colonize a blastocyst even if they were older and yet surprisingly express characteristics in an appropriate tissue-specific manner (for example, some carcinoma cells incorporated into the dermis of a white mouse and expressing the fur color of the dark donor mouse). These same cells would cause lethal tumors if injected into an adult mouse. Besides the relevance of this experiment for cancer, it also offered the possibility that one might introduce new genes into the germ line.

A similar set of experiments was done by Beatrice Mintz at the Fox Chase Cancer Center at the Institute of Cancer Research in Philadelphia. The resulting mouse was a mosaic having two genomes that expressed themselves concurrently. Offspring of these chimeric mice could give rise to transgenic mice. Mintz also demonstrated that the viral DNA segments injected into the blastocysts of mice could be found in the genome of adult mice and their progeny.

By 1980, most groups were using direct microinjection of DNA and injecting copies of a specific DNA segment with the proper flanking sequences into the nuclei of recently fertilized eggs. The latter technique was used by Brinster's group to transfer the human growth hormone gene to mouse embryos. In 1979 and 1980, Brinster and his coworkers, utilizing direct microinjection, tried five different genes: rabbit beta-globin, SV40 virus, HSV thymidine kinase, a sea urchin histone, and Xenopus 5S RNA (ribonucleic acid). Only thymidine kinase showed evidence of incorporation and expression. Brinster's group found that having the appropriate RNA polymerase present in addition to other sequences was important. Brinster and Palmiter attached the proper flanking sequences to the mouse metallothionein (MT) gene and then fused a gene for the enzyme thymidine kinase (TK) and its promoter to the MT gene. They then fused this MT-TK gene to the gene for human growth hormone. This triple gene was inducible, assayable, had the proper insertion sequences, and might even be expressed. The mouse metallothionein portion would, they hoped, switch on the thymidine kinase enzyme when the cell was exposed to a heavy metal, such as cadmium. They observed the induced expression of the thymidine kinase in their first group of newborn mice in 1981. This experiment demonstrated that one could use control genes and insertional DNA sequences from species that were not closely related in order to introduce and express transferred foreign genes. Antibody to this new gene inhibited 95 percent of the thymidine kinase, confirming that the incorporated gene was being controlled by MT's regulator gene.

Between December, 1980, and December, 1981, several groups reported transferring new genes from one animal to another, resulting in the production of transgenic animals. The most dramatic experiment of these was by Brinster and Palmiter, in which human growth hormone gene was fused to the MT-TK gene to create a line of giant mice (they grew to twice their normal size). Later, Robert Hammer from Brinster's laboratory used the same procedure to correct a genetic defect in a line of dwarf mice, proving that gene therapy was possible on mouse embryos. Growth hormone acts on most cells (maybe all), and therefore tissue specificity, was not an issue in this case, although it might be for other genes. Tissue-specific expression would be desirable for most genes, such as the immune response genes or hemoglobin genes. This has been demonstrated also for an immunoglobin kappa gene by Brinster's group, which is working on the temporal controls for tissue-specific expression of other genes. The ultimate goal is to insert single copies of genes and have them expressed only in the proper tissue at the right time.

Impact of Event

The ability to maintain embryos in culture and to alter their genome has provided investigators with a significant opportunity to enhance the knowledge of how living systems develop. The technique of transferring DNA to developing cells will advance basic embryology and open up the possibility of genetic therapy. The techniques described represent the next step after a defective gene has been found. Since this work was accomplished, the genes for several genetic diseases have been localized. Scientists can make the products of some of those genes utilizing biotechnology and, in some cases, can alleviate or ameliorate undesirable phenotypic expressions. Manipulating the phenotypes of plants and animals could improve food output or allow animals to be more resistant to disease. One might even be able to save endangered species.

In the field of medicine, the knowledge of gene regulation and control would aid in the fight against cancer, altered immune conditions, congenital conditions, and possibly some of the consequences of aging. In combination with in vitro fertilization, the possibility of eliminating congenital defects exists. For example, if both parents are carriers of a trait, 75 percent of the offspring will be producing some aberrant form of that protein, and the severity of the disease will be related to the amount of abnormal protein (or lack of) produced. Adding "good" genes might reduce the amount of abnormal peptide incorporated into the desired protein and/or provide more of the good protein.

Experiments indicate that control genes are genetic switches and that many are also generic—that is, capable of switching on any gene attached to it. The transfer of these control genes may, in some cases, be all that is necessary to correct some genetic defects or to alter the phenotype. Using control genes sensitive to particular triggers in the environment, one could maintain some kind of control over the gene after it has been inserted into the genome so as to have it expressed when appropriate.

The aforementioned accomplishments have stimulated a rash of new experiments, as measured by their citation frequency. The successes of these researchers in mammalian gene transfer have stimulated other investigators. The early experiments were high-risk, low-probability experiments, and they have led to an exponential increase in experiments in this realm.

Bibliography

Brinster, Ralph, and Richard Palmiter. "Introduction of Genes into the Germ Line of Animals." *The Harvey Lectures* 80 (1986): 35. A well-written, concise, and not too technical overview of gene transfer in animals, including landmark experiments, technical difficulties, and prospects for its application in medicine.

Burns, George, and Paul Bottino. *The Science of Genetics.* New York: Macmillan, 1989. An introductory text in genetics. Good discussion of the new genetic procedures, including DNA fingerprinting.

Gurdon, John, and Douglas Melton. "Gene Transfer in Amphibian Eggs and Oocytes." *Annual Review of Genetics* 15 (1981): 63. Excellent review by a pioneer in genetic material transfer between animals. Well referenced. Contains a thorough discussion of techniques and status of the field at the time. For the general reader.

Lewin, Benjamin. *Genes IV.* New York: Oxford University Press, 1990. A detailed upper-level text in molecular genetics, well referenced and illustrated, with a glossary and summaries at the end of each chapter. Informative.

Mange, Arthur, and Elaine Mange. *Genetics: Human Aspects.* 2d ed. Sunderland, Mass.: Sinauer, 1990. This book is for nonmajors who want a one-semester course in genetics. Covers the basic concepts from gene structure to population genetics, with emphasis on human conditions. Has an excellent bibliography, name index, and subject index.

Palmiter, Richard, and Ralph Brinster. "Germline Transformation in Mice." *Annual Review of Genetics* 20 (1986): 61. A good historical review of gene transfer in mice covering both the authors' work as well as that of others. Well referenced and well written.

Rothwell, Norman. *Understanding Genetics.* 4th ed. New York: Oxford University Press, 1988. Excellent coverage of gene transfer in animals. In general, a very good text for the biology major. Includes review questions and problems. Bibliography found at the end of each chapter, but no references.

Scangos, George, and Charles Bieberich. "Gene Transfer into Mice." In *Advances in Genetics.* Vol. 24 in *Molecular Genetics of Development*, edited by John Scandalios. New York: Academic Press, 1987. A well-referenced review of gene transfer in mice specifically for the serious student.

Stine, Gerald. *The New Human Genetics.* Dubuque, Iowa: Wm. C. Brown, 1989. An excellent text for the allied health student or premed student. Light treatment of quantitative genetics and heavy treatment of clinical genetics. Well illustrated with references, summaries, problems, and a glossary.

William D. Niemi

Cross-References

Watson and Crick Develop the Double-Helix Model for DNA (1951), p. 1406; Nirenberg Invents an Experimental Technique That Cracks the Genetic Code (1961), p. 1687; Kornberg and Coworkers Synthesize Biologically Active DNA (1967), p. 1857; Cohen and Boyer Develop Recombinant DNA Technology (1973), p. 1987; The First Successful Human Embryo Transfer Is Performed (1983), p. 2235; Murray and Szostak Create the First Artificial Chromosome (1983), p. 2251; Willadsen Clones Sheep Using a Simple Technique (1984), p. 2273; A Gene That Can Suppress the Cancer Retinoblastoma Is Discovered (1986), p. 2331; The Search Continues for the Gene That Begins Male Development (1987), p. 2346; Erlich Develops DNA Fingerprinting from a Single Hair (1988), p. 2362.

BELL LABORATORIES SCIENTISTS ANNOUNCE A LIQUID-JUNCTION SOLAR CELL OF 11.5 PERCENT EFFICIENCY

Category of event: Chemistry
Time: May-June, 1981
Locale: Murray Hill, New Jersey

One of a long series of solar cells developed by Bell Laboratories, this particular cell raised the hope that liquid-junction devices could compete in efficiency with solid-state devices

Principal personages:
ADAM HELLER (1933-   ), a member of Bell Laboratories technical staff
BARRY MILLER (1933-   ), a member of Bell Laboratories technical staff and longtime collaborator with Heller in the development of solar cells
F. A. THIEL (1930-   ), a member of Bell Laboratories technical staff and codeveloper of the liquid-junction solar cell

Summary of Event

Efforts to harness the sun's energy for human purposes are as old as history. Today's technology of solar panels and passive solar heating is but a development of the animal skin hung at night over the hut's open window. The solar furnace is a high-technology form of the burning glass, known for centuries. The sun as heat source has long been used.

A silicon crystal by itself can absorb solar energy to produce an electron and a hole, but these recombine quite rapidly and do no useful work. Yet, a thin p-silicon layer on an n-silicon base (or vice versa) creates a potential surface that holds electron and hole apart long enough that the electron can be drawn off to flow through an external circuit. This is, at the simplest, how a solar cell works. Such a cell, employing crystalline silicon, can have a solar-to-electrical energy conversion efficiency of about 20 percent. Other materials can push efficiency close to 30 percent. The difficulty is cost: Pure, carefully "doped" crystals are expensive. Polycrystalline materials are cheaper but much less efficient.

One solution to this problem that appeared promising for a time was that of combining the semiconductor device with a chemical half-cell consisting of oxidized and reduced species in solution in water or other media, with an inert electrode-like platinum to carry current from the cell. This would give a number of advantages. The dissolved species could be oxidized and reduced forms of some inorganic ion such as vanadium could be used. This would give some control over the cell's voltage, as the concentration of dissolved ions alters the voltage of the half-cell. The potential difference between solid and solution could make separation of electrons and holes cleaner and more complete, suppressing unproductive recombination. This would lead to more efficient production of electricity in an external circuit. Moreover, if the liquid half-cell contains species that can be reduced or oxidized to usable products, these can be removed and used as needed.

Use of the sun for other kinds of energy is more recent. Only a century ago, it was found that light shining on substances such as zinc or cadmium sulfides induces a flow of electrons that can be made to do work in an external electrical circuit or be stored in a battery. The sunlight has been used to bring about chemical reactions, notably the production of hydrogen gas, which can be stored and used later as a fuel. By processes like these, the sun's energy can be saved for use even when it is not shining brightly.

Utilization of solar energy for direct production of electricity requires cells made of semiconductors. These are materials in which valence (bonding) electrons are not normally free to migrate and carry an electric current, but can be excited to a conducting state with a small amount of extra energy. The standard example of a semiconductor is the element silicon, in which a bonding electron can be raised to a conducting energy state, leaving behind a positive "hole" in the body of the silicon crystal. Both electrons and holes can act as charge carriers in conduction of current.

Taking this a step further, if the silicon is "doped" with a small amount of phosphorus, the extra electron in the phosphorus atom (compared with the number in silicon) is a conducting electron, and the result is n-silicon (negative silicon, because of the extra electrons). Similarly, a p- (for positive) silicon can be made by doping with aluminum, which is missing an electron compared with silicon. Such a material conducts electricity via the holes, which migrate within the body of the material as electrons do in n-silicon. Compounds of a Group III and a Group V element (such as gallium arsenide and indium phosphide) can also be made to show n- and p-behavior. Other products are possible, including fuel gases such as methane and ethane from inexpensive starting materials such as acetic acid.

At the beginning of the 1980's, the production of electricity was the major goal. More than a decade of development of liquid-junction cells culminated in the liquid-junction solar cell, announced in 1981 by Adam Heller, Barry Miller, and F. A. Thiel, all of Bell Laboratories (one of the major research centers in the area), in a publication in *Applied Physics Letters*. Technically, their cell is a p-indium phosphide/vanadium(II)-vanadium(III)-hydrochloric acid/carbon cell. Its 11.5 percent efficiency was sufficient to raise the hope that liquid-junction cells might be competitive with all-solid-state solar cells, which were well above this efficiency at the time, but only in the expensive crystal form. Other liquid-junction cells exceeded 14 percent efficiency, but only with artificial light sources. By the mid- to late 1980's, however, solid-state crystalline cells of silicon and other elements reached efficiencies near 30 percent, and even solid-state polycrystalline and amorphous silicon, in inexpensive thin films (the kind used in solar-powered calculators), could produce electricity in the 10 to 15 percent efficiency range. It is worth noting that all the liquid-junction work was done with crystalline semiconductors; the cost-cutting thin-film materials were not evaluated. The net result was that, for electricity production, the liquid-junction cells took a back seat to solid-state devices by the late 1980's, and research on liquid-junction devices has come to a near standstill. Tandem solid-state cells (two piggy-back cells absorbing at different wavelengths in the visible spectrum) achieved more than 30 percent conversion efficiency by the early 1990's. In addition to the double-absorption feature, these cells relied on concentration of sunlight by mirror and lens systems.

Scientists were determining if liquid-junction cells could be used for production of chemical materials with solar energy. By their nature, these must have a liquid portion containing dissolved chemical materials. The dissolved species are oxidized or reduced to usable forms by the electron flow generated by the semiconductor part of the device. Many problems arise, the most prominent being that when the semiconductor forms an anode, it consumes itself because the holes (positive centers) that gather at the surface of the anode are powerful oxidizing agents (electron removers), and if diffusion processes do not bring chemical materials from the solution to the surface quickly enough, the holes oxidize the adjacent portion of the anode. Methods have been developed to suppress this phenomenon but not to eliminate it.

Reduction reactions at the cathode are more successful because the cathode is more stable, but even at this point chemical limitations take over. Easily reducible species such as hydrogen ion react at the semiconductor cathode, but many metals are more difficult to reduce and do not yield to solar-induced reactions. Hydrogen production is, in fact, the most promising of the solar chemical reactions, particularly as the product, hydrogen gas, can be easily stored and used as a fuel. Research in this area continues, but at a slow pace.

Impact of Event

Given that research on liquid-junction solar cells has fallen off and solid-state devices are receiving nearly all the attention in solar generation of electricity, Heller, Miller, and Thiel's cell in itself is not an important landmark. Even the stacked cells of gallium arsenide and crystalline silica that broke the 30 percent efficiency barrier in 1988 have only a symbolic impact on the effort to use the sun's energy to generate electricity, particularly as they are far from ready for commercial application.

Taken collectively, however, these cells and a host of others in four decades of development have an impact that is more than merely symbolic. They represent the steady progress toward solar electricity as a power source competitive with grid-distributed commercial electricity. Solar electricity was first considered as a commercial possibility after the oil crisis in 1973. At that time its cost, with existing technology, was about fifteen dollars per kilowatt hour. By the late 1980's, that figure had fallen to thirty cents per kilowatt hour, within striking distance of conventionally generated power, at six to twelve cents.

Already, solar electricity is in use in remote places where power lines cannot be brought in because of distance or expense: telecommunications relay stations, remotely operated lighthouses, and most particularly, the various space probes, vehicles, and orbiters. In some cases, solar electricity is inexpensive enough to be fed into the power grid at peak demand times when the price paid by power companies is high. One such installation is in operation at Carissa Plains, California, built by ARCO Solar, Inc., and Pacific Gas and Electric Company as a profit-making venture. It produces 7.2 megawatts peak power, enough for the needs of a fair-sized town.

Although the Carissa Plains station is the largest built in the 1980's, it is not the only one in the United States. Many others can be found in California and in Alabama. Other nations have built facilities with capacities ranging from 100 kilowatts or so up to a megawatt. Japan has led in the research and development of these technologies, but other nations have kept close pace: Australia and New Zealand, Germany and other European countries, even the Sahel desert region in northern Africa, have solar power stations. A powerful attraction of these installations in remote areas is that, unlike coal- or oil-fired turbine systems, the solar power stations require very little maintenance and have operating lives of twenty or thirty years before replacement is necessary. Estimates of solar power's share of the world market by the year 2000 range from 10 to 40 percent.

Bibliography

"The Bright, Wet Look for Solar Cells." *Science News* 124 (December 10, 1983): 376. One of the few mentions in the popular science press of liquid-junction cells at the time when they were regarded as competitive with solid-state.

DeMeo, Edgar A., and Roger W. Taylor. "Solar Photovoltaic Power Systems: An Electric Utility R and D Perspective." *Science* 224 (April 20, 1984): 245-251. A thorough discussion of solar electric power from the commercial standpoint, with projections over two decades. Cost and output projections; some discussion of technologies.

Hamakawa, Yoshihiro. "Photovoltaic Power." *Scientific American* 256 (April, 1987): 86-92. A sound and thorough discussion of the use of semiconductors as solar cells, covering both theory and materials technology. Economics of solar energy, some discussion of existing installations.

Heller, Adam. "Hydrogen-Evolving Solar Cells." *Science* 223 (March 16, 1984): 1141-1148. Best discussion in this set of references of production of chemicals by liquid-junction solar cells. Very knowledgeable, if rather technical exposition of chemistry of cells. Large bibliography, well connected to points made in article.

Heller, Adam, Barry I. Miller, and F. A. Thiel. "11.5% Solar Conversion Efficiency in the Photocathodically Protected P-Indium Phosphide/Vanadium (3+) Ion-Vanadium (2+) Ion-Hydrogen Chloride/Carbon Semiconductor Liquid Junction Cell." *Applied Physics Letters* 38, no. 4 (1981): 282-284. Announcement of the cell that is the subject of this article. The authors' principal innovation appears to have been an oxidative surface treatment of the indium phosphide semiconductor.

Hubbard, H. M. "Photovoltaics Today and Tomorrow." *Science* 244 (April 21, 1989): 297-304. Excellent coverage of both economics and technology of solar electricity production. Large bibliography.

Maugh, Thomas H., II. "Catalysis in Solar Energy." *Science* 221 (September 30, 1983): 1358-1361. "Fuels from Solar Energy: How Soon?" *Science* 222 (October 14, 1983): 151-153. Simple and lucid explanations of both solid-state and liquid-junction cells and how they are used in producing electricity and bringing about chemical reactions.

Pool, Robert. "Solar Cells Turn 30." *Science* 241 (August 19, 1988): 900-901. Announcement of Sandia National Laboratories' tandem solar cell that broke the 30 percent conversion efficiency "barrier." Careful description of the construction of a cell.

Rabani, J., ed. *Photochemical Conversion and Storage of Solar Energy.* Jerusalem: Weizmann Science Press, 1982. Articles contributed by major workers in the field, covering every aspect of the material indicated by the title. Technical.

Robert M. Hawthorne, Jr.

Cross-References

Elster and Geitel Devise the First Practical Photoelectric Cell (1904), p. 208; Einstein Develops His Theory of the Photoelectric Effect (1905), p. 260; Thomson Wins the Nobel Prize for the Discovery of the Electron (1906), p. 356; Millikan Conducts His Oil-Drop Experiment (1909), p. 443; Esaki Demonstrates Electron Tunneling in Semiconductors (1957), p. 1551; Manabe and Wetherald Warn of the Greenhouse Effect and Global Warming (1967), p. 1840; Heeger and MacDiarmid Discover That Iodine-Doped Polyacetylene Conducts Electricity (1977), p. 2063; The First Commercial Test of Fiber-Optic Telecommunications Is Conducted (1977), p. 2078; Solar One, the Prototype Power Tower, Begins Operation (1982), p. 2216; The British Antarctic Survey Confirms the First Known Hole in the Ozone Layer (1985), p. 2285.

CASSINELLI AND ASSOCIATES DISCOVER R136A, THE MOST MASSIVE STAR KNOWN

Category of event: Astronomy
Time: June, 1981
Locale: Washburn Observatory, University of Wisconsin, Madison

The discovery of the supermassive starlike object, R136a, presented important theoretical problems concerning stellar size and stability

Principal personages:
> JOSEPH PATRICK CASSINELLI (1940-265), an American astronomer known for his studies on the formation of very massive stars using X-ray and radio frequencies
> JOHN SAMUEL MATHIS (1931-370), an American astronomer known for his work on the theoretical properties of interstellar dust
> BLAIR DEWILLIS SAVAGE (1941-557), an American astronomer who has studied the properties of the interstellar medium and whose work led to the discovery of a corona around the Milky Way

Summary of Event

One of the interests of stellar astronomers is the search for large stars and how they evolve. The Orion nebula has been observed for protostars, stars not yet at the point of undergoing nuclear fusion for energy. Infrared images revealed a strange object embedded in the densest center portion of the nebula. The object, called a Becklin-Neugebauer object (BN), was measured at 200 astronomical units (distance from the earth to the sun) across, but with a surface temperature of only 600 Kelvins.

Betelgeuse, also in the constellation of Orion, is a cool red star in the late stages of stellar evolution. One of the brightest objects in the night sky, Betelgeuse is very luminous, in spite of its low temperature. Astronomers have calculated its diameter as one thousand times that of the sun. If Betelgeuse was placed at the center of the solar system, Earth would be engulfed below its surface. Betelgeuse is classified as a red supergiant and must be nearing the end of its lifetime.

The 30 Doradus, or Tarantula nebula, in the Large Magellanic Cloud has been of ongoing interest to astronomers since data were obtained in 1978 from the International Ultraviolet Explorer (IUE) satellite. The 30 Doradus nebula is the most luminous region of ionized hydrogen in the local group of galaxies. It appears illuminated by an intense source equal to as many as one hundred blue giant-type stars. This nebula was the center of attention in February of 1987, when the magnificent supernova of 1987 occurred to its southwest. The central region of the nebula is dominated by a peculiar hot object or objects with a strong stellar wind (rapid moving outflow of gas). In 1977, it was suggested that R136a, the brightest and bluest component of the radiation, might be a massive star. R136a was the first extragalactic object studied in high resolution with the IUE satellite. Observations were made initially to obtain information about the nature of the gas in the Milky Way halo.

Observational data from the years 1978 to 1980 were assembled by the research team of Joseph Patrick Cassinelli, John Samuel Mathis, and Blair DeWillis Savage. Small- and large-aperture cameras were used to collect both low- and high-resolution spectra. The small-aperture camera gave a good overall view of the shape of the spectrum from 120 to 300 nanometers (a nanometer is one billionth of a meter), showing the relative positions of all the absorption and emission lines. The large-aperture camera was used to study in detail the individual spectrum lines.

An ionized helium line was found in R136a that is similar to that found in Wolf-Rayet stars, which are very hot and have broad emission lines. The helium and a nitrogen emission line were observed as broadened with flat bottoms. Emission is caused by shells of gas that the star ejects into space. This was evidence of a very large mass loss by stellar winds.

There were other indications from spectral studies that R136a was very hot; an ionized silicon emission line that is strong in stars with temperatures less than 50,000 Kelvins was weak in the spectrum of R136a. This was consistent with data from R136a showing a significant contribution to ultraviolet radiation, which is expected from a very hot star.

On the basis of the earlier data, the research team reported in 1981 that the radiation appears to come from a single superluminous object with a radius one hundred times, and a mass equal to twenty-five hundred times, that of the sun. They believed this energy was consistent with hydrogen fusion processes in its interior.

Later observations included data from the IUE covering the period from 1980 to 1982. The overall quality of the data was improved by applying an averaging process to the low- and high-resolution spectra. Additional spectral lines were observed, including ionized oxygen and iron. The oxygen lines were present only in hot blue stars of class 04 or younger age, suggesting a very hot star or collection of stars.

Ultraviolet variability was studied over the four-year period for evidence of the physical characteristics of R136a. No significant changes in the stellar absorption lines or spectral lines were observed during this time. This suggests that the star or stars do not vary in radiation output as do some of the more massive variable stars.

Radio measurements of the rate of ionization of the 30 Doradus nebula have led to an estimated nebular electron temperature of about 12,000 Kelvins. To produce that much ionization would require approximately one hundred hot 05-type stars or sixty 04-type stars. Measurements show that the ionized regions have a much higher temperature when compared with other nebulas of the Large Magellanic Cloud and Milky Way galaxy. The researchers concluded that a luminous object in the interior of the 30 Doradus nebula was responsible for this ionization.

The technique of speckle interferometry was developed in 1979, allowing for the first time direct measurement of the diameter of the planet Pluto. This technique involves taking a series of photographic images to freeze out the effects of turbulence in Earth's atmosphere, thereby improving resolution. Earlier measurements made from speckle interferometry showed what could be a cluster of stars having a diameter of less than 1,000 astronomical units. Interestingly enough, however, a secondary component appeared that was dimmer and slightly separated from the primary source of radiation.

When the analysis of the complete set of data from 1978 to 1982 was finished, the research team concluded that the radiation from R136a was dominated by emission from a supermassive star or a small group of supermassive stars. The following properties were derived from theory using the complete data set, which resulted in modifying the original size estimates somewhat: a radius equal to fifty times that of the sun, a mass equal to 2,100 times the mass of the sun, and a luminosity equal to 60 million times that of the sun. The researchers also believed that the deduced properties were compatible with the theory of internal structure for supermassive stars. It was argued that such a star might form by ordinary stellar collapse if the right types of interstellar dust were present or perhaps by the coalescence of several stars in a cluster of high stellar density. Such a star would be expected to have a lifetime of some 2 million years.

Impact of Event

The existence of massive stars such as red or blue supergiants has been known for some time. A supermassive star such as R136a exceeded the mass of the most massive stars by a factor of twenty times. There are theoretical problems associated with very large stars, including the formation, stability, lifetime, and upper mass limits. The upper mass limits for a star to form and be vibrationally stable are not known from observations or theory.

An alternative explanation to the single star hypothesis for R136a is that perhaps the radiation is coming from a compact star cluster instead. To explain the spectral lines and the overall radiation intensity (continuum) as a result of R136a would require a cluster of as many as thirty hot blue 03 class stars and fifteen Wolf-Rayet-type stars. According to the Wisconsin astronomers, this would be an unusual mixture of star types because the 03-type stars are very young, but the Wolf-Rayet stars are of a later spectral type and are older. As a rule, stars within the same cluster are assumed to have about the same age, forming from gas and dust that is gravitationally contracting.

An important development regarding the supermassive star hypothesis occurred in 1988 when two German astronomers, R. Neri and Michael Grewing at the University of Tübingen, were using advanced speckle interferometry. R136a was now reported as resolved into at least five stars. Another luminous blue object in the Large Magellanic Cloud, previously known as one of the most massive and luminous stars, was resolved into a cluster of six lesser stars. Later in 1988, Eta Carinae, a very luminous star in the Milky Way galaxy, was resolved into four components, each less than 0.03 second of arc apart.

These advanced techniques used a charge-coupled device (CCD) exposed to red light and subjected to extensive image processing. This technique improved the light-gathering methods that were one of the serious problems facing the earlier researchers. The reduction of radiation in this region of the sky is called extinction and is caused by the high concentrations of gas and dust in the nebula. The effect of the extinction is to make the brightness dimmer or reddened by the Milky Way foreground dust that is between the object and the observer.

The 1988 discoveries could have an important impact on the scale of distances in the universe. In galaxies as far away as 10 million light-years, the brightest stars within the galaxy are used as distance calibrators. The magnitudes of such stars are compared to those in the brightest clusters of the Milky Way. Nearby clusters are less likely to be mistaken for single stars than the more distant ones. If the stars are really multiples, their distances would tend to be underestimated. This would, in turn, lead to smaller values for the Hubble constant (relates the velocity of recession of galaxies to their distance) and would give larger estimates for the size and age of the universe.

Bibliography

Cassinelli, Joseph P., John S. Mathis, and Blair D. Savage. "Central Object of the 30 Doradus Nebula, a Supermassive Star." *Science* 2212 (June, 1981): 1497-1501. Using data from the International Ultraviolet Explorer satellite, reports the first discovery of a peculiar hot massive object. Presents arguments regarding the nature of R136a—whether it could be a single supermassive star or a cluster of stars.

Cooke, Donald A. *The Life and Death of Stars.* New York: Crown, 1985. A superbly illustrated book with color plates of nebulas that may be stellar birthgrounds and nurseries. Presents the Great Orion nebula as a case study. Diagrams show the interiors of stars and mechanisms of heat transport within a star.

Kippenhahn, Rudolf. *100 Billion Suns.* Translated by Jean Steinberg. New York: Basic Books, 1983. The author and director of the Max Planck Institute for Astrophysics, using humor and insight, describes discoveries in stellar astronomy, including his own. Illustrations show the evolutionary paths of both stars and star clusters in different stages of their lifetimes.

Savage, Blair D., Edward L. Fitzpatrick, Joseph P. Cassinelli, and Dennis C. Ebberts. "The Nature of R136a, the Superluminous Central Object of the 30 Doradus Nebula." *The Astrophysical Journal* 273 (October, 1983): 597-623. Presents additional data subsequent to the initial discovery. Discusses the theoretical problems associated with the supermassive star hypothesis.

Time-Life Books, eds. *Stars.* Alexandria. Va.: Time-Life Books, 1989. Uses dramatic multicolored diagrams to illustrate fast-burning giant stars and the convective heat transport mechanism from the core outward. Discusses high mass stars and how they evolve from protostars.

Michael L. Broyles

Cross-References

Hertzsprung Notes the Relationship Between Color and Luminosity of Stars (1905), p. 265; Hertzsprung Describes Giant and Dwarf Stellar Divisions (1907), p. 370; Hertzsprung Uses Cepheid Variables to Calculate the Distances to Stars (1913), p. 557; Russell Announces His Theory of Stellar Evolution (1913), p. 585; Michelson Measures the Diameter of a Star (1920), p. 700; Supernova 1987A Corroborates the Theories of Star Formation (1987), p. 2351.

THE IBM PERSONAL COMPUTER, USING DOS, IS INTRODUCED

Category of event: Applied science
Time: August 12, 1981
Locale: Boca Raton, Florida

International Business Machines (IBM) entered the personal computer market-place, giving credibility and new direction to the microcomputer revolution

Principal personages:

TOM WATSON, SR. (1874-1956), the founder of IBM Corporation, who set corporate philosophy and marketing principles

FRANK CARY, the chief executive officer of IBM at the time the decision was made to market a personal computer

JOHN OPEL, a member of the Corporate Management Committee

GEORGE BELZEL, a member of the Corporate Management Committee

PAUL RIZZO, a member of the Corporate Management Committee

DEAN MCKAY, a member of the Corporate Management Committee

WILLIAM L. SYDNES, the leader of the original twelve-member design team

Summary of Event

International Business Machines (IBM) was formed in 1914 when a banker persuaded three companies to combine and form the Computer-Tabulating-Recording Corporation. IBM has become the world's leading computer company. Many consider IBM to be a symbol of institutionalized excellence, progress, and profit. After IBM failed to enter the minicomputer market in the late 1960's and early 1970's, many wondered if IBM would enter the emerging microcomputer market. When IBM failed to enter the minicomputer market, it left room for new companies like Digital Equipment Corporation and Data General to make large profits in an open market. In the late 1970's, Apple Computer, Radio Shack, and several others proved that there was a new lucrative market in microcomputers. IBM executives saw the potential of this market. In order to make an impact, IBM had to move quickly because time was of the essence.

Many at IBM doubted that the bureaucracy could be made to move with the events unfolding outside the company. Apple Computer was beginning to make inroads at large IBM accounts. IBM stock began to stagnate on Wall Street. A 1979 *Business Week* article asked: "Is IBM just another stodgy, mature company?" The personal computer market was expected to grow more than 40 percent in the first part of the 1980's, but it was clear, also, that IBM would have to make some internal changes in order to accomplish the task of bringing a competitive personal computer to the market. IBM's initial investigations into the production and marketing of personal computers produced grim results. The estimates on the mass purchase of parts for the project ran higher than the potential price it could bring on the market.

The decision to build and market the personal computer (PC) was made by the Corporate Management Committee (CMC). CMC members included chief executive officer Frank Cary, John Opel, George Belzel, Paul Rizzo, Dean McKay, and three senior vice presidents. In July of 1980, Cary gave the order to proceed. He wanted the PC to be designed and built within a year. The CMC approved the initial design of the PC one month later. Twelve engineers, with William L. Sydnes as the leader, were appointed as the design team. At the end of 1980, the team had grown to 150.

For many years, IBM had been a company set in its ways. In many ways, it remains so. Typically, IBM sticks to traditions set up by its founder Tom Watson, Sr. It was clear that a competitive microcomputer could be produced using only nontraditional methods. Most of the components of the PC had to be produced outside IBM. Microsoft won the contract to produce the PC's disk operating system (DOS) and the BASIC language that is built into the PC's read only memory (ROM). Intel Corporation's sixteen-bit 8088 was chosen as the central processing unit (CPU) chip in the PC. Outside programmers were solicited to write software for the PC; this strategy would not have been heard of, within IBM, only ten years earlier. IBM approached ComputerLand and Sears to handle the sales and service for the PC. IBM saw that Apple Computer had successfully used a network of franchises to market its computers. Training ComputerLand and Sears to handle the service broke a seventy-year tradition of in-house servicing of products. IBM reorganized in fall, 1981. This reorganization shifted the focus of the entire organization to a product-driven operation. It also reduced the amount of internal competition between divisions. IBM hired the firm of Lord, Geller, Frederico, and Einstein, Inc., of New York to design a media campaign for the new personal computer. Readers of magazines and periodicals saw Charlie Chaplin advertising the PC. The PC was delivered on schedule to the public on August 12, 1981. The price of the basic "system unit" was $1,565. A system with 64 kilobytes of random access memory (RAM) and a 13-centimeter single-sided disk drive, holding 160 kilobytes, and a monitor was priced about $3,000. A system with color graphics, a second disk drive, and a dot matrix printer cost about $4,500. It was the combination of IBM's "radical" use of outside resources, internal organization, and high-level support from management that made the PC possible.

While nearly everyone agreed that IBM had done everything right, acceptance of the PC was slow at first. Many familiar application packages had been adapted to the PC and were available when it was introduced. VisiCalc from Personal Software—the program that is credited with making the microcomputer revolution—was one of the first available. Other packages included a comprehensive accounting system by Peachtree Software, and a word processing package called Easywriter from Information Unlimited Software. IBM had software that would allow the PC to be used as a "dumb terminal" on larger computer systems. As the selection of software grew, so did sales. In the first year after its introduction, the IBM PC went from a zero market share to 28 percent of the market. The success of the PC has not been IBM's alone. Many hundreds of companies were able to produce software and hardware for the PC. The basic PC soon was upgraded to a system with 256 kilobytes of memory and a 13-centimeter double-sided drive capable of storing 360 kilobytes.

The PC software market grew rapidly. Within two years, powerful products such as Lotus Corporation's 1-2-3 spreadsheet had come to market. Many believe that Lotus 1-2-3 is the program that caused the PC to take off in business. Many companies produced add-on boards that plugged into the PC's expansion slots. Up to 640 kilobytes of memory could be added to the PC using expansion boards from companies such as Quadram and AST. Businesses found the PC to be a powerful tool. As new uses were found, programs grew, and required PCs to have more memory. The PC was expandable; this meant that when the needs of the user outgrew the PC, the PC could grow. It was not uncommon to find people who had purchased memory expansion boards so that they could build huge spreadsheets in Lotus 1-2-3. The PC has survived because of its expansion capability. When a new graphics board is introduced, it can be added to any PC.

With the increasing popularity of the PC, the market for information about the PC grew. A whole industry of computer magazines about personal computers was created. They offered reviews of new hardware and software products, tips and hints for many popular programs, and advertising for the multitude of companies manufacturing software and hardware for the PC.

IBM has continued to upgrade the PC. In 1983, the PC/XT was introduced. It had more slots and a fixed disk offering 10 million bytes of storage for programs and data. Many of the companies that made expansion boards found themselves able to make whole PCs. An entire range of PC-compatible systems was introduced to the market, many offering features that IBM did not include in the original PC. The original PC has become a whole family of computers sold by both IBM and other companies. The hardware and software continue to evolve; each generation offers more computing power and storage capacity with a lower price tag.

Impact of Event

It is fair to say that IBM's entry into the personal computer marketplace continues to affect the entire computer marketplace. Personal computers took over the function of many programs on traditional mainframes. Management information systems departments—the people who ran the mainframes in most companies—had to embrace the PC or face being phased out in favor of this new technology. The PC, in conjunction with a spreadsheet program and word processing software, gave anyone within a company, including the corporate executive, the ability to track budgets, run simulations, and prepare reports. IBM proved that it could respond to the computer market. IBM also proved that by using resources outside the company itself it could make a cost-effective entry in a new market. This was a major break from the original sales and management rules as set down by Watson when he founded IBM.

IBM's entry into the microcomputer market gave microcomputers credibility. Apple Computer's earlier introduction of its computer did not win wide acceptance with the corporate world. Apple Computer did thrive within the educational marketplace. IBM's name already carried with it much clout, because IBM was a successful company. Apple Computer represented all that was great about the "new" microcomputer, but the IBM PC carried IBM's stability and success.

IBM set the standard for personal computers that came later. IBM chose as its character set the American Standard Code for Information Interchange (ASCII) in spite of the fact that IBM's mainframes used the proprietary Extended Binary Coded Decimal Interchange Code (EBCDIC). The PC accessed memory in 8-bit chunks called bytes. Other microcomputers did this, too. Most mainframes and minicomputers had word accessible memory. More than one character would usually fit in a word. Making the smallest addressable chunk of memory the same size as a character made it easy to manipulate textual information as well as numbers.

IBM coined the term "personal computer" and its acronym "PC." PC is now used almost universally to refer to the microcomputer. It also had great significance with the consumers of the PC that had previously used a large mainframe computer that had to be shared with the whole company. This was their personal computer. This was important to many PC buyers, as the company mainframe was perceived as complicated and slow. The PC owner now had complete control.

It took several years for the IBM Personal Computer to make an impact throughout the growing personal computer marketplace. IBM had created a new market, an exploding one, for the first business personal computer. Much of the credit must go to the companies that wrote software for the PC. Other companies, such as Digital Equipment Corporation, tried to enter the personal computer marketplace, but lacked the backing of the popular software houses. These "other" computers were not attractive to the marketplace because they lacked the feature that IBM PC owners now took for granted: a large base of software from which to choose. It took time for IBM and other companies to build this base of software. Computer companies that failed to come out with a strong base of software went bankrupt. Many companies now produce or sell systems that are software and hardware compatible with the IBM PC.

Bibliography

Burton, Kathleen. "Anatomy of a Colossus." *PC Magazine* 1, no. 9 (January, 1983), no. 10 (February, 1983), no. 11 (March, 1983). This series of articles is a well-written history of the IBM corporation. The second installment describes what IBM had to overcome to develop the PC. It offers a different perspective from that of earlier articles, since it was written more than a year after the release of the personal computer.

Evans, Christopher. *The Micro Millennium*. New York: Viking Press, 1980. This book shows the impact the computer revolution, especially the microcomputer, has had on society. Even though it was written before the IBM PC was introduced, it makes many accurate predictions of the changes caused in society by computers. It predicts a future in which people could work from home using a personal computer with a phone connection back to the company, a future in which the printed word is dead. It offers a good view of the effects that computer technology could have.

Fertig, Robert T. "Is IBM the Work in Apple's Future?" *PC Magazine* 3 (February, 1984): 152-156. This article describes the worries in the industry about the well-being of Apple Computer in the face of IBM's challenge.

McMullen, Barbara E., and John F. McMullen. "Apple Charts the Course for IBM." *PC Magazine* 3 (February, 1984): 122-130. This article shows the relationship between Apple Computer and IBM. IBM borrowed the strategy that Apple used to market its personal computer and took the lead in the microcomputer market.

Sculley, John, with John A. Byrne. *Odyssey: Pepsi to Apple, A Journey of Adventure, Ideas, and the Future.* New York: Harper & Row, 1987. This book gives many insights into the competition between Apple Computer and IBM. It presents a view from Apple's perspective of IBM's movement into the personal computer market. It shows how IBM followed Apple's lead in the microcomputer market. IBM made Apple's strategy work better than it did for Apple. It gives a perspective on the mistakes that Apple believed it made in competing with IBM.

Staples, Betsy. "IBM Personal Computer: The Big Blue Giant Makes Its Move." *Creative Computing* 11 (November, 1981): 14-18. This article is one of the first reviews of the IBM Personal Computer. Staples believes that IBM had done everything right. *Creative Computing* magazine was one of the first computer magazines to embrace the new microcomputer technology.

Watson, Thomas J. *Father, Son & Co.: My Life at IBM and Beyond.* New York: Bantam Books, 1990. A very good history of IBM. Written by the son of the founder of IBM, it describes the philosophy that made the company what it is, and how the philosophy still plays a role in the operation of IBM. Depicts the relationship between the founder of IBM, his son, and the company. Describes the successes of IBM and examines its shortcomings.

Daniel P. Boehlke

Cross-References

Backus' IBM Team Develops the FORTRAN Computer Language (1954), p. 1475; Hopper Invents the Computer Language COBOL (1959), p. 1593; Kemeny and Kurtz Develop the BASIC Computer Language (1964), p. 1772; The Floppy Disk Is Introduced for Storing Data Used by Computers (1970), p. 1923; The Microprocessor "Computer on a Chip" Is Introduced (1971), p. 1938; Apple II Becomes the First Successful Preassembled Personal Computer (1977), p. 2073; IBM Introduces a Personal Computer with a Standard Hard Disk Drive (1983), p. 2240.

CLEWELL CORRECTS HYDROCEPHALUS
BY SURGERY ON A FETUS

Category of event: Medicine
Time: September, 1981
Locale: Denver, Colorado

Clewell performed surgery on a hydrocephalic fetus through fetoscopy, by placing a shunt to allow the removal of fluid from the brain before birth, leading to more normal brain development

Principal personages:

WILLIAM H. CLEWELL (1943-   ), an obstetrician-gynecologist who guided studies on correcting prenatal hydrocephalus without opening the uterus of the mother

JOHN B. NEWKIRK, a bioengineer who designed the shunt device used by Clewell in treating fetal hydrocephalus

MICHAEL R. HARRISON (1944-   ), a pediatric surgeon at the University of California, San Francisco, who studied prenatal treatment of induced hydrocephalus in fetal lambs and monkeys

MARIA MICHEJDA, a physician at the National Institutes of Health who performed experiments on induced hydrocephalus in fetal monkeys using a screw-in device to relieve fluid and pressure

Summary of Event

In the late 1970's and early 1980's, fetal medicine was expanding in several fields. Treatment of the fetus had been slowly becoming possible over the previous decades as new techniques were developed for recognizing and managing anatomical and physiological problems. Fetal blood transfusions were begun in the 1950's as a treatment for erythroblastosis fetalis, the destruction of fetal red blood cells by the immune system attack of an Rh-negative mother against her Rh-positive fetus. This allowed the birth of babies who would otherwise probably have died in utero. Fetal surgery also was being done by the early 1980's by opening the mother's abdomen and uterus and operating on the fetus, usually after removing the affected portion of the fetus from the uterus. Such a major operation would nearly always necessitate a cesarean delivery later. Emphasis was being placed on trying to develop techniques to correct abnormalities with the least invasiveness for both the mother and the fetus.

A major improvement in the recognition and treatment of fetal aberrations was the result of technological improvements in ultrasound imaging and the ability to interpret what was visualized by this process in the fetus. Diagnosis of brain enlargement in the fetus by ultrasound became possible in the 1970's. A fetus with hydrocephalus, literally "water in the head," develops ventriculomegaly, or enlarged brain ventricles, the cavities that contain cerebrospinal fluid. This abnormality produces thinning of the cerebral lobes and skull enlargement that generally causes difficulties in normal delivery. Mental and physical retardation most often accompany the physical deformities of the head. Fluid buildup occurs as a result of blockage of the pathway by which cerebrospinal fluid normally enters the bloodstream. It was recognized that if another way could be found to drain off the excess cerebrospinal fluid, perhaps the head enlargement and cerebral thinning could be prevented and a more normal child could be born.

In 1981, William H. Clewell and his team of physician-scientists developed a means of treating this problem by placement of a shunt in the head of an affected fetus, allowing the excess fluid to drain into the surrounding amniotic fluid. At the time of this innovative treatment, Clewell had previous experience in treating fetuses, having transfused blood numerous times over the previous four years into fetuses that were affected by erythroblastosis fetalis. The precise placement of the transfusion needle into the fetal abdomen was possible with guidance by ultrasound.

The parents of the hydrocephalic fetus that Clewell treated by shunt placement had a previous son born with congenital hydrocephalus. This was caused by the presence of a defective X-linked gene for aqueductal stenosis, or narrowing of the passage by which fluid is normally removed from the brain. The second pregnancy was being carefully followed, and it was recognized that the male fetus definitely was affected at twenty-one weeks of gestation, about halfway through the pregnancy. The parents did not wish to consider an abortion and wanted treatment of the fetus to limit or prevent damage caused by the hydrocephalus.

Clewell and a team of two other obstetricians, two radiologists/ultrasonologists, a neurosurgeon, and a bioengineer were involved in this treatment. A miniature shunt was designed by John B. Newkirk, the bioengineer, to be placed in the skull of the fetus as a drainage route for excess fluid. It was constructed of medical-grade Silastic tubing with a valve that would limit the rate of outflow of cerebrospinal fluid and prevent backflow of amniotic fluid into the brain. The shunt was put into place in the fetal skull at twenty-three weeks of gestation, when the ultrasound measurements showed that ventriculomegaly was increasing.

Although this process was a surgical procedure, there was no need to open up the mother's abdomen and uterus to treat the fetus. The shunt had been designed so that it could be passed into the fetal head through a blunt needle with a pointed stylet attached. The needle and stylet were clearly visible with ultrasound and could be guided indirectly by looking at the real-time images on the computer screen. It was thus possible to use local anesthesia to allow a small incision in the skin and abdominal fascia of the mother. The uterus was entered only with the needle, and the shunt was passed through the needle into the head of the fetus. The ventricles could be visualized clearly with ultrasound, and shunt placement was similar to that used in an infant, although this placement was done by looking at the screen rather than at the infant's head to place the tubing. The needle and stylet were withdrawn, leaving the shunt correctly in place, as determined by ultrasound. The fetus showed no distress at the procedure, and only sporadic uterine contractions that soon stopped were produced as a result of the treatment. The process was much less invasive than surgery that would allow the fetal head to be seen directly for shunt placement, and the risk of fetal loss from surgery-induced premature birth was reduced.

As a result of the shunt placement, the size of the fetus' ventricles decreased and cerebral thickness increased over several weeks. Measurements through thirty-two weeks of gestation showed that the shunt was performing as desired. Examination at thirty-four weeks, however, showed increased ventricular size, presumed to be the result of obstruction of the shunt. The mother was treated with a drug to induce maturation of the fetal lungs, because the normal time of delivery is about forty weeks. The baby was delivered at thirty-four weeks' gestation by cesarean section. The shunt was found to be completely blocked by growth of tissue. The vigorous, active infant was examined soon after birth by ultrasound and computerized tomography (CAT scan), both of which confirmed the need for a shunt. The original prenatal shunt was removed, and a new shunt was put in place to divert fluid into the baby's peritoneal cavity. When examined at thirteen weeks after birth, the child showed some physical delay, but by eighteen weeks, he showed improved strength and coordination and had learned to smile.

Impact of Event

In the early 1980's, there were numerous investigations dealing with the prenatal treatment of abnormal fetuses. The use of ultrasound was replacing fluoroscopy (X rays) as a means of visualizing the movement of structures within the body, particularly in the developing fetus. Technology was producing greater resolution in ultrasound, so more detail could be distinguished in earlier fetuses. Research in animal systems, particularly the fetal monkey and fetal lamb, was leading to greater familiarity with the development of hydrocephalus and possible ways to treat it. In these animal studies, however, it was common to pull the fetal head out of an incision in the uterus, both to induce hydrocephalus artificially and then later to treat it. Michael R. Harrison and a group at San Francisco were involved in these studies, placing shunts in lambs and monkeys with hydrocephalus induced by chemical treatment with a teratogen. Maria Michejda and Gary D. Hodgen at the National Institutes of Health had developed a screw-in vent for use in treating induced hydrocephalus in fetal monkeys. Their process again required cutting into the uterus, and the results were disappointing because the chemical produced other neural tube defects as well as hydrocephalus, and no treated monkeys were fully normalized by the release of excess fluid.

A relatively noninvasive clinical treatment for hydrocephalus used before Clewell's procedure involved repeatedly puncturing the fetal skull and brain to draw off fluid, a process called cephalocentesis. Jason C. Birnholz and Frederic D. Frigoletto had tried this approach in early 1981, withdrawing fluid from a fetus six times between twenty-five and thirty-two weeks of gestation. After delivery at thirty-four weeks, the child had a developmental level of only six months at sixteen months of age, suggesting that the treatment was not able to alleviate all the effects of hydrocephalus.

Clewell's procedure was recognized and referenced in numerous papers on prenatal hydrocephalus following his report of the process. It does not appear to have had much effect on perinatal management of hydrocephalus, however, and there was no immediate increase in the use of the treatment. This resulted from the fact that many cases of recognizable prenatal hydrocephalus are accompanied by other abnormalities, and a simple treatment, such as shunting, can correct only the physical condition of ventricular enlargement. While this can be important in certain cases, such as the X-linked hereditary condition of aqueductal stenosis in Clewell's patient, it is not as useful in other forms of hydrocephalus. This is shown in a case report by Frigoletto and colleagues in November, 1982, in which they used a shunt similar to Clewell's to treat a fetus that had other anomalies as well as hydrocephalus. The baby died at five and one-half weeks of age of cardiac arrest, apparently caused by these other anomalies. These investigators recommended several conditions under which prenatal treatment for hydrocephalus would be advisable: when detection of hydrocephalus is too early to wait for birth and postnatal shunting; simple obstructive hydrocephalus, not associated with other major developmental problems of the brain or other systems; no chromosomal aberrations; progressive ventricular enlargement; and a team of physicians representing different disciplines involved in treatment.

Clewell and others reported on the evaluation and management of ventriculomegaly in 1985, describing the fetal surgery approach as experimental. Reasons to include a fetus or to exclude one from this treatment were listed. Inclusion would be possible only if there were progressive ventricle enlargement and cerebral thinning, gestational age at shunting less than thirty-two weeks, a single fetus, and no other significant detectable abnormalities. Exclusion from consideration would occur in the case of twins, gestational age great enough that birth could soon occur and treatment could then take place, or the presence of other anomalies. Five pregnancies were discussed in this report in which shunts were placed. Two of the resulting children were normal at two and three years of age, two were severely retarded at three years, and the last fetus was electively aborted after the shunt was placed. These results suggest that, in some cases, the treatment completely corrects the effects of hydrocephalus. In another paper in 1985, Clewell and coauthors again stressed the poor prognosis for most hydrocephalus detected prenatally. Studies done by a team at Yale University Medical School, led by F. A. Chervenak in 1985, showed that shunts performed there were not able to prevent severe retardation in the infants born after shunting.

Additional severe abnormalities often are present in fetuses that show ventriculomegaly, even if the abnormalities are not detectable before birth. With early detection and fully informed decisions by parents, often the chosen course with fetal hydrocephalus is elective abortion. In a small number of cases, Clewell's approach to prenatal treatment works very well, but in the majority of cases of ventriculomegaly, death or severe retardation occurs. Careful selection is necessary to determine which fetuses may benefit from this treatment. It may take several more years before Clewell's procedure can be reasonably evaluated to allow for testing the criteria for patient selection and to follow up the results of children treated. In the mid-1980's, modifications of the procedure and mechanical corrections that help prevent shunt obstruction or displacement have produced a better technique that will be able to help at least some children to develop normally after this treatment.

Bibliography

Chervenak, Frank A., Glenn Isaacson, and John Lorber. *Anomalies of the Fetal Head, Neck, and Spine: Ultrasound Diagnosis and Management*. Philadelphia: W. B. Saunders, 1988. Written for the physician who deals with ultrasound and neurological disorders; provides much information to the lay reader as well. Covers the normal anatomy of the brain and spinal cord of the fetus, and shows how it is envisioned by the use of ultrasound. Includes photographs of neural structures as they develop at different stages of gestation. Discusses diagnosis and prognosis of fetal hydrocephalus, with a very brief mention of shunt placement for prenatal treatment.

Clewell, William H., et al. "Placement of Ventriculo-Amniotic Shunt for Hydrocephalus in a Fetus." *The New England Journal of Medicine* 303 (October 15, 1981): 955. A very short communication that first announced the event. Written while the pregnancy was still in progress.

_____, et al. "A Surgical Approach to the Treatment of Fetal Hydrocephalus." *The New England Journal of Medicine* 306 (June 3, 1982): 1320-1325. Describes the team of physicians that developed and placed the shunt to treat fetal hydrocephalus, as well as the shunt itself and the procedure of placement. Includes numerous sonograms to show reduction of fluid in the ventricles after shunt placement. A photograph shows the shunt still in place shortly after birth of the child.

Frigoletto, Fredric D., Jr., Jason C. Birnholz, and Michael F. Green. "Antenatal Treatment of Hydrocephalus by VentriculoAmniotic Shunting." *JAMA* 248 (November 19, 1982): 2496-2497. A case report that briefly describes a shunt treatment that failed to work, followed by a second shunt on the same fetus that did allow fluid drainage. Includes recommendations for guidelines to be used in selecting cases for treatment.

Harrison, Michael R., Mitchell S. Golbus, and Roy A. Filly. "Management of the Fetus with a Correctable Congenital Defect." *JAMA* 246 (August 14, 1981): 774-777. Contains a useful set of guidelines for assessing the means of management of congenital malformations of the fetus. Prenatal hydrocephalus is included in the group that may require intervention in utero, which may include the use of a catheter, or shunt.

_____. *The Unborn Patient: Prenatal Diagnosis and Treatment*. Orlando, Fla.: Grune & Stratton, 1984. A major resource covering prenatal anatomical and physiological disorders, although out of date for current methods. Discusses fetoscopy and fetal sampling methods, as well as congenital hydrocephalus. Includes numerous photographs and sonograms to show what could be seen in ultrasound observations of the affected fetuses.

Levene, Malcolm I., Michael J. Bennett, and Jonathan Punt, eds. *Fetal and Neonatal Neurology and Neurosurgery.* Edinburgh: Churchill Livingstone, 1988. A large volume that covers the subject in depth for physicians. Includes numerous chapters by various contributors on the normal and abnormal aspects of neurological development in the fetus and infant. A section on imaging shows how to interpret ultrasound images. A section on defects of the neural tube discusses diagnosis and management of hydrocephalus, and two chapters on fetal ventriculomegaly and infantile hydrocephalus are covered under neurosurgery.

Michejda, Maria, and Gary D. Hodgen. "In Utero Diagnosis and Treatment of Non-Human Primate Fetal Skeletal Anomalies: I. Hydrocephalus." *JAMA* 246 (September 4, 1981): 1093-1097. Discusses research in the experimental production of fetal hydrocephalus in monkeys, and the subsequent treatment of the disorder by placement of a screw-type shunt or vent to allow outflow of cerebrospinal fluid from the enlarged brain ventricles.

Jean S. Helgeson

Cross-References

Röntgen Wins the Nobel Prize for the Discovery of X Rays (1901), p. 118; Blalock Performs the First "Blue Baby" Operation (1944), p. 1250; Bevis Describes Amniocentesis as a Method for Disclosing Fetal Genetic Traits (1952), p. 1439; Donald Is the First to Use Ultrasound to Examine Unborn Children (1958), p. 1562; Hounsfield Introduces a CAT Scanner That Can See Clearly into the Body (1972), p. 1961; Brown Gives Birth to the First "Test-Tube" Baby (1978), p. 2099; Daffos Uses Blood Taken Through the Umbilical Cord to Diagnose Fetal Disease (1982), p. 2205; The First Successful Human Embryo Transfer Is Performed (1983), p. 2235.

COLUMBIA'S SECOND FLIGHT PROVES
THE PRACTICALITY OF THE SPACE SHUTTLE

Category of event: Space and aviation
Time: November 12-14, 1981
Locale: 222 kilometers above Earth

Circling above the globe, the world's first reusable spacecraft opened the future to
payloads and experimenters to whom space was previously inaccessible

Principal personages:
JOE H. ENGLE (1932-   ), a United States Air Force colonel and STS 2
   commander
RICHARD H. TRULY (1937-   ), a Navy captain and STS 2 pilot
JOHN W. YOUNG (1930-   ), an STS 1 commander who became the
   first person to fly into space six times by being the commander of the
   STS 9 mission
ROBERT L. CRIPPEN (1937-   ), a Navy captain and STS 1 pilot

Summary of Event

The concept of a reusable spacecraft has been around for as long as there have been dreams of space flight. Thus, it was not surprising when the X-15 high-speed research aircraft program was developed in 1953. The X-15 would test the principles of launching a manned spacecraft such as a rocket and having it return to Earth and land like an airplane. The aircraft was drop-tested from under the wing of a large carrier aircraft, much like its predecessors, the X-1 and its variants, and the X-2. The modified Boeing B-52 bomber took the X-15 to a predetermined drop altitude and released it. The X-15's powerful rocket engine then propelled it to the edge of space and nearly six times the speed of sound.

Later, a slightly larger version of the X-15 would be placed on the nose of a Titan booster and launched into space. After deploying a small payload or conducting experiments, the X-15 would reenter the atmosphere and glide to a landing on the dry lake bed at Edwards Air Force Base, California. Budgetary cutbacks by Congress led to the reduction of the X-15 to a purely atmospheric test vehicle and, after 199 successful missions, the X-15 program came to a halt in 1968. During the X-15 program, a number of experiments used ablative materials (ones that remove heat by melting and carrying the heat from the vehicle). These tests led to the development of the heat-soak tiles currently used on the space shuttle orbiters.

In October, 1957, the Air Force began the X-20 Dyna-Soar (for Dynamic Soaring) program, which combined the efforts of three earlier projects to develop a reusable space plane. Out of a conference that opened on October 15 at the National Aeronautics and Space Administration's (NASA) Ames Research Center in northern California came three proposals for manned spacecraft. The first was a blunt reentry shaped craft—what essentially emerged as the Mercury spacecraft. A second design was for a lifting body—one that had a modest lift-to-drag ratio and permitted limited maneuverability during reentry. The third design was that of a flat-bottom hypersonic glider. This third type was chosen for the X-20.

Like the X-15, the X-20 would be launched atop a Titan booster. It would carry experiments or other payloads, including deployable satellites, into low-Earth orbit, and then glide back to Earth. Like the X-15 program, Dyna-Soar was destined to be made extinct by congressional budget cuts.

In the wake of the successful manned lunar landings in 1969, NASA proposed a choice of three long-range goals for the space program: first, an $8- to $10-billion per year program involving a manned Mars expedition, a space station in lunar orbit, and a fifty-person Earth-orbiting space station serviced by a reusable ferry, or space shuttle; second, an intermediate program, costing less than $8 billion per year, that deleted the lunar-orbiting space station, but kept the other elements; finally, a relatively modest $5-billion-per-year program that would embrace an Earth-orbiting space station and the space shuttle as its link to Earth.

Congress and the White House rejected all three as too expensive. By the spring of 1971, NASA was determined to get at least the space shuttle, since it would be the first step if any of the other programs got approval. Failing to impress the Office of Management and Budget that the proposed shuttle program would be cost-effective, NASA turned to the Department of Defense (DOD). Under NASA's plan, the DOD would provide a large portion of the funding for the shuttle, and in return, the shuttle would be used for military missions in addition to its scientific ones. In addition, the shuttle would become the only launch vehicle in NASA's fleet, thereby replacing expendable boosters and saving billions of dollars. Congress approved it, and plans were drawn for a shuttle system. A number of aerospace contractors submitted different designs, but the general concept involved a large, winged, liquid-fueled reusable booster with a smaller, winged, reusable orbiter. The orbiter would be carried aloft on the back of the booster. Before reaching orbit, the booster would separate and glide back to the launch site. The orbiter would be placed in low-Earth orbit, where it would conduct experiments, deploy satellites, or perform other on-orbit tasks. After three to seven days, the orbiter would fire its engines, reenter the atmosphere, and glide to a landing at the launch site. An air-breathing jet engine could be deployed to give the orbiter additional range, if necessary.

Budgetary cuts reduced the ideal space transportation system to the four-part system it became. Essentially, the shuttle system consists of an orbiter with three main engines to be used during launch; a large, external fuel tank for the engines; and a pair of strap-on solid rocket boosters to get the stack off the launch pad. The orbiter can be reused, as well as the major components of the solid rocket boosters, which parachute back to the water after they run out of fuel. Only the external tank is discarded. The space shuttle is launched from the same launch complexes and launch platforms from which the Apollo Moon-landing missions began. Extensive modifications allowed for the differences in the vehicles.

By 1977, the design and development phase of the program was complete, and a series of drop tests were conducted using an orbiter identical to later flight versions, but which was incapable of space flight. *Enterprise* (named for the starship from the television series *Star Trek*) was placed on top of a modified Boeing 747 airliner. It was carried to a predetermined altitude and released. The glide characteristics of the orbiter could be determined, and hands-on experience landing at speeds close to those expected on orbital missions could be gained.

On December 29, 1980, the first operational space shuttle, *Columbia*, was rolled to launch pad 39A in preparation for its first flight. This was to be a manned flight, and it marked the first time that a manned space vehicle would be flown without the benefit of unmanned test flights—another result of budgetary constraints.

The STS 1 mission blasted off from the Kennedy Space Center on April 12, 1981, twenty years to the day after the first astronaut was launched into space by the Soviet Union. The STS 1 commander was John W. Young; the pilot was Robert L. Crippen. The two-day "shakedown cruise" showed that the systems worked and that the orbiter could make the conversion from rocket to orbiter to glider without enormous problems. The only major concept yet to be tested was that of reusability. The only way to find out was to fly *Columbia* on its second planned space mission.

The STS 2 mission was scheduled to be launched on November 4, 1981. The STS 2 commander was Joe H. Engle; the pilot was Richard H. Truly. A problem with one of *Columbia*'s auxiliary power units (APUs), however, could not be resolved in time to meet the launch deadline for the day. The three APUs provide hydraulic power to the orbiter for movement of the main engines and aerosurfaces, as well as other systems. The APU 1 and 3 lubrication oil systems had to be flushed and the launch was pushed back to November 12. At 59.887 seconds past 10:09 A.M. eastern standard time, the era of the reusable shuttle vehicle began. All phases of the launch proceeded normally, and *Columbia* was placed into a 222-kilometer-high orbit. The mission had a planned duration of approximately five days, but a fuel cell failure prior to five hours into the mission resulted in a decision to shorten the flight to the preplanned minimum mission of about fifty-four and one-half hours. During the shortened mission, more than 90 percent of the high-priority flight tests were completed successfully. *Columbia* carried the OSTA-1 (for NASA's Office of Space and Terrestrial Applications) experiments pallet, which included several Earth observation experiments. The Development Flight Instrumentation, used to monitor *Columbia*'s systems during flight, showed that the orbiter functioned as planned. The Remote Manipulator System's 15-meter robot arm was first flown on this mission. On later missions, it would be used to deploy and retrieve large payloads.

Columbia landed on Runway 23 at the Edwards Air Force Base dry lake bed on November 14 at 6:23 P.M. The concept of a reusable manned spacecraft had been proven feasible.

Impact of Event

In a "throw-away" society where everything is used once, then tossed aside, the idea of building a spacecraft that could be used a hundred times was pure science fiction until the space shuttle. Prior to the Space Transportation System, a satellite or probe was launched into space and, if it got there successfully, the probe lived its life until it ran out of fuel, electrical power, or funding. Then, it was discarded for a newer model. Sometimes it was replaced before it died. By building a vehicle that could carry large payloads to low-Earth orbit, it was possible to go to those old satellites and either repair them or bring them back to Earth. The cost of such a vehicle would be prohibitive if it, too, was used once, then discarded. The ideal craft for this job was one that reused all or most of its components and needed only to be resupplied and refueled: the space shuttle. During the waning days of Apollo, NASA wanted to build an Earth-orbiting space station from which it could conduct manned and unmanned exploratory missions. A shuttle vehicle would be used to take materials and personnel to and from the space station. NASA was told it could have one or the other. The space shuttle was their choice since it could be used later to build a space station. In addition, the savings and revenue generated by having this shuttle transport commercial payloads would more than pay for itself and the station.

After being given the go-ahead for the space shuttle, the congressional budget strings were tight. As funds were being cut, so were the plans for a completely reusable space vehicle. Soon, the launch vehicle was reduced to one that was part reusable, part nonreusable. With the second flight came the proof that it was indeed practical and soon it would be an operational vehicle. By the end of 1985, twenty-three space shuttle missions had been flown with more or less complete success. Funding for a space station was being approved by a Congress that had some confidence in NASA, despite occasional launch delays and minor vehicle problems. Congress had enough confidence in the Space Transportation System that it permitted two of its members to take trips aboard the vehicle.

Then, on January 28, 1986, seven crew members aboard the shuttle *Challenger* were killed when their vehicle disintegrated one minute and thirteen seconds into the flight. Shortly thereafter, unmanned Titan and Delta launch vehicles were destroyed during launch. NASA's space program was at a standstill. In the ensuing weeks and months, much criticism was leveled at NASA. Continuous cutbacks, coupled with increasing demands, placed the space agency in a difficult position.

The significance of the second flight of the space shuttle can never be overlooked. Two astronauts placed their lives in the hands of the engineers and scientists who said it was feasible. Out of it came the world's first reusable space vehicle.

Bibliography

Allen, Joseph P., with Russell Martin. *Entering Space: An Astronaut's Odyssey.* New York: Stewart, Tabori & Chang, 1984. The story of an astronaut's journey in space, told in conjunction with a study of the American exploration of space. Filled with colorful prose and spectacular photographs of space travel, including hundreds of color photographs taken during the manned Gemini, Apollo, Skylab, Apollo-Soyuz, and space shuttle programs. Includes a brief bibliography and an index.

Baker, David. *Space Shuttle.* New York: Crown, 1979. Covers the early days of the Space Transportation System and how the shuttle evolved into its present configuration. Includes several artists' conceptions and photographs of models of the earlier versions. Contains a preview of a flight of the shuttle to orbit and how it will be used to carry cargo to space.

Furniss, Tim. *Space Shuttle Log.* London: Jane's, 1986. A concise work about the Space Transportation System. The first chapter traces the development and testing of the system, detailing the earlier designs of the vehicle and some of the research aircraft that paved the way for the orbiter. A listing of future flights includes the ill-fated STS 51-L mission. Includes many black-and-white photographs.

Jenkins, Dennis R. *Rockwell International Space Shuttle.* Arlington, Tex.: Aerofax, 1989. Discusses the history and origins of the Space Transportation System. Includes specifications and diagrams of early designs and of several vehicles used to test the final version. Focuses on the finer details of the orbiter and its systems. Drawings and black-and-white photographs. Cutaway drawings permit an in-depth study of the components. A good research book.

National Aeronautics and Space Administration. *STS-2 Orbiter Mission Report.* Washington, D.C.: Government Printing Office, 1982. The official report for the second mission of *Columbia* during the Orbiter Flight Test program. Details the performance of each of the vehicle's main systems—orbiter, solid rocket motors, and external tank. Includes a complete biomedical, trajectory, and flight control evaluation. Black-and-white photographs, line drawings, charts, and tables.

Otto, Dixon P. *On Orbit: Bringing on the Space Shuttle.* Athens, Ohio: Main Stage, 1986. Chronicles the Space Transportation System from its inception through the *Challenger* accident. Discusses the early design concepts for the shuttle and some of the factors that influenced its evolution. Gives a brief description of a typical shuttle mission. Several black-and-white photographs from each flight and a portfolio of color pictures in the center section are included.

Russell R. Tobias

Cross-References

Tsiolkovsky Proposes That Liquid Oxygen Be Used for Space Travel (1903), p. 189; Goddard Launches the First Liquid Fuel Propelled Rocket (1926), p. 810; The Germans Use the V-1 Flying Bomb and the V-2 Goes into Production (1944), p. 1235; The First Rocket with More than One Stage Is Created (1949), p. 1342; Sputnik 1, the First Artificial Satellite, Is Launched (1957), p. 1545; The United States Launches Its First Orbiting Satellite, Explorer 1 (1958), p. 1583; Shepard Is the First United States Astronaut in Space (1961), p. 1698; Glenn Is the First American to Orbit Earth (1962), p. 1723; The First Humans Land on the Moon (1969), p. 1907; Skylab Inaugurates a New Era of Space Research (1973), p. 1997.

BAULIEU DEVELOPS RU-486, A PILL THAT INDUCES ABORTION

Category of event: Medicine
Time: 1982
Locale: Paris, France

Baulieu developed RU-486, a progesterone antagonist, as an early contragestin drug in humans, thus giving rise to success and controversy

Principal personages:
ÉTIENNE-ÉMILE BAULIEU (1926-   ), a French biochemist and endocrinologist at the University of Paris who won the Lasker Prize in 1989
GEORGES TEUTSCH, a French chemist in the department of endocrinology at Roussel Uclaf who was in charge of the research team that synthesized RU-486
ALAIN BÉLANGER, a postdoctoral student on Teutsch's team who actually synthesized RU-486 and other similar compounds
DANIEL PHILIBERT, a physicist and pharmacologist in endocrinology who evaluated the binding characteristics of the compounds and found that RU-486 blocked progesterone

Summary of Event

Étienne-Émile Baulieu literally rescued a steroid (derivative of cholesterol) compound, RU-486 (trade name Mifepristone), from certain oblivion at the drug company Roussel Uclaf. He went on to show that it is an effective and safe drug to use as a contragestive agent (prevention of or disruption of implantation of the embryo in the uterus) within the first forty-nine days after a missed menstrual period. As a consultant for the company, he understood the significance of a compound that blocked the action of progesterone (a steroid and a naturally occurring hormone) and bound tightly to the receptor (molecule that acts as a link to obtain or prevent activity) for progesterone.

In 1980, at Roussel Uclaf, Alain Bélanger, in the chemistry laboratory of Georges Teutsch, synthesized several steroids that bound to steroid receptors. Their main interest was studying alterations in the steroids that affect binding capacity. Teutsch and Daniel Philibert had established a research project looking for glucocorticoid (a steroid hormone) antagonists. These compounds would compete with the natural steroid, bind to the receptor, and thus cover it up, but not activate it. The most potent of the group was RU-486, as it was able to block the actions of a very strong synthetic glucocorticoid, dexamethasone. Philibert found in animal studies that the compound also blocked progesterone activities and bound tightly to the progesterone receptor.

Baulieu convinced the company to test the product in fertility control tests, largely because it bound tightly enough to the receptor to outcompete successfully the natural steroid, progesterone. Many tests were performed on rabbits, rats, and monkeys, showing that even in the presence of progesterone, RU-486 prevented the formation of secretory tissue in the uterus, could induce a premature menstrual cycle, and could terminate a pregnancy. Since it was shown to be nontoxic even in high doses, in October of 1981, Baulieu began testing with human volunteers. By 1985, large-scale testing was done in France, Great Britain, Holland, Sweden, and China. There was an 85 percent success rate administering a relatively low dose of RU-486 orally. There was complete expelling of the embryo and all the endometrial surface. If a low dose of a prostaglandin (a hormonelike substance that acts on smooth muscle present in the uterus to enable contraction) is given forty-eight hours later, this raises the success rate to 96 percent. The side effects were minimal, and the low dose of RU-486 did not interfere with the actions of the necessary glucocorticoids in the body. It acts as an antagonist of glucocorticoids only at much higher doses. As reported in June, 1990 (*Scientific American*, pages 42 to 48), there had been forty thousand terminations of pregnancy done with RU-486 and a prostaglandin. In 1990, the French government started subsidizing the cost of using RU-486.

In a detailed study reported in the March, 1990, issue of *The New England Journal of Medicine*, Baulieu and others found that, with one dose of RU-486, followed in thirty-six to forty-eight hours by a low dose of prostaglandin, 96 percent of the 2,040 women had a complete abortion with few side effects. One percent failed to abort, and 3 percent had an incomplete expulsion or excessive bleeding and required surgical intervention. The women were kept in the clinic for four hours after receiving the injection or suppository of prostaglandin to watch for side effects, which were usually mild and included nausea, vomiting, abdominal pain, and diarrhea. Fewer than 2 percent of the women complained of any further side effects when they returned for a checkup; these included the side effects mentioned, plus dizziness, headache, rash, and, rarely, a fever. They used two different prostaglandins and found that the higher doses of one of them caused a faster expulsion but also caused more pain and a longer period of bleeding. The frequency of bleeding problems was the same in the study as in the most used surgical procedure for the early stages of pregnancy termination. In animal studies, there are reports of harm to the fetuses from RU-486 and prostaglandins. A preliminary study of the effects of RU-486 on early monkey embryos, with and without progesterone, shows high tolerance in most cases. There is also a report of a woman given a high dose of prostaglandin at seven weeks who gave birth to a child with abnormal digits and an enlarged head. She had been told to call back if there was no expulsion, but she did not. It is not absolutely certain that there were no other drugs or harmful products taken during the early pregnancy.

The French government approved RU-486 in September, 1988, for use in state-controlled clinics. In October, 1988, Roussel Uclaf withdrew it from the market because of pressures and anticipated boycotts of the company products. After petitions and boycott threats from the members of the World Congress of Obstetrics and Gynecology present at a large meeting, and from other sources, the French government (which has a 36 percent interest in Roussel Uclaf) ordered them to resume distribution of the drug. By fall, 1989, between one-fourth and one-third of the early abortions in France were done using RU-486 and a prostaglandin. Neither Roussel Uclaf, its partners, nor any other drug company has volunteered to produce and distribute the drug outside France. The testing for approval of the drug was completed in Great Britain and in Holland, but the parent company, Hoechst AG, either has not applied for the license to market it or has withdrawn the application. In the United States, testing is not allowed using government funds, but some private money has financed tests that support and extend those in Europe.

Although there are few side effects, which are mild, if the drug is not given properly and followed by the prostaglandin, there could be rare serious problems such as incomplete abortions and severe bleeding. This seems more likely to occur if a black market arises because of the unavailability of the drug in some countries. The drugs may then be given without correct supervision.

There are other potential uses for RU-486 in the treatment of breast cancers that require progesterone for growth and possibly for other tumors when the cells have steroid receptors on their surfaces. RU-486 is also being investigated for use in treatment of glaucoma to lower pressure in the eye that may be caused by an increased level of steroid hormone, Cushing's syndrome to keep a patient alive until surgery, skin wounds to promote healing, and the cervix at birth to soften it and thus aid in normal delivery. The most promising of all the uses may be as a real contraceptive, that is, prevention of ovulation without the addition of estrogen, a potential cancer-causing agent. There have also been tests to use RU-486 as a pill more like the current birth control pill, a combination of estrogen and progesterone which the user takes for a certain number of days and which then induces menstruation whether there was a fertilization or not, thus preventing implantation in the uterus because the nutritive lining would be shed.

Impact of Event

Baulieu's discovery that RU-486, in conjunction with a dose of a prostaglandin, will allow early abortions to be performed without surgery, major side effects, and publicity has added to the polarization of groups for and against the right to end an unwanted pregnancy. It has been approved for use in France and China; there have been threats against the producers and loud outcries from some groups on both sides of the issue in the United States. The cause of the controversy, namely, a new drug that will terminate unwanted pregnancies within the first seven weeks, will not go away as there are now other very similar drugs being tested worldwide. On the occasion of renewal of distribution of RU-486, the French Minister of Health Claude Evin said, as noted in *The New York Times* of October 29, 1988: "From the moment Government approval for the drug was granted, RU-486 became the moral property of women, not just the property of the drug company."

The World Health Organization and others have estimated that annually up to 200,000 women lose their lives (mostly in undeveloped countries) because of botched abortions. The regime of RU-486 and a prostaglandin would still require professional supervision and at least two visits to a clinic, but the strict sterility and surgical skills would be needed only in the 4 percent who do not respond fully to this method. There is much discussion within the medical profession that RU-486 is a beneficial drug that is withheld from many countries not only for medical reasons but also for political and possibly financial and emotional reasons. The World Health Organization has the right to commandeer the drug and make it available to poor countries for cost; they are waiting until even more testing ensures that taking such a drastic step is completely safe.

There is an opinion expressed in popular magazines and newspapers that if the drug is to become available in the United States, some small company that will not be affected by boycotts will have to produce it backed by funds from nongovernmental sources. Some articles noted that if the groups desiring availability of RU-486 would present as strong and united a front as the opposition, there would also be more push to test and distribute it in the United States. If RU-486 becomes the drug of choice for other needs such as treatment of breast cancer, then perhaps it will be marketed more widely.

Clearly, this drug is controversial not because of side effects or improper testing of women in developing countries (the early birth control pill would be an example of a drug of this type) but because of a prior division of opinion: whether choice should exist for ending unwanted pregnancies. So far, both family planning agencies and women's health groups are solidly behind the effort to test and market RU-486 worldwide.

Bibliography

Baulieu, Étienne-Émile. "RU-486 as an Antiprogesterone Steroid: From Receptor to Contragestion and Beyond." *JAMA* 262 (October 6, 1989): 1808-1814. A somewhat technical article, but written for a wide audience. Several diagrams help to illustrate meanings. Background information, chemical structures of steroids, mechanism of action explained, discussion of testing, terminology explained, and other uses are discussed. Many references listed.

Cherfas, Jeremy. "Étienne-Émile Baulieu: In the Eye of the Storm." *Science* 245 (September 22, 1989): 1323-1324. Some background on Baulieu and his general work in endocrinology and, more specifically, searching for molecules similar to progesterone. Short, but interesting view of Baulieu and that area of science research.

Fraser, Laura. "Pill Politics." *Mother Jones* 13 (June, 1988): 31-33. Good basic information on the drug testing and the politics of its distribution. Much discussion on why it is not available in the United States. Some information on contraceptive methods in general and a comparison with the potential of RU-486. Some discussion of its use as a contragestive and what happens when it is given.

Halpern, Sue M. "RU-486: The Unpregnancy Pill." *Ms.* (April, 1987): 56-59. Some grandiose predictions about RU-486, but gives information and asks critical questions about why it does not work in some women. Other aspects of possible potential harm to women discussed. Some discussion of use as a contraceptive agent.

Palca, Joseph. "The Pill of Choice?" *Science* 245 (September 22, 1989): 1319-1323. A combination of much of the information that appeared in other sources about the controversies, the reasons that it will be a boon especially to developing countries, the history of its development, and some discussion on its immediate future. Written for the general audience. Other articles in the issue are referred to, although one is very technical.

Ulmann, André, Georges Teutsch, and Daniel Philibert. "RU 486." *Scientific American* 262 (June, 1990): 42-48. Aimed at the general reader, with many diagrams and a few pictures of the models of the steroid molecules. A history of the synthesis of RU-486 and related compounds and how they work. Basic human reproduction described and illustrated. Some discussion of the testing studies. One page on other uses. Four references are listed.

Judith E. Heady

Cross-References

Birth Control Pills Are Tested in Puerto Rico (1956), p. 1512; The Plastic IUD Is Introduced for Birth Control (1960's), p. 1629.

CECH DEMONSTRATES THAT RNA
CAN ACT AS AN ENZYME

Category of event: Biology
Time: 1982
Locale: Boulder, Colorado

Cech demonstrated that RNA can act as an enzyme to catalyze biochemical reactions, thus providing evidence of the process of chemical evolution

Principal personages:
THOMAS CECH (1947-   ), a research professor of the American Cancer Society who initiated the investigations into catalytic RNA
ARTHUR J. ZAUG, a colleague of Thomas Cech during the initial stages of RNA research
TAN INOUE, a collaborator with Thomas Cech who later showed the ability of RNA to catalyze biochemical reactions

Summary of Event

In the 1920's, Aleksandr Ivanovich Oparin and John Burdon Sanderson Haldane independently proposed that the early Earth atmosphere lacked oxygen but contained an abundant amount of hydrogen-containing compounds, such as ammonia, methane, water vapor, hydrogen gas, hydrogen cyanide, carbon monoxide, carbon dioxide, and nitrogen. They both proposed that these gases spontaneously combined in the presence of energy. There was no lack of energy on the surface of the early earth because of volcanic eruptions, lightning, and ultraviolet radiation. It was in this type of environment—a reducing atmosphere without oxygen present—Oparin and Haldane hypothesized, that life began.

This model was tested, in 1953, by Stanley Miller, a graduate student at the University of Chicago. He built a system of interconnecting tubes and flasks designed to simulate the primitive Earth atmosphere and primordial ocean. After a week, he analyzed the results and found simple organic compounds, such as organic acids and amino acids (building blocks of proteins). This experiment paved the way for other experimenters. Utilizing different mixtures of gases, other scientists have produced virtually all the organic building blocks necessary for life and found in cells, such as nucleotides, sugars, and fatty acids.

Once all the building blocks were formed, the next important step was to link these simple molecules into long chains, or polymers. An example of polymerization would be the linking of amino acids to form a long chain called a protein. Another polymer would be the polynucleotide, a long chain of single nucleotides.

There are two types of nucleotides, deoxyribonucleic acid (DNA) and ribonucleic acid (RNA), which are very similar in structure. They differ in that DNA contains the pentose sugar deoxyribose, while RNA contains the pentose sugar ribose. Ribose has an —OH (hydroxyl group) instead of —H (hydrogen atom) at the number 2 carbon atom. DNA also contains the four nucleotide bases, adenine (A), guanine (G), cytosine (C), and thymine (T), while RNA contains the same nucleotide bases as DNA except that uracil (U) replaces thymine.

Many hypotheses suggest the early polymers may have been formed by different mechanisms. One method to concentrate the organic molecules would be the process of evaporation, which would leave a high concentration of organic molecules that are necessary for polymerization. Another possibility would be clay particles in the soil. Clay has charges that attract and adsorb ions and organic molecules to its surface, thus bringing molecules so close to one another that they could polymerize into long chains. The adsorbed metal ions could also provide a site for the formation of polynucleotides. Once the polynucleotides had formed, they could then act as a template specifying a complementary sequence of new polynucleotides. This is the result of the preferential bonding of certain nucleotides to one another, such as adenine with uracil or thymine, and guanine with cytosine. Presently, this simple mechanism accounts for the transfer of genetic information from cell to cell and generation to generation.

The process of polymerization of nucleotides is slow and relatively ineffective and would be hindered from occurrence under the conditions found on the primitive earth. Even the clay and metal ions would have been slow. Presently, enzymes, which are proteins, function to catalyze (speed up the biochemical reactions involved in) the formation of polynucleotides. They, however, would not have been present in this prebiotic solution.

A discovery in 1981 by Thomas Cech and Arthur J. Zaug indicated how the early polynucleotides might have been replicated. RNA was thought to be a simple molecule, but now this appears not to be the case. Research with the ciliated protozoan *Tetrahymena thermophila* showed the existence of an RNA molecule with catalytic activity. Ribosomal RNA (rRNA) is synthesized as large molecules, which is then spliced to the correct size. In *Tetrahymena*, the surprise came when this reaction was found to occur without the presence of proteins to catalyze the reaction. The only requirement for the reaction to occur was magnesium ions and the nucleotide guanosine triphosphate (GTP). This was a surprising result because in 1981 the current dogma in science stated that enzymes are proteins and catalyzed all the reactions of a cell. Cech later showed that the RNA molecule contained the catalytic activity to splice itself. This self-splicing mechanism resembled the activity of an enzyme and Cech coined the term "ribozyme" to describe the RNA enzyme. Ribozyme is distinguished from an enzyme because it works on itself, which is unlike other enzymes that work on other molecules. Later, Cech studied the properties of the ribozyme and found it similar to enzymes in that it accelerated the reaction and was highly specific. In addition, the three-dimensional structure, as in enzymes, was found to be critical in the activity of the ribozyme. Cech's research showed that if the ribozyme was put in a solution that prevented folding, then the ribozyme showed no catalytic activity, similar to any other enzyme.

The mechanism still needed refinement. Knowing that the folding was important in the catalytic activity aided in discovering the process. Cech and colleagues, Brenda L. Bass, Francis X. Sullivan, Tan Inoue, and Michael D. Been, discovered that the folding was essential in creating binding sites for GTP. This also activated the phosphate group and increased the likelihood for splitting the RNA molecules. Information on the exact mechanism of activation of guanosine is still unclear, but the reaction catalyzed by the ribozyme was speeded up by a factor of 10 billion. Thus, the ribozyme was established as having many enzymelike properties, such as accelerating the reaction and having a three-dimensional structure like an enzyme. It was observed that the ribozyme kept acting on itself. A true catalyst is not converted in the reaction, and the ribozyme was altered in the reaction. In 1983, this distinction also appeared. Zaug and Cech began working with *Tetrahymena* so that a shortened form of the RNA intron could work as a true enzyme.

Another surprise came out of this research. As the ribozyme acted as an enzyme by splicing another RNA chain, it also was synthesizing a nucleotide polymer of cytosine. Not only was the ribozyme acting as a splicer but also it was behaving as a polymerase enzyme by synthesizing chains of RNA molecules that were up to thirty nucleotides long. Since then, other researchers have strung together strings of nucleotides up to forty-five nucleotides long. These results lead to the implication that RNA can duplicate RNA genes. In 1982, sequences of RNA molecules, introns (intervening sequences in genes), were found to be similar in different types of cells, such as fungi (yeast and *Neurospora crassa*) and protozoans. Remarkably, this discovery of a self-splicing RNA molecule was also found in a bacterial virus. This was a startling discovery since fungi, protozoans, and viruses are thought to be only very distantly related. A conserved sequence implies an essential function even in the face of evolutionary divergence. This indicates the ribozyme may have evolved relatively early in the evolution of life.

Impact of Event

Thomas Cech's work leads to the implication that RNA can duplicate RNA. These conclusions have a profound impact on the theory of the origin of life and chemical evolution. The results of RNA having catalytic activity support the theory that the role of RNA might have been as the primordial genetic material. These functions have now been taken over by DNA and proteins.

In the scheme of the evolution of life, it is thought that RNA was the primordial genetic material. Both RNA and DNA can store genetic information, but only RNA, as shown by Cech, can act as a catalyst to speed up chemical reactions. With Cech's discovery of splicing of RNA by RNA, this implies that proteins that may have been in existence might not have been needed for gene duplication. The self-splicing of RNA can be considered a primitive form of genetic recombination, since new combinations of RNA sequences are thereby created. Thus, the first genes are thought to be composed of RNA. RNA genes that were combined in a molecule that provided useful products could be at an advantage in the primordial mix.

Cech's discovery that RNA can catalyze reactions also has led researchers to speculate that RNA might catalyze other reactions. RNA has enough structure that it might be possible to catalyze other biochemical reactions. RNA may not have a high rate of activity, but even a modest rate and some specificity would be faster than no enzyme at all. This would be valuable early in the evolution of life.

If primitive cells, which were surrounded by a membrane, contained these ribozymes, they would have been at a selective advantage over other such cells. Thus, the primitive genetic material would be duplicated and passed to other cells. In addition, these RNA molecules could also bind amino acids in close proximity to allow the amino acids to combine into short polypeptides. These polypeptides could then act as a primitive enzyme, and if they aided the cell in replication and survival of RNA, then the cell could split and pass the genes on to other cells.

Laboratory experiments cannot establish without a doubt how chemical and cell evolution took place. Cech's work, however, does establish a plausible scenario in which RNA might have been the primordial genetic material and enzyme. These functions have been taken over by DNA and proteins, but they are linked together by RNA. How life started still remains a mystery, but the pieces of the puzzle are coming together.

Bibliography

Alberts, Bruce, et al. *Molecular Biology of the Cell*. 2d ed. New York: Garland, 1989. An advanced biology text; however, chapter 1, "The Evolution of the Cell," is well written and easy to understand. The authors cover the sequence of events on the possible evolution of the cell and Cech's impact on the theory of chemical evolution. A good list of references is found at the end of the chapter.

Amato, I. "RNA Offers a Clue to Life's Start." *Science News* 135 (June 17, 1989): 372. A short article that discusses the research indicating that RNA can copy itself without assistance. It is nontechnical and summarizes the developments up to 1989 in this area of research.

Campbell, Neil A. *Biology*. 2d ed. Redwood City, Calif.: Benjamin/Cummings, 1990. An introductory college textbook that is geared for the biology major but can be easily understood by the high school student. Chapter 24, "Early Earth and the Origin of Life," covers chemical evolution and the possible sequence of events. References are found at the end of the chapter for further reading. The book also contains interviews with leading scientists in the field. Includes a glossary, diagrams, and cross-references to other chapters.

Cech, Thomas. "Ribozyme Self-Replication." *Nature* 339 (June 15, 1989): 507-508. A short article but technical. The introduction and conclusions, however, are easily read and understood for their implications of RNA's role in the evolution of life.

_____. "RNA as an Enzyme." *Scientific American* 255 (November, 1986): 64-75. A nice review of the research that Cech has done and the implications of his results. Parts of the article are easier to read than others.

_____. "Self-Splicing RNA: Implications for Early Evolution." In *International Review of Cytology*, edited by D. C. Reanney and Pierre Chambon. Vol. 93. San Diego: Academic Press, 1985. An advanced source, yet the introduction and conclusions are easy to understand. The article chronologically follows the development of Cech's research and reasoning.

Lonnie J. Guralnick

Cross-References

Hopkins Discovers Tryptophan, an Essential Amino Acid (1900), p. 46; Pauling Develops His Theory of the Chemical Bond (1930), p. 926; Krebs Describes the Citric Acid Cycle (1937), p. 1107; Ochoa Creates Synthetic RNA (1950's), p. 1363; Lipmann Discovers Acetyl Coenzyme A (1951), p. 1390; Watson and Crick Develop the Double-Helix Model for DNA (1951), p. 1406; Miller Reports the Synthesis of Amino Acids (1953), p. 1465; Barghoorn and Coworkers Find Amino Acids in 3-Billion-Year-Old Rocks (1967), p. 1851; Kornberg and Coworkers Synthesize Biologically Active DNA (1967), p. 1857.

DeVRIES IMPLANTS THE FIRST JARVIK-7 ARTIFICIAL HEART

Category of event: Medicine
Time: 1982
Locale: University of Utah, Salt Lake City, Utah

DeVries implanted the first Jarvik-7 artificial heart in a patient with end-stage heart disease

Principal personages:
WILLIAM CASTLE DEVRIES (1943-   ), a surgeon at the University of Utah in Salt Lake City and later at Humana Hospital in Louisville, Kentucky
ROBERT JARVIK (1946-   ), the chief developer of the Jarvik-7, after whom the device was named, and president of Symbion, Inc., the manufacturer of the Jarvik-7 heart
BARNEY CLARK (1921-1983), a Seattle dentist and the first recipient of the Jarvik-7 artificial heart

Summary of Event

In 1982, Barney Clark was diagnosed by his cardiac specialist, William Castle DeVries, as having only hours left to live. The decision was made to implant the recently developed Jarvik-7 heart in Clark. The Food and Drug Administration (FDA), which has authority over the use of medical "devices," had already given DeVries and his associates permission to perform a total of seven Jarvik heart implants for permanent use. The operation was performed on Clark and was initially considered successful. Widespread news coverage heralded this first triumph of the scientific community to support or replace a severely dysfunctioning heart with a total artificial heart. The media considered this achievement a major event in the history of cardiac care: It seemed DeVries had proved that an artificial heart had been developed that was almost as good as a human heart.

The Jarvik-7 heart, designed and produced by researchers at the University of Utah in Salt Lake City and named after chief developer Robert Jarvik, is one of the most well-known advances in so-called spare-parts medicine. Essentially an air-driven pump made of plastic and titanium, the expensive bionic device is the size of a human heart. It consists of two hollow chambers of polyurethane and aluminum, each containing a flexible plastic diaphragm. Two plastic hoses, routed through incisions in the abdomen, carry compressed air to the heart to pump blood through the pulmonary artery to the lungs and the aorta to the body. A disadvantage of the Jarvik-7 is that patients had to be tethered by 2-meter hoses to a large, cumbersome air compressor, which then had to be wheeled from room to room along with the patient.

Shortly after Clark's surgery, DeVries went on to implant the device in several other patients with end-stage heart disease. For the time being, all patients survived the operation. As a result, DeVries was soon offered a position at Humana Hospital in Louisville, Kentucky. Humana offered to pay for his first one hundred implant operations. Unfortunately, in the three years following DeVries' earliest operations with the Jarvik-7, doubts and criticism plagued the earliest version of the artificial heart. Of the people who, by then, had received the plastic and metal device as a permanent replacement for their own diseased hearts, three had died and four suffered disabling strokes. The FDA asked Humana Hospital and Symbion, Inc., the manufacturer of the medical device, for detailed histories of the artificial-heart recipients.

Infections were diagnosed as the cause of each patient's condition. Life-threatening infection, or "foreign body" response, is a major limitation in the extended use of any artificial organ. The Jarvik heart, with its metal valves, plastic body, and velcro attachments, provided a receptive area for particularly virulent strains of bacteria resistant to even the most potent antibiotics. Also, the intensive support required to sustain the artificial heart recipients through surgical complications compromised their immune systems and predisposed them to infections that eventually affected the device itself.

By 1988, researchers concluded that severe infection was virtually an inevitable occurrence and that it was a function of how long each patient had to rely on the artificial heart for circulatory support. Consequently, experts recommended the use of the device be limited to thirty days. Long before this research had been completed, however, heart specialists and medical ethicists had questioned whether the artificial heart experiment was worth continuing. They suggested that it was time to go back to the research phase and to experiment on animals.

When the artificial heart program was first started, the question was simply if it would support life. Soon, however, it was not known if the research should continue. In fact, a heated controversial debate grew around the use of the Jarvik-7. Some ethicists argued that the fact that people were going to die anyway did not justify any human experimentation. Extension of life was considered an ambiguous benefit when it allows a patient to live somewhat longer, but burdened by pain, distress, or incapacity. At times, DeVries claimed it was worth the price for patients to reach another birthday; at other times, he admitted that if he thought a patient was going to spend the remainder of his life in a hospital, he would think twice about performing the implant.

Also, there was an ethical debate that centered on the question of "informed consent"—that is, the patient's understanding that going ahead with a procedure not only has a high risk of failure but also may leave the patient miserable even if it succeeds. Getting informed consent from a dying person is a delicate proposition, because, understandably, the patient will resort to anything. Therefore, one must consider if the ordeal is worth the risk for such patients, if the suffering is justifiable, and who should make the decision: the patient, researchers, or federal agencies.

Another question was that of cost. It had to be decided if high-cost, high-technology solutions such as the Jarvik heart should be financed or if more should be allocated to preventive measures. Although it was estimated once that a total artificial heart implant could be performed for about $150,000, actual estimates for DeVries' patients totaled up to $1 million each.

The medical community was criticized for involving a private corporation in basic research. They called this "for-profit medicine." Humana, a multibillion-dollar corporation, offered to pay for the first one hundred heart implants. Critics saw this not only as a priceless public relations ploy but also as a get-rich scheme, since the corporation could land a potential windfall starting with the hundred and first patient. Although the surgical techniques involved could not be patented, the expertise Humana would develop could give it a virtual monopoly on a market worth an estimated $5 billion.

DeVries' motives and interests were also questioned: Should the surgeon who performs the implantation own twenty-seven thousand shares of the company that makes the device? Controversies raged on all aspects of the Jarvik-7 experiment. Finally, in December of 1985, an advisory panel of experts recommended that the FDA permit the experiment to continue—but only with close monitoring on a case-by-case basis. Not long after, when experts recommended use of the device be limited to thirty days, the Jarvik heart began to be thought of as a "bridge" transplant, while patients awaited living donor hearts.

In the next few years, pneumatically powered mechanical hearts similar to the Jarvik were employed as a temporary method of circulatory support (or bridge) to buy time while a suitable donor heart was identified and cardiac transplantation was performed. Meanwhile, research on the results of these implantations demonstrated that the less time the prosthesis remained in place, the better the results. Unfortunately, infection continued to be a serious and, at times, lethal accompaniment of staged heart replacement.

Concurrent with the efforts to develop these mechanical devices, cardiac transplantation became a reality. By the end of 1985, an international registry reported that 2,577 patients had received cardiac transplants and that four-year survival among those heart recipients was 76 percent. The successes of cardiac transplantation led to a waning of interest in the Jarvik and other artificial hearts as a long-term support device and to a demand for donor hearts that rapidly exceeded the supply. Subsequently, the unavailability of donor hearts for patients in desperate need of immediate transplantation gave rise to the use of the Jarvik as a temporary bridge to transplantation.

By 1988, major improvements were made on the partial and total artificial hearts being developed. First, these were powered by implantable compact electric motors rather than by external consoles, which did away with the need for the clumsy pneumatic power line. With these devices, electrical energy was transmitted across the intact skin using inductive coupling. This markedly reduced the risk of infection. Moreover, the bulky external power unit required for the pneumatic hearts was replaced with a portable battery of house-current energy supply to enhance the comfort and mobility of patients.

Impact of Event

Experience with the Jarvik-7 made the world keenly aware of how far the medical community is from the dream of the implantable permanent mechanical heart. Patients as well as physicians soon realized it was not the panacea it was once thought to be. Nevertheless, it was a bionic breakthrough in the field of spare-parts medicine, which pushed back the boundaries of conventional medicine. For centuries, the problem of duplicating nature's handiwork with human technology has challenged scientists. The Jarvik-7, however clinically successful it was or was not, represented an advance in the relatively new field of artificial body parts. Other human-made components include heart valves and synthetic blood vessels, and artificial inner ears that enable profoundly deaf patients to improve their understanding of speech. The permanent artificial heart experiment is not concluded. A second generation of devices overcame some of the limitations of the Jarvik by using electric rather than pneumatic drives and taking advantage of new miniaturization and pump technologies. These newer hearts showed some technical promise, but the larger social and ethical questions remain.

The Jarvik-7 raised many difficult ethical questions about the consent process, for-profit medicine, and the like. Indeed, it demonstrated the confusion of American medical experimentation. DeVries' patients have forced one to ask whether medical experimentation has become an acceptable form of euthanasia. For example, one might ask if it is acceptable for such patients to volunteer for experiments that could hasten their deaths. Barney Clark's widow testified that her husband wanted to make a contribution to the artificial heart program.

Long ago, it was once proper to experiment on criminals and the insane, but today one wonders if the quest for a formula to avoid the inevitability of death makes it acceptable to try questionable regimens on the aged and terminally ill. Society as a whole has yet to come to terms with many such issues brought to light by Dr. De Vries' first implantation of an artificial heart. Although this medical experiment in borrowing time had its sobering limitations, it demonstrated its usefulness as an emergency stopgap measure for patients with end-stage heart disease when used as a bridge to a human-heart transplant. The bridge procedure's success rate, however, lags significantly behind that for normal transplants. Moreover, there is another ethical concern: Patients with mechanical hearts may be given priority over others waiting for donor hearts, which are in desperately short supply.

In the long run, some type of permanent artificial implant may be necessary if the growing demand for replacement hearts is to be met. The early efforts of DeVries with the Jarvik and other work with bridge implants have served as a step toward the realization of the artificial heart dream.

Bibliography

Adler, Jerry. "I Have Him Back Again: The Controversial Operation Behind Him, the World's Second Artificial Heart Recipient Faces a New Life." *Newsweek* 104 (December 10, 1984): 74. An excellent summary of the Jarvik controversy. Includes a diagram explaining how the Jarvik-7 works and a sidebar on for-profit medicine titled "Operating for Profit."

Cowart, Virginia S. "Artificial Heart Can Be Reassuring Backup for Surgeons, Says Pioneer." *JAMA* 259 (February 12, 1988): 785. An interview with DeVries that highlights benefits of the Jarvik. Details the experience of one patient with Jarvik followed by a human heart transplant.

Lyons, Albert S., and R. Joseph Petrucelli II. "Cardiology and Cardiac Surgery." In *Medicine: An Illustrated History.* New York: Harry N. Abrams, 1978. This massive volume is a highly readable and interesting survey of the history of medicine. Heavily illustrated and includes a selected bibliography.

Pierce, William S. "Permanent Heart Substitution: Better Solutions Lie Ahead." *JAMA* 259 (February 12, 1988): 891. This article highlights the Jarvik heart's contribution to ongoing research and its effectiveness as a bridge to cardiac transplantation. Pierce describes the improvements being made on artificial hearts. Includes an excellent bibliography.

Rice, Louis B., and Adolf W. Karchmer. "Artificial Heart Implantation: What Limitations Are Imposed by Infectious Complications?" *JAMA* 259 (February 12, 1988): 894-895. A thorough discussion of the primary limitation of artificial hearts as well as of ventricular assist pumps. Bibliography included.

Nan White

Cross-References

Einthoven Develops the Forerunner of the Electrocardiogram (1900's), p. 41; Carrel Develops a Technique for Rejoining Severed Blood Vessels (1902), p. 134; Abel Develops the First Artificial Kidney (1912), p. 512; McLean Discovers the Natural Anticoagulant Heparin (1915), p. 610; Drinker and Shaw Develop an Iron Lung Mechanical Respirator (1929), p. 895; Gibbon Develops the Heart-Lung Machine (1934), p. 1024; Blalock Performs the First "Blue Baby" Operation (1944), p. 1250; Favaloro Develops the Coronary Artery Bypass Operation (1967), p. 1835; Barnard Performs the First Human Heart Transplant (1967), p. 1866; Gruentzig Uses Percutaneous Transluminal Angioplasty, via a Balloon Catheter, to Unclog Diseased Arteries (1977), p. 2088.

COMPACT DISC PLAYERS ARE INTRODUCED

Category of event: Applied science
Time: 1982-1983
Locale: Japan, Europe, and the United States

The introduction of compact disc players, devices for playing digitally encoded music, created a new level of sound quality and rejuvenated the music industry

Principal personages:
HARRY NYQUIST (1889-1976), the Bell Laboratories physicist who formulated the sampling theorem that formed the basis of digital audio recording systems
CLAUDE ELWOOD SHANNON (1916-   ), the Bell Laboratories mathematician who created the science of information theory
JOSEPH FOURIER (1768-1830), the French mathematician whose work established the theory of digital audio

Summary of Event

Ever since Thomas Alva Edison invented the phonograph in 1876, people have been captivated by the ability to hear popular music in their homes. A steady string of improvements followed, and when affordable mass-produced flat disc records became available in the first half of the twentieth century, the recording industry entered the American mainstream. The introduction of magnetic tape recorders, stereo, and cassette tapes in the decades after World War II created an even larger market. As the technologies associated with the audio industry improved, however, new recording mediums became a priority. When the Japanese electronics giants who supplied much of the audio equipment began to look for a replacement recording medium in the 1960's, they turned to digital audio.

Traditional sound recording had been done in analog, a continuous waveform. The analog pattern of a song was encoded on the record, and the record player and receiver turned this information back into recognizable sound. Digital recording was more complicated, yet yielded better results. In digital recording, instantaneous samples of an analog wave are taken at set intervals, and the information is then stored as binary code; binary code is the representation of traditional numerical quantities in base 2, or a system of 0's and 1's. Binary coding had been used for some time in the computer industry and thus, technologically, the story of digital audio roughly parallels that of the computer. Theoretically, however, digital audio had been anticipated for some time.

The earliest form this took was in the work of the French mathematician Joseph Fourier. Fourier's theorem stated that discrete samples taken of a continuous waveform could reproduce the original waveform exactly, given that enough samples were taken. An example of what Fourier meant can be shown given a simple sine wave.

The analog method of recording the wave would be to record every point along the wave. The digital method of recording, as outlined in Fourier's theorem, would be to take readings of value, in this case, the amplitude of the wave, at regular intervals. Fourier said that from these discrete samples, one could reproduce the wave exactly, with no loss of information (or in this instance, sound). It was not until 1928, however, that anyone developed a formula for telling how many samples one needed to take to be able to use Fourier's theorem. At a meeting of the American Institute of Electrical Engineers, a physicist with Bell Laboratories presented a paper entitled "Certain Topics in Telegraph Transmission Theory." The physicist, Harry Nyquist, had been studying telegraph signaling, with the aim of optimizing the signaling rate. He found that in the terms of the sine wave, if a guarantee was needed that no error would enter into a reproduction, one needed to take S samples per second to reproduce a wave with a bandwidth of S/2, or a number twice the bandwidth. Given that the average upper limit of human hearing was a bandwidth of 20 kilohertz, this meant that forty thousand samples a second would guarantee error-free reproduction of sound. To give some room for error and to mesh with existing television technology, a sampling rate of 44,100 was established when the compact disc was introduced.

Nyquist, however, certainly never envisioned the compact disc when he derived his theorem. The technological realities were still far in the future. Theoretical developments that would be utilized later continued to be made. The most important of these was the 1948 work of another Bell Laboratories researcher, the mathematician Claude Elwood Shannon. In a paper entitled "A Mathematical Theory of Communication," Shannon outlined an entirely new science—information theory. He said that what communications lines carried were units of information that could be measured, transmitted, and tested. The terms that explained the new theory were the terms of binary coding. Information basically explained how noise in the form of random errors could corrupt the flow of information, now measured in bits (a unit of value 0 or 1), and it suggested ways in which this random error could be reduced or even eliminated. The error-correction codes that would find their way into compact disc encoding were developed in 1960 by two Massachusetts Institute of Technology researchers, Irving S. Reed and Gustave Solomon.

Several technological breakthroughs were still needed before the compact disc player could even become a dream. Developments in the computer industry, coupled with the development of the transistor and later the integrated circuit, or microchip, made optical data storage possible. The development of a working laser in 1960 meant that research into optical storage could begin. By the late 1960's, the Japanese firms NHK Technical Research Institute and Sony Corporation had working digital recorders, and the Dutch conglomerate NV Philips had begun pursuing optical storage. During the early 1970's, the digital recorders found their way into the recording studio, and soon several companies were using digital master recordings to produce their analog long-playing records (LPs).

The technical problems of a laser-read optical disc were solved—and solved for the low-cost consumer market—in 1978, when several companies, led by Philips' Magnavox subsidiary, produced working laser videodisc systems. These LP-sized discs stored an hour of analog video on each side. They proved to be commercial failures, primarily because of the coming of the videocassette recorders, but they proved the basic underlying practicalities of compact disc technology.

By the end of the 1970's, there were at least nine separate prototype digital audio systems. Still stung from the format wars associated with both the videodisc and the videocassette, two of the largest players, Philips and Sony, agreed in 1979 to work together to produce a standard format. By acting together and licensing the basic technology, thirty-five electronics manufacturers officially adopted the Philips/Sony standard in 1980. Two years later, in 1982, the compact disc player was introduced into Japanese and European markets, followed in 1983 by that of the United States.

The features of the compact disc (CD) readily showed their difference from the standard analog LPs. The CD was much smaller than the LP; it measured 120 centimeters in diameter, was thinner at only 1.2 millimeters, and weighed only about 14 grams. Its composition was different as well. The CD had a base composed of a type of hard Plexiglas, into which the information was pressed. On top of that came a reflective layer of aluminum or silver, and on top of that a protective layer of acrylic resin, onto which the labeling could be screened.

The information was stored in the form of "pits" sunk into the base material. Although all the pits were the same width, they differed in length, as did the plain regions between them. Each pit and plain represented one binary number value, which contained not only the original sound but also the error-checking and error-correcting schemes. The information was read when the tightly focused low-intensity laser passed over the spiral track of pits and plains, with each diffracting the light differently. The differing diffraction patterns were then translated into binary. Finally, a converter changed the binary information back into an analog waveform. Digital audio had arrived.

Impact of Event

The immediate impact of the CD and the CD player was not overly spectacular. The plans for digital audio disc technology had been publicly known for several years before its introduction. In fact, were it not for a recession, the CD player could have been introduced a year earlier. Second, there was debate about whether the new medium was actually as good as it was billed. Some claimed that it brought with it its own kind of distortion; others claimed that it lacked the "warmth" of the analog recordings. Perhaps the greatest reason that the CD player failed to leap immediately into the limelight, however, was the lack of CDs to play in them. When the players were introduced, there were only two manufacturing plants for CDs in the world, one in Hanover, West Germany, and another in Japan.

Within a few days, however, the situation had changed. New manufacturing plants, including one in Terre Haute, Indiana, greatly increased the number and variety of available CDs. Listeners were generally impressed with the quality of the sound.

The price of individual CDs, initially somewhat high, came down steadily. Their size made them easier to use and store, and their near indestructibility and promise of a long lifetime offered an end to the need to replace worn or damaged records.

Although it was evident at first, the introduction of the CD standard was the death blow for the LP. With retailers unwilling to support three standards for long, the CD's success meant that fewer and fewer LPs would be released. By the end of the 1980's, LPs, which had earlier slipped behind sales of cassette tapes, were reduced to a trickle. Many new albums were not even released in an LP version. Even the nationwide Record Bar retail chain made plans to change its name.

Sales of both players and discs steadily increased. In the first year after its introduction, thirty thousand CD players were sold in the United States alone, accompanied by 800,000 discs. By 1984, 230,000 players had been sold. By 1986, the figure was 3 million.

New technologies, however, made CDs a bit precarious. Plans were being made for a digital audio tape (DAT), which was to be to the traditional cassette what CD was to the LP, as well as being recordable. CD-ROM, a first cousin of the CD which stored computer data, was introduced in 1984, and plans for other types of CD mediums for the computer were being made. The Japanese were even envisioning technologies that might make the CD obsolete by the twenty-first century. Whatever its ultimate fate, CD technology succeeded in opening up new paths for both the electronics industry and the consumer and guaranteed that, while nothing ever stays the same, it would at least sound that way.

Bibliography

Berger, Ivan, and Hans Fantel. *The New Sound of Stereo.* New York: New American Library, 1986. Aimed primarily at the consumer. Some elementary coverage of compact discs. Photographs, illustrations, index.

Brewer, Bryan, and Edd Key. *The Compact Disc Book: A Complete Guide to the Digital Sound of the Future.* San Diego: Harcourt Brace Jovanovich, 1987. Excellent popular look at compact disc technology. Includes some history, along with consumer information. A good introduction to the topic. Photographs, illustrations.

Millman, S., ed. *A History of Engineering and Science in the Bell System: Communications Sciences (1925-1980).* New York: AT&T Bell Laboratories, 1984. A house history written by the technical staff. Nothing on compact discs, but excellent coverage of the work of Nyquist and Shannon. Semipopular, but difficult. Excellent references. Photographs, illustrations, index.

Pohlmann, Ken C. *The Compact Disc: A Handbook of Theory and Use.* Madison, Wis.: A-R Editions, 1989. A good introduction to the topic. Somewhat more technical than the Brewer and Key book (see above), but without the consumer bent. Good historical section, with some minor errors. Illustrations, index.

Ranada, David. "Digital Debut: First Impressions of the Compact Disc System." *Stereo Review* 48 (December, 1983): 61-70. Review article looks at the first compact disc players sold in the United States. Explains the theories involved and outlines the historical development. Very readable. Photographs.

Schetina, Erik S. *The Compact Disc*. Englewood Cliffs, N.J.: Prentice-Hall, 1989. Short, but generally quite readable. Aimed at the hobbyist. Illustrations, index.

Sweeney, Daniel. *Demystifying Compact Discs: A Guide to Digital Audio*. Blue Ridge Summit, Pa.: TAB Books, 1986. Popular look at compact discs by a *Stereo Review* writer. Maintains the magazine's flavor. See historical development. Photographs, poorly indexed.

George R. Ehrhardt

Cross-References

Johnson Perfects the Process to Mass-Produce Disc Recordings (1902), p. 138; Fessenden Perfects Radio by Transmitting Music and Voice (1906), p. 361; The Principles of Shortwave Radio Communication Are Discovered (1919), p. 669; Armstrong Perfects FM Radio (1930), p. 939; Shockley, Bardeen, and Brattain Discover the Transistor (1947), p. 1304; Sony Develops the Pocket-Sized Transistor Radio (1957), p. 1528; The First Laser Is Developed in the United States (1960), p. 1672; The Cassette for Recording and Playing Back Sound Is Introduced (1963), p. 1746; Optical Disks for the Storage of Computer Data Are Introduced (1984), p. 2262.

DAFFOS USES BLOOD TAKEN THROUGH THE UMBILICAL CORD TO DIAGNOSE FETAL DISEASE

Category of event: Medicine
Time: 1982-1983
Locale: Paris, France

Daffos used ultrasound to guide needle placement into the umbilical cord to withdraw pure fetal blood for rapid prenatal diagnosis of disease in the fetus

Principal personages:

FERNAND DAFFOS (1947-   ), a French obstetrician and gynecologist who guided studies on obtaining fetal blood using ultrasound to localize umbilical vessels

MARTINE CAPELLA-PAVLOVSKY, an obstetrician-gynecologist who assisted Daffos in using sonography for fetal blood collection

FRANÇOIS FORESTIER, a biologist who oversaw the laboratory work on fetal blood drawn by Daffos' procedure

D. V. I. FAIRWEATHER, a London professor of obstetrics who collected fetal blood for prenatal diagnosis of inherited blood disorders

Summary of Event

The team of Fernand Daffos, Martine Capella-Pavlovsky, and François Forestier developed a new and improved technique for collection of fetal blood samples in 1982, which they reported after numerous successes in 1983. Pure samples of fetal blood were obtained from the large vein in the umbilical cord, using a long, twenty-gauge spinal needle that was guided by real-time ultrasound imaging. The transducer of the ultrasound apparatus was held immobile at the point of needle entry on the mother's abdomen, and it showed the location of the needle with respect to the placenta, fetus, and umbilical cord. Blood was collected from a point on the cord about 2.5 centimeters from its connection to the placenta, into a syringe attached to the needle. Between 1 and 2 milliliters of blood was collected in each case. The new procedure was a dramatic improvement in fetal blood collection, as compared to other methods in use at the time for prenatal diagnosis of disease in the fetus.

Prenatal diagnosis of numerous genetic defects can be done by various means, including laboratory examination of blood, skin, or amniotic fluid cells from the fetus. For different suspected disorders, different methods and tissues work better. A common diagnostic procedure in use since the 1950's is amniocentesis, in which a sample of amniotic fluid surrounding the fetus is withdrawn by a needle. Cells from the fluid are grown for several weeks in the laboratory and then karyotypes are prepared from the cells or other diagnostic tests are performed. There is a specific time span in which collection must occur, with best results at about sixteen to eighteen weeks of gestation. Repeat sampling for karyotyping is not possible later if the first cells fail to grow, and results are not available until weeks after the amniocentesis.

Direct examination of the fetus by a process called fetoscopy also began in the 1950's, first through the dilated cervix and then across the uterine wall. The fetoscope is a fairly large instrument, about a centimeter in diameter, used to provide light and a means of observing the fetus. Blood collection using this direct visualization allowed the prenatal diagnosis of various hemoglobinopathies (hereditary disorders of the blood such as thalassemia or sickle-cell anemia) in the early 1970's and hemophilia, Duchenne muscular dystrophy, and other hereditary disorders in the late 1970's.

The fetoscopic procedure depends on ultrasound to visualize the location of the fetoscope, the collection device, and various parts of the fetus. Ultrasound, or sonography, uses directed high-frequency sound waves that are bounced off tissues and collected. Reflected sound collected by the transducer device is interpreted as light and dark areas on a computer screen, and moving internal tissues can be visualized. As technology has improved the resolution of these systems, it has become possible to see very small features of fetal tissues within the uterus.

One drawback to fetoscopy for fetal blood collection in the early 1980's was that the fetoscope could be rigid with a narrow diameter or flexible with a wide diameter. An optimal instrument for this use would be both flexible and of a small diameter; this has since been developed. In addition, fetoscopy is an invasive procedure, requiring surgery. A cut must be made in the mother's abdomen and uterine wall to allow the admission of the fetoscope, and a larger cut will increase the risk of inducing labor. A large cut is required to allow room for the sampling device that must accompany the fetoscope for blood collection. Because of the requirement for surgery, fetoscopy cannot be used in some cases. In many cases, the placenta—the organ of gas and nutrient exchange between the mother and fetus—is located on the uterine wall in such a way that it blocks access by the fetoscope. It would not be possible to cut through the placenta safely to allow entry of the instrument. Another disadvantage is that fetoscopy for blood collection can be performed most successfully only within the narrow time window between eighteen and twenty-one weeks of gestation. Risks associated with this procedure include premature labor and delivery, rupture of the fetal membranes without delivery, infection of the mother or fetus, maternal or fetal hemorrhage, and injury to the fetus. In 1984, the reported miscarriage rate associated with fetoscopy was about 7.5 percent, with another 1.8 percent loss in the perinatal period as a result of fetoscopy. The success rate for obtaining blood by this method, however, was 95 percent at that time.

Another means of obtaining blood from the fetus still in use in the early 1980's was placentesis, puncture of the placenta to remove fluid. This older procedure generally was being replaced by fetoscopy. Although not requiring use of the fetoscope, placentesis involved ultrasound. Sonography was used to locate the placenta, where maternal and fetal blood accumulate for the exchange processes. This less invasive procedure involved passing a long needle into the placenta to withdraw blood from it. Unfortunately, samples containing from 2 percent fetal blood to 100 percent fetal blood have been reported using this technique, and maternal blood contamination occurred in a reported 76 percent of the samples. Although the maternal blood cells can be removed selectively from the sample, the procedure was considered seriously flawed in this respect. The risk of miscarriage following placentesis was also higher than that for fetoscopy, because of tissue disruption in the placenta as the blood was being withdrawn.

Daffos and his colleagues' new technique removed many of the problems associated with prenatal diagnosis by amniocentesis, placentesis, and fetoscopy. The narrow 10- or 13-centimeter-long needle used in this method could be placed into the vein either through the uterine wall or through the placenta, allowing greater adaptability for use with different placenta placements. The procedure was easier to perform than fetoscopy; in sixty-two of the early cases reported, the first blood collection attempt was successful, and another four cases were successful at the second attempt twenty-four hours after the first. Blood could not be obtained in only two pregnancies out of the first sixty-eight sampled, giving an early failure rate of only 3.1 percent. Dilution of the blood sample occurred in two cases, but maternal blood contamination was not seen. In three cases, the fetus was sampled a second time after an interval of a few days, and the hematocrit and hemoglobin levels showed no significant loss of blood from the first sampling. In all the procedures, only one abdominal wall puncture was needed to enter the amniotic cavity, and in most cases, the procedure was completed within ten minutes with a single puncture of the umbilical cord. In some cases, the cord slipped away from the needle, causing a longer procedure time, and in two cases, it was necessary to penetrate the cord two or three times to obtain a sample. Because this procedure is less invasive, requiring only entry of the needle instead of a surgical incision, the risks to mother and fetus were reduced drastically compared to fetoscopy, and pregnancy loss was not observed. The gestational age range for the procedure was much greater than that for fetoscopy, between seventeen and thirty-two weeks in reports of the first sixty-six successful cases, as compared to eighteen to twenty-one weeks for fetoscopy. During development of this procedure, blood was drawn from fetuses that were scheduled for elective abortion or for prenatal diagnosis of toxoplasmosis, rubella, or Duchenne muscular dystrophy. Other uses were added later.

This new fetal blood sampling procedure combined the simplicity of placentesis with even better results than fetoscopy. Multiple samples could be taken over a period of several days with no apparent harm to the fetus, which could not be done with fetoscopy. Fetal blood cells were evaluated more easily than cells collected in a similar noninvasive way through amniocentesis. In contrast to the fibroblast cells obtained by amniocentesis, blood cells are grown in culture for only two or three days rather than for several weeks before being karyotyped to show chromosomal structure.

Impact of Event

Fetal blood sampling was an accepted procedure for prenatal diagnosis at the time of Daffos' development of umbilical cord puncture. Clinical use of fetoscopy was replacing placentesis in treatment centers in England, Europe, and the United States. Of particular importance was the detection of hemoglobinopathies by this method, as done by D. V. I. Fairweather's group in London. While Daffos' procedure was clearly superior to fetoscopy, Fairweather's group continued to use their more familiar method through 1985. A paper published in 1985 stated that "fetoscopy is now regarded as the method of choice for diagnostic fetal blood sampling," even though fetal loss was relatively high, particularly in cases where a repeat procedure was required. Daffos' method was not discussed in Fairweather's paper.

The procedure developed by Daffos, Capella-Pavlovsky, and Forestier was independently developed in Sweden in 1984. P. G. Lindgren and Bo S. Lindberg published a paper in 1985 describing their development of ultrasound-guided puncture of the umbilical cord in rhesus monkeys to collect fetal blood. The Daffos papers again were not referenced. Daffos and his group continued to use the procedure for numerous applications, showing its versatility. Small amounts of fetal blood were used to provide information about infection, genetic diseases caused by abnormal genes or chromosomes, and normal developmental processes of growth and blood production in the fetuses studied.

An interesting use of this fetal blood collection procedure was reported in 1985, in the prenatal diagnosis of genetically produced high cholesterol levels, called familial hypercholesterolemia. The standard means of prenatal diagnosis of this disease is through amniocentesis. A French couple had a previous child homozygous for familial hypercholesterolemia, with cholesterol levels about eight times normal. Prenatal testing by amniocentesis of the mother's second pregnancy gave ambiguous results. It was not possible to repeat the amniocentesis, so Daffos' group sampled for fetal blood, and the assay for cholesterol was performed on the blood. Diagnosis of a homozygous fetus enabled the parents to decide on elective abortion, and direct examination of the tissues showed that the fetus was indeed affected. The patient's third pregnancy was tested in the same way by fetal blood sampling, and again a homozygous fetus was aborted. Many genetic diseases can be determined from fetal blood sampling, and it may be expected that, as this technique becomes more widespread, the waiting period between fetal sampling and diagnosis will be greatly reduced. One effect of such prenatal diagnosis is that parents who might not want to risk having an affected child can use prenatal diagnosis to allow them the chance to have a normal child and to abort an affected fetus.

Application of Daffos' procedure may be used also in pharmacological studies of chemical or drug movement between the mother and fetus. Repeated fetal blood samples or intravenous injections can be obtained using ultrasound to guide the needle to the umbilical vein, as reported by Daffos' group in 1984.

Studies on fetal growth control processes were done by Daffos' group using blood left over after collection to test for possible fetal infection by maternal transmission of disease. Various growth factors were examined by immunoassay, leading to the conclusion that fetal growth in midpregnancy does not appear to depend on the same hormonal mechanisms seen in later growth in infants and children.

The large number of research and clinical reports from Daffos and his colleagues has underscored the usefulness of this means of fetal blood collection. The widespread use of the technique makes it possible to produce diagnostic results quicker and at less risk to mother and fetus than with previous methods.

Bibliography

Daffos, Fernand, Martine Capella-Pavlovsky, and François Forestier. "Fetal Blood Sampling via the Umbilical Cord Using a Needle Guided by Ultrasound: Report of Sixty-Six Cases." *Prenatal Diagnosis* 3 (1983): 271-277. A longer report that discusses the use of a needle visualized by ultrasound to enter the umbilical cord and collect fetal blood uncontaminated by maternal blood or amniotic fluid. Includes a discussion of the results of the blood evaluations.

_____. "A New Procedure for Fetal Blood Sampling in Utero: Preliminary Results of Fifty-three Cases." *American Journal of Obstetrics and Gynecology* 146 (August 15, 1983): 985-987. A short report that describes the procedure used by Daffos and his group to obtain fetal blood samples from the umbilical cord. A sonogram and drawing illustrate how it was performed.

Forestier, François, et al. "Hematological Values of 163 Normal Fetuses Between Eighteen and Thirty Weeks of Gestation." *Pediatric Research* 20 (1986): 342-346. Gives normal blood values for various cells and chemicals in fetuses of different gestational ages, as determined through the process of umbilical cord blood collection. Interesting as a research use of the method developed by Daffos and his colleagues.

Harrison, Michael R., Mitchell S. Golbus, and Roy A. Filly. *The Unborn Patient: Prenatal Diagnosis and Treatment*. Orlando, Fla.: Grune & Stratton, 1984. Gives a general overview of the state of fetal diagnosis at about the time of Daffos' development of fetal blood collection. Discussions of amniocentesis, genetic counseling, and fetoscopy and fetal sampling are interesting.

Kurjak, Asim, ed. *The Fetus as a Patient*. Amsterdam: Exerpta Medica, 1985. Contains the papers presented at an international symposium on the fetus as patient in June, 1984. Several topics are of interest in regard to prenatal diagnosis and the moral, ethical, and legal aspects of what can be done for or to the fetus. Covers genetic abnormalities.

Tortora, Gerard J., and Nicholas P. Anagnostakos. *Principles of Anatomy and Physiology*. 6th ed. New York: Harper & Row, 1990. A college textbook on human anatomy and physiology; useful in helping to visualize the fetus and the umbilical connection to the placenta. Shows fetal circulation through the umbilical cord in the chapter on cardiovascular vessels and routes, and gives further information on the fetus in the chapter on development.

Jean S. Helgeson

Cross-References

Blalock Performs the First "Blue Baby" Operation (1944), p. 1250; Bevis Describes

ASTRONOMERS DISCOVER AN
UNUSUAL RING SYSTEM OF PLANET NEPTUNE

Categories of event: Astronomy, and space and aviation
Time: 1982-1989
Locale: New Zealand; Chile; Pasadena, California

By use of ground-based telescopes and the Voyager 2 spacecraft, astronomers learned that Neptune has rings made of dustlike particles, as well as a unique incomplete ring

Principal personages:

ANDRÉ BRAHIC, an astronomer at the University of Paris and head of the observing team at New Zealand

WILLIAM B. HUBBARD (1940-   ), an astronomer at the University of Arizona and head of the observing team at the Cerro Tololo Inter-American Observatory

E. F. GUINAN, an astronomer at Villanova University and head of the observing team at European Southern Observatory

Summary of Event

In 1979, the Voyager 1 space probe discovered a ring of material around the planet Jupiter. This meant that of the solar system's four gas giant planets—Jupiter, Saturn, Uranus, and Neptune—all but Neptune were known to have rings. Most astronomers, thus, expected to find rings at Neptune also.

The search for Neptunian rings was done by viewing stellar occultations. Similar to an eclipse, an occultation occurs when a planet obscures a star as seen from Earth. By watching the behavior of the starlight as it vanishes and reappears, scientists can determine the planet's precise diameter as well as information about its atmosphere. If, minutes before or after passing behind the planet, the star should pass behind one of the planet's moons or rings as well, the star will wink out momentarily—even if the occluding object is too faint to be seen from Earth. The rings of Uranus were discovered in this manner. A stellar occultation by Neptune happens only about once a year.

In 1968, E. F. Guinan of Villanova University had observed an occultation of Neptune in New Zealand. The primary record of the event was accidentally lost, but a secondary record had been made on punched cards. These cards were not as precise as the primary record, which had been made on a paper strip chart, so they were set aside and ignored for more than ten years. When the cards were finally examined, they were discovered to be surprisingly informative; they showed that the star had flickered distinctly before disappearing behind Neptune. The fact that the starlight fluctuated rather than went out completely suggested one of two things: that a moon had partly blocked the star—an extremely unlikely occurrence, considering the tiny disk of such a moon as seen from Earth—or that the star had passed behind a diffuse ring. Guinan announced the discovery of rings at Neptune in 1982.

In 1981 and 1983, more Neptunian occultations were observed, but neither provided conclusive evidence of rings. Another occultation took place on July 22, 1984. This one was viewed by two astronomical teams in Chile about 100 kilometers apart. André Brahic of the University of Paris led the group at the European Southern Observatory in Cerro La Silla, while William B. Hubbard of the University of Arizona headed the group at the Cerro Tololo Inter-American Observatory. Both astronomers reported the same results: Before passing behind Neptune, the star flickered for about one second. This could not have been caused by a satellite; the observatories had seen the occultation at the same time from slightly different angles. This required either the absurd coincidence of two moons simultaneously passing the star in the views of the respective telescopes or the existence of a stream of debris orbiting Neptune—a ring. Once again, data suggested that Neptune had a ring system; but this one had a surprising twist. After the star passed behind the ring and then Neptune, it should have flickered again as it passed the ring a second time; however, there was no flicker. Somehow, the ring did not totally encircle the planet; the stream of particles formed only a portion, or arc, of a ring.

Such a finding should have been impossible. Ring particles closest to a planet orbit much faster than those farther away, because gravity is stronger closer to the planet. As a result, a ring arc would be unstable: It would spread itself eventually into a full ring. Even if the arc were of recent origin—such as a moon coming too close to Neptune and being torn apart by the planet's gravity—theoreticians calculated that it would become a ring in only three to ten years. Yet, when Hubbard, Brahic, and their colleagues compared previous occultations that had provided no evidence of a ring, they concluded that their ring arc "encircled" only about 10 percent of Neptune. To astronomers, it seemed more likely that the ring arcs were millions of years old—perhaps as old as the solar system itself—than the possibility that the fragments had been discovered during the few years in which they would exist.

Several scientists searched for mechanisms that could maintain the arcs. Most of these theories relied on the gravitational effects of small moons called "shepherds," like those known to orbit close to the rings of the other gas giants. The gravity of such a moon guides, or "shepherds," nearby ring particles into new positions, changing their orbits and the shape of the ring. Shepherd moons can keep a thin, dense ring from expanding over time into a wide, diffuse one; another such moon may split one ring into two by clearing out a gap down its center. Theorists such as Jack Lissauer of the State University of New York believed that one or more Neptunian shepherds could keep the stream of ring fragments from spreading either forward and back along their orbit or from side to side. This phenomenon had a precedent. In 1980 and 1981, the two Voyager probes examined the ring system of Saturn. Two incomplete "ringlets" were found among the many thin ringlets that made up the rings. The Voyager scientists believe that the ringlets were caused by undiscovered shepherds too small for the scientists to see.

As Voyager 2 approached Neptune for its 1989 close flyby, the mission astronomers viewed its progress from the Jet Propulsion Laboratory in Pasadena, California, hoping that the spacecraft's high-resolution cameras would be able to solve the mystery of the peculiar partial rings. Instead, Voyager 2 made two surprising discoveries on its arrival. The first was that Neptune had six small moons never seen before; however, none of them was in the necessary orbit to have any effect on the ring arcs. No shepherds were found. The second discovery was that in addition to ring arcs—now known to be three—Neptune had full rings as well. These rings are too faint and diffuse to be detected on Earth, even by occultation. In fact, they are too faint to be seen by a viewer at Neptune. It was only by computer enhancement of the Voyager images that they were discovered at all.

There are four Neptunian rings, made primarily of tiny particles best described as dust or smoke. The three ring arcs, made of rocky particles perhaps as large as a fist, are embedded in the outermost ring, strung together in one 35-degree sector. Two rings are a few kilometers thick and are 62,900 and 53,200 kilometers in radius, respectively. Two more are broader; one has a radius of 41,900 kilometers, and the other extends from the inner thin ring out to nearly 59,000 kilometers. Also circling Neptune is a broad sheet of fine material called the "plateau," which may overlap all the rings and is so diffuse that it is barely visible in enhanced Voyager photographs that show clearly the other very faint rings.

Impact of Event

Each time the Voyager probes examined a planet, they saw a ring system unlike any of the others. Jupiter has a single thin ring, while Saturn has thousands forming a band more than 66,000 kilometers wide. Saturn's rings are bright enough to be seen in archaic telescopes, while the meter-sized boulders that make up Uranus' eleven thin rings are blacker than coal dust. Neptune's rings are made of invisible smoke, dust, and streams of debris that cannot be explained fully yet. As Carolyn Porco of the University of Arizona and the Voyager imaging team has said, astronomers are "puzzled" by this wide variation; they have begun to look for a single theory of ring creation and mechanics that can explain all four systems.

Voyager images of the cratered and fractured surfaces of the moons of the outer planets have led scientists to believe that these moons are frequently bombarded, sideswiped, and careened into new orbits by passing comets from beyond Pluto's orbit. Such violence can easily turn a satellite into a ring of debris.

This, however, does not solve the mystery of the ring arcs—That is, if they are newly created (astronomically speaking) or if they have circled Neptune for billions of years. If the ring arcs are a long-lasting feature, astronomers need to determine what holds them together. Also, if the culprit is a shepherd moon, it is puzzling why Voyager 2 did not reveal it. The Voyager scientists believe that Saturn's broken ringlets could be influenced by moons too small for Voyager to see. It may be that the same holds true of Neptune.

If the ring arcs were formed recently—perhaps since 1979—and are transitory, then the timing of the arcs' creation was extremely fortunate for Voyager, because experts believe that the arcs would persist for less than a decade. If this was the case, however, astronomers are wondering what Guinan recorded in 1968.

One theory advanced by Larry Esposito of the University of Colorado is that recently formed rings are not uncommon. As moons and planetoids are reduced to boulder-sized rocks, those fragments are eroded constantly by micrometeorites, radiation, gravitational interactions, and collisions between particles. Eventually, these particles become dust-sized and are more likely to disperse out of orbit into the planet's clouds or out into space. Rings can be "rejuvenated," according to Esposito, by a recent collision or near-collision between a satellite and another body. This suggests that Neptune's thin dust rings are the oldest rings in the solar system and that the arcs are becoming the newest.

Another of Esposito's theories is that rings, although stable, are fluid. The debris may shift continually and reshape itself as a result of random collisions and sideswipes between ring particles and shepherd moons, so that the ring does not maintain a uniform thickness and density indefinitely. Ring arcs and incomplete ringlets, then, are a transitory stage in the life of a planet's rings.

Bibliography

Burgess, Eric L. *Uranus and Neptune: The Distant Giants.* New York: Columbia University Press, 1988. Despite the title, concentrates primarily on Uranus because of the book's publication between the Voyager encounters of Uranus and Neptune. Contains some history of Neptunian research and the itinerary of the "upcoming" Voyager flyby. For general reading.

Kerr, Richard A. "Voyager Finds Rings in Need of Rejuvenation." *Science* 245 (September 29, 1989): 1451. Part of a special section on Neptune, this article discusses the theory that a planet's rings are born, decay, die, and are replaced by new rings. Informative piece for all readers.

_____. "Why Neptunian Ring Sausages?" *Science* 245 (September 1, 1989): 930. Larry Esposito is a proponent of the theory that planetary rings are continually being shaped and reshaped by gravity and impacts with the bodies. This short article is a sidebar to a piece about the Voyager encounter with Neptune and its moon Triton. For a wide audience.

Littmann, Mark. *Planets Beyond: Discovering the Outer Solar System.* New York: John Wiley & Sons, 1988. This entertaining and informative general-readership book examines the discovery and exploration of Uranus, Neptune, Pluto, and a hypothetical tenth planet. Numerous sidebars explain important concepts without interrupting the narrative; some of the sidebars are first-person accounts by the researchers involved. The book describes the early evidence of Neptunian rings, but regrettably was published only one year before the Voyager encounter and falls short of including that information.

Miller, Ron, and William K. Hartmann. *The Grand Tour: A Traveler's Guide to the*

Solar System. New York: Workman, 1981. This general-readership book contains fascinating paintings and an unusual format. It examines briefly all bodies in the solar system at least 1,600 kilometers in diameter, in size order from Jupiter down to the largest asteroids. The book was published too early to include mention of Neptune's rings, but there is artwork of the similar Uranian system from a rare angle: underneath the rings on the planet itself.

Smith, B. A., et al. "The Neptune Ring System." *Science* 246 (December 15, 1989): 1431-1437. One of a group of articles by the Voyager 2 science team about the findings from Voyager's encounter with Neptune. It is probably the most complete and up-to-date information available on Neptune's rings and arcs, and the theories about their nature. Unfortunately, it does not address the question of their origin. Other articles in this group examine Neptune's chemistry, weather, moons, and the like. Highly technical; written for scientists, but suitable for college students.

Shawn Vincent Wilson

Cross-References

Astronomers Discover the Rings of the Planet Uranus (1977), p. 2068; Voyager 1 and 2 Explore the Planets (1977), p. 2082; The First Ring Around Jupiter Is Discovered (1979), p. 2104.

SOLAR ONE, THE PROTOTYPE POWER TOWER, BEGINS OPERATION

Category of event: Earth science
Time: April, 1982
Locale: Daggett, California

This experimental pilot plant provided new opportunities for technology improvements to be incorporated into future commercial-sized plants

> *Principal participants:*
> SOUTHERN CALIFORNIA EDISON COMPANY, the company that had a major role in the design and construction of nonsolar facilities
> U.S. DEPARTMENT OF ENERGY, the organization that provided major funding, maintained on-site technical staff, and tested the performance of the components
> LOS ANGELES DEPARTMENT OF WATER AND POWER, the company that provided the technical and operating support
> CALIFORNIA ENERGY COMMISSION, the organization that participated in project reviews

Summary of Event

"On April 12, 1982, at 3:09 P.M., the Solar One turbine/generator was synchronized to the SCE (Southern California Edison) system for the first time, marking the dawning of a new age of electrical power generation." This optimistic statement about the start-up of Solar One came from the Communications Department of SCE. The facility—a cooperative effort between the U.S. Department of Energy, SCE, the Los Angeles Department of Water and Power, and the California Energy Commission—was an experimental pilot plant. Generating electricity for consumers to use was only part of its purpose. It also provided opportunities to study technologies that can be included in future commercial-sized solar powered electric generating plants.

The concept underlying Solar One is to "gather" the sunlight that falls over a large area and concentrate it onto a small area. In this small area, the intense heat of the concentrated light will be used to boil water, produce steam, and then use the steam to drive a turbine that, in turn, drives an electric generator. This method of producing electric power is similar to more conventional methods, except for the way in which the steam is generated. In conventional electric generating plants, coal or oil is burned to obtain the heat needed to produce steam. Also, in some plants, atomic reactors are used to obtain the necessary heat. If solar energy could be harnessed to do the job, there would be some significant benefits. Coal is a good energy source, but burning coal often produces air pollutants that are undesirable. Oil is also a very good source of energy, and it is cleaner than coal. Oil reserves, however, are rapidly being depleted. Atomic energy was once thought of as the power source of the future, but concerns about safety have slowed the development of atomic energy in the United States. Hence, solar energy may well be very important in meeting future energy needs.

Solar One consists of an array of heliostats (mirrors) that focus sunlight on a centrally located receiving tower, often referred to as the "power tower." Here, steam is generated and turns a turbine that then drives an electrical generator. During periods when excess steam is available, it is used to heat oil in a thermal storage tank. Later, steam can be generated by using heat from the oil in the storage tank. While this description gives the general idea behind the project, it does not give any feeling for the immensity and complexity of the undertaking. For example, each heliostat is about the size of a two-car garage, and there are 1,818 of them.

There are several major areas of technical investigation for Solar One. These include evaluation of mirror performance, boiler performance, heat storage, materials and equipment, and environmental impact.

Because the heliostats are a major component of the system, more details about them are important to an understanding of the overall working of Solar One. Each heliostat consists of a second surface (silver-backed glass) mirror, bonded to an aluminum honeycomb core 6.4 centimeters thick. This is bonded to a steel pan and the edges are sealed. The mirror is slightly curved to have a focal length of about 305 meters. Each heliostat is mounted on a support pedestal that is 3.05 meters tall and 0.51 meter in diameter. This pedestal houses the electronic controls that allow the mirror to track the sun; it is mounted on a reinforced concrete foundation that is 3.05 meters deep and almost 1 meter in diameter.

Each computer-controlled heliostat is a sun-tracking mirror. Sunlight strikes the heliostat and is reflected to the receiver/boiler that is situated at the top of the 91-meter power tower. There, the heat is absorbed and used to convert water to very high temperature steam. In general, the collector system has operated without major problems. After a few months of operation, however, the reflectivity of the mirrors has degraded from the original 91 percent reflectivity to about 72 percent reflectivity. An experimental wash program was begun. In this program, pressurized, demineralized water was used to rinse off the heliostats, returning the reflectivity to a high level. About one minute is needed to wash one heliostat.

The heliostats are turned in a stow position (mirror surface down) at night, at times when wind velocity exceeds 72 kilometers per hour, and during other types of adverse weather conditions. They can withstand winds of 80 kilometers per hour in the vertical position and 145 kilometers per hour in the stow position without structural damage. Normally, the heliostats are "awakened" prior to sunrise and moved from the stow position to a standby position. When all the heliostats are focused on the receiver/boiler, it becomes very hot and glows brightly.

The power tower is 91 meters high. The top 14 meters is the receiver, where the sunlight is focused. The receiver is 7 meters in diameter and consists of twenty-four panels. The function of the panels is to provide a vessel in which water can be converted to superheated steam by the energy from the heliostat field. Inside each panel are seventy tubes of a special alloy with a melting point of about 1,430 degrees Celsius. The water used is extensively demineralized and flows vertically upward through the receiver panels, where it is converted to steam at a temperature of about 515 degrees Celsius and a pressure of about 1,550 pounds per square inch. The life expectancy of these panels is greater than thirty years. The steam is then directed to a conventional turbine/generator on the ground.

The thermal storage unit is a vertical cylindrical tank filled with a sand and rock mixture through which caloria oil passes. Caloria oil is a petroleum-based oil that is about the consistency of mineral oil when it is cold. To charge the storage unit, steam from the receiver flows into a heat exchanger near the top of the tank. In this heat exchanger, the steam loses some of its heat to the caloria oil. The oil is then introduced at the top and flows downward through the tank. As the oil flows through the rock and sand mixture, it transfers its heat to the rock and sand. The process continues until the thermal storage unit is fully charged—that is, until the entire contents of the tank are at a temperature of about 300 degrees Celsius. The stored energy can be used to generate steam to run the turbine for a limited period when sunlight is not available. The output of the station, however, is much less when it is being operated from stored energy rather than from sunlight.

Solar One has some operational requirements that make it quite different from conventional power stations. In a conventional plant, systems are turned on and run for very long periods of time. In fact, they are shut down only when absolutely necessary. In this way, the output of the station is maximized and the thermal cycling of the equipment is minimized. At Solar One, systems were started up and shut down every day. This placed extreme thermal stress on the equipment. The expansion and contraction that resulted from great changes in temperature (for example, the receiver panels expand about 15 centimeters vertically and 2.5 centimeters horizontally when heated up to their operational temperature) have taken their toll. Three major equipment failures occurred that were attributed to thermal stress. These include warping of one of the receiver boiler panels, failure of valves, and numerous leaks in the receiver and thermal storage system. Some modifications of equipment and of start-up procedures were made to minimize thermal stress. Also, consideration was given to operating the thermal storage system at night to keep all systems at near-normal temperatures and ready for resumed full operation at sunrise.

Impact of Event

There is little argument that whenever electricity is generated on a large scale, there is bound to be some environmental impact. Solar One is no exception. There is little air pollution from this type of electricity-generating plant, and there is little, if any, water pollution. The impact of the solar power tower concept is on the use of land. This type of system requires a large area. Solar One is a relatively small facility, but even so, it covers 321 hectares. Another potential problem is that systems such as Solar One are most effective in sunny, arid, ecologically fragile desert biomes, where there may not be enough water for use in cooling towers to recondense spent steam. Water usage at Solar One is about 100 acre feet per year. Information about long-term effects on the flora and fauna of the region is not readily available.

Solar One was conceived and built as a pilot plant and, after being in operation for about six and one-half years, it was shut down in September of 1988. The closing does not mean that the plant was a failure. On the contrary, Solar One demonstrated that the central receiver concept works as expected. It was expensive to build (about $1.4 billion), but the excessive costs resulted mainly from the fact that components had to be specially built. The information provided by its operation, such as data on maintenance of heliostats and the failure of materials and equipment, will be valuable in the next generation of power towers. There is little doubt that more applications of this concept will be made. In fact, several similar plants that have benefited from knowledge gained from the operation of Solar One would be placed in operation in the southwestern part of the United States. In addition, the concept would be explored in several other countries, including Spain, Italy, France, and Japan.

Bibliography

Bartel, J. J., Sandia Labs, and Solar One Visitor Center. "Solar 10 MWe Pilot Plant Fact Sheet." May, 1983, 1-26. Contains detailed information about all aspects of Solar One, including specifications, costs, and principal participants. One section, called "Solar One Startup," prepared by the Research and Development and Corporate Communications departments of Southern California Edison, gives a good explanation of operating procedures and some problems encountered.

Bylinski, George. "Getting Energy with Mirrors (Solar One Project)." *Fortune* 104 (November 2, 1981): 114-119. A summary of the concept of the power tower, accompanied by excellent photographs that show very well the size and complexity of the project.

Metz, William D. "Solar Thermal Energy: Power Tower Dominates Research." *Science* 197 (July 22, 1977): 353-356. Written before Solar One was built; discusses various ways to make use of solar energy. At the time, the power tower concept was favored, and this article describes how four major, competing contractors envisioned the project. Also describes potential problems.

ReVelle, Charles, and Penelope ReVelle. "Solar Energy." In *The Environment: Issues and Choices for Society.* 2d ed. Boston: Willard Grant Press, 1984. This chapter, in an environmental science text for college-level students, describes several applications of solar energy, including specific reference to Solar One as a viable alternative energy source.

Schefter, Jim. "Solar One: Sun, to Heat, to Electricity." *Popular Science* 221 (October, 1982): 114. A brief article that describes, in general, how Solar One works. Easy-to-understand language.

David W. Appenbrink

Cross-References

Steinmetz Warns of Pollution in *The Future of Electricity* (1908), p. 401; Geothermal Power Is Produced for the First Time (1913), p. 547; Callendar Connects Industry with Increased Atmospheric Carbon Dioxide (1938), p. 1118; The World's First Breeder Reactor Produces Electricity While Generating New Fuel (1951), p. 1419; Bell Telephone Scientists Develop the Photovoltaic Cell (1954), p. 1487; The United States Opens the First Commercial Nuclear Power Plant (1957), p. 1557; Manabe and Wetherald Warn of the Greenhouse Effect and Global Warming (1967), p. 1840; Bell Laboratories Scientists Announce a Liquid-Junction Solar Cell of 11.5 Percent Efficiency (1981), p. 2159; The British Antarctic Survey Confirms the First Known Hole in the Ozone Layer (1985), p. 2285; The Chernobyl Nuclear Reactor Explodes (1986), p. 2321.

THE FIRST COMMERCIAL GENETIC ENGINEERING PRODUCT, HUMULIN, IS MARKETED BY ELI LILLY

Category of event: Medicine
Time: May 14, 1982
Locale: Indianapolis, Indiana

Eli Lilly and Company developed an industrial method to supply enough genetically engineered human insulin (Humulin) to meet the needs of the millions of diabetics worldwide

Principal personages:
> IRVING STANLEY JOHNSON (1925-   ), an American zoologist and vice president of research at Eli Lilly Research Laboratories who was involved in the development of Humulin
> RONALD E. CHANCE (1934-   ), an American biochemist who was involved in design and purification of Humulin at Eli Lilly Research Laboratories

Summary of Event

Carbohydrates (sugars and related chemicals) are the main dietary food and energy source for humans. Most carbohydrate eaten is polymers of the sugar glucose, like plant starch and its animal equivalent, glycogen. In affluent countries, more than 50 percent of the dietary intake of calories is composed of carbohydrates. In underdeveloped countries, dietary carbohydrate content is even higher, reaching 70 to 90 percent of the caloric intake. Therefore, the appropriate disposition of dietary carbohydrate—more than a pound each day—is essential to the quality and length of human life. Normally, most dietary carbohydrate is rapidly used (metabolized) to produce the energy needed to run the body. Excess carbohydrate eaten is either converted to fat, for storage, or stored as the glucose polymer glycogen. Body glycogen—about a pound in most adult humans—is broken down to produce energy when needed. Several diseases of carbohydrate metabolism prevent normal carbohydrate disposition and cause health problems in humans.

The most frequently seen disease of carbohydrate metabolism is diabetes mellitus, usually referred to as diabetes. It is found in more than 70 million people worldwide. Diabetes is caused by either insufficient body levels or underutilization of the pancreatic hormone insulin. The diabetes caused by insufficient insulin—juvenile-onset diabetes—begins early in life. It is treated by administration of insulin. Milder diabetes caused by insulin underutilization—maturity-onset diabetes—begins later in life, mostly in obese people. It is not treatable by insulin therapy. (All further discussion of diabetes here will be limited to juvenile-onset diabetes.)

Uncontrolled cases of diabetes are very dangerous because they can lead to deterioration of the eyes and eventual blindness, to kidney dysfunction, to severe damage to blood vessels and other cardiovascular problems (for example, atherosclerosis), to a reduced life expectancy, and to coma followed by death. It is these complications that explain why diabetes is the third most common killer in the United States. The basis for most of the problems caused by diabetes is the high glucose levels in the blood. For example, cataracts often form in diabetics, as excess glucose is deposited in the lens of the eye.

Major symptoms of diabetes include continual thirst, excess urination, and large amounts of sugar in blood (hyperglycemia) and urine (glycosuria). In fact, the term "diabetes mellitus" means excessive excretion of sweet urine, and measurement of the glucose content of urine is one way to diagnose the disease. A more reliable test for diabetes, however, is examination of the blood glucose content and its response to glucose intake, the so-called glucose tolerance test (GTT). People given a GTT are first told to fast overnight and then given about one-quarter pound of glucose (dissolved in water) to drink.

Blood glucose levels are measured shortly before glucose administration and every thirty to sixty minutes during the next four to six hours. In nondiabetics, the glucose levels observed during a GTT do not rise above 180 milligrams per 100 milliliters. Furthermore, they drop to about 100 milligrams per 100 milliliters within two hours, as the glucose is assimilated because of pancreatic insulin production. In juvenile-onset diabetics, the blood glucose levels rise much higher and do not drop at normal rates. Furthermore, glucose from the test dose appears in the urine of most such diabetics.

Until the 1920's, control of the symptoms and consequences of juvenile-onset diabetes was possible only by the use of diets that severely restricted carbohydrate intake. These diets were only moderately successful. Then, Sir Frederick G. Banting and Charles H. Best succeeded in preparing purified insulin from animal pancreases and gave it to patients. At that time, the use of insulin gave juvenile-onset diabetics their first chance to live normal life spans. The endeavors of Banting and his co-workers won for them the 1923 Nobel Prize in Physiology or Medicine.

The usual treatment for juvenile-onset diabetes quickly became the routine injection of insulin, isolated by drug companies, from the pancreases of cattle and pigs slaughtered by the meat-packing industry. Most diabetics survive well on the animal insulins. Unfortunately, use of the animal insulins has two drawbacks: First, about 5 percent of diabetics are allergic to animal insulins; they can experience severe allergic reactions. Second, and more important, the world supply of animal pancreases is limited by and dependent on fluctuation of the demand for meat foods. In this respect, it alarmed many that this supply declined sharply from 1970 to 1975, while the number of diabetics continued to increase. Consequently, the search for a better, less limited supply of insulin began.

At that time, study of insulin, from the pancreases of people who donated their bodies to science, showed human insulin to be nonallergenic and thus preferable to animal insulin. Therefore, it became apparent that a suitable means for chemical or biological preparation of human insulin would solve the problems both of insulin allergy and of the fluctuating insulin supply. This endeavor became a major goal of pharmaceutical research. Eli Lilly and Company was the first pharmaceutical house to achieve success, and on May 14, 1982, it filed a new drug application with the Food and Drug Administration (FDA) for the human insulin preparation they called Humulin.

Human, beef, and pork insulins are similar, small proteins—that is, amino acid polymers or polypeptides—that differ slightly from the amino acids they contain. Each of these proteins is composed of two short amino acid chains (the A and B chains) joined together chemically by so-called disulfide bonds. The strategy used to make Humulin relies heavily on the methodology called genetic engineering (or recombinant DNA—deoxyribonucleic acid—research). The basis for this methodology is the arduous, but deceptively simple-sounding, ability of the modern biotechnology to incorporate the genetic message for a desired polypeptide into the genetic information of a chosen microorganism; to grow the microorganism on an industrial scale; and finally, to isolate and purify the desired polypeptide from the microorganism.

The current commercial method for producing Humulin (devised in collaboration by scientists at Genentech, Inc., and Eli Lilly) is described by Eli Lilly spokesman Irving Stanley Johnson in *Science* (1983). It uses the bacterium *Escherichia coli* as the carrier microorganism. Two modified strains of the bacterium are produced by genetic engineering and grown on an industrial scale. The first strain makes the A chain, and the second strain makes the B chain. After the bacteria are harvested, the A and B chains are isolated and purified separately. Then, the two chains are combined chemically by linking them together with disulfide bonds. Finally, repurification yields Humulin. This insulin preparation was shown to be chemically, biologically, and physically identical to the human insulin isolated from cadaveric human pancreases by Eli Lilly's Ronald E. Chance and coworkers in *Diabetes Care* (1981).

Impact of Event

In 1982, genetically engineered human insulin was approved for pharmaceutical use by the FDA and equivalent regulatory agencies around the world. This insulin preparation, marketed under the trade name Humulin, is medically important because it generates a reliable supply of insulin and because it does not contain an allergenic. In addition, Humulin manufacture is important because it is a test case proving the industrial viability of recombinant DNA research.

Humulin manufacture guarantees a reliable, easily increased supply of insulin for the growing number of insulin-requiring diabetics around the world. The importance of Humulin availability is best clarified by consideration of problems foreseen in the 1970's, when slaughterhouse beef and pork pancreases were the sole source of insulin. It became clear then that the supply of available pancreases had plateaued, while the number of insulin-requiring diabetics (4 million cases in industrial nations) was rising steadily. This led to fear that an insulin shortage would occur by 1990; Humulin availability has ended this fear.

Many members of the biomedical community view the lack of allergic reactions from Humulin as being less important than its promised abundance. They state that this results from the fact that a lack of allergenicity will help probably only the 3 to 4 percent of diabetics who experience severe allergic responses to beef and pork insulin. In contrast, it is stressed that the real treatment problems seen with insulin-dependent diabetics are: how to obtain adequate amounts of insulin to serve all of them and how to get the hormone into patients in the most appropriate way. Finally, it cannot be emphasized too strongly that a prime aspect of the importance of Humulin production is the fact that it was the first genetically engineered industrial chemical. It thus began an era wherein recombinant DNA technology could be viewed as a viable source of pharmaceuticals, agricultural chemicals, and other important industrial products. It seems likely that gains from such efforts will produce understanding of cancer and other disease processes, as well as leading to methodologies that will provide adequate amounts of food in a world where productive land areas are decreasing, while the population increase continues.

Bibliography

Chance, Ronald E., Eugene P. Krieff, James A. Hoffmann, and Bruce H. Frank. "Chemical, Physical, and Biologic Properties of Biosynthetic Human Insulin." *Diabetes Care* 4 (March/April, 1981): 147-154. This detailed technical paper describes the chemical, physical, and biological properties of Humulin and shows its chemical, physical, and biological equivalence to insulin isolated from human pancreases. Tests that show the biological equivalence of Humulin and pork insulin are also described. Forty-one references on development and testing of Humulin are included.

Chenault, Alice. "Diabetes." In *McGraw-Hill Encyclopedia of Science and Technology.* 6th ed. New York: McGraw-Hill, 1987. This brief exposition of diabetes presents solid but simple explanations of juvenile-onset and maturity-onset diabetes. Some common medical problems caused by diabetes (for example, kidney damage, eye damage, and atherosclerosis) are described. In addition, possible causes of diabetes and its treatment are discussed. Contains useful sources of additional information.

Johnson, Irving S. "Human Insulin from Recombinant DNA Technology." *Science* 219 (February, 1983): 632-637. Johnson, an Eli Lilly Research Laboratories vice-president, describes Humulin development; the basis for commencement of the project; aspects of regulation of recombinant DNA research; the preparation and testing of Humulin; and the impact of recombinant DNA research on the pharmaceutical industry, including hopes for the future. Twenty-one references cover scientific and regulatory aspects.

Lehninger, Albert L. *Principles of Biochemistry.* New York: Worth, 1982. Chapters 24 and 25 of this excellent college biochemistry text cover aspects of insulin biochemistry and diabetes that include information on diabetes diagnosis and treatment, the pancreas, and the production of insulin. Useful diagrams showing the structure of insulin and the glucose tolerance test are included.

Orci, Lelio, Jean-Dominique Vassali, and Alain Perrelet. "The Insulin Factory." *Scientific American* 236 (September, 1988): 85-94. This article describes insulin biosynthesis, processing, and secretion. Very useful to readers who wish to understand the nature of insulin and provides background for dealing with the concept of insulin as a protein composed of A and B chains that can be taken apart and put back together.

Smith, Emil L., et al. "The Pancreatic Islets." In *Principles of Biochemistry: Mammalian Biochemistry.* New York: McGraw-Hill, 1983. Coverage of insulin is strong, including insulin synthesis by islet cells; insulin actions; and effects of insulin on sugar, lipid, and protein metabolism. The mechanism of insulin uptake by cells it affects, mechanisms of insulin action, and regulation of its secretion are also described. The information here is more technical than that presented in the Lehninger text (cited above).

Sanford S. Singer

Cross-References

Bayliss and Starling Discover Secretin and Establish the Role of Hormones (1902), p. 179; Banting and Macleod Win the Nobel Prize for the Discovery of Insulin (1921), p. 720; The Artificial Sweetener Cyclamate Is Introduced (1950), p. 1368; Du Vigneaud Synthesizes Oxytocin, the First Peptide Hormone (1953), p. 1459; Sanger Wins the Nobel Prize for the Discovery of the Structure of Insulin (1958), p. 1567; Kornberg and Coworkers Synthesize Biologially Active DNA (1967), p. 1857; Cohen and Boyer Develop Recombinant DNA Technology (1973), p. 1987; The Artificial Sweetener Aspartame Is Approved for Use in Carbonated Beverages (1983), p. 2226.

THE ARTIFICIAL SWEETENER ASPARTAME IS APPROVED FOR USE IN CARBONATED BEVERAGES

Category of event: Chemistry
Time: 1983
Locale: Washington, D.C.

The U.S. Food and Drug Administration approved the use of l-aspartyl-l-phenylalanine methyl ester (aspartame) for use in carbonated beverages

Principal personages:
ARTHUR H. HAYES, JR. (1933-   ), a physician and commissioner of the U.S. Food and Drug Administration
JAMES M. SCHLATTER (1942-   ), an American chemist
MICHAEL SVEDA (1912-   ), an American chemist and inventor
LUDWIG FREDERICK AUDRIETH (1901-   ), an American chemist and educator
IRA REMSEN (1846-1927), an American chemist and educator
CONSTANTIN FAHLBERG (1850-1910), a German chemist

Summary of Event

People have sweetened food and beverages since before recorded history. The use of honey as a sweetener dates from biblical times; other sweeteners include maple syrup (used by the American Indians), cane syrup (South Pacific), and, more recently, granulated sugar and corn syrup. Adding these ingredients enhanced the taste of many foods and acted as a preservative. The most widely used sweetener is sugar, or sucrose. The only real drawback to the use of sucrose is that it is a nutritive sweetener: In addition to adding a sweet taste, it adds calories. Because sucrose is readily absorbed by the body, an excessive amount can be life-threatening to diabetics. This fact alone would make the development of nonsucrose sweeteners attractive.

There are three common nonsucrose-type sweeteners currently in use around the world: saccharin, cyclamates, and aspartame. Saccharin was the first of this group to be discovered, in 1879. Constantin Fahlberg synthesized saccharin based on the previous experimental work of Ira Remsen using toluene. This product was found to be three hundred to five hundred times as sweet as sugar, although some people could detect a bitter aftertaste. With the help of a relative, Fahlberg founded Fahlberg, List & Co., and in 1886, Fahlberg started the production and distribution of this artificial sweetener.

In 1944, the chemical family of cyclamates was discovered by Ludwig Frederick Audrieth and Michael Sveda. Although these compounds are only thirty to eighty times as sweet as sugar, there was no detectable aftertaste. By the mid-1960's, cyclamates had replaced saccharin as the leading nonnutritive sweetener in the United States. Although this compound is still in use throughout the world, on October 1969, the U.S. Food and Drug Administration (FDA) removed it from the list of approved food additives because of tests that indicated possible health hazards.

Aspartame, or l-aspartyl-l-phenyl-alanine methyl ester, is the latest in artificial sweeteners that are derived from natural ingredients—in this case, two amino acids, one from milk and one from bananas. Discovered by accident in 1965 by American chemist James M. Schlatter when he licked his fingers during an experiment, aspartame is 180 times as sweet as sugar. In 1974, it was approved for use in dry foods such as gum and cereal and as a sugar replacement. Shortly after its approval for this limited application, the FDA held public hearings on the safety concerns raised by John W. Olney, a professor of neuropathology at Washington University in St. Louis. There was some indication that aspartame, when combined with the common food additive monosodium glutamate, caused brain damage in children. These fears were confirmed, but the risk of brain damage was limited to a small percentage of individuals with a rare genetic disorder. At this point, the public debate took a political turn: Senator William Proxmire charged FDA Commissioner Alexander M. Schmidt with "misfeasance," or public misconduct. This controversy resulted in aspartame being withdrawn from the marketplace in 1975.

In 1981, the new FDA commissioner, Arthur H. Hayes, Jr., reapproved aspartame for use in the same applications: as a tabletop sweetener, a cold-cereal additive, in chewing gum, and other miscellaneous uses. In 1983, it received final approval for use in carbonated beverages, its largest application to date. Later safety studies revealed that children with a rare metabolic disease, phenylketonuria, could not ingest this sweetener without severe health risks because of the presence of phenylalanine in aspartame. This condition occurs in individuals who have only a single copy of the phenylalanine hydroxylase gene instead of two copies. This results in a rapid buildup of phenylalanine in the blood. Laboratories simulated this condition in rats and found that high doses of aspartame inhibited the synthesis of dopamine, a neurotransmitter, which can lead to an increase in the frequency of seizures. There was no direct evidence, however, that aspartame actually caused seizures in these experiments.

Many other chemical compounds are being tested for use as sugar replacements, the sweetest being an analog of aspartame [N-(4-nitrophenylcarbamoyl)-L-aspartyl-L-phenylalanine methyl ester]. This compound is seventeen thousand to fifty-two thousand times sweeter than sugar.

Aspartame has a tendency to break down when exposed to heat, and for this reason, the sweetener is not used in products that require baking. Even at room temperature (25 degrees Celsius), aspartame will break down; it has a half-life of 262 days. In aqueous solutions, it is most stable in a pH of 4, the acidity of vinegar.

It should be noted that although aspartame is currently marketed by G. D. Searle & Co., the largest supplier of this product to the U.S. market, the patent for the manufacturing of aspartame expired in 1987.

Impact of Event

The business fallout from the approval of a new low-calorie sweetener occurred over a short span of time. In 1981, sales of this artificial sweetener by Searle were $74 million. In 1983, sales rose to $336 million and exceeded half a billion dollars the following year. These figures represent more than 2,500 tons of this product. In 1985, 3,500 tons of aspartame were consumed. Clearly, this product's introduction was a commercial success for Searle. During this same period, the percentage of reduced-calorie carbonated beverages containing saccharin declined from 100 percent to 20 percent in an industry that had $4 billion in sales. Universally, consumers preferred products containing aspartame; the bitter aftertaste of saccharin was rejected in favor of the new, less powerful sweetener.

The health risks associated with aspartame are not inconsequential. As much as 2 percent of the U.S. population has a genetic defect that hinders the efficient metabolism of phenylalanine, one of the two amino acids that make up aspartame. Phenylketonuria, a disease normally limited to infants, can lead to mental retardation and seizures. The FDA has established 50 milligrams per kilogram of body weight as an "acceptable daily intake" for aspartame; the average 340-gram can of diet soda contains 200 milligrams. With moderate consumption of diet beverages, cereal, chewing gum, and other low-calorie products, it is very possible for individuals to exceed this guideline. At these levels of use, some researchers have found that test subjects, with the genetic defect that reduces the body's ability to break down phenylalanine, exhibit slower brain wave patterns. It was also found that aspartame will lower blood pressure in laboratory animals. These mixed signals caused ambivalence among regulators.

There is a trade-off in using these products. The FDA found evidence linking both saccharin and cyclamates to an elevated incidence of cancer. Cyclamates were banned in the United States because of those studies, and the FDA tried to ban saccharin for the same reason. Public resistance to this measure caused the FDA to back away from its position. The rationale was that because diabetics and overweight people had higher health risks because of the presence of granulated sugar and corn syrup in carbonated beverages, the slight chance of contracting cancer was a risk many would choose to take. The total domination of aspartame in the soft-drink market proves the earlier assumption.

Bibliography

Allman, W. F. "Aspartame, Some Bitter with the Sweet." *Science 84*, 5 (July/ August, 1984): 14. An article on the phenylalanine risk to individuals with a slight genetic defect. This article questions the benefits when viewed against the serious health risks to some individuals.

Gissen, J. "Beginner's Luck." *Forbes* 133 (May 21, 1984): 240. An interview with the head of G. D. Searle's NutraSweet division, Robert Shaprio. Shaprio identifies marketing plans for future production increases.

Smith, J. R. "Aspartame Approved Despite Risk." *Science* 213 (August 28, 1981): 986-987. This article outlines the controversy surrounding Hayes's decision to approve aspartame. It also profiles the product's main antagonist, John Olney.

_____. "Sweet Talk." *Scientific American* 167 (September, 1987): 16. A nontechnical review of the health risks associated with aspartame, including results of tests done on human volunteers by Paul Spiers of Beth Israel Hospital in Boston.

Teitelman, R. "Bittersweet." *Forbes* 134 (August 27, 1984): 36-37. An in-depth look at the business considerations and problems that were facing G. D. Searle one year after the approval of aspartame in soft drinks. Contains an inset on other sweeteners that eventually may provide NutraSweet with competition for the artificial sweetener market.

Charles A. Bartocci

Cross-References

Bayliss and Starling Discover Secretin and Establish the Role of Hormones (1902), p. 179; Banting and Macleod Win the Nobel Prize for the Discovery of Insulin (1921), p. 720; The Artificial Sweetener Cyclamate Is Introduced (1950), p. 1368; Sanger Wins the Nobel Prize for the Discovery of the Structure of Insulin (1958), p. 1567.

RUBBIA AND VAN DER MEER ISOLATE
THE INTERMEDIATE VECTOR BOSONS

Category of event: Physics
Time: 1983
Locale: European Laboratory for Particle Physics (CERN), Geneva, Switzerland

Rubbia and van der Meer isolate the intermediate vector bosons, the particles that carry the weak nuclear force, providing confirming evidence for the unified electro-weak theory

Principal personages:

CARLO RUBBIA (1934-   ), an Italian physicist and head of the Euro-
    pean Laboratory for Particle Physics (CERN) who discovered weak
    neutral currents and the bosons, W^+, W^-, and Z^0 particles

SIMON VAN DER MEER (1925-   ), a Dutch physicist who shared the
    1984 Nobel Prize in Physics with Rubbia for the discovery of the
    intermediate vector bosons

Summary of Event

One of the largest research projects in contemporary physics has been the study of the forces that act between the elementary particles of matter. The experiment that resulted in the discovery of the W^+, W^-, and Z^0 particles, or the "intermediate vector bosons," provided at least partial confirmation for a theory about two of these forces. Before the experiment can be discussed, however, an understanding of the four forces in nature is essential.

The gravitational force is the weakest of the four forces, yet any particle with mass feels its effects. The gravitational force acts over large distances and is always attractive. It holds the universe together, keeps objects from floating off Earth's surface, and keeps Earth in orbit around the sun.

The electromagnetic force acts on electrically charged particles such as electrons and protons and is much stronger than the gravitational force. This force can act over an infinite distance and can be either repulsive or attractive. This latter fact can be confirmed by moving two magnets into close proximity—like poles (charges) repulse each other, while opposites attract.

The other two forces are nuclear forces, a strong and a weak version, which act over very small distances. The weak force is responsible for the phenomenon of radioactive decay in radioactive materials; the strong force acts between the protons and neutrons that constitute the nucleus of the atom. Thus, the strong force holds the nucleus together. Until 1967, there were four fairly well understood theories explaining each of the four forces. In that year, Abdus Salam of Imperial College, London, and Steven Weinberg of Harvard University, and, independently, Sheldon L. Glashow of Harvard University, developed the first of the so-called unified theories. The idea is that rather than having four theories to explain four forces, one unified theory would explain them all. The theory, which came to be known as the Weinberg-Salam electro-weak theory (W-S), had as an experimentally testable consequence the existence of three massive particles known as the "intermediate vector bosons." Such theories are known as quantum field theories and describe the forces between two or more particles as the exchange of a third intermediate particle between them. In other words, rather than conceptualizing the force as simply a continuous field, the force is conceived as the exchange of discrete "quanta." In the electromagnetic force, a massless particle known as the "photon" is exchanged; it has been known to exist since Albert Einstein first discussed the quantized nature of light in 1905. In the W-S theory, the intermediate vector bosons—in addition to the photons—carry the newly termed electro-weak force.

If the W-S theory was correct, an experiment should have been possible that could detect the intermediate vector bosons. This was the Carlo Rubbia experiment conducted at CERN. The theory predicted the existence of three particles: the W^+, the W^-, and the Z^0. They carry positive, negative, and neutral electrical charges, respectively. Because of their relatively heavy masses—eighty to ninety times that of the proton—experimental apparatus that could generate extremely high energies would be required to isolate the particles.

Two large experiments were conducted at CERN, called UA-1 and UA-2 for Underground Area One and Underground Area Two. The difference between the two experiments rests in the type of target used for the detection of the bosons, yet both employed the huge proton accelerator, the Superproton Synchrotron (SPS) at CERN. Rubbia's first step in the experiment was to modify the SPS from simply a proton accelerator (a large magnetized ring that accelerates protons to very high energies) to a proton-antiproton collider (an apparatus that would accelerate protons and the particle of identical mass and opposite charge—the antiproton—at high energies for the purpose of smashing them together and detecting the residue of the collision, which often includes new particles). In theory, the intermediate vector bosons should be among the collision products of a proton-antiproton collision.

In 1976, Rubbia faced the problem that new proton-antiproton accelerators would not be in operation until the mid-1980's. One such machine, the United States' Fermilab's Tevatron, would not come on-line until 1985 to 1986. As a solution, Rubbia spearheaded the renovation of the CERN SPS, converting the proton accelerator into a collider. Simon van der Meer, formerly in charge of magnetic power at CERN, supervised the conversion, which was approved on Rubbia's recommendation in 1978 at a cost of $100 million. Van der Meer's most important contribution was the design of the new antiproton storage ring. In the new proton-antiproton collider, when a beam of accelerated protons is shot at a copper target, a relatively small number of antiprotons are produced. These particles are collected and stored in the Antiproton Accumulator until several billion have accumulated. At this time, they are sent back into the main ring of the accelerator, while protons are introduced simultaneously, but in the opposite direction. Eventually, at energies approaching 300 billion electronvolts, the two opposing beams are allowed to collide. Detectors are placed at the point of collision to record the products of the reaction.

By late 1981, the new collider was in operation and produced energies nearly twenty times greater than the energies that the unmodified SPS could generate. Rubbia enlisted almost two hundred experimental and theoretical physicists to take part in the project. The team at UA-1 employed a target to observe collision products that weighed almost 2,000 tons and cost $20 million. The target's size could be attributed to some future planning on the part of the Rubbia team. Not only could this target detect the intermediate vector bosons but also it could, under the proper conditions, detect the more elusive Higgs particle, another prediction of W-S theory. This detector can measure the energy of particles by measuring the curvature of their paths in the presence of a large magnetic field. UA-2 employed a target one-tenth the size of UA-1's, which was designed exclusively for the intermediate vector boson search. Experimental runs on both targets began in early 1982 after some crucial trial-and-error adjustments on the newly constructed collider.

In order to detect any subatomic particle, experimenters do not look for them directly—they are far too small for even the most powerful microscopes. Instead, the targets are embedded with sophisticated electrical equipment that can detect the properties attributed to the particles. The search for the intermediate bosons is one such case. Theoretically, after a highly energetic proton-antiproton collision, if any bosons are produced, their high mass and equivalently (according to Einstein's theory of relativity) high energies would make them very short-lived particles; they would decay quickly into other, less interesting particles. As a result, Rubbia and his team were forced to look for specific patterns on their targets that would indicate the presence of the bosons. Each of the three bosons has a unique decay pattern, according to the W-S theory.

The Z^0 particle, being electrically neutral, would decay into two less massive particles with opposite charge, such as an electron and a positron. Physicists can recognize easily the tracks of positrons and electrons on a detector, and a highly energetic pair of them traveling in opposite directions would provide evidence for the Z^0 boson. Each of the W particles carries a charge; therefore, their decay products must have an equivalent sum charge. For example, a W^+ carries an electrical charge of $+1$. Its decay products might consist of a positron, which carries a $+1$ charge, and an electron neutrino, which has no charge. Because neutral particles such as electron neutrinos cannot be imaged in electrical detectors, evidence of the W^+ would consist of a single, highly energetic positron. By January of 1983, Rubbia and his team were reporting success.

By the end of the summer of 1983, Rubbia and 134 other physicists had collaborated in an experiment that isolated all three of the intermediate vector bosons in very close agreement with the predictions of the W-S theory. UA-1 reported fifty-five events which they maintain produced both W^+ and W^- particles, while UA-2 reported thirty-five events. Both reported a mass of approximately 81 billion electronvolts, plus or minus 2 billion electronvolts; the W-S theory predicted 83 billion electronvolts. UA-1 reported six Z^0 events, while UA-2 reported three, at masses that were also in very close agreement with the experiment. This confirmation of the W-S theory further justified the Nobel Committee's awarding of the 1979 prize in physics to Salam, Weinberg, and Glashow for the theory. In 1984, Rubbia and van der Meer shared the Nobel Prize for their experimental achievement.

Impact of Event

The impact of the isolation of the intermediate vector bosons has taken on significant scientific and technological dimensions. Scientifically, the discovery had the immediate effect of confirming the predictive success of the Weinberg-Salam electroweak unified theory. In addition, it has provided the impetus for further work in unifying the other forces, the gravitational and the strong nuclear force. Several attempts to unify the electro-weak theory with a theory of the strong nuclear force have met with some limited success, although a unification of all four forces has yet to be accomplished.

One particularly striking effect of the experiment was the discovery of so-called weak neutral currents. What is so striking is that this discovery occurred before the discovery of the bosons. The experiments that led to the discovery of weak neutral currents were actually a fortunate consequence of some of the early studies of the intermediate vector bosons. A weak neutral current is one way in which the presence of the Z^0 particle can be detected. In 1973, experiments conducted on weak interactions revealed that charge was preserved before and after weak interaction experiments, indicating that a neutral particle must be exchanged in the interaction. That particle, according to W-S theory, was the Z^0. Rubbia, together with David B. Cline and Alfred Mann, made this discovery in experiments conducted at Fermilab. The result not only gave an early vote of confidence for the new W-S theory but also spurred Rubbia to continue the search for the intermediate vector bosons at CERN.

Technologically, the success of the boson search was a victory to those physicists who favored the use of proton accelerators in high-energy physics experiments. The Rubbia results have been instrumental in the U.S. Department of Energy's push to develop the massive superconducting supercollider, a proton-antiproton collider that will attain energies at least forty times those of CERN's collider. The results also struck a blow to proponents of electron-positron colliders, who claimed that proton-antiproton machines create too much interference to detect such short-lived particles as the intermediate vector bosons. Rubbia's results indicated that this was clearly not the case, although Rubbia has favored using both types of machines, and CERN now has in operation an electron-positron collider that has been operational since 1989. A similar device was scrapped in midconstruction at the United States' Brookhaven Laboratory in favor of the much more powerful superconducting supercollider.

Bibliography

Cline, David B., Carlo Rubbia, and Simon van der Meer. "The Search for Intermediate Vector Bosons." *Scientific American* 246 (March, 1982): 48-59. A rather technical treatment of the design and purpose of the Rubbia experiment. Readers should not be intimidated by the technical aspects, for the prose is elegant and the illustrations are quite helpful, especially in understanding the interaction between the apparatus employed and the theory that predicted the existence of the bosons. Both aspects of the experiment are explained thoroughly.

Galison, Peter. "Ending a High Energy Physics Experiment." In *How Experiments End*. Chicago: University of Chicago Press, 1987. This book is primarily an account of three historical events in twentieth century physics: the discovery of the muon, the determination of the gyromagnetic ratio, and the discovery of weak neutral currents. Galison's discussion of the discovery of weak neutral currents stresses the interdependence of theory and experimental apparatus in the interpretation of experimental results. Provides excellent historical background to the W-S theory and stresses the expanding role of engineering and technology in high-energy physics.

Hawking, Stephen W. "Elementary Particles and the Forces of Nature." In *A Brief History of Time: From the Big Bang to Black Holes*. New York: Bantam Books, 1988. This book is perhaps the finest piece of popular science ever produced. Hawking is a leader in the fields of physics and cosmology, and the chapter on particles and forces explains the idea of unified theories in intuitively accessible terms without recourse to mathematical equations. For those who are interested only in the theoretical aspect of unified field theories, this book is an excellent starting point.

Robinson, Arthur. "CERN Vector Boson Hunt Successful." *Science* 221 (August, 1983): 840-842. This brief article presents the results of the Rubbia experiment and discusses many of its implications for unified theories in physics. It is nontechnical in its presentation and amply illustrated.

Taubes, Gary. *Nobel Dreams: Power, Deceit and the Ultimate Experiment*. New York: Random House, 1986. Taubes tells the personal story of Rubbia's work, which culminated in the 1984 Nobel Prize in Physics. The book discusses many of the facets of high-energy physics not seen in scientific papers, such as the tremendous monetary commitments necessary for the design, construction, and ultimate completion of experiments, and the personal conflicts that develop when more than one hundred physicists work on the same high-stakes project.

William J. McKinney

Cross-References

Lawrence Develops the Cyclotron (1931), p. 953; Cockcroft and Walton Split the Atom with a Particle Accelerator (1932), p. 978; Gell-Mann Formulates the Theory of Quantum Chromodynamics (QCD) (1972), p. 1966; Georgi and Glashow Develop the First Grand Unified Theory (1974), p. 2014; The Tevatron Particle Accelerator Begins Operation at Fermilab (1985), p. 2301; The Superconducting Supercollider Is Under Construction in Texas (1988), p. 2372.

THE FIRST SUCCESSFUL HUMAN EMBRYO
TRANSFER IS PERFORMED

Category of event: Medicine
Time: January-October, 1983
Locale: Torrance, California

An embryo from a donor woman was transferred to the uterus of an infertile woman, enabling her to give birth to a genetically unrelated infant

> *Principal personages:*
> JOHN E. BUSTER (1941-   ), a professor of obstetrics and gynecology who headed the team that performed the first human embryo transfer
> MARIA BUSTILLO (1951-   ), an assistant professor of obstetrics and gynecology who worked with Buster on the first successful human embryo transfers
> HORACIO B. CROXATTO, a Chilean physician who showed that a human egg could be extracted from the uterus without surgery

Summary of Event

The 1978 announcement that a test-tube baby had been born caused many to wonder whether a new era in reproductive biology had begun. After all, if a woman could have her eggs fertilized in a dish and then transferred to her uterus, then maybe it would be possible to transfer her fertilized eggs to another woman. The notion that some women would become breeders for others was reminiscent of George Orwell's futuristic novel, *Nineteen Eighty-Four* (1949).

Although the public was not aware of it, these possibilities were not so remote. The necessary technology had been developing in the laboratory for almost a century. In 1890, a Cambridge physiologist, Walter Heape, first produced the birth of young from an unrelated mother. He surgically removed two embryos from a pregnant Angora rabbit and placed them into the reproductive tract of a Belgian hare. The hare gave birth to six young—four hares and two rabbits. Over the years, embryo transfer had been tried successfully by scientists in several laboratories and had been successful in fourteen different species.

Farmers, particularly, were quick to recognize the potential of this technology. By the early 1980's, embryo transfer was widely used in the cattle industry, with more than ten thousand calves a year produced in this way. A prize cow would be artificially inseminated with sperm from a superior bull and, after a few days, the embryo would be flushed out and inserted into the uterus of a genetically less valuable cow. Whereas a top milk-producing cow could naturally give birth to about ten calves in her lifetime, with embryo transfer, some cows have produced as many as fifty offspring a year. Conservation biologists also exploited this technique. To increase the reproductive rate of an endangered species, they transferred embryos to a more common species for gestation. In 1982, an ordinary cow gave birth to a rare, wild Asian ox at the Bronx Zoo in New York City.

There was little question that the procedure of embryo transfer could be performed eventually on humans as well. In 1972, Horacio B. Croxatto and his team in Chile reported that a human egg could be safely flushed out of the uterus. They suggested that "egg transfer may have a place in the future as an alternative for the infertile couple. . . ." Concerns about the ethical implications of embryo transfer hindered research on humans. Even after the success of in vitro fertilization in England, the National Institutes of Health refused to fund research dealing with human embryos.

A solution was suggested by two brothers, Richard Seed, founder of a cattle-breeding company, and Randolph Seed, a surgeon. In 1980, they replicated Croxatto's work, in this case flushing out a fertilized human egg. Their idea was to fund research on human embryo transfer with money from investors willing to risk their capital on a new venture. The company they created, Fertility and Genetics Research, provided a grant to a team headed by John E. Buster and Maria Bustillo at Harbor-University of California, Los Angeles Medical Center in Torrance, California. The goal of this group was to develop a nonsurgical technique to transfer a fertilized egg from the uterus of one woman to that of another woman.

To assemble a group of egg donors, they placed an advertisement in the local newspapers: "Help an infertile woman have a baby. Fertile women ages 20-35 willing to donate an egg. Similar to artificial insemination. No surgery required. Reasonable compensation." Of four hundred respondents, forty-six women were selected after medical and psychological screening.

The doctors were treating fourteen women who were unable to conceive because of blockage of the oviduct, the tube where an egg and sperm usually meet. Conventional surgical therapy had not helped them. As each donor woman reached the fertile point of her menstrual cycle, the physicians called in one of the infertile women who was similar in general appearance and stage of her menstrual cycle. The infertile woman's husband provided sperm, which was injected into the uterus of the donor. Under natural conditions, a fertilized egg enters the uterus but does not attach to the wall of the uterus until five days after fertilization takes place. The donor women were recalled to the hospital five days after they had been inseminated. A small amount of fluid was injected into the uterus and then sucked out, bringing the embryo with it. The embryo was then placed through the narrow cervical opening into the uterus of the infertile woman. From January through October of 1983, twenty-nine inseminations were performed, from which twelve embryos were transferred. Two of these developed into normal pregnancies.

Early in 1984, the first embryo-transferred baby was born. In a year that was filled with comparisons to Orwell's classic novel, the press was quick to cover the story of a woman giving birth to a baby conceived in another woman's womb. Another baby born the same year caused a similar stir. An Australian woman was twenty-five years old when doctors diagnosed her infertility as resulting from premature menopause. Although her ovaries were not functioning, her uterus seemed normal. A second woman, undergoing surgery for in vitro fertilization, agreed to donate one of her eggs to the first woman. After the egg was fertilized in vitro with sperm from the recipient's husband, it was transferred to the woman's uterus. She was treated with hormones usually produced by the normal ovary, and eventually gave birth to a baby that was not genetically related to her. The physicians who performed the procedure, Peter Lutjen, Alan Trounson, and their associates at Monash University, referred to the procedure as "ovum transfer." Technically, a fertilized egg, or ovum, is called a "blastocyst" until the second week of pregnancy, when the word "embryo" is used. For this reason, many people refer to the Buster and Bustillo technique as ovum transfer as well.

Physicians were generally pleased with the success of these new methods of treating infertility. In the work of Buster and Bustillo, however, a second problem was raised, that of financial profit. The members of Fertility and Genetics Research sought to recoup their investment by obtaining a patent on an instrument that had been developed in the research, a catheter used to retrieve and transfer the embryo. To the consternation of most medical practitioners, they also sought a patent on the technique used in the transfer. Patenting a medical technique means that users would be required to pay a fee to the company, raising concern that privacy would be compromised. The vision of medicine as a profit-making venture was abhorrent to many.

Buster went on to study the possibility that embryo transfer might be used for prenatal diagnosis of genetic disease. The other investigators continued their work as medical practitioners, researching and treating infertility in women.

Impact of Event

Like other advances in reproductive technology, embryo transfer was hailed by infertile couples who might benefit from the technology. Its potential use includes treatment of women with scarred oviducts, premature menopause, or congenital malfunctioning of the ovaries, as well as women who risk passing on a serious genetic disease if their own eggs are used. Yet, the technique is not without risk. For the donor woman, there is the chance that the embryo will implant in the uterus before it can be washed out, leading to an unwanted pregnancy. There is also the risk of infection for both the donor and the recipient. These risks can be minimized by using an egg fertilized in vitro, rather than in a donor woman.

The technique of embryo transfer brought into focus some ethical and legal questions. For the first time, the genetic mother and the gestational mother were not the same person. That raised the question of who should be considered the legal mother. Also, a genetic mother might decide to keep the baby herself. In general, the woman who carries the baby throughout pregnancy is considered the legal mother.

Perhaps the most significant impact of embryo transfer was in the discussion it generated regarding the commercialization of medical technology. Although a small number of patents had been granted in the past for certain medical techniques, these were related generally to some specific surgical advance. The general consensus in the medical community was that this situation did not qualify for a patent. Although the catheter used by the California team was patented, other workers were able to develop similar catheters that were equally as effective. In 1986, Leonardo Fonmigli reported a third embryo-transfer baby born in their clinic in Italy.

It is too early to tell how extensively embryo transfer will be used to treat infertility. Possibly, ovum transfer, with its lesser risk for both women, will be more appealing in most cases. Nevertheless, the 1984 birth of a genetically unrelated infant sparked the imagination of the world with its potential for a new reproductive technology.

Bibliography

Andrews, Lori B. "Legal Aspects of Assisted Reproduction." In *In Vitro Fertilization and Other Assisted Reproduction*, edited by Howard J. Jones and Charlotte Schrader. New York: New York Academy of Sciences, 1988. Written by a research fellow of the American Bar Foundation who describes the legal permissibility of embryo research in the United States, permissibility of paying an egg donor, laws regarding the screening of donors, assigning legal parenthood, and payment for the service. Well written, with a bibliography.

Austin, C. R. *The Mammalian Egg.* Springfield, Ill.: Charles C Thomas, 1961. A detailed account of information available at the time on mammalian eggs. The final chapter on the manipulation of the egg includes discussion of studies on in vitro fertilization and embryo transfer. The first appendix summarizes ninety-eight reports published since 1890 in which eggs were transferred between individuals, including pigs, cows, sheep, rats, mice, and rabbits.

Glover, Jonathan, et al. *Ethics of New Reproductive Technologies.* DeKalb: Northern Illinois University Press, 1989. A report to the European Commission on some of the ethical issues raised by embryo transfer and other reproductive technologies. A thoughtful book on the possible social, ethical, and political ramifications of the different ways humans may have children.

Robertson, John A. "Ethical and Legal Issues in Human Egg Donation." *Fertility and Sterility* 52 (1989): 353-363. Written by a professor of law; discusses questions raised by embryo transfer. Topics include welfare of offspring, donor anonymity, rights of donor and recipient to rear an infant, donation of eggs by relatives, paying egg donors, ownership of human eggs, research on donor eggs, and freezing eggs. References.

Walters, Leroy. "Ethical Aspects of the New Reproductive Technologies." In *In Vitro Fertilization and Other Assisted Reproduction*, edited by Howard J. Jones and Charlotte Schrader. New York: New York Academy of Sciences, 1988. Summarizes the statements produced by ethical committees from six countries on issues such as embryo freezing and egg donation. Includes references. Provides a source of information on the use of ovum transfer to treat infertility.

Judith R. Gibber

Cross-References

Ivanov Develops Artificial Insemination (1901), p. 113; Bevis Describes Amniocentesis as a Method for Disclosing Fetal Genetic Traits (1952), p. 1439; Donald Is the First to Use Ultrasound to Examine Unborn Children (1958), p. 1562; Brown Gives Birth to the First "Test-Tube" Baby (1978), p. 2099; Clewell Corrects Hydrocephalus by Surgery on a Fetus (1981), p. 2174; Daffos Uses Blood Taken Through the Umbilical Cord to Diagnose Fetal Disease (1982), p. 2205.

IBM INTRODUCES A PERSONAL COMPUTER
WITH A STANDARD HARD DISK DRIVE

Category of event: Applied science
Time: March 8, 1983
Locale: Boca Raton, Florida

IBM Corporation was the first manufacturer to offer a personal computer with a large-capacity disk drive (hard disk) system as standard equipment, giving desktop computers large storage capabilities

Principal personages:
ALAN SHUGART, an engineer who first developed the floppy disk as a means of mass storage for mainframe computers
PHILIP D. ESTRIDGE, the director of IBM's product development facility in Boca Raton, Florida, where the IBM PC and PC XT were designed and manufactured
THOMAS J. WATSON, JR. (1914-   ), the chief executive officer of IBM Corporation during the development of the first PC and the PC XT

Summary of Event

When International Business Machines (IBM) introduced its first microcomputer, called simply the IBM PC (for "personal computer," the name by which all such single-user systems are now generally known), the occasion represented less a dramatic invention than the confirmation and legitimization of a trend begun some years before. A number of companies had introduced personal computers before IBM; one of the best known at that time was Apple Corporation's Apple II model, for which desirable software for business and scientific use was quickly developed. Nevertheless, the personal computer was quite expensive and was often looked upon as an oddity, not as a common tool. Under the leadership of its chairman, Thomas J. Watson, Jr., giant mainframe computer manufacturer IBM decided to develop the PC. A PC design team headed by Philip D. Estridge was assembled in Boca Raton, Florida, and it quickly developed its first, pace-setting product. It is an irony of history that IBM anticipated selling only a hundred thousand or so of these machines, mostly to scientists and technically inclined hobbyists. Instead, IBM's product sold exceedingly well, and its design parameters as well as its operating system became standards.

The original PC represented a consolidation of computing technology available at the time of its introduction. In its most basic configuration, it addressed the need to store data by offering a port (connection) for the use of a cassette recorder as a means of mass storage; a floppy disk drive capable of storing approximately 160 kilobytes of data was initially only an extra cost option. While home hobbyists were accustomed to using a cassette recorder as a means of data storage, such storage was far too slow and awkward for professional use in business and science; consequently, virtually every IBM PC sold was equipped with at least one 5.25-inch floppy disk drive. In general, the PC's specifications differed little from those of other personal computers such as the Apple II.

All computers require memory of two sorts in order to carry out their tasks. One type of memory is main memory, or random access memory (RAM), the memory used by the processor to store data it is using while operating. Physically, the type of memory used for this function is typically built of silicon-based integrated circuits, which have the advantage of speed (to allow the processor to fetch or store the data quickly) but the distinct disadvantage of losing or "forgetting" data when the electric current that sustains them is turned off. Further, such memory generally is relatively expensive. Thus, another type of memory—long-term storage memory—was developed, also known as mass storage. Mass storage devices include magnetic media (tape or disk drives) and optical media (such as the compact disk storage device, CD-ROM). While the speed with which data may be retrieved from or stored in such devices is rather slow compared to the processor's speed or RAM speed, a disk drive—the most common form of mass storage used in PCs—has the virtue of storing relatively large amounts of data in a small space. Above all, it does so quite inexpensively.

Early floppy disk drives (so called because the magnetically treated material on which data are recorded is made of a very flexible plastic substance) held 160 kilobytes of data using only one side of the magnetically coated disc (a normal, double-spaced, typewritten page represents about 2 kilobytes of information). Later developments increased storage capacities to 360 kilobytes by using both sides of the disc and, with increasing technological ability, 1.44 megabytes (or millions of bytes). While such capacities seem large, the needs of business and scientific users soon outstripped available space. In contrast, mainframe computers, which are typically connected to large and expensive tape drive storage systems, could store gigabytes (or millions of megabytes) of information. Since even the mailing list of a small business or a scientist's mathematical model of a chemical reaction easily could require greater storage potential than early PCs allowed, the need for a mass storage device that allowed very large files of data was clear.

The answer was the hard disk drive, also known as a fixed disk or Winchester drive (IBM uses the term "fixed disk drives," reflecting the fact that, in their most usual configuration, the disk itself is not only rigid but also permanently mounted in its drive unit). IBM had invented the notion of a fixed, hard magnetic disk as a means of storing computer data around 1955 and, under the direction of Alan Shugart in the 1960's, the floppy disk was developed as well. As the engineers of IBM's facility in Boca Raton further refined the idea of the original PC to define the PC XT, it became clear that chief among the needs of users was the availability of larger mass storage options. The decision was made to add a 10-megabyte hard disk drive to the PC.

The availability of such a large amount of disk storage space required modifications to IBM's PC-DOS operating system as well. A single 5.25-inch floppy disk drive then held a maximum of 360 kilobytes of data, so keeping track of individual files was not difficult. Yet, a 10-megabyte hard drive (the standard issue size in the original XT series) was the equivalent of almost thirty smaller disks—finding one specific file in a large list of files could become confusing for the user. Thus, in cooperation with IBM, Microsoft enhanced its DOS to include options specifically geared toward managing many files on large-volume disk drives. In particular, commands allowing users to create a tree-structured directory system were added so that the 10 megabytes (modern drives can offer ten or one hundred times as much space) could be divided into smaller units (or "branches" of the main or "root" directory), each with a name given by the user. Commands allowed users to create these directories (MD, or "make directory"), remove them (RD, or "remove directory") or change (CD, or "change directory") quickly between various directories and subdirectories. Since programs now could reside in differently named parts of the single large drive, the ability to define a "path" was added as well. A "path statement," usually included as a part of a basic startup file that sets individually tailored operating parameters when the machine is first started, allows the operating system to seek programs in directories other than the one currently in use. Later, another command, "append," was included to allow the locations of nonprogram files in other directories as well.

Another change made for the PC XT was the ability of the computer to "boot" (start its operating system) from its hard disk, thus freeing users from the need to swap various floppy disks to start up their machine and its programs. A small industry grew quickly to answer a perceived problem this convenience offered. While in the past a user could safely and simply lock up floppy disks in a drawer or cabinet and thereby prevent unauthorized use of them, the hard drive was a fixed, permanently mounted unit within the PC. As a result, one's data were made easily available to others. Numerous software and hardware manufacturers offered means of encrypting data or otherwise securing it from prying eyes.

On March 8, 1983, less than two years after the introduction of the first PC, IBM introduced the PC XT. Like the original, it was an evolutionary design, not a revolutionary one. The inclusion of a hard disk drive, however, signaled general acceptance of mass storage devices in future personal computers.

Impact of Event

Above all else, any computer provides a means for storing, ordering, analyzing, and presenting information. If the personal computer is to become the information appliance some have suggested it will be, the ability to manipulate very large amounts of data will be of paramount concern. Random access memory offers the greatest speed in allowing a processor to store or fetch data, but its expense—especially in the light of the constantly growing amount of data created—makes it an unlikely prospect for all but the most advanced scientific tasks. On the other hand, hard disk drive technology was greeted enthusiastically in the marketplace, and the demand for hard drives has seen their numbers increase even as their quality increases and their prices drop.

It is easy to understand one reason for such eager acceptance: convenience. Floppy-bound computer users find themselves frequently changing (or "swapping") their disks in order to allow programs to find the data they need. Indeed, a floppy disk limits the size of a data file to that which can fit on a single floppy. A hard drive allows users to keep the equivalent of thirty or more floppies "in" their machines, quickly available. Moreover, because hard drives generally do not allow one to change the actual disk (a misnomer; actually the typical hard drive consists of multiple platters and read/write heads that transfer data to and from each individual platter), the drive unit can be optimized for maximum speed. Thus, hard disk drives are capable of finding files and transferring their contents to the processor much more quickly than a floppy drive. A user may thus create exceedingly large files, keep them on hand at all times, and manipulate data more quickly than with a floppy.

The distinction was made earlier between main memory, used by the processor to carry out its work, and mass storage, of which the hard drive is an example. In fact, the distinction has blurred somewhat since the introduction of the hard disk drive. Since such drives can fetch their data quite quickly (though not as quickly as RAM allows), some software manufacturers have created programs that use data files (called "overlay" files) as a substitute for main memory. Thus, a spreadsheet program need not keep all of itself in main memory but simply fetches from the hard disk overlay file whatever additional data are needed to perform a specific function. Indeed, some programs for the IBM PC XT (and compatible computers made by other manufacturers) are so large that they can be used only on computers equipped with a hard disk drive. While a hard drive is a slow substitute for main memory, it allows users to enjoy the benefits of larger memories at significantly lower cost. That the introduction of the IBM PC XT with its 10-megabyte hard drive was a significant milestone in the development of the PC can be measured by the simple fact that few PCs sold for serious use are without a hard drive.

Bibliography

Curran, Susan, and Ray Curnow. *Overcoming Computer Illiteracy: A Friendly Introduction to Computers.* New York: Penguin Books, 1983. Provides a basic introduction to personal computers used in the period around the introduction of the original IBM PC and the PC XT. Includes discussion of PC fundamentals, types of software, and suggested uses.

Ditlea, Steve, ed. *Digital Deli: The Comprehensive, User-Lovable Menu of Computer Lore, Culture, Lifestyles, and Fancy.* New York: Workman, 1984. Despite its seeming randomness of organization, this unusual compendium provides a useful history of computer development, while focusing on the use of personal computers. As its subtitle suggests, it is filled with interesting and useful discussion about the computer industry as well.

Fallow, Allan. *Understanding Computers: Memory and Storage.* Alexandria, Va.:

Time-Life Books, 1987. Provides a very clear discussion of how modern mass storage devices work (including, along with disk drives, tape drives and optical media). Illustrations are a strong feature. Includes a bibliography, as well as a glossary of terms related to storage and memory functions.

Kean, David W. *IBM San Jose: A Quarter Century of Innovation*. San Jose, Calif.: International Business Machines Corporation, 1977. Provides somewhat technical discussion of design and implementation of the floppy disk drive, which, inasmuch as it was developed after the hard disk drive, shares the same basic operating principles. Well documented.

Shore, John. *The Sachertorte Algorithm and Other Antidotes to Computer Anxiety*. New York: Viking Press, 1985. Provides a useful introduction to the concepts of computing with psychological insight into how people first come to computers and how they learn to use them; amusing yet accurate representation not only of hardware basics but also of programming.

Sobel, Robert. *IBM vs. Japan: The Struggle for the Future*. New York: Stein & Day, 1986. While concentrating on the similarities and differences in corporate styles in IBM and in Japanese organizations, this text provides a useful and revealing early history of the origins of the IBM PC series.

Joseph T. Malloy

Cross-References

UNIVAC I Becomes the First Commercial Electronic Computer and the First to Store Data on Magnetic Tape (1951), p. 1396; "Bubble Memory" Devices Are Created for Use In Computers (1969), p. 1886; The Floppy Disk Is Introduced for Storing Electronic Data (1970), p. 1923; Apple II Becomes the First Commercially Successful Personal Computer (1977), p. 2073; IBM Personal Computer Using DOS Is Introduced (1981), p. 2169; Optical Disks for the Storage of Data Are Introduced (1984), p. 2262.

THE FIRST TRACKING AND DATA-RELAY SATELLITE SYSTEM OPENS A NEW ERA IN SPACE COMMUNICATIONS

Category of event: Space and aviation
Time: April 4, 1983
Locale: Geostationary orbit

The Tracking and Data-Relay Satellite System replaced NASA's ground-tracking stations, making possible nearly continuous communication with most U.S. spacecraft in Earth orbit

Principal personages:

ROY BROWNING, the TDRSS program manager for NASA who directed the agency's involvement in the project

EDWIN A. COY, vice president and TDRSS program manager of Spacecom, the corporation that built the TDRSS with NASA

JOE WELLENS, an engineer who was program manager of TDRSS for TRW Space Systems; the contractor that built the spacecraft and ground facilities under contract to Spacecom

Summary of Event

The National Aeronautics and Space Administration (NASA) became seriously interested in communications satellites in geostationary orbit as a solution to the problem of frequent tracking and communication lapses inherent in the earth-based tracking system in use at the time. In the mid-1960's, NASA began research and development toward what became known as the Tracking and Data-Relay Satellite System (TDRSS) in the early 1970's. Roy Browning was appointed program manager by NASA for the TDRSS project.

Up to that time, spacecraft in low Earth orbit were able to communicate with ground control centers only if they were above the horizon over strategically placed antennas in NASA's Ground Spaceflight Tracking and Data Network (GSTDN). Horizon-to-horizon passage at any tracking station lasted only a few minutes, and political, geographic, and economic limitations on the number and placement of such facilities meant that communications were never more than sporadic. Typically, they totaled less than twenty minutes out of every ninety-minute orbit. The space-based tracking system offered fewer total installations required to maintain communications, and all key ground installations could be located on United States soil.

In December of 1976, NASA signed a joint endeavor agreement with the Space Communications Company (Spacecom) to develop the TDRSS. Edwin A. Coy was selected program manager of Spacecom for TDRSS. The new system was to be built around three large Tracking and Data-Relay Satellites (TDRS's) and a single ground terminal in White Sands, New Mexico, and was intended to begin limited operation in 1979. Spacecom selected TRW Space Systems and the Harris Corporation to build the hardware for the system. Joe Wellens was the engineer who was appointed program manager of TRW Space Systems for TDRSS.

Three TDRS spacecraft were to operate in geostationary orbit over the equator. At an altitude of 35,680 kilometers, they would orbit eastward at a speed matching exactly the rotation of Earth on its axis, so that each would remain over the same point on Earth's surface indefinitely. From a vantage point 150 times higher than the spacecraft it served, a single TDRS could maintain communications with a low-orbiting satellite as the latter traveled halfway around the world. Two predetermined orbital "duty stations" (TDRS-East and TDRS-West), spaced 130 degrees of longitude apart, would let the satellites observe a minimum of 80 percent of the orbit of any spacecraft dependent upon them. The choice of orbital location allowed both duty stations to "see" the ground terminal at White Sands. A third position, midway between the other two, would be used for a "standby" TDRS.

Tracking and Data-Relay Satellites are the largest and heaviest spacecraft designed for geostationary orbit and the largest privately owned communications satellites ever built. Each one weighs 2,120 kilograms and measures 17.4 meters across. TDRS is the first communications satellite able to handle all three main frequencies employed by scientific and defense spacecraft, manned spacecraft, and commercial satellites: the S, Ku, and C bands. A single TDRS spacecraft's capacity as a conduit for communications traffic is enormous. Digital and analog data, along with audio and video signals, flow simultaneously at transmission rates so great that the entire contents of a twenty-volume encyclopedia could be transmitted in one second. Like water moving through a bent pipe, the communications signals enter TDRS from one direction and leave in another. Signals originating on Earth involve an "uplink" to a TDRS and a "forward link" to the user spacecraft. Signals originating on the user spacecraft involve a "return link" to a TDRS and a "downlink" to the receiving antennas on Earth.

Two large (4.8-meter-diameter) dish antennas aboard the TDRS are used for high data-rate Ku- and S-band transmissions to and from other spacecraft. Fabricated of molybdenum mesh and plated with 24-karat gold, each can be used by only one spacecraft at a time, but can receive up to 300 megabits (300 million electrical impulses) per second on the return link. Their forward link transmission capability is 25 megabits per second. Both antennas are independently steerable by ground command, so that each follows the spacecraft it is communicating with as it passes below. A phased array antenna consisting of thirty elements is used for low data-rate S-band transmissions and is available for two simultaneous forward links and ten simultaneous return links per TDRS. All communications between TDRS and Earth (both uplink and downlink) pass between the satellite's 2-meter dish antenna called the Space Ground Link and one of the three 18.3-meter dishes at the White Sands Ground Network.

The spacecraft requires 1.7 kilowatts of electrical power, obtained from two solar cell arrays extended at right angles to the large dish antennas. Batteries store excess electrical power so that the satellite can continue to operate during the hours-long periods when it, like the ground point below, experiences night. The batteries are contained in a hexagonally shaped core module, along with more than 40 transponders and the spacecraft's attitude control system and hydrazine propellant.

Compactly folded up to fit into the shuttle payload bay, a TDRS and its two-stage Inertial Upper Stage (IUS) booster occupy most of the total space available and weigh 16,783 kilograms. In addition, a large piece of payload support equipment called a tilt table is required to deploy the TDRS/IUS once in orbit. The tilt table supports the weight of the payload during ground handling and launch. Once in space, it elevates the forward end of the TDRS/IUS at a steep angle to the body of the shuttle to facilitate deploying the payload by a spring-activated mechanism in the tilt table.

TDRS-1 was carried into orbit on April 4, 1983, as the major payload of the sixth shuttle mission, which was also the maiden flight of the *Challenger* orbiter. Lift-off from Cape Canaveral occurred at 1:30 P.M. eastern standard time. Once in the intended 280 kilometer-high orbit, the deployment of TDRS-1 was the mission's first priority. At 11:52 P.M., the TDRS/IUS was pushed free of the *Challenger* and the shuttle maneuvered a safe distance away. Less than one hour later, at 12:27 A.M. on April 5, the IUS first stage ignited, placing the spacecraft into a transfer orbit with an apogee (highest point) at geostationary altitude and a perigee (lowest point) at the altitude of the space shuttle. The IUS second stage was to have circularized that orbit at geostationary altitude, but problems developed soon after its ignition at 5:45 A.M. Eighty seconds into the planned 105-second burn, the IUS veered off course suddenly and telemetry was lost. Flight controllers commanded emergency separation of the two vehicles, but the $100 million TDRS-1 continued tumbling out of control at thirty revolutions per minute in a useless egg-shaped orbit. Quick work by flight controllers stabilized TDRS-1 before its batteries died, allowing the solar panels to be deployed and giving engineers several weeks to develop and test a plan to try and salvage the mission.

On May 2, the spacecraft was commanded to begin firing a pair of small attitude control thrusters for periods of up to three hours per orbit to slowly maneuver the satellite into its intended orbit. The process required weeks, because the thrusters could generate less than a kilogram of thrust apiece. TDRS-1 finally reached its 35,680-kilometer orbital altitude on June 29, 1983, and was subsequently moved to the TDRS-East duty station, located above the equator at 41 degrees west longitude. It went into service on October 17, 1983. Meanwhile, problems with the IUS booster, the TDRS spacecraft, and with the shuttle caused serious delays in the launch of TDRS-2 and 3. Expected to go into service within months of its sister craft, TDRS-2 was finally manifested for flight in January of 1986 and was lost in the explosion of *Challenger*. The TDRSS network was considered so vital that a replacement for TDRS-2 was the first payload carried by the shuttle when flights resumed on September 29, 1988. It was placed in the TDRS-West position at 171 degrees west longitude. Two flights later, on March 13, 1989, TDRS-3 was put in orbit. Each of the spacecraft has an expected operating life of ten years.

Impact of Event

The communications revolution on Earth has linked individuals to information, regardless of where the information might be located physically, thereby increasing the speed and accuracy with which sound decisions can be made. The Tracking and Data-Relay Satellite System is an analogous development in the space program of the United States. It enhances substantially the performance and utility of manned and unmanned spacecraft in low Earth orbit by maintaining almost constant contact with them and providing very high-capacity data throughput.

Many modern spacecraft could not operate if they were still bound by the limitations of the former GSTDN system. Landsat Earth resources satellites utilize an instrument called a thematic mapper, which collects so many data about each ground site imaged that the former means of communication could never have kept up with the information coming from only one such satellite. The Hubble Space Telescope alone requires about thirty minutes of high data-rate communications per orbit for scientific and operational information related to its mission. Other major users of TDRSS among the unmanned satellite family will be: Solar Mesosphere Explorer, Cosmic Background Explorer, Earth Radiation Budget Experiment, Gamma Ray Astronomy Observatory, Geopotential Research Mission, and the Upper Atmosphere Research Satellite. All are expected to be in operation by the year 2000.

The space shuttle is a major user of TDRSS services. The operation and safety of such a complex spacecraft and the variety of routine tasks carried out simultaneously by crews of up to seven individuals require many channels of information to be provided almost around the clock. Routine communications demands are escalated significantly by certain payloads and activities. One example is the European Space Agency's Spacelab, which is a sophisticated orbiting laboratory carried inside the payload bay for operational periods of a week or more. Spacelab's capacity to generate scientific data is so great that on its first mission—only six weeks after TDRS-1 became operational—more space-to-ground data were transmitted than on all thirty-nine previous U.S. manned missions. Most of its experiments cannot operate successfully without TDRSS support. Extravehicular activity (EVA) by shuttle astronauts places an added burden on communications links, also because of the dangers involved and the complexity of the tasks the crew performs, such as repairing disabled spacecraft and assembling structures.

America's space station *Freedom*, planned to be permanently manned and fully operational before the end of the twentieth century, will expand upon the demands made by today's spacecraft. With three large manned research modules, crew accommodations for eight astronauts, and numerous unmanned experiments mounted externally, it will, from a communications standpoint, be the equivalent of several Spacelabs and many unmanned satellites in one package. When fully operational, *Freedom* is expected to generate as much information flow as all the current TDRSS users together.

Bibliography

Covault, Craig. "Loss of TDRS-A Averted by Joint Action." *Aviation Week and Space Technology* 118 (April 11, 1983): 19-21. A detailed account of the events that occurred, which nearly resulted in the loss of the first TDRS spacecraft, including a timelined breakdown of the period immediately following second-stage ignition of the faulty Inertial Upper Stage (IUS) booster. Comments and decisions of some of the key flight controllers are revealed. (TDRS-A was the designation of the satellite that became TDRS-1 when it successfully went into operation on October 17, 1983.)

Faget, Max. "Tracking and Communications." In *Manned Space Flight*. New York: Holt, Rinehart and Winston, 1965. Faget describes the general functions of tracking and communications and the specific problems and solutions developed for Projects Mercury and Apollo. Careful study of figure 6.3, which includes a map of the location of the Manned Space Flight Network (MSFN) tracking centers around the world and a listing of their capabilities, shows many of the serious limitations of an Earth-based tracking network.

"First Flight of SS Challenger." *Space World* 234-235 (June/July, 1983): 4-6. A staff report on the STS 6 mission, which carried the first TDRS into orbit. Readers will find a brief day-by-day summary of mission activities and a discussion of the problems that developed because of the malfunction of the satellite's IUS booster.

Froelich, Walter. *The New Space Network: The Tracking and Data Relay Satellite System*. NASA EP-251. Washington, D.C.: Government Printing Office, 1986. This twenty-eight-page booklet is the layperson's single source of information about the overall system and its various components. Written in nontechnical language and illustrated with numerous color photographs and clear diagrams.

Smith, Bruce A. "TDRS Thrusters Readied for Geostationary Shift." *Aviation Week and Space Technology* 118 (May 9, 1983): 16-17. One of several articles in this journal during the spring of 1983 that detail the week-by-week efforts to raise TDRS-1 from the useless orbit into which it was placed by the malfunction of its IUS booster. These discussions explain general engineering considerations related to raising the satellite to a geostationary orbit and reveal the profound impact of the problems on NASA's future plans. Suitable for the technically literate reader.

Thomas, Shirley. *Satellite Tracking Facilities: Their History and Operation*. New York: Holt, Rinehart and Winston, 1963. Useful for the reader wishing a fuller understanding of the problems of providing guidance and communications to spacecraft in low Earth orbit from Earth-based tracking stations. Describes the two networks originally used by NASA, the MSFN and the Spaceflight Tracking and Data Network (STDN), later combined into the Ground Spaceflight Tracking and Data Network (GSTDN). Suitable for a general readership and is indexed and referenced with footnotes.

Richard S. Knapp

Cross-References

Tiros 1 Becomes the First Experimental Weather Reconnaissance Satellite (1960), p. 1667; Echo, the First Passive Communications Satellite, Is Launched (1960), p. 1677; Glenn Is the First American to Orbit Earth (1962), p. 1723; Telstar, the First Commercial Communications Satellite, Relays Live Transatlantic Television Pictures (1962), p. 1728; The Orbital Rendezvous of Gemini 6 and 7 Succeeds (1965), p. 1803; Skylab Inaugurates a New Era of Space Research (1973), p. 1997; _Columbia_'s Second Flight Proves the Practicality of the Space Shuttle (1981), p. 2180; Spacelab 1 Is Launched Aboard the Space Shuttle (1983), p. 2256.

MURRAY AND SZOSTAK CREATE
THE FIRST ARTIFICIAL CHROMOSOME

Category of event: Biology
Time: September, 1983
Locale: Boston, Massachusetts

Murray and Szostak invented a working artificial chromosome to study natural chromosome behavior, which ultimately proved to be an invaluable tool to recombinant DNA technology

Principal personages:

JACK W. SZOSTAK (1952-   ), a British-born Canadian citizen who became a professor at Harvard Medical School with research interests in genetic recombination and ribosomes

ANDREW W. MURRAY (1956-   ), a graduate student working in Szostak's laboratory at the time the first artificial chromosome was invented

MAYNARD VICTOR OLSON (1943-   ), a professor of medical genetics who developed techniques for cloning human DNA into yeast by means of artificial chromosome vectors

Summary of Event

Advancements in medicine often rest on the wings of achievements in basic biology and technology. The artificial chromosome dually distinguishes itself by giving biologists insight into the fundamental mechanisms by which cells replicate as well as by its role as a tool in genetic engineering technology. Since its invention in 1983 by Andrew W. Murray and Jack W. Szostak, working at the Dana-Farber Cancer Institute in Boston, Massachusetts, the importance of the artificial chromosome was quickly appreciated by scientists and its value to medicine exploited.

In order to appreciate the benefits of an artificial chromosome, one must first be aware of the role natural chromosomes play in the living cell. Chromosomes are essentially carriers of genetic information, that is, they possess the genetic code which is the blueprint for life. In higher organisms, the number and type of chromosomes that a cell contains in its nucleus is characteristic of the species. For example, each human cell has forty-six chromosomes, while the garden pea has fourteen and the guinea pig has sixty-four. The chromosome's job in a dividing cell is to replicate and then distribute one copy of itself into each new daughter cell. This process, referred to as mitosis or meiosis, depending upon the actual mechanism by which this occurs, is of supreme importance to the continuation of life, and errors may have disastrous consequences. Indeed, most errors in chromosomal inheritance are lethal and result in the truncation of the whole cell line containing the error. This, ironically, limits the damage that would otherwise result if these abnormal cells proliferated. More dangerous are abnormal cell lines that are not sufficiently damaged to cause cell death but rather continue to thrive and cause disease. The most famous example of this is Down's syndrome, the most common congenital disease in humans. It is characterized by atypical facial features such as a moon face, low set ears, widely separated eyes, mental retardation, and frequently serious heart malformations. This syndrome occurs because the chromosomes fail to segregate competently into the proper cells during development, which results in the appearance of an extra chromosome twenty-one in every cell. Hundreds of other genetic diseases have also been identified in which cells have either too few or too many chromosomes or chromosomes that are damaged during segregation.

Chromosomes are primarily composed of two raw materials, protein and the nucleic acid deoxyribonucleic acid (DNA), which is the actual carrier of genetic information. DNA is a duplex of molecular chains twisted around each other to form a spiral staircase. Each strand of this duplex is a polymer made up of the chemical bases adenine (A), guanine (G), thymine (T), and cytosine (C) attached together chemically like links on a chain. The precise arrangement of these bases forms an elaborate code in which every group of three bases determines a specific amino acid that will be incorporated into a protein. Any group of bases large enough to code for one entire protein is called a gene, the total collection of which ultimately determines an organism's physical characteristics. The bases on one strand of the DNA molecule are paired with the bases on the opposite strand according to strict rules: A binds with T and G binds with C; therefore, the strands are complementary and the base sequence of one strand can be deduced from the base sequence on the opposite strand.

In 1953, when James D. Watson and Francis Crick discovered the structure of DNA, for which they won the 1962 Nobel Prize in Physiology or Medicine, it was immediately apparent to them how the double helical nature of DNA might explain the mechanism behind cell division. During DNA replication, the chromosome unwinds to expose the thin threads of DNA. The two strands of the double helix separate, and each acts as a template for the formation of a new complementary strand, thus forming two complete and identical chromosomes that can be distributed subsequently to each new cell. This distribution process, referred to as segregation, relies on the chromosomes being pulled along a microtubule framework in the cell called the mitotic spindle.

An artificial chromosome is a model of a natural chromosome, designed in the laboratory, which possesses only those functional elements its creators desire. In order to be a true working chromosome, however, it must, at minimum, maintain the machinery to carry out replication and segregation. The technique of studying nature by making a simplified model and subsequently refining it is an approach more common to the physicist than the biologist. This is caused by the complexity of biological systems that make it difficult to discern and isolate the essential features of the system. Nevertheless, by the early 1980's, Murray and Szostak had recognized the possible advantages of having a simple, controlled model to study chromosome behavior, since there are several inherent difficulties associated with studying chromosomes in their natural state. Since natural chromosomes are large and have poorly defined structures, it is almost impossible to sift out for study those elements that are essential for replication and segregation and those that are dispensable. Previous methods of altering a natural chromosome and observing the effects is difficult because the cells containing that altered chromosome usually die. Furthermore, even if the cell survives, analysis is complicated by the extensive tracts of genetic information the chromosome carries. Artificial chromosomes are simple, having known components even if the functions of those components are poorly understood. In addition, since artificial chromosomes are extra chromosomes carried around by the cell like a parasite, their alteration does not kill the cell.

Prior to the synthesis of the first artificial chromosome, the essential functional elements necessary to accomplish replication and segregation had to be identified and harvested. One of the three chromosomal elements thought to be required is the origin of replication, the site at which the synthesis of new DNA begins. The relatively weak interaction between DNA strands at this site facilitates their separation, making possible—with the help of appropriate enzymes—the subsequent replication of the strands into sister chromatids. The second essential element is the centromere, a thinner segment of the chromosome that serves as the attachment site for the mitotic spindle. Sister chromatids are pulled into diametric ends of the dividing cell by the spindle apparatus, thus forming two identical daughter cells. The final functional element is a repetitive sequence of DNA located at each end of the chromosome, called a telomere, which is needed to protect the terminal genes from degradation.

In the late 1970's, a group of researchers at Stanford University, Kevin Struhl, Dan Stinchcomb, Stewart Scherer, and Ronald W. Davis, discovered short sequences of yeast DNA thought to be origins of replication, since they could replicate independently of chromosomal DNA. These sequences, called plasmids, segregated poorly, however, because of the lack of a centromere, thus resulting in both copies remaining with the mother during cell mitosis. The yeast centromere was cloned in 1980 by Louise Clarke and John A. Carbon of the University of California at Santa Barbara. It is a stretch of DNA which, when inserted into plasmids, causes them to segregate correctly in 99 percent of cell divisions. The last chromosomal element to be cloned was the telomere, which was accomplished in 1982 by Szostak and Elizabeth H. Blackburn of the University of California at Berkeley.

With all the functional elements at their disposal, Murray and Szostak proceeded to construct their first artificial chromosome. Once made, this chromosome would be inserted into yeast cells to replicate, since these cells are relatively simple and well characterized but otherwise resemble cells of higher organisms. In addition, inheritance errors are rare in yeast. Construction begins with a commonly used bacterial plasmid, a small, circular, autonomously replicating span of DNA. Enzymes are then called upon to create a gap in this cloning vector into which the three chromosomal elements are spliced. In addition, genes that confer some distinct trait such as color to yeast cells are also inserted, thus allowing one to detect which cells have actually taken up the new chromosome subsequent to their incubation together. Although their first attempt resulted in a chromosome that failed to segregate properly, by September, 1983, Murray and Szostak had published in the prestigious British journal *Nature* their success in creating the first artificial chromosome.

Impact of Event

The artificial chromosome has had a tremendous impact on the understanding of chromosome behavior and the mechanism of heredity. Murray and Szostak set out to answer three questions. First, they wanted to test the validity of the commonly held belief that replication origins, centromeres and telomeres, are the only essential items necessary for competent chromosome segregation. This was proved by demonstrating competent chromosomes after the removal of all segments of DNA except for the functional elements. Second, they wanted to learn more about how the mitotic spindle attaches to the centromere. Third, they wanted to learn how the errors in segregation that give rise to diseases such as Down's syndrome arise. One of their discoveries is that chromosomes must be greater than a certain minimum length, about 100,000 base pairs; otherwise, they will segregate randomly. A possible explanation for this finding is that chromosomes will not remain attached to the mitotic spindle unless they are under tension. This tension is created when sister chromatids are entwined about each other as they are being pulled to opposite sides of the cell. The shorter the chromosome is, the fewer molecular interactions it has with its homolog and the less avidly it will bind to it. The implication is that anything which breaks a chromosome, such as a mutagen, might shorten it enough to cause problems in segregation and genetic disease.

One of the most exciting aspects of the artificial chromosome is its application to recombinant DNA technology, which is the creating of novel genetic materials by combining segments of DNA from various sources. For example, the yeast artificial chromosome can be used as a cloning vector. That is, a segment of DNA containing some desired gene is inserted into an artificial chromosome and is then allowed to replicate in yeast until large yields of the gene are produced. David T. Burke, Georges F. Carle, and Maynard Victor Olson at Washington University in St. Louis have pioneered the technique of combining human genes with yeast artificial chromosomes and have succeeded in cloning large segments of human DNA. Although amplifying DNA in this manner has been done before, using bacterial plasmids as cloning vectors, the yeast artificial chromosome has the advantage of being able to hold much larger segments of DNA, thus allowing scientists to clone very large genes. This is of great importance, since the genes that cause diseases such as hemophilia and Duchenne's muscular dystrophy are enormous in size. The most ambitious project for which the yeast artificial chromosome is being used is the national project to clone the entire human genome.

Bibliography

Murray, Andrew W., and Jack Szostak. "Artificial Chromosomes." *Scientific American* 257 (November, 1987): 62-68. A brief, up-to-date account of the artificial chromosome and its applications to biological problems. Written by the scientists who invented the artificial chromosomes. This is the best article for the nonscientist; includes several clarifying illustrations as well as a bibliography for the interested reader.

_____. "Construction of Artificial Chromosomes in Yeast." *Nature* 305 (September, 1983): 189-193. The original, landmark paper that started it all. Intended for the scientist, this article explains in detail how to construct artificial chromosomes and provides original data and discussion of results. Numerous references to the scientific literature are provided.

Perbal, Bernard. *A Practical Guide to Molecular Cloning.* New York: John Wiley & Sons, 1988. A detailed and functional guide on all aspects of molecular cloning. A book for the advanced reader interested in the practical details of genetic engineering and recombinant DNA technology.

Stryer, Lubert. *Biochemistry.* 3d ed. New York: W. H. Freeman, 1988. An extremely good introductory textbook considered a standard for students of biology, biochemistry, or medicine. Although it fails to discuss artificial chromosomes specifically, chapters 4 and 5 are essential for understanding DNA's role in the mechanism of heredity. Chapter 6 provides an up-to-date introduction of recombinant DNA technology.

Watson, James D. *The Double Helix: A Personal Account of the Discovery of the Structure of DNA.* New York: Atheneum, 1968. A personal narrative on the origin of the Watson-Crick model of DNA, which revolutionized biology and earned for them the 1962 Nobel Prize in Physiology or Medicine. Written for the lay reader, this book is a classic that shows how scientific discoveries occur. It is filled with anecdotes and photographs of several eminent scientists involved in the discovery.

Kenneth S. Spector

Cross-References

Sutton States That Chromosomes Are Paired and Could Be Carriers of Hereditary Traits (1902), p. 153; Morgan Develops the Gene-Chromosome Theory (1908) p. 407; Johannsen Coins the Terms "Gene," "Genotype," and "Phenotype" (1909), p. 433; Avery, MacLeod, and McCarty Determine That DNA Carries Hereditary Information (1943), p. 1203; Watson and Crick Develop the Double-Helix Model for DNA (1951), p. 1406; Cohen and Boyer Develop Recombinant DNA Technology (1973), p. 1987; A Human Growth Hormone Gene Transferred to a Mouse Creates Giant Mice (1981), p. 2154.

SPACELAB 1 IS LAUNCHED ABOARD
THE SPACE SHUTTLE

Category of event: Space and aviation
Time: November 28, 1983
Locale: Low Earth orbit

Developed and funded in Europe at a cost of approximately one billion dollars, Spacelab 1 marked Europe's entry into the manned spaceflight endeavor

Principal personages:

JOHN W. YOUNG (1930-   ), the commander of STS 9, who was the chief of the NASA Astronaut Office

OWEN K. GARRIOTT (1930-   ), a NASA mission specialist aboard Spacelab 1 who was a veteran of the Skylab 3 crew

BREWSTER H. SHAW (1945-   ), the pilot of STS 9, who made his first spaceflight as a member of the *Columbia* Spacelab 1 crew

ROBERT A. PARKER (1936-   ), a NASA astronaut and astronomer who made his first spaceflight as a mission specialist aboard Spacelab 1

BYRON K. LICHTENBERG (1948-   ), a biomedical engineer who was selected as the American payload specialist aboard Spacelab 1

ULF MERBOLD (1941-   ), a physicist who was selected as the European payload specialist aboard Spacelab 1 and the first foreign citizen to fly in an American spacecraft

MICHAEL L. LAMPTON (1941-   ), an astronomer who served as backup American payload specialist

WUBBO OCKELS (1946-   ), a nuclear physicist who served as backup European payload specialist

Summary of Event

In 1969, the United States invited the member nations of the European Space Conference to participate in the National Aeronautics and Space Administration's (NASA) "Post-Apollo Program." The ten nations that made up the Conference (West Germany, Belgium, Denmark, Spain, France, England, Italy, The Netherlands, Switzerland, and Austria) responded with the idea of a space laboratory whose operators could be expert scientists rather than professional astronauts. This proposal was adopted by the European Space Conference ministers and became an official program of the European Space Research Organization (ESRO) in August, 1973. ESRO's responsibilities were assumed later by the European Space Agency (ESA), which was created by merging ESRO and another European space agency in 1975. ESA was responsible for funding, developing, and building Spacelab, while NASA was responsible for launch and operations.

Spacelab is a modular facility designed to ride into space in the payload bay of the space shuttle and remain there for periods of a week or more while conducting a wide range of manned and unmanned research tasks. The principal elements are a pressurized module, in which scientists work, and one or more U-shaped equipment platforms, called pallets, on which experiments requiring direct exposure to space can be mounted. The pressurized module consists of two segments, each 4 meters in diameter, the maximum width that the space shuttle's payload bay can accommodate. A "core segment" 4.26 meters long contains the data processing and utilities support, and a 2.74-meter-long "experiment segment" provides added work space and room for additional equipment racks, if needed. The pressurized module can be flown as a core segment only, called the short module, or in combination with the experiment segment, called the long module. The configuration selected for Spacelab 1 consisted of the long module and one pallet. In this arrangement, Spacelab offered 22 cubic meters of pressurized volume for research and had a mass of 11,539 kilograms.

More than four hundred proposals were received in response to a joint NASA/ESA "Announcement of Opportunity" inviting research investigations for the first Spacelab mission. Narrowing the field down to the seventy-two that eventually included the mission's "flight experiments" involved judgments on both the scientific merit of the proposals and the suitability of the experiments for flight aboard the Spacelab and the shuttle. Scientists from eleven European nations plus Canada, the United States, and Japan were principal investigators of the flight experiments selected for the Spacelab 1 mission.

The experiments were grouped into five general categories, which reflected the multidisciplinary focus of the mission. The Atmospheric Physics and Earth Observations experiments group examined in detail the composition, temperature, and motion of Earth's atmospheric gases, and tested a high-resolution mapping camera and an all-weather radar for gathering Earth-surface data of scientific and commercial value. The Space Plasma Physics experiments group studied the character of Earth's ionosphere—a plasma envelope situated between the atmosphere and the magnetosphere—where the sun's energy and Earth's environment interface. The Solar Physics and Astronomy experiments determined how constant the sun's energy output is and detected deep-sky objects visible only in the ultraviolet and X-ray wavelengths. A Materials Sciences and Technology experiments group conducted pilot studies in crystal growth, fluid physics, chemistry, and metallurgy in the microgravity environment. Medical and biological experiments included the Life Sciences experiments group, which assessed the influence of microgravity on various living systems and life functions, and the adaptation of various organisms to the space environment.

The primary goal of the Spacelab 1 mission was to verify the spacecraft's performance through a broad variety of scientific experiments, chosen especially to test flight and ground support systems and personnel. A secondary goal was to demonstrate, through a program of actual experiments, the exciting capabilities of Spacelab's instruments and methods for performing scientific research impossible in any other setting. The final goal was to demonstrate the feasibility of cooperative space research projects undertaken by international teams of scientists.

Launch of Spacelab 1 aboard the STS 9 mission was originally scheduled for September 30, 1983, a date that would have ideally satisfied the requirements of the astronomy experiments for a new moon because the skies would be as dark as possible, and the metric camera's requirement for a time of year when ground illumination was well distributed over the entire globe, hence an equinox. The first Tracking and Data-Relay Satellite System (TDRSS-1), a communications satellite considered essential for Spacelab's high data-rate communications needs, however, was delayed in getting into operation, forcing a one-month postponement of the launch. After rescheduling the mission for late October, technicians discovered a problem with the shuttle's solid rocket boosters and the launch had to be delayed an additional month.

Once the countdown finally got under way, it proceeded smoothly, and *Columbia* thundered into space from Launchpad 39-A on November 28, only milliseconds off the 11:00 A.M., eastern standard time, opening of the next available launch window. *Columbia* circled Earth every ninety minutes in an orbit 240 kilometers high and inclined 57 degrees to the equator. This "high-inclination orbit" was favored for the fact that it overflew a large percentage of the inhabited portions of Earth. Spacelab 1 was activated two hours and thirty minutes after launch, and the crew entered the pressurized module one hour later. The heavy load of experimental work dictated a very regimented flight plan. Spacelab 1 was operated around the clock for 231 hours by dividing the crew into two teams alternating in twelve-hour shifts. John W. Young, Robert A. Parker, and Ulf Merbold constituted the Red Team, while Brewster H. Shaw, Owen K. Garriott, and Byron K. Lichtenberg made up the Blue Team.

The first day of the mission was devoted to life science experiments, with particular attention to the crew's vestibular and general physiological response to zero-gravity. The focus of this research was on the causes of space adaptation syndrome, the troublesome "space sickness" that affects most astronauts during their first hours in space. Following up on research conducted ten years earlier aboard Skylab, the experiments revealed the nature of a suspected coupling between the eye and the vestibular system, leading to revisions in existing medical theories. Day two included a verification test of Spacelab's ability to withstand prolonged exposure to space cold. Days three and four involved further heat and cold verification tests. Days five through eight were dedicated to Spacelab experiments. The mission was judged to be going so well, and the use of onboard consumables was so conservative, that by day four there was discussion of extending the mission one additional day. The "bonus day" option was approved by mission managers on day six, allowing the crew to catch up on a few experiments that had been missed.

Spacelab 1 performed with remarkably little mechanical trouble. On the first day, a Remote Acquisition Unit (RAU) associated with collecting data from the pallet developed a malfunction and thereafter operated only when the shuttle's payload bay was shaded from the sun, but the problem was not serious. Of much greater concern was the High Data Rate Recorder (HDRR), which malfunctioned on the fourth day, threatening to destroy the mission's ability to save scientific data when the shuttle was out of range of TDRSS. Crewmen discovered that one of the tape drive capstans simply needed lubrication, after which the machine functioned normally for the remainder of the flight.

Crew egress from the pressurized module came at nine days, seventeen hours, and five minutes into the mission. *Columbia* returned to the main runway at Edwards Air Force Base at 3:47 P.M. Pacific standard time after a flight of ten days, seven hours, and forty-seven minutes, during which it circled Earth 166 times.

Impact of Event

Prior to Spacelab, the opportunities for an independent scientist to conduct complicated research involving access to the space environment were infrequent. Experiments had to be simple enough to be operated by astronauts who were not specialists in the research area and who had only very limited time to give to each experiment, or they had to be automated so that they could operate without human attention. Neither approach allowed much opportunity for a researcher to interact with an experiment in progress or to modify the experiment in response to the data or any problems that might develop. The only alternative was for scientists to become astronauts, but the extraordinarily long training time involved presented a serious obstacle.

Spacelab 1 demonstrated the best solution yet to this dilemma. The powerful and flexible orbital laboratory was complemented by a special crew category called payload specialist, and by a unique management interface between NASA and ESA spaceflight professionals on one side and the many scientists involved in the mission on the other. It amounted to a user management team and was called the Investigator Working Group (IWG).

The normal Spacelab mission crew of six includes two payload specialists, who are research scientists who have undergone enough NASA flight training to be able to live and work safely and productively aboard the space shuttle but not to operate it. Most of their preparation, called "mission-dependent training," is associated with the experiments themselves and is provided by the mission experimenters rather than NASA or ESA.

After the selection of the flight experiments was decided, the principal investigators of each of the chosen experiments were invited to participate in the IWG. This committee had responsibilities for guiding the incorporation of the many individual experiments into a payload and coordinating and communicating the needs of the user scientists to the mission manager. The IWG is also the body that selected the payload specialists and their backups; the backup American payload specialist was Michael L. Lampton, and the backup European payload specialist was Wubbo Ockels.

The ability to have highly trained scientists aboard Spacelab missions and strong involvement by the principal investigators in mission management decisions is important. These features allow the on-the-ground experimenters to exercise an onboard presence during the mission, which is essential in order to react appropriately and efficiently to new research opportunities and to problems that may occur.

The timely completion and very successful performance of the European-built Spacelab equipment, together with the high level of international scientific and management cooperation exhibited during all phases of the Spacelab 1 mission, have provided substantial confidence in the ability of NASA and ESA to forge a strong partnership in future space exploration. Reflights of Spacelab are a periodic feature of the shuttle flight schedule, even as both ESA and NASA plan more advanced orbiting laboratories to be incorporated in future space stations. ESA's Columbus laboratory—a direct outgrowth of Spacelab—will begin as a dependent module of the U.S. space station *Freedom* but will permit expansion to enable it eventually to operate independently.

Bibliography

Baker, Wendy. *NASA: America in Space.* New York: Crescent Books, 1986. Included in this chronicle of the technology, the people, and the events from almost three decades of U.S. space exploration are a general description of Spacelab and capsule summaries of missions 1, 2, 3, and D-1. This reference is useful for younger readers and helps put the Spacelab program into the context of space research endeavors in general. The book is generously illustrated with more than two hundred color photographs from NASA archives, but the quality of reproduction is disappointing.

Chappell, Charles R., and Karl Knott. "The Spacelab Experience: A Synopsis." *Science* 225 (July 13, 1984): 163-165. A report by two of the Spacelab 1 project managers, which presents highlights of the mission, based on a symposium conducted at the Marshall Space Flight Center in March of 1984. It gives a layperson's overview of the scientific results. Almost the entire issue is devoted to Spacelab 1, although most of the articles are too technical for the general reader.

Dooling, Dave. "Getting Ready for Spacelab 1." *Space World* 228 (December, 1982): 4-9. This article was prepared a year before the mission actually flew and discusses the integration of the hardware and the final training of the crew. Dooling draws on a meeting of the IWG to provide a number of first-person quotes, which gives an interesting insight into the enthusiasm prevailing in both the scientific community and in the two space agencies involved. Black-and-white photographs show Spacelab being assembled and tested.

_____. "Mission Report: STS-9." *Space World* 242 (February, 1984): 16-17. A short article that has more to do with the performance of the space shuttle *Columbia* and the experimental equipment than with the actual findings from the mission. At the time the article went to press, the results of the experiments were not yet available. Mention is made of a little-known fire that broke out in the aft end of the shuttle upon landing, but extinguished itself before the astronauts or flight controllers were aware of it.

Garriott, Owen K., Robert A. R. Parker, Byron K. Lichtenberg, and Ulf Merbold. "Payload Crew Members' View of Spacelab Operations." *Science* 225 (July 13, 1984): 165-167. Assesses the flight from the perspective of those who executed it.

Written by the scientists-astronauts aboard Spacelab 1, it features their view of their role, an assessment of how it compared with the role of the Skylab astronauts, and their perception of what the future of scientists in space might be. Suitable for general readers.

Longdon, Norman. *Spacelab Data Book*. Paris, France: European Space Agency, 1983. Presents a thorough overview of the Spacelab program, including historical background, management structure, crew selection and training, and the technical details of the spacecraft. Written for the nonspecialist and abundantly illustrated with color photographs and high-quality technical art, it is easy for the general reader to gain a deep understanding of the program. The book was written prior to the Spacelab 1 mission, so it does not contain specific mission results.

Silvestri, Goffredo, et al. *Quest for Space*. New York: Crown, 1987. Offers brief discussions of the general features and purposes of the spacecraft of all nations. It was written by Europeans, and its coverage of European space initiatives is particularly good. The narrative on Spacelab is adequate for students, and the excellent technical illustrations (in color) by Amedio Gigli would complement readings of more advanced texts.

Waldrop, M. Mitchell. "Spacelab: Science on the Shuttle." *Science* 222 (October 28, 1983): 405-407. A fresh and fascinating discussion of how the Spacelab program evolved and how it fits into the context of the overall space shuttle program. It is a story not told in public relations pieces and press releases about Spacelab. Full of frank quotations from individuals involved, the article conveys both the hopes and the frustrations associated with the project.

Richard S. Knapp

Cross-References

Skylab Inaugurates a New Era of Space Research (1973), p. 1997; *Columbia*'s Second Flight Proves the Practicality of the Space Shuttle (1981), p. 2180; The First Tracking and Data-Relay Satellite System Opens a New Era in Space Communications (1983), p. 2245; The First Permanently Manned Space Station Is Launched (1986), p. 2316.

OPTICAL DISKS FOR THE STORAGE OF COMPUTER DATA ARE INTRODUCED

Category of event: Applied science
Time: 1984
Locale: United States, Europe, and Asia

After many years of development, optical disk technology, promising great strides in storage capacity and reliability as well as competition for magnetically based disk media, was made available commercially

Principal participants:

IBM CORPORATION, Armonk, New York, the largest information technology company in the world, which provided direction in the early work in optical storage

SONY CORPORATION, Tokyo, Japan, a major electronics company that collaborated with NV Philips on research and development of optical data storage devices

NV PHILIPS, INC., Eindhoven, The Netherlands, the company that collaborated with Sony on research, development, and setting standards in the optical disk storage market

APPLE COMPUTER, INC., Cupertino, California, a company that provided leadership in the use of optical disks

NEXT, INC., Palo Alto, California, the company that pioneered the use of erasable optical drives

GROLIER ELECTRONIC PUBLISHING, INC., Danbury, Connecticut, the company that transferred an entire encyclopedia to a single compact disk

Summary of Event

Data storage using optically based methods is an exciting yet evolutionary step in mass storage technology. The excitement results from the numerous significant advantages that optical disks offer over traditional magnetic storage media; the emergence of the optical disk is, however, evolutionary, indicating the degree of overlap with the dominant magnetically based technology. It is thus helpful to compare and contrast optical disks with magnetic disks and tape.

Virtually all computers use two types of data storage media: magnetic tape and magnetic disk. Magnetic tape consists of a continuous thin plastic ribbon coated with iron oxide. It is typically packaged in a plastic reel and is analogous to the tape reels found on reel-to-reel tape recorders. Growing in popularity for microcomputers are smaller tape cartridges that are analogous to cassette tapes. A magnetic tape is mounted on a tape drive (similar in function to a tape recorder/player) that has a read/write head that creates or reads magnetized sections of iron oxide on the tape. When input is required from the tape, the read/write head detects magnetized spots of left or right orientation corresponding to the binary digits 0 or 1 (all computer information is encoded using the base-two numbering system known as binary) on the tape and converts the sequencing of these spots into electrical signals, which are then interpreted by the computer. When output to the tape is desired, the computer sends the proper sequence of electrical signals to the read/write head, which then magnetizes a stretch of tape. The use of a read/write head in some form is common to all the storage media considered in this article. The relevant feature for comparison to disk technology, both magnetic and optical, is the method of data access. Magnetic tape passes by the read/write head in the forward or reverse direction. Information can thus be accessed only in a sequential fashion.

The second and more common type of secondary data storage medium is the magnetic disk. It comes in two varieties: removable and fixed. The ubiquitous floppy diskette in its 5.25- and 3.5-inch standard sizes dominates the microcomputer media market. It consists of a flexible circular piece of plastic coated with a magnetizable iron oxide. This thin platter, analogous to a phonograph record, is enclosed in a paper or plastic jacket for protection with a small oblong opening cut into the jacket for access by the read/write head. The main advantage of floppy diskettes over fixed hard disks is their portability. The trade-off is in capacity and speed; fixed disks offer significantly more storage space, and the information that is stored can be accessed by a computer in about one-tenth of the time it takes for a floppy "disk." Fixed or hard disks are similar to floppy disks in operation but differ in construction and performance. Fixed disks usually consist of one or more metal platters coated with the same magnetizable material used with floppy disks but are sealed permanently in a cabinet in order to create a dust-free environment.

Information is written to or retrieved from floppy and fixed disks by a read/write head that skims over the surface of the spinning disk. Data are organized on the surface of the disk in concentric rings called tracks and can be accessed rapidly by radially positioning the read/write head to the proper track and then reading or writing data along a track as the disk spins underneath. This is known as random access and is considerably faster than the sequential mode used for magnetic tape media. All optical disks for computer data storage use random access.

Although IBM experimented with optical disk storage in the mid-1960's, it was not until the advent of the semiconductor laser that the idea became practical. In the late 1960's, Sony Corporation and NV Philips, Inc., entered agreements on cooperative research in the optical disk arena. The first commercial product realized from this joint venture was the laser videodisk in the late 1970's. The laser videodisk is analogous to the phonograph record (LP) in that information—in this case, video rather than audio or computer data—is recorded in a spiral track in analog, rather than digital, fashion. Information in analog form can take on any value, whereas digital information can be only one of two values: 1 or 0. The videodisk player used a laser stylus to play back prerecorded information.

The next commercial product to be launched from this merger was the compact disc (CD) in 1982. Videodisk technology was adapted to store audio information in a digital format. The move to the digital format was not only pivotal in producing higher-quality sound but also meant that more sound could be embedded in a small, 12-centimeter platter than could be found on the analog tracks of a larger phonograph record. The success of this product is readily evident from a cursory visit to any music store.

The natural extension of this idea to the computer world emerged in 1984, when CD-ROMs were made commercially available. CD-ROM is an acronym for Compact Disk-Read Only Memory. Computer data, prerecorded on optical disks, could be read but not written. Audio compact discs and CD-ROM disks are fabricated in the same manner. A high-power laser is used to burn pits in the recording-medium layer of a master disk. Information is retrieved from the disk by reading the sequencing of "pit" and "land" (the space between the "pits") areas using a lower-powered laser that is focused to a small spot on a single circular tract. As the disk spins at a speed of up to 3,600 revolutions per minute (rpm), the laser beam from the optical head shines on the pit and land regions, resulting in variations in the intensity of the reflected light. The reflected light is then captured by an optical system using lenses and focused onto an electrical device known as a photodetector. The photodetector detects the light intensity fluctuations and translates them into electrical signals which, in turn, are translated into video, audio, or computer text by means of electrical circuitry. Another significant event for computer data storage that took place in 1984 was the introduction of WORM optical disks. WORM is an acronym for Write Once Read Many. WORM drives allow the computer user to write to the optical disk once, read what was written many times, yet not erase what has been written. Every time information is saved to the disk, it resides there permanently.

The ultimate goal is to merge the best characteristics of the optical and magnetic disk technologies—that is, create a fast, high-capacity disk on which one could both read and write repeatedly. Such a goal was attained in 1988, when the first erasable optical disks became available. The achievement required a fusion of magnetic and optical physics and represents a remarkable feat in applied science. The erasable, or magneto-optical, disk is similar to the CD-ROM disk in construction but differs in the recording medium. Rather than burning pits in a metallic coating, the read/write head laser heats spots in the magneto-optical medium, and then a weak magnetic field is applied to magnetize the region of the spot. The direction of magnetization, analogous to the metal arrow in a compass, is either up or down (representing a 0 or a 1) and lies perpendicular to the plane of the platter. To read data on the disk, the laser is run at lower power. When the laser beam is reflected off the magnetized regions, it experiences a change in its physical characteristics. This phenomenon, known in physics as the Kerr effect, is detectable by a special optical system. In the case of the magneto-optical medium, the change occurs in one direction if the magnetization is up (for example, representing a 1) and in another direction if the magnetization is down. Again, the sequencing of the transitions between up and down states corresponds to the sequencing of 1's and 0's in the binary words that include all computer data.

Impact of Event

Advances in secondary storage technology for computer data have followed the major trend that characterizes the short history of the computer industry—miniaturization. This beneficial trend in hardware of packing more devices, circuits, and information in an ever-shrinking piece of silicon substrate is proceeding at a rapid pace and is evident particularly with the advent of the optical disk.

Conventional tape technology provides low capacity and long data access times. Newer tape formats (such as VHS and DAT) improve on both measures. Floppy disks, while providing portability, are inferior in data storage capacity. Hard disks are faster than floppies and provide greater data storage capacity. None of the tape or magnetic disk technologies can, however, achieve the performance of optical media. CD-ROM, WORM, and magneto-optical disks offer portability and thus security (the disks are removable), reliability (there is no wear on the media, since access is accomplished by a laser that sits relatively far from the surface of the platter compared to the read/write head for a magnetic disk or tape), storage capacity (a single 5-inch CD can store as much as 275,000 pages of text), and all for considerably less cost per bit of data.

The most compelling application for CD-ROMs is as a publishing medium. This type of optical disk has found multiple uses in government and industries (ranging from insurance companies to national laboratories) where large data storage is needed. Imagine storing the contents of an entire set of an encyclopedia on one compact disk. Indeed, this was one of the first applications for CD-ROM. The twenty-volume Grolier's *Academic American Encyclopedia* resides on a mere 20 percent of the surface of a single disk. Also imagine that one could have access to a particular topic in seconds in that encyclopedia and then locate in the same quick fashion all relevant and related topics. In addition, color pictures and animation, video stills and films, accompanied by stereo sound, stored on the same disk, would be available immediately for the topic of inquiry. One can easily see the exciting pedagogical potential and dramatic entertainment value in this technology. Companies marketing laser videodisks and CD-ROMs have had limited commercial success in this arena thus far but are finding renewed life as they embellish the products to appeal to a broader audience. Optical storage media are having a direct impact as the world of entertainment and computing coalesce under the banner of interactive multimedia. Apple Computer, Inc., and NeXT, Inc., in California, have provided leadership in this developing realm. Indeed, in 1989, NeXT, Inc., became the first microcomputer manufacturer to include a removable magneto-optical drive as standard equipment. Apple Computer is also a leader in the use of optical media for education.

Although magnetic disk and tape technology is still very much alive in research and development laboratories and will certainly continue to compete in price and performance with optical disks, there is little doubt that the inherent advantages of optical media eventually will enable it to dominate the data storage marketplace.

Bibliography

Brand, Stewart. *The Media Lab: Inventing the Future at MIT*. New York: Viking, 1987. A popular presentation of the exciting research being conducted in the futuristic media laboratory at MIT. Chapter 2 contains a short discussion on new media directly relevant to this article. Successfully conveys the excitement and potential benefits of technologies based on interactive media.

Buddine, Laura, and Elizabeth Young. *The Brady Guide to CD-ROM*. Englewood Cliffs, N.J.: Prentice-Hall, 1987. A standard reference to the world of CD-ROM and optical technology. Takes the reader through discussions of the technology, applications, industry standards, and software. Very well written and accessible. Includes many figures and diagrams.

Hecht, Jeff, and Dick Teresi. *LASER: Supertool of the 1980s*. New York: Ticknor & Fields, 1982. A book in the same vein as *The Media Lab* (cited above) that presents all of the glitter of lasers. There are some references to optical media, but this work appeared at the same time audio compact discs first came to market. Much discussion on current as well as potential laser applications.

Kryder, Mark H. "Data-Storage Technologies for Advanced Computing." *Scientific American* 243 (October, 1987): 116-125. An excellent review of the state-of-the-art in disk storage technology. Part of a dedicated issue on the topic of the next revolution in computers. Includes good discussion on the physics underlying magneto-optical disk technology with superb diagrams and photographs.

Lambert, Steve, and Suzanne Ropiequet, eds. *The New Papyrus: CD-ROM*. Bellevue, Wash.: Microsoft Press, 1986. Classic visionary work on what the future holds for this revolutionary medium. Includes descriptions of hardware and media, marketing advice, case studies, and tutorials on handling audio, video, and text. Microsoft Corporation is a leading developer of microcomputer software and is now working on DVI, a new standard in the interactive optical media market.

White, Robert M. "Disk-Storage Technology." *Scientific American* 243 (August, 1980): 138-148. An excellent review of the state-of-the-art in disk storage technology. The primary focus of the article is the explanation of the physics behind magnetic disk technology; therefore, it is directed toward readers with some high school physics background. Although somewhat dated, its predictions about future developments were on target.

Paul G. Nyce

Cross-References

UNIVAC I Becomes the First Commercial Electronic Computer and the First to Use Magnetic Tape (1951), p. 1396; Bubble Memory Devices Are Created for Use in Computers (1969), p. 1886; The Floppy Disk Is Introduced for Storing Data Used by Computers (1970), p. 1923; The IBM Personal Computer, Using DOS, Is Introduced (1981), p. 2169; Compact Disc Players Are Introduced (1982), p. 2200; IBM Introduces a Personal Computer with a Standard Hard Disk Drive (1983), p. 2240.

SIBLEY AND AHLQUIST DISCOVER A CLOSE HUMAN AND CHIMPANZEE GENETIC RELATIONSHIP

Category of event: Anthropology
Time: 1984
Locale: Yale University, New Haven, Connecticut

Sibley and Ahlquist showed that human and chimpanzee DNA is 99 percent identical, making chimpanzees the closest living relative to humans, with a 5-million-year-old common ancestor

Principal personages:

CHARLES GALD SIBLEY (1917-   ), a professor of biology, ornithologist, and taxonomist who developed techniques for reconstructing evolutionary trees from DNA comparisons among bird species and applied those techniques to primate evolutionary relationships

JON AHLQUIST, a professor of biology who collaborated with Sibley on the reconstruction of avian and primate evolutionary lineages from DNA comparisons among species

VINCENT M. SARICH (1934-   ), a professor of anthropology who used molecular analyses of proteins and reached the same conclusions as Sibley and Ahlquist

ALLAN C. WILSON (1918-   ), a professor of biochemistry who collaborated with Sarich

Summary of Event

The consensus among anthropologists before 1980 was that humans evolved from a common ancestor with apes approximately 15 to 20 million years ago, and the earliest humans were represented by Ramapithecine fossils dating back to that period. *Ramapithecus* was considered to be an early human because its jawbone fragments were more rounded like the human jaw and less like the rectangular ape jawbone. New evidence emerged, however, during the 1980's that indicated a much more recent human origin from a common ancestor with apes. Charles Gald Sibley and Jon Ahlquist reported that human and chimpanzee deoxyribonucleic acid (DNA) is approximately 99 percent identical, which is typical of species that are much closer than humans and chimpanzees were supposed to be. Sibley and Ahlquist suggested also that chimpanzees are more closely related to humans than gorillas, which was surprising since it was assumed, based on appearance, that the opposite was the case. Regardless of whether chimpanzees are closer to humans or gorillas, Sibley and Ahlquist's results show that all three species have separated from one another relatively recently, possibly as recently as 5 million years ago.

Sibley and Ahlquist based their conclusions on comparisons between human DNA and DNA from chimpanzees and gorillas. DNA, the chemical form of genes, has a "double helix" molecular structure, similar in shape to a ladder twisted into a spiral. The two sides of the ladder correspond to two strands of DNA wound around each other, and the rungs correspond to chemical bonds holding the two strands together. The strands of the double helix consist of chemical subunits called nucleotides, and all genes are constructed from only four different kinds of nucleotides. The chemical names of the nucleotides are adenine, guanine, cytosine, and thymine, abbreviated A, G, C, and T, respectively. An average gene consists of a continuous linear strand of hundreds of nucleotides, and it is the exact number and order of nucleotides that makes each gene and each species' genetic code unique. The entire set of human genes contains approximately 3 billion nucleotides. Sibley and Ahlquist discovered that approximately ninety-nine out of every one hundred nucleotides are exactly the same between the genetic codes of humans and chimpanzees.

A universal feature of DNA is that the bonds holding the two strands of nucleotides together can connect only a G on one strand with a C on the other, and vice versa, or an A with a T; no other combinations are allowed. The attractions between G and C and between A and T are so specific that these nucleotides attract each other spontaneously to hold together the two strands of the double helix. Sibley and Ahlquist used this specific binding of DNA strands to compare the DNA of different species. They started by heating the DNA until the two strands of the double helix separated. Normally, as the DNA cools, the two strands bind back together, since every G matches up with its original C partner on the opposite strand, and every A matches with a T. Sibley and Ahlquist compared DNA between species by melting apart the DNA of both species, mixing them together, and measuring how well they bind to each other as the mixed pool of DNA strands cools. If the DNA of two species is identical, the nucleotides of the two strands will match perfectly and bind strongly. If the DNA of two species is 1 percent different, then ninety-nine out of one hundred nucleotides will match, and the mixed DNA will reassociate with 99 percent the strength of identical DNA.

The strength of binding between the two strands of DNA is measured by determining the temperature required to melt them apart. If more nucleotides match, the binding is tighter, and it takes a higher temperature to separate them. If fewer matches are made, there is weaker binding across the two strands. Mismatched DNA has weaker binding and melts apart more easily and thus separates at a lower temperature. Overall, each one degree drop in melting temperature when the DNA of two species is mixed corresponds to approximately 1 percent mismatching between the two species' genetic codes.

Since human and chimpanzee DNA mixed together separate at about one degree below the temperature at which human DNA separates from itself, there is approximately a 1 percent difference between the human and chimpanzee genetic codes. Sibley and Ahlquist estimated also that the genetic difference between humans and gorillas was approximately one-third greater than the difference between humans and chimpanzees. Their discovery of the small difference between the genes of humans and chimpanzees reveals more than a close genetic relationship between the two species. It also provides an independent estimate, compared to the fossil record, of the time since the two species diverged from a common ancestor.

Estimates of the ages of fossils are based on radiometric dating. The age of a fossil is calculated by measuring the amount of radioactive decay in the fossil since it was formed and using the rates of decay, which are constant for a specific element, to calculate the fossil's age. Since many radioactive elements decay at very slow rates over millions or billions of years, it is possible to date fossils that are millions or billions of years old. Fossil records are the only remains of extinct species, but species that are alive now can be thought of as "living fossils," because all organisms carry DNA in their cells that is inherited from distant ancestors. Humans and chimpanzees, for example, have been evolving as separate species only long enough for 1 percent of human DNA to change. If humans had been separated earlier from chimpanzees, the DNA would differ by a greater percentage, since it would have had more time to change. The DNA of living species, in short, represents a "molecular clock" that carries a record of the divergence of all related species from their common ancestors. The exact time that two species have been evolving apart can be calculated if the rate at which changes in DNA evolve can be determined. Sibley and Ahlquist estimate that overall, the total DNA of a species changes at the rate of approximately 1 percent every 5 million years. The rate of 5 million years of evolutionary divergence per 1 percent difference in DNA comes from two sources: One is direct measurement of the DNA mutation rate; the second is calibration of DNA comparisons against the fossil record in species, where good fossil records of ancestral origins exist and living descendants have also survived whose DNA can be analyzed. The fossil record, for example, of related flightless birds such as ostriches in Africa and South American rheas indicates that they separated approximately 80 million years ago. Their DNA is approximately 16 percent different, which corresponds to about 5 million years of independent evolution per 1 percent change in DNA. A number of similar calibrations all yield comparable answers. Since human and chimpanzee DNA is approximately 1 percent different, it follows that this difference represents about 5 million years of divergent evolution from a common ancestor.

Impact of Event

Anthropologists were reluctant to accept Sibley and Ahlquist's conclusions. Molecular taxonomists argued for a recent divergence of humans from apes, but most anthropologists believed in a very ancient human origin. The two schools of thought might still be divided except for the 1979 discovery of new fossils showing Ramapithecines to be ancestors of orangutans, not humans. The original Ramapithecine fossils were only jawbone fragments, but the new, more complete, specimens were nearly identical to modern orangutans and showed that orangutans were well on the way toward their present form 15 million years ago. This revision left the fossil record without any ancient human ancestors until the relatively recent "Lucy" (*Australopithecus afarensis*), who dates back to 3.8 million years ago. With evolutionary reconstructions based on fossils and DNA comparisons suddenly in agreement, the consensus among anthropologists shifted toward accepting a common human and chimpanzee ancestor approximately 5 million years old.

The demonstration by Sibley and Ahlquist of the close genetic relationship between humans and chimpanzees was anticipated by earlier evidence from investigators such as Vincent M. Sarich and Allan C. Wilson, who demonstrated extremely close molecular resemblances among humans, chimpanzees, and gorillas, and significant differences between these three species and other apes. Previous studies, however, were based on only one or a few specific proteins or genes, or else comparisons of total DNA that were not exact enough to date the origins of different primate species. Sibley and Ahlquist, however, had spent years carrying out approximately twenty thousand comparisons of DNA among bird species, and used their expertise to conduct the most definitive study of primate DNA comparisons to date.

The 1980's were an exciting period for the study of human evolution. The picture of human evolution is more complete and more secure since it is based on two consistent but independent reconstructions from fossils and DNA comparisons. Although 5 million years of evolutionary divergence between humans and chimpanzees is consistent with the fossil record, debate continues among molecular taxonomists about the exact calibration of DNA differences to evolutionary time. Some argue that 1 percent DNA difference corresponds to as little as 2 to 3 million years of divergence, while others estimate 6 to 8 million years. Further DNA comparisons and refinement of techniques can provide increasingly precise estimates, and additional fossil discoveries can confirm or challenge the predictions of molecular taxonomists.

The primary and profound change that occurred in the view of human evolution during the 1980's is the recognition of a much closer evolutionary relationship between humans and great apes than had been previously recognized. Ten years of progress in the study of human evolution reduced the evolutionary distance between humans and apes by more than 10 millon years to a mere 5 million years. The realization that 5 million years is a short time in the history of life, and that chimpanzees and humans remain approximately 99 percent genetically identical, indicates that biologically, chimpanzees are nearly human and humans are barely different from them.

Bibliography

Campbell, Bernard G. *Human Evolution*. 3d ed. New York: Aldine, 1985. An introductory college-level book by an anthropologist who presents an objective, comprehensive, and detailed overview of human evolution. The 475-page book includes an index, glossary, extensive bibliography, and an appendix listing human fossil discoveries.

Gribbin, John, and Jeremy Cherfas. *The Monkey Puzzle*. London: Bodley Head, 1982. A well-written account of human and primate evolution revised to reflect the evidence from molecular taxonomy. The molecular evidence is reviewed extensively from the very beginnings of such work up to, but not including, the work of Sibley and Ahlquist. This book is very readable, and includes biographical information about the scientists involved. Includes an index and further references.

Lewin, Roger. "DNA Reveals Surprises in Human Family Tree." *Science* 226 (December, 1984): 1179-1182. A research news article in the leading American scientific journal. Lewin describes Sibley and Ahlquist's results and techniques in more detail, but in relatively nontechnical language. This article is written in response to Sibley and Ahlquist's original paper, and includes a sense of the initial reactions of the scientific community.

_____. *Human Evolution*. New York: W. H. Freeman, 1984. A brief, general introduction to human evolution. This book includes a survey of fossil and molecular evidence, including Sibley and Ahlquist's work, and a discussion of cultural evolution. Contains a bibliography and an index. Clearly written in a nontechnical style.

_____. *In the Age of Mankind*. Washington, D.C.: Smithsonian Books, 1988. A beautifully illustrated book, with a contemporary and comprehensive summary of human evolution, including the molecular evidence, fossil evidence, and cultural evolution. An index is included, and the photographs are excellent.

_____. "My Close Cousin the Chimpanzee." *Science* 238 (October, 1987): 273-275. A research news article reflecting the status of Sibley and Ahlquist's discovery three years after it was first reported. Lewin describes the general consensus in support of Sibley and Ahlquist's main conclusions and additional evidence supporting their views. Since by this time the close relationship of humans to apes has been established, Lewin focuses on the question of whether chimpanzees are more closely related to humans or gorillas.

Pilbeam, David. "The Descent of Hominoids and Hominids." *Scientific American*, March, 1984, 84-96. A well-written account of the changing view of human evolution as a result of both the molecular evidence and the reassignment of *Ramapithecus* from the human to orangutan lineage, with an emphasis on the fossil evidence. This article is especially significant because Pilbeam was originally one of the strongest supporters of the view that humans diverged from apes 20 or more million years ago, but shifts his views here to support the more recent 5 million year divergence.

Bernard Possidente, Jr.

Cross-References

Boule Reconstructs the First Neanderthal Skeleton (1908), p. 428; Zdansky Discovers Peking Man (1923), p. 761; Dart Discovers the First Recognized Australopithecine Fossil (1924), p. 780; Watson and Crick Develop the Double-Helix Model for DNA (1951), p. 1406; Leakey Finds a 1.75-Million-Year-Old Fossil Hominid (1959), p. 1603; Nirenberg Invents an Experimental Technique That Cracks the Genetic Code (1961), p. 1687; Simons Identifies a 30-Million-Year-Old Primate Skull (1966), p. 1814; Anthropologists Discover "Lucy," an Early Hominid Skeleton (1974), p. 2037; Hominid Fossils Are Gathered in the Same Place for Concentrated Study (1984), p. 2279; Scientists Date a *Homo sapiens* Fossil at Ninety-two Thousand Years (1987), p. 2341.

WILLADSEN CLONES SHEEP
USING A SIMPLE TECHNIQUE

Category of event: Biology
Time: 1984
Locale: Cambridge, England

Willadsen improved on a previous technique of cloning large mammals, developing a simple, easily used new method of producing identical sheep for use in experimentation

Principal personages:
> STEEN M. WILLADSEN, a veterinarian who developed both an early complex method and a simplified method for dividing early sheep embryos to produce two or more identical animals
> R. A. GODKE, an animal physiologist who collaborated with Willadsen in the development of a simple procedure for production of identical sheep twins
> C. POLGE, an animal physiologist who assisted Willadsen in his studies of embryo division and growth in cows
> J. P. OZIL, a French scientist and head of the team working on splitting cow embryos to produce identical twins
> STEVEN A. VOELKEL (1957-   ), a reproductive physiologist who worked on splitting sheep and cattle embryos, producing the first bisected embryo twin lambs in 1984

Summary of Event

Steen M. Willadsen and R. A. Godke reported in 1984 on their development of a new simplified technique for the rapid production of split embryos in sheep. This new procedure improved upon a previous splitting method first developed by Willadsen in 1979, which required several steps occurring over about three days. As with earlier techniques in use at the time, such splitting produced two separate embryos from a single embryo collected from a mother ewe. The resultant half embryos, called demi-embryos, were then implanted into the mother's or a different ewe's reproductive tract to allow for the development of identical twin lambs. The new procedure was considerably simplified, as it required only a single surgery on the mother, and the two demi-embryos were reimplanted into the same ewe within about an hour of the removal of the original embryo.

Work had been done since the 1950's in removing embryos from both laboratory animals and large domestic animals and then alternatively replacing them, transferring them to new host mothers, or freezing them for storage and later transfer. In the 1970's, this research led to the development of methods of separating the individual cells (blastomeres) from early two-cell, four-cell, or eight-cell embryos to allow the production of multiple individuals from the same original embryo. Such production of identical twins, triplets, and higher multiples is a means of artificial cloning. Work in this area was important to reproductive physiologists interested in how embryos develop, to researchers in other fields who needed twins for experiments in various areas, and to farmers and ranchers who wanted to increase the number of specifically bred animals.

In 1979, Willadsen published the description of his original method for micromanipulation of sheep embryos to divide an embryo in half at the two-cell stage. He first treated female sheep with hormones to cause superovulation, the production of more than the normal number of ova. The sheep were mated, and Willadsen surgically removed fertilized ova in the second day of the estrus cycle, when each embryo contained only two cells. Using a microscope, he held the embryo stationary with a capillary pipet and tore into the zona pellucida, the tough, clear outer coating that protects a mammalian egg. Willadsen separated the two cells and placed each into a new empty zona that had been collected previously. He then coated the demi-embryos with agar, a jellylike material derived from seaweed, to protect them within the torn zonas. A second layer of agar was used to coat the first, producing a small plug containing the two demi-embryos from a single original embryo. Each cylinder produced in this manner was transferred into the oviduct of a recipient ewe that was synchronized at the same point in its estrus cycle as the mother, and the oviducts were ligated (tied off) to prevent the loss of the agar plug. After two and one-half days, the embryos were flushed out of the host mother's oviduct and evaluated microscopically. Small hypodermic needles were used to manipulate the embryos out of the agar, and the freed structures were transferred to other synchronized ewes. Sixteen sets of monozygotic (identical) embryo pairs were produced in this way, and both embryos of each set were transferred into the same host mother. Ten of the ewes carried the resultant pregnancies to term, resulting in the birth of five single lambs (when only one of the demi-embryos survived) and five sets of monozygotic twin lambs.

Willadsen also found in 1980 and 1981 that embryos could be split in half or in quarters at both the four-cell and eight-cell stages of development, and the resultant demi-embryos were able to produce normal offspring. This was shown by production of identical quadruplets in sheep in 1980 and in cattle (with C. Polge) in 1981. For this work, the same agar technique was used, with the embryo divided into halves or quarters. Identical twins, triplets, or quadruplets were produced in these studies. Single cells (blastomeres) derived from eight-cell sheep embryos also were able to develop into lambs, but the recovery rate was very low, and either no offspring or only one lamb resulted from each of numerous attempts.

Other investigators in this area were also working on producing cloned animals. J. P. Ozil and his colleagues in France thought that Willadsen's agar-coating method was more complex than necessary and devised a technique that eliminated that aspect of the procedure. In experiments with cattle, they used later-stage multicellular embryos, called gastrulas, and omitted the protective agar-coating step. With no agar around the embryo, it was not necessary to retrieve them three days later to scrape off agar so that implantation could take place in the host mother's uterus. In this process, the demi-embryos were separated microsurgically and then placed in empty zonas. These "naked" demi-embryos were inserted directly into the synchronized host-mother recipients, with resultant rates of pregnancy (64 percent) and of twinning (67 percent) that were lower than Willadsen's rate in sheep (75 percent), but still reasonably high.

Willadsen decided in 1984 to try the later, multicellular embryos as the source of demi-embryos. Rather than collecting embryos at day two of the estrus cycle, in the two-cell stage, or day three, in the four-cell or eight-cell stage, the new method reported by Willadsen and Godke harvested embryos at day six, seven, or eight of the cycle. At day six, a sheep embryo is in the morula stage of development, and at day seven or eight, it is in the blastocyst stage. At these stages, there are hundreds of small cells, and the zona pellucida has disintegrated, no longer necessary for protection. Two specific areas develop in the blastocyst, an inner cell mass that becomes the actual individual, and the outer layer (trophectoderm) that forms fetal membranes that attach to the mother's uterus. In Willadsen's new procedure, these embryos were collected surgically and then cut in half using a capillary pipet and a fine glass needle. The hollow ball of cells was held so that the inner cell mass was at the top, and division was accomplished by passing the needle into the ball and moving it up and down to divide the embryo, cutting it against the pipet. While the two demi-embryos were not exactly the same, each contained about the same amount of inner cell mass and trophectoderm, enough to produce a new individual. The mother ewe from which the original embryo was taken remained under anesthesia for approximately an hour while the procedure was performed, then the two resultant demi-embryos were replaced into her uterus. This rapid removal, bisection of the embryo, and replacement was much more efficient than the previous method. Of the eighteen ewes implanted with a pair of demi-embryos, the pregnancy continued to term in sixteen ewes (85 percent pregnancy). Seven produced single lambs, one produced nonidentical twins (one lamb was from an embryo that had not been recovered and split), and eight produced identical twins (50 percent twinning). All twin production was from the bisection of blastocysts, with no twins developing from morula-stage embryos. Unpublished studies in cattle by Willadsen and colleagues also showed high viability of similarly split and transferred embryos.

Impact of Event

A major reason for the importance of Willadsen's original technique was that it allowed the use of genetically identical animals in experimental studies, so that differences in response to particular treatments could be assumed to be the result of the treatment and not of genetic differences between test organisms. It had been estimated that the use of one pair of twins in an experiment in cattle, as the experimental animal and its untreated control, was the informational equivalent of using twenty-two unrelated animals under the same experimental conditions. The birth of identical twins in cattle and sheep occurs very infrequently, in less than 1 percent of births, and such births occur at random in the population. This experimental means of producing identical twins from specific animals and specific matings was able to increase the accuracy and efficiency of experiments and greatly reduce the number of animals needed to obtain useful results. This was of importance in controlling the time and cost elements of animal experimentation.

Willadsen recognized that his earlier technique was rather complicated for general use in the burgeoning market for production of twin cattle and sheep, because production was desired on the ranch as well as in the laboratory. Use of the agar-coating method required that the user be experienced in embryo transplantation and in micromanipulation for division of the original embryo.

The publication of a simplified new procedure provided a significant improvement over the previous method, making it possible for even relatively inexperienced users to produce reasonably successful results in twin production. The procedure developed by Willadsen opened up the production of identical twins in sheep and cattle to many more research investigations and agricultural uses. Not only did it become possible to produce twins for use in experiments but also it was easier to produce many offspring of a desirable mating. Through superovulation of the mother before mating, cloning of the resultant embryos, and implantation of the demi-embryos into host mothers, the actual mother could then be used repeatedly for ovum production, rather than being pregnant herself. Many offspring could be produced in this way for use in experiments or for agricultural applications, such as further breeding or milk or meat production.

Another use of the technique was that one of a pair of demi-embryos could be transferred to a host mother immediately, while the other was placed in frozen storage and later thawed and implanted. One of Willadsen's papers contains a picture of identical twins produced in such a way, with one of them two weeks old and the other two and one-half months old. Such methods would allow an experiment to be carried out on genetically identical individuals at the same time, but with the individuals at different ages.

The embryo-splitting technique is widely used now in agriculture. Researchers such as Steven A. Voelkel and Godke in Louisiana began using it in 1984 in studies on cattle, sheep, and pigs. Voelkel was involved in development of a modification of the technique that allowed automatic splitting of the embryo using a micromanipulator. At Southwestern Louisiana University's animal research center, the method has also been applied to the embryos of rhesus monkeys in an attempt to develop a colony consisting of identical twins for use in research. Since monkeys are closely related to humans, such a colony would be of great help in studies related to human health, such as development of malaria and AIDS (acquired immune deficiency syndrome) vaccines.

Bibliography

Bavister, Barry D., ed. *The Mammalian Preimplantation Embryo: Regulation of Growth and Differentiation in Vitro.* New York: Plenum Press, 1987. Discusses embryonic development in a variety of animal species. No specific discussion of twin production, but good background information on in vitro embryo culture in the two appendices. Chapter 12 is of particular interest on the growth of early embryos of domestic animals, mostly pigs.

Beier, H. M., and H. R. Lindner, eds. *Fertilization of the Human Egg in Vitro.* Berlin: Springer-Verlag, 1983. Contains a paper by Willadsen and Fehilly on the developmental capacity of separated cells from two-, four-, and eight-cell sheep embryos.

Brackett, Benjamin G., George E. Seidel, and Sarah M. Seidel. *New Technologies in Animal Breeding.* New York: Academic Press, 1981. Of general interest in animal breeding; also contains a chapter on "Parthenogenesis, Identical Twins, and Cloning in Mammals" that references work done by Willadsen and coworkers from 1979 to 1981. Discusses other cloning methods, also.

McKinnell, Robert Gilmore. *Cloning: A Biologist Reports.* Minneapolis: University of Minnesota Press, 1979. Written by a professor of cell biology; provides a clear scientific explanation for the general reader of what is involved in cloning. Frogs, the highest animals that had been cloned at the time the book was written, are discussed in depth. The information may be applied to sheep cloning as well, with some modifications.

Ozil, J. P., Y. Heyman, and J. P. Reynard. "Production of Monozygotic Twins by Micromanipulation and Cervical Transfer in the Cow." *Veterinary Record* 110 (February 6, 1982): 126-127. Discussion of splitting of cattle embryos accompanied by pictures of the resultant identical twins. The authors call their process useful at the practical level, as compared to Willadsen's more difficult laboratory techniques of the early 1980's.

Willadsen, S. M. "A Method for Culture of Micromanipulated Sheep Embryos and Its Use to Produce Monozygotic Twins." *Nature* 277 (January 25, 1979): 298-300. Willadsen's first paper on splitting sheep embryos; discusses his original technique that involved coating the split segments with agar for protection.

Willadsen, S. M., and R. A. Godke. "A Simple Procedure for the Production of Identical Sheep Twins." *Veterinary Record* 114 (1984): 240-243. Presents Willadsen's simplified procedure for splitting an early sheep embryo. Describes the collection, splitting, and return of the embryos to the original or host mother, as well as the results.

Willadsen, S. M., and C. Polge. "Attempts to Produce Monozygotic Quadruplets in Cattle by Blastomere Separation." *Veterinary Record* 108 (March 7, 1981): 211-213. Describes the techniques and results of a procedure for dividing cow embryos into four parts and attempting to produce a calf from each quarter embryo. Cow embryos are noted to be easier to work with than sheep embryos split using the same technique.

Jean S. Helgeson

Cross-References

The U.S. Centers for Disease Control Recognizes AIDS for the First Time (1981), p. 2149; A Human Growth Hormone Gene Transferred to a Mouse Creates Giant Mice (1981), p. 2154; The First Successful Human Embryo Transfer Is Performed (1983), p. 2235; Sibley and Ahlquist Discover a Close Human and Chimpanzee Genetic Relationship (1984), p. 2267.

HOMINID FOSSILS ARE GATHERED IN
THE SAME PLACE FOR CONCENTRATED STUDY

Category of event: Anthropology
Time: April 6-September 9, 1984
Locale: The Bronx, New York

The gathering of most important hominid fossils, along with a simultaneous symposium, led to the focusing of theories concerning human ancestry into two primary hypotheses

> *Principal personages:*
> JOHN VAN COUVERING, an associate of Micropaleontology Press, American Museum of Natural History
> ERIC DELSON, an associate of the Department of Vertebrate Paleontology, American Museum of Natural History
> IAN TATTERSALL, an associate of the department of anthropology, American Museum of Natural History

Summary of Event

On April 6, 1984, fifty-seven of the world's leading paleoanthropologists gathered at the New York Museum of Natural History to study forty of the most important early human fossils found up to 1984. Contrary to popular belief, the fossil remains of early human and near-human beings are extremely rare, owing to the very special set of circumstances that must occur for a fossil to form and be preserved and the great difficulties involved in finding those that do exist. These few precious relics of human antiquity are scattered among the world's great museums, carefully guarded by museum curators. Consequently, scientists and scholars who study and write about the evolution of the human species have never had the opportunity to study firsthand the evidence on which their discipline is based. The symposium at New York's Museum of Natural History marked the only time in history that most of the important fossils relating to early humans and humanlike creatures have been available for the direct study by the world's most foremost authorities on early forms of humans.

"Anthropology," the study of humankind, has various subdisciplines that include the study of human biology, the study of human cultural evolution, and the study of human adaptation. "Paleoanthropology" refers to the study of ancient varieties of human species and species ancestral to humans. Paleoanthropologists base their studies on the fossil remains of ancient hominids (a hominid is any member of the taxonomic family Hominidae, which includes modern human beings and ancestors). Paleoanthropologists are concerned particularly with untangling the complex path of human evolution. Obviously, these scientists labor under a severe handicap because they are unable to examine firsthand most of the fossils upon which their discipline is based. The organizers of the symposium at the New York Museum of Natural History designed their program to give paleoanthropologists access to the most important early hominid fossils.

Eric Delson, Ian Tattersall, and John Van Couvering, all associated with the New York Museum of Natural History, originated the idea for the symposium in April, 1979. In addition to their desire to gather as many important hominid fossils as possible for examination by paleoanthropologists, they had another motive for organizing the symposium. By their own admission, Delson, Tattersall, and Van Couvering shared with other evolutionary scientists alarm at the growing influence and visibility of so-called creation science—those who insist that God created all existing species, including humankind. Advocates of creation science contend that the evidence for their position is at least as convincing as the evidence for evolution and should be included in high school and college biology texts. The organizers of the symposium proposed that after paleoanthropologists had had an opportunity to view and examine the hominid fossils, they should be displayed to the public at the New York Museum of Natural History. Such a display would go far, the organizers hoped, toward dispelling in the minds of the general public what they considered the obscurantist contentions of creation scientists.

After securing the permission and cooperation of the directors of the museum for their proposed symposium and exhibit, Delson, Tattersall, and Van Couvering tackled the many problems associated with its realization. Before even approaching the curators of the museums around the world where the fossils they hoped to acquire were permanently housed, the project originators had to secure adequate funding for the proposed symposium and public display. At a meeting in New York in 1981, Delson, Tattersall, and Van Couvering, along with four noted American paleoanthropologists (David Pilbeam of Harvard University, Elwyn Simmons of Duke University, and Clark Howell and Desmond Clark of the University of California at Berkeley) presented their proposals to the heads of four private foundations which support physical anthropology (the Wenner-Gren Foundation, the L. S. B. Leakey Foundation, the National Geographic Society, and the Foundation for Research into the Origins of Man). The representatives of those organizations agreed to furnish the bulk of the funding needed for the project. With the funding arranged, the project coordinators turned their attention toward establishing security measures that would assure the safety of the priceless human remains they proposed to bring to New York. Those security measures had to take into consideration not only precautions against accidents in shipment, display, and storage but also protection from deliberate acts of violence by those who might want to destroy the fossils for political reasons. After working out seemingly foolproof security procedures, the project organizers began contacting the museums of the world where the most important hominid fossils were located. They were pleasantly surprised by the eagerness with which most curators accepted their invitations to bring the fossils to New York. Eventually, fifty-seven of the most eminent paleoanthropologists from around the world accepted invitations to participate and most indicated a willingness to prepare scientific papers for presentation at the symposium. The working out of all the details took nearly three years.

At the very last moment before the opening of the program, curators of three of the twenty-five museums who had agreed to participate in the project reconsidered. Despite those disappointments, the project was a success. From April 6 to 9, the paleoanthropologists were able to examine and compare the fossil ancestors of contemporary humanity gathered from various parts of the world during the one hundred-plus year history of paleoanthropology. From April 9 to 13, scientists participated in a symposium entitled "Paleoanthropology: The Hard Evidence" at which they presented papers that greatly clarified the current status of scientific theory concerning the evolution of modern humans and the evidence on which those theories are based. From April 13 to September 9, 1984, the fossils were made available to the public in an exhibit entitled "Ancestors: Four Million Years of Humanity." When the exhibit ended, almost a half-million people from around the world had viewed the evidence upon which contemporary understanding of human origins is based.

Impact of Event

The "Ancestors" project had many far-reaching results. One of the most important outcomes was the emergence of a consensus in paleoanthropological circles concerning the main lines of human evolution. Before the symposium part of the project, paleoanthropologists from around the world seemed to be too busy engaging one another in often acrimonious disputes concerning obscure points within their discipline to give any clear explanation of their findings to the general public, which found their arguments virtually incomprehensible. There emerged from the symposium an account of the major lines of human evolution on which probably most scientists who study the origins of humankind can agree, though many would dispute various minor issues. The account is as follows.

Four to five million years ago in Africa, a hominid called *Australopithecus afarensis* evolved from creatures ancestral to both humans and contemporary apes. *Australopithecus afarensis* was small, averaging around 1.2 meters in height, and had a brain only slightly larger than that of the modern chimpanzee, but it walked completely upright and its teeth much more closely resembled the teeth of modern humans than they resembled the teeth of modern apes. During the succeeding 3 million years, several other species evolved from *Australopithecus afarensis*, including *Australopithecus africanus* and *Australopithecus robustus*, both of which eventually became extinct, and *Homo habilis*, which was directly ancestral to modern man. The first members of *Homo habilis* appeared about 2.2 million years ago and may have spread outside the continent of Africa to as far away as southeast Asia and Java.

In Africa and perhaps southeast Asia, another human ancestor called *Homo erectus* evolved from and replaced *Homo habilis* around 1.5 million years ago. *Homo erectus* (some famous specimens of which include so-called Java man and Peking man) was much larger in size than *Homo habilis*, averaging almost as tall as modern humans. His brain was much larger than the brain of *Homo habilis*, though not as large as the average brain size of the human species. *Homo erectus* spread throughout Africa, Europe, and Asia, only to be replaced around 400,000 years ago in many locales by *Homo sapiens neanderthalensis* (popularly known as Neanderthal man), whose members apparently evolved in Europe. Concurrently with *Homo sapiens neanderthalensis*, who was not directly ancestral to modern humans, there evolved in Africa humans' immediate ancestor, *Homo sapiens* (often referred to by anthropologists as archaic or primitive *Homo sapiens*), which replaced *Homo erectus* throughout the regions not dominated by *Homo sapiens neanderthalensis*. Nevertheless, members of *Homo erectus* survived in some parts of the world until about 250,000 years ago, at which time the species apparently became extinct.

Homo sapiens neanderthalensis was shorter on the average and had a much more massive bone structure than modern man. His brain size was well within the range for modern humans, but his skull was much more slanted (virtually no forehead) and his pronounced brow ridges were rounded rather than straight as in previous hominids. In addition, Neanderthal man had several skeletal features markedly different from *Homo sapiens sapiens*. Neanderthal man used fire and buried his dead with considerable ritual.

Archaic *Homo sapiens* (some famous specimens of which include Steinheim man from Germany and Swanscombe man from England), on the other hand, were much more similar in most respects to modern human beings than was Neanderthal man. The stature of *Homo sapiens* was nearly that of modern man, and his brain was only slightly smaller on average. He had very little in the way of a forehead or chin, and his bone structure was more massive than that of his modern descendants. Archaic *Homo sapiens* used fire and evolved a very distinctive stone tool culture. Also, he may have built permanent villages.

Perhaps 60,000 to 100,000 years ago, modern man (*Homo sapiens sapiens*) evolved in Africa. By 35,000 years ago, *Homo sapiens sapiens* had replaced both *Homo sapiens neanderthalensis* and archaic *Homo sapiens*, becoming the only living hominid.

In addition to clarifying contemporary theories concerning human ancestry so that it can be understood by nonprofessionals, the "Ancestors" project had two other important results. First, the project dealt a powerful blow to creation science. The assembly and display of the hard evidence for human evolution made it very difficult for creation scientists to explain away beings whose remains clearly reveal them to have been closely related to but not identical with modern humans. Also, creation scientists had made much of the often well-publicized disputes between paleoanthropologists, arguing that even the "evolutionists" themselves were divided about the validity of evolutionary theories. The symposium made it clear that although paleoanthropologists might dispute obscure points about the evolutionary path of humankind, they agreed on major points, something that many members of the general public had not understood prior to the well-publicized symposium.

Finally, the "Ancestors" project created a precedent for international cooperation in the study of human origins. The disputes between paleoanthropologists in the years before often took on distinctly nationalist and political overtones, which were extremely disruptive to progress in the field. The project showed the many advantages that would accrue from international cooperation rather than competition and, it is hoped, will usher in a new age of progress in the search for human origins.

Bibliography

Day, Michael H. *Guide to Fossil Man*. Chicago: University of Chicago Press, 1986. The most complete catalog of hominid fossils available. Contains photographs and illustrations that greatly aid the nonprofessional in understanding the main points of human evolution.

Delson, Eric, ed. *Ancestors: The Hard Evidence*. New York: A. R. Liss, 1985. Contains a brief account of the genesis of the "Ancestors" project, along with all the papers presented at the concurrent paleoanthropological symposium. Contains color photographs of the fossils assembled in New York, with a complete history of each one and an evaluation of its significance.

Else, James G., and Phyllis C. Lee, eds. *Primate Evolution*. New York: Cambridge University Press, 1986. This account of the evolution of the order Primates, which includes hominids and pongids (the modern apes) as well as other species, will aid the interested student in untangling the often confusing theories concerning the emergence of human species from earlier forms. The authors draw heavily on material from the "Ancestors" symposium for their account of hominid evolution. Contains many illuminating photographs and illustrations.

Feder, Kenneth L., and Michael Alan Park. *Human Antiquity*. Mountain View, Calif.: Mayfield, 1989. An excellent account of the current state of knowledge concerning human origins and early hominid forms in terms understandable by most people. Contains excellent photographs and illustrations and an outstanding bibliography. The authors' interpretations were influenced quite obviously by the papers presented at the symposium.

Reichs, Kathleen J., ed. *Hominid Origins*. Washington, D.C.: University Press of America, 1983. The confusion within and without professional paleoanthropological circles concerning human origins before the "Ancestors" symposium will become readily apparent to the readers of this book.

Weaver, Kenneth F. "The Search for Our Ancestors." *National Geographic* 178 (November, 1985): 561-629. One of the best popular treatments of ancient hominids and human evolution available. Replete with the splendid photographs and illustrations that readers of National Geographic have come to expect. Inspired and heavily influenced by the "Ancestors" program.

Paul Madden

Cross-References

Boule Reconstructs the First Neanderthal Skeleton (1908), p. 428; Zdansky Discovers Peking Man (1923), p. 761; Dart Discovers the First Recognized Australo-

THE BRITISH ANTARCTIC SURVEY CONFIRMS
THE FIRST KNOWN HOLE IN THE OZONE LAYER

Category of event: Earth science
Time: 1985
Locale: Halley Bay, Antarctica

In 1985, scientists of the British Antarctic Survey reported the first long-term trend data in total column ozone over Antarctica, confirming the presence of major ozone depletion in the stratosphere

Principal personages:
JOSEPH C. FARMAN, an English meteorologist
F. SHERWOOD ROWLAND (1927-   ), a physical chemist who was the codiscoverer of the chlorofluorocarbon/ozone effect
MARIO JOSÉ MOLINA (1943-   ), a physical chemist who worked with Rowland on the chlorofluorocarbon/ozone effect

Summary of Event

As has been long known from astronomical observations, stratospheric ozone acts as an effective shield or filter to screen out most short wavelength ultraviolet radiation in sunlight. Ozone is formed in the stratosphere by reaction of atomic oxygen with diatomic molecular oxygen via photolysis driven by absorption of solar ultraviolet radiation. As a trace element, ozone accounts for only about 0.0001 percent of all oxygen in the earth's atmosphere, with the highest (only marginally larger) concentrations occurring at altitudes between 12 and 25 kilometers.

Although it was only in the late 1960's that wider scientific and public attention was focused on ozone in connection with the greenhouse effect and degradation of the stratospheric ultraviolet, the first studies and scientific conferences on the nature of the ozone layer date to the late 1920's. The French physicist Marie-Alfred Cornu in 1878 first suggested that ultraviolet portions of the solar spectrum lost in telescopic observations was caused by chemical absorption by the earth's atmosphere. In 1880, specific ozone bands were discovered, extending from nearly 2,100 to 3,200 angstroms. In 1917, Lord Rayleigh's experiments gave conclusive proof of ultraviolet radiation absorption by the "ozonosphere." Thereafter, many questions arose regarding the precise height and geographic and temporal variations of the stratospheric ozone layer. Lord Rayleigh's observations of the spectrum of the rising and setting sun initially located most ozone between 40 and 60 kilometers above sea level. In 1921, Charles Fabry's spectrophotometric tests and balloon sondeing in 1934 refined this estimate to nearly its present-known limits.

Many researchers and government and public interest leaders since the late 1960's feared that even slight changes in the ozone layer, caused by industrial pollution, would result eventually in large changes in the amount of damaging ultraviolet radiation at the earth's surface. Increasing medical evidence substantiated that more than 90 percent of skin cancer is associated with solar exposure in the ultraviolet radiation region. In addition to eye disease, increases in exposure to ultraviolet radiation have been widely reported to be detrimental to plant and marine biota growth and development. In response to these concerns, beginning in 1975 and subsequently, the National Research Council of the American National Academy of Sciences initiated a number of reports on the topic of how changes in stratospheric ozone concentrations may affect public health and welfare.

In the 1975 paper by F. Sherwood Rowland and Mario José Molina ("ChloroFluoro-Methanes in the Environment," *Reviews of Geophysics and Space Physics* 13, pages 1-36), scientific and public attention was focused to the specific modes of chemical catalysis by which industrial emissions enter and remain in the upper atmosphere to break down ozone concentrations. Because of their great chemical stability, chlorofluoromethanes (and chlorofluorocarbons such as Freon) are not removed from the atmosphere after discharge until they diffuse to stratospheric or even tropospheric altitudes. A typical chlorofluorocarbon molecule survives for approximately seventy-five years in the atmosphere before being decomposed by sunlight into its constituent chlorine atoms. Free chlorine is released in the presence of ultraviolet radiation. Free chlorine and nitrous oxide enter into one or more catalytic cycles where even small chlorofluoromethane concentrations can destroy large ozone quantities. Rowland and Molina argued that adding man-made substances that destroy ozone create new (im)balances between a complex and often delicate network of production and removal processes that govern total ozone abundance.

Since the late 1950's, surface and atmospheric ozone measurements had been conducted sporadically in connection with local pollution studies and more continually at Antarctic and Texas monitoring stations. Measurements of the total quantity of ozone above a unit area of the earth's surface (called total column ozone) is the standard reference measure for assessing ozone levels. Early measurements were made using potassium iodide based electrochemical sensors, not always of certain reliability. From about 1975, however, most ozone measurements have been undertaken using ultraviolet photometry.

At first, it was recognized that a dominant feature of near-surface ozone distribution is its annual variations caused by seasonal circulation cycles. The circumpolar regions were known to be highly variable in upper atmospheric conditions, so although some prior studies had suggested small ozone depletion levels over Antarctica, these were neither a matter of experimental concern nor factored into numerical models of atmospheric chemistry. Shortly after Rowland and Molina's work, researchers (1976) published in *Science* a report on an apparent eleven-year variation in polar ozone and stratospheric ion chemistry, attempting to relate changes in concentration level with natural changes in solar ultraviolet radiation output.

Following the El Chichón (Mexico) 1982 volcanic eruption, a 40 percent increase in stratospheric column-free chlorine was detected between 20 and 40 degrees latitude, which was then proposed as having possible effects on local decreases in ozone concentration. In 1984, it was reported that extremely low total ozone amounts of the Japanese Syowa Research Station in Antarctica existed from September to October in 1982 and 1983, based on ozone balloon sonde observations. Likewise, another study in 1984 reported 125 ozone-sonde results for the period from 1966 to 1971, which showed broad ozone maxima in surface and altitude readings, together with a long winter minimum at higher levels. Nothwithstanding the meteorologic interest in stratospheric ozone depletion at circumpolar latitudes (since the poles are far from pollution sources), because of the error bars and apparent rarity in space and time of these observations, these reports apparently generated little interest in the scientific community.

As Joseph C. Farman recounted for the English journal *New Scientist* (November 12, 1987, pages 50-54), as early as 1982, the British Antarctic Survey had noticed anomalous depletions in upper-layer ozone over their coastal base at Halley Bay, which had begun ozone monitoring in 1957. In particular, the measured ozone abundance over sectors of the Antarctic measured in 1985 were only half those in 1975. Because of the age of their measuring instruments, Farman returned the Dobson spectrophotometer to England for checking and recalibration, and shortly thereafter had additional experts fly to Antarctica to double-check the findings. Even with new instruments, the results were confirmed, and published in the journal *Nature* on May 16, 1985.

Initial scientific reactions were largely skeptical. One major objection was that Farman's data represented only point (and not wider areal) measurements. Another reason for disbelief was the fact that neither the precision Total Ozone Mapping Spectrometer and the less-accurate Solar BackScattering Ultraviolet Recorder aboard the U.S. *Nimbus*-7 satellite, nor the ground station at McMurdo Sound, had shown any excessive ozone depletion. Nevertheless, when a NASA (National Aeronautics and Space Administration) meteorology group reexamined the latter data bases, it was discovered that the satellites had recorded the same ozone hole as the English, but had misprocessed the data because the computer algorithms had been flagged to disregard "unrealistic" results. Further independent studies were carried out in 1985 by East German scientists, confirming that the ozone hole occurred over almost half of Antarctica in winter and above the Weddell Sea in early spring.

Impact of Event

The sudden discovery of a widescale major ozone "sink" at the South Pole was unexpected by the atmospheric and environmental sciences communities and their computer models. Major short period changes in stratospheric ozone concentration implied to many that the ozone layer was somehow being influenced by one or more processes not previously recognized or accounted for. As a direct result of scientific and governmental concern, early in 1986, independent studies set forth theoretical proposals for ozone-destroying chemistry involving stratospheric polar clouds that could act to accelerate ozone depletion. With high-altitude winter temperatures below -85 degrees Celsius, scarce water vapor over Antarctica can condense to form thin, high ice clouds. It was proposed that the ice surfaces of these clouds could notably facilitate conversion of relatively inert forms of chlorine into active chlorine.

During the astral spring of 1986, the United States National Science Foundation funded the first national ozone experiment, including a fully instrumented ozone station at McMurdo Sound, and high-altitude ER-2 and DC-8 flights from Punta Arenas, Chile, over Antarctica to heights of up to 19 and 10 kilometers, respectively. The abundance of chlorine monoxide at 18 kilometers measured by the ER-2 was sufficient alone to account for the observed rate of ozone decrease. During the second national ozone experiment in spring, 1987, it was found that the ozone hole not only had become deeper but also was associated with unusually high levels of chlorine dioxide inversely correlated with ozone levels. The 1987 South Pole ozone hole showed more than 97 percent ozone depletion at several altitudes between 12 and 20 kilometers, with most ozone disappearance occurring within a few weeks during late August and September, reaching a minimum value in mid-October. Independent measurements by other teams from East Germany, England, and Japan confirmed these findings, consistent with the postulated ice catalytic reactions. For the first time, from another experiment there was also indications of an analogous though lesser low-ozone zone over the North Pole between February and March.

Theoretical models and experimental observations have given largely different dependences of the concentration of chlorine on altitude in the upper stratosphere, a particularly critical discrepancy since these are heights at which ozone is most sensitive to perturbations caused by chlorofluorocarbon pollution. Most computer models of the stratosphere predict that the largest expected reductions in atmospheric ozone caused by chlorofluorocarbons should occur near altitudes of about 40 kilometers, from a localized (polar) region. Although agreement between observed and computed values of free chlorine atomic oxygen is adequate for heights below 35 kilometers, discrepancies remain at higher altitudes, as well as with 1989 satellite data showing the South polar ozone hole's size as being nearly twice that of the Antarctic continent.

Most of the physics involved in these atmospheric chemistry prediction algorithms is reasonably well understood. Nevertheless, large error bars for concentrations of some chemical species remain a problem, because the more than thirty-year database of largely ground-based measurements gives inadequate resolution and is subject to other uncertainties. Concentrations of chlorofluorocarbons and nitrous oxide appear to be increasing on a global basis at rates of 1 to 2 percent to more than 7 percent per year, but the precision and geographic validity of these rates are the subject of much debate. Further concerns with modeling are the large number and complex interlinkages and feedback loops for many processes. These make it difficult, for example, to sort out unambiguously the real long-term trends in stratospheric ozone loss caused by cumulative industrial pollution from fluctuations in natural pollution caused by volcanic dust and from the quasi-biennial oscillations caused by perturbations in polar atmospheric flow patterns.

A technique for analyzing trace gases in air bubbles trapped in polar ice offers the opportunity of accurately determining past atmospheric gas concentrations for periods from ten to thousands of years, and may be able to improve inputs to atmospheric chemistry models. Initial results show that present excess concentrations in Antarctic ice cores are caused entirely by modern industrial emission. The question of improving the chemistry and thermohydrodynamics of the computer models themselves, however, is a more difficult problem. Rowland and Molina suggested in 1975 that an atmospheric catastrophe is not only possible but also possible within a surprisingly short period of time. It remains to be seen whether the 1985 ozone hole discovery was made in sufficient time to understand and reverse the phenomenon.

Bibliography

Berger, Wolfgang H., and Laurent D. Laberyrie, eds. *Abrupt Climatic Change: Evidence and Implications*. Boston: D. Reidel, 1987. A broad account. Includes a critical assessment of atmospheric numerical modeling capabilities.

Fisher, David E. *Fire and Ice: The Greenhouse Effect, Ozone Depletion, and Nuclear Winter*. New York: Harper & Row, 1990. Places the polar ozone hole into the total context of long-term catastrophic effects of air pollution.

Hare, Tony. *The Ozone Layer*. New York: Watts, Franklin, 1990. A somewhat alarmist and selective view. Offers a good overall treatment for the general reader in nontechnical language. Illustrations.

Jones, Robin Russell, and Tom Wigley, eds. *Ozone Depletion: Health and Environmental Consequences*. New York: A. R. Liss, 1989. An accurate discussion of human medical consequences of ultraviolet radiation.

Roan, Sharon. *Ozone Crisis: The Fifteen Year Evolution of a Sudden Global Emergency*. New York: Wiley, 1989. The history of stratospheric ozone depletion.

Worrest, Robert C., and Martyn M. Caldwell, eds. *Stratospheric Ozone Reduction, Solar Ultraviolet Radiation, and Plant Life*. New York: Springer-Verlag, 1986. Examines possible effects of increased ultraviolet radiation on natural and agricultural growth and mutation.

Zerefos, C. S., and A. Ghazi. *Atmospheric Ozone*. Boston: D. Reidel-Kluwer, 1985. One of the most comprehensive collections documenting the state of knowledge immediately before the ozone hole discovery.

Gerardo G. Tango

Cross-References

Bjerknes Publishes the First Weather Forecast Using Computational Hydrodynamics (1897), p. 21; Teisserenc de Bort Discovers the Stratosphere and the Troposphere (1898), p. 26; Steinmetz Warns of Pollution in *The Future of Electricity* (1908), p. 401; Fabry Quantifies Ozone in the Upper Atmosphere (1913), p. 579; Callendar Connects Industry with Increased Atmospheric Carbon Dioxide (1938), p. 1118; Tiros 1 Becomes the First Experimental Weather Reconnaissance Satellite (1960),

CONSTRUCTION OF THE WORLD'S LARGEST TELESCOPE BEGINS IN HAWAII

Category of event: Astronomy
Time: 1985
Locale: Mauna Kea, Hawaii

Using a new technology of mirror construction and computer alignment, the telescope provided a quantum leap in the abilities of ground-based, visible-light astronomy

Principal personages:
JAMES ROGER PRIOR ANGEL (1941-   ), a British-born engineer, astronomer, and major innovator in telescope mirror fabrication
JACQUES BECKERS (1934-   ), a Dutch-born astronomer in charge at various times of major American and European telescope development programs

Summary of Event

In 1985, a newly organized consortium called the California Association for Research in Astronomy (CARA) announced plans of radical design for a large new telescope facility, to be located on the slopes of the Mauna Kea volcano in Hawaii, about 4,100 meters above sea level. The site offered some of the best observing conditions in the world. A gift of $70 million from the W. M. Keck Foundation provided major funding for construction.

Since the mid-1970's, when Soviet engineers decided to build a reflecting telescope with a mirror diameter of almost 600 centimeters, prevailing opinion had been that the size of ground-based optical telescopes had reached practical limits. The Soviet mirror was not very successful. Conventional methods of mirror preparation, using fused silicates for material and precision machinery to grind the parabolic surface on the mirror, impose problems that compound themselves as mirror diameter increases and simply became impossible for the ponderous Soviet telescope, which weighs more than 50 tons. The previous record holder for size, the 508-centimeter Hale reflector at Mount Palomar Observatory in California, completed in 1948, had demanded many technical innovations, including patterns of wafflelike indentations to reduce the enormous weight of the single piece of Pyrex comprising the mirror. It required several attempts to manufacture a usable blank mirror disk for Hale and months of grinding and polishing of the parabolic surface resulting in wastage of tons of extremely expensive Pyrex.

The engineers of the Keck telescope promised to overcome the physical and financial obstacles to larger mirrors with new engineering. Instead of a single mirror, the Keck instrument would align thirty-six hexagonal mirrors, each approximately 1.8 meters in diameter, into a single array with light-gathering power equal to a single mirror with a diameter of 10 meters (roughly four times the gathering power of the

Hale telescope). The mirrors will be arranged in a mosaic pattern reminiscent of a honeycomb (or, as one engineer noted, old-fashioned bathroom floor tiles). Positioning of each mirror segment will be controlled by computers throughout observing sessions, so that they focus their reflected light on a single, secondary mirror as a composite image. Computers also continuously adjust thirty-six support points under each mirror segment, as well as alignment of the telescope support structure, to minimize thermal and gravitational deformation. This technology will permit each segment to be a scant 7.5 centimeters in thickness, almost waferlike by the standards of earlier mirrors of comparable diameter. The weight of all thirty-six segments will be about equal to that of the single primary mirror in the Hale telescope.

The hexagonal shape of the mirror segments raised problems with regard to conventional fabrication, grinding, and polishing techniques designed for circular mirrors. Surfaces of the mirror segments must be configured for specific locations in the array, and differing surface and grinding formulas must be used for particular positions. Engineers met this challenge by developing a process called stressed mirror polishing, which applies predetermined forces to the mirror edges to achieve appropriate levels of distortion during polishing. When the forces are released, the parabolic surface of the mirror relaxes into desired shape. Construction problems of segmented mirrors are more than offset by their advantages over single mirrors. In addition to overcoming the weight problem, damage to the mirror surface may be repaired, or the mirror segment replaced, fairly inexpensively, whereas serious damage to a primary mirror (the size of the one in the Hale instrument) would be a major scientific disaster and almost unthinkably expensive at current prices. Equipment for aluminizing and polishing small mirrors is much less costly and easy to obtain; for the giant primary mirrors of the Hale instrument, customized machinery had to be built. Analysis of the mechanical and structural characteristics of the telescope frame and mounting assembly may be completed, and changes made if required, with only a few of the mirror segments installed.

Technology for the Keck telescope borrows from an earlier experiment called the Multi-Mirror Telescope (MMT), a six-segment mirror installed at the Mount Hopkins Observatory in Arizona in 1979. Announcement of building plans for the Keck Observatory came, in fact, amid a rush of technological developments which collectively breathed new life into ground-based visual astronomy and may hold the promise of achievements undreamed of in the mid-1970's. The major competition to the MMT-Keck technology came from British-born engineer James Roger Prior Angel and his team at the Steward Observatory Mirror Laboratory at the University of Arizona. In 1983, Angel's team began work on a series of new techniques for manufacturing reflector mirror disks. They were convinced that rigid mirrors much larger than 5 meters could be fabricated successfully and that, in the long run, their simplicity would be far more desirable in isolated observatories than the extremely complex computer programs and analog devices required for segmented mirror systems.

Traditionally large telescope mirrors are made from fused quartz or silicate glass. Angel's team decided to use borosilicate, a Pyrex-like material, in molten form.

Ventilating devices were installed in the furnace to remove fumes as the material heated to minimize the danger of impurities turning up in the finished mirror blank, a problem that had turned others away from borosilicate material. The first attempt at a borosilicate mirror was a 3.5-meter experiment, in which a revolving furnace heated the material to molten form over a honeycomb support structure of carefully machined fiberboard hexagonal cores. Spinning the furnace as it held molten material allowed the team to create the parabolic surface for the mirror by centrifugal force. (This process also allows deeper and more accurate parabolic surfaces than may be obtained by conventional grinding processes. Since a deeper parabolic curve means a shorter telescope focal length, the technique should allow larger mirrors to be installed in observatories of modest size.)

By 1989, the Angel team had cast a 6.5-meter blank destined to replace the six-mirror MMT array on Mount Hopkins. Encouraged by these early results, the National Science Foundation elected to provide major funding. Barring unforeseen difficulties with the process, mirrors 8 meters in diameter could be ready for installation by the mid-1990's.

Another initiative came from the Dutch-born astronomer Jacques Beckers, founding director of the MMT project at Mount Hopkins. Under his direction, plans developed for a massive project called the National New Technology Telescope (NNTT), in which the experience gained with MMT would be applied to a foursome of 8-meter mirrors possibly cast by the Angel method. Such an instrument would increase the light-gathering capacity of ground-based astronomy by an order of magnitude over that of the Hale telescope. Unable to persuade federal officials to fund NNTT adequately, Beckers resigned from the program in April, 1988, to join the European Southern Observatory (ESO) project at La Silla in Chile. Here, another New Technology Telescope was in the works, a 3.6-meter testbed for a gargantuan Very Large Telescope (VLT) project with a mirror diameter of 16 meters. The VLT will have a light-gathering capacity very close to that of the NNTT design for which Beckers could not obtain funding in the United States. The ESO project borrowed yet another innovation in telescope design developed in the 1980's by a research team at the University of Texas. Known as "active optics," the system depends upon a flexible rather than a rigid mirror —a useful analogy would be a soft rather than a hard contact lens—with very extensive computer mediation controlling large numbers of adjustable pads at frequent intervals on the back surface of the mirror. The flexible mirror, like the Angel mirrors, can be thinner and lighter than conventional mirrors of the Hale variety.

Impact of Event

In the last quarter of the twentieth century, ground-based visual astronomy, widely perceived in 1975 as a technological dead end and an increasingly marginal pursuit, instead embarked upon a period of unprecedented technological progress and daring engineering. As the Keck Observatory becomes fully operational, larger mirrors are produced by the Angel Team and construction begins on the European VLT. The public will become aware of astronomical breakthroughs unlike any since early in the twentieth century. (Many of the new technologies are transferable to orbiting platforms, such as the Hubble Space Telescope stationed in 1990.)

The stiff competition among nations and research teams presumably will establish new conventions and standards of large telescope manufacture and may result also in changed world leadership in astronomical sciences. Despite the considerable economies realized by new methods of mirror manufacture and fabrication, major observatory projects have become extremely expensive. Computer support, global communication networks, staff salaries, and the great difficulties of getting personnel and materials to increasingly far-flung observatory sites, all confront governments and scientific groups with a challenge. The good fortune of the Keck Observatory project in obtaining private funding is the exception rather than the rule. National governments must be prepared to fund these projects or risk falling behind the breakneck pace of the space sciences.

The stakes could not be higher in terms of knowledge to be gained. The largest telescopes of mid-century design, with considerable ingenuity, can detect objects some 8-10 billion light-years distant from Earth. New instruments, such as the Keck telescope, should be able to detect phenomena up to 15 billion light-years or more from Earth. Thus, they will bring astronomers to the very frontiers of the universe, in the sense that light observed from these immensely distant sources must have been emitted at around the time of the beginning of the universe. The new telescopes promise to be not only instruments of observation but also potential keys to the most basic questions of cosmology.

Bibliography

Asimov, Isaac. *Eyes on the Universe: A History of the Telescope.* Boston: Houghton Mifflin, 1975. Useful general work. A classic statement of the opinion, common among astronomers and science writers in the mid-1970's, that ground-based visual astronomy had reached its practical limits on mirror size and sensitivity and that the future of visual astronomy lay only in space.

Cornell, James, and John Carr, eds. *Infinite Vistas: New Tools for Astronomy.* New York: Charles Scribner's Sons, 1985. Written in a conversational style by a number of respected general science authors, this work explores both space and ground-based technologies, including new types of mirror fabrication and how they work together to open new paths for research. Suggestions for further reading are very well chosen.

Faber, S. M. "Large Optical Telescopes—New Views into Space and Time." *Annals of the New York Academy of Sciences* 422 (1982): 171-179. Summarizes technological developments that revolutionized prospects for ground-based astronomy and contrasts relative convenience of even the most complex earthbound instruments.

Fischer, Daniel. "A Telescope for Tomorrow." *Sky and Telescope* 78 (September, 1989): 248-252. Synopsis of the 3.6-meter New Technology Telescope at the European Southern Observatory in Chile and how its technology relates to the multi-faceted attempts to construct ever larger telescopes.

Fisher, Arthur. "Spinning Scopes: Making Larger Reflecting Telescope Mirrors in a Rotating Furnace." *Popular Science* 231 (October, 1987): 76-81. Summary of the Angel technology for mirror fabrication, how it has revolutionized the outlook for firm mirrors, and possible future projects.

Henbest, Nigel. "The Great Telescope Race." *New Scientist* 119 (October 29, 1988): 52-59. General discussion of the profusion of large-mirror telescope technology in the 1980's. Speculates that, within twenty years, one or another of the technologies may emerge as a standard.

Krisciunas, K. *Astronomical Centers of the World.* Cambridge, England: Cambridge University Press, 1988. An important and useful survey of the large number of new or rapidly expanding observatories in the late twentieth century, including extensive coverage of developments in the Southern Hemisphere. Treats "seeing" conditions at such locations as Cerro Tololo and Mauna Kea and the growing threat of light and atmospheric pollution near older centers in the American Southwest.

McCoy, Jan. "Angel Builds 'Em Bigger." *Sky and Telescope* 76 (August, 1988): 128-129. Excellent, illustrated synopsis of the Angel process for mirror fabrication using a spinning furnace and honeycombed compartments.

Nelson, Jerry. "The Keck Telescope." *American Scientist* 77 (March/April, 1989): 170-176. Progress report on the great segmented-mirror telescope at Mauna Kea announced in 1985 as the world's largest light-gathering surface.

Sinnott, Roger W. "The Keck Telescope's Giant Eye." *Sky and Telescope* 80 (July, 1990): 15-22. Summary article with numerous excellent illustrations as telescope neared completion. Includes discussion of preliminary testing using four mirror segments.

Ronald W. Davis

Cross-References

Hale Establishes Mount Wilson Observatory (1903), p. 194; Hale Oversees the Installation of the Hooker Telescope on Mount Wilson (1917), p. 645; Schmidt Invents the Corrector for the Schmidt Camera and Telescope (1929), p. 884; Reber Builds the First Intentional Radio Telescope (1937), p. 1113; Hale Constructs the Largest Telescope of the Time (1948), p. 1325; NASA Launches the Hubble Space Telescope (1990), p. 2377.

JEFFREYS DISCOVERS THE TECHNIQUE OF GENETIC FINGERPRINTING

Categories of event: Biology and medicine
Time: March 6, 1985
Locale: University of Leicester, Leicester, England

Jeffreys produced "fingerprints" of human DNA, which are completely specific to an individual and can be applied directly to issues of human identification such as establishing family relationships and identifying criminals

> *Principal personages:*
> ALEC JEFFREYS (1950-), an English geneticist, active in biochemical genetics and in human and animal developmental genetics
> VICTORIA WILSON (1950-), a coworker of Jeffreys at the University of Leicester, England
> SWEE LAY THEIN (1951-), a biochemical geneticist at John Radcliffe Hospital, Oxford, England

Summary of Event

In 1985, Alec Jeffreys, a geneticist at the University of Leicester in England, developed a method of deoxyribonucleic acid (DNA) analysis that provides a visual representation of the human genome. Jeffreys' discovery had an immediate, revolutionary impact on problems of human identification, especially the identification of criminals. Whereas earlier techniques, such as conventional blood typing, provide evidence that is merely exclusionary, DNA fingerprinting provides positive identification. For example, under favorable conditions, the technique can establish with virtual certainty whether a given individual is or is not a murderer or a rapist. The applications are not limited to forensic science; DNA fingerprinting can also establish definitive proof of parenthood (paternity or maternity), and it is invaluable in providing markers to map disease-causing genes on chromosomes. In addition, the technique is utilized by animal geneticists to establish paternity and to detect genetic relatedness between social groups.

DNA fingerprinting (also referred to as genetic fingerprinting) is a sophisticated technique that must be executed carefully to produce valid results. The technical difficulties arise partly from the complex nature of DNA. DNA, the genetic material responsible for heredity in all higher forms of life, is an enormously long, double-stranded polymeric molecule composed of four different units called bases. The bases on one strand of DNA pair with complementary bases on the other strand. A human being contains twenty-three pairs of chromosomes; one member of each chromosome pair is inherited from the mother, the other from the father. Each chromosome has a continuous stretch of double-stranded DNA containing 50 million to 500 million base pairs. The order, or sequence, of bases forms the genetic message, called the genome. Scientists did not know the sequence of bases in any sizable stretch of DNA prior to the 1970's because they lacked the molecular tools to cleave DNA into fragments that could be analyzed. This situation changed with the advent of biotechnology in the mid-1970's.

The door to DNA analysis was opened with the discovery of bacterial enzymes called DNA restriction enzymes. A restriction enzyme binds to DNA whenever it finds a specific short sequence of base pairs (analogous to a code word), and it cleaves DNA at a defined site within that sequence. A single enzyme finds millions of cutting sites in human DNA, and the resulting fragments range in size from tens of base pairs to hundreds or thousands. The fragments are separated in order of their length by a process called gel electrophoresis, in which an electrical field separates smaller and larger DNA fragments on a porous gel. Fragment size can be estimated by comparing the position in a gel of a fragment with that of marker DNA pieces of known size. The fragments are transferred from the gel to a membrane filter, where they are exposed to a radioactive DNA probe, which can bind to specific complementary DNA sequences in the fragments. X-ray film sandwiched to the membrane detects the radioactive pattern. The developed film, called an autoradiograph, shows a pattern of DNA fragments, which is similar to a bar code and can be compared with patterns from known subjects. Only fragments that bind labeled probe DNA can be seen in the autoradiograph; the other fragments are invisible.

The uniqueness of a DNA fingerprint depends on the fact that, with the exception of identical twins, no two human beings have identical DNA sequences. Of the 3 billion base pairs in human DNA, many will differ from one person to another. Many of these sequence variations result from base pair changes that create or destroy a cleavage site for a restriction enzyme and thereby cause variation in fragment length from one individual to the next. The variation can be detected by the use of labeled probe DNA sequences that are complementary to sequences located near restriction cutting sites. This type of probe reveals whether a certain cleavage site is present or absent in a given sample of DNA. Because cleavage sites are either present or absent, however, this type of probe is only moderately sensitive, and the resulting fragment band pattern, or autoradiograph, will not be absolutely unique for an individual.

In 1985, Jeffreys and his coworkers, Victoria Wilson at the University of Leicester and Swee Lay Thein at the John Radcliffe Hospital in Oxford, discovered a vastly more powerful type of probe, powerful enough to produce a DNA fingerprint. Jeffreys had found previously that human DNA contains many multirepeated minisequences called minisatellites. Minisatellites consist of sequences of base pairs repeated in tandem, and the number of repeated units varies widely from one individual to another. Every person, with the exception of identical twins, has a different number of tandem repeats and, hence, different lengths of minisatellite DNA. By virtue of this difference, DNA fragments, which vary in length correspondingly, can be generated by cleaving DNA at a restriction site close to minisatellite DNA. The complexity of minisatellite DNA, and the multiple repeat options of the tandem sequences, enable probes to minisatellites to be very sensitive in differentiating one person from another. By the use of two labeled probes to detect two different minisatellite sequences, Jeffreys obtained a unique fragment band pattern that was completely specific for an individual.

The power of the technique derives from the law of chance, which indicates that the probability (chance) that two or more unrelated events will occur simultaneously is calculated as the multiplication product of the two separate probabilities. Jeffreys used two different probes to produce autoradiographs, which contained a total of thirty-six significant bands. He calculated that the probability of two unrelated people having an identical band was one in four, or one-fourth; hence, the chance of two people having thirty-six identical bands is the fraction one-fourth multiplied by itself thirty-six times. The resulting fraction is extremely small—less than 1 in 10 trillion. Given the population of the world, it is obvious that the technique can distinguish a person from anyone else in the universe. Jeffreys called his band patterns DNA fingerprints because of their ability to individualize. As he stated in his landmark research paper, published in the English scientific journal *Nature* in 1985, probes to minisatellite regions of human DNA produce "DNA 'fingerprints' which are completely specific to an individual (or to his or her identical twin) and can be applied directly to problems of human identification, including parenthood testing."

Impact of Event

The impact of genetic fingerprinting was immediate and broad-ranging. Genetic fingerprinting provides a powerful method for establishing family relationships in paternity (and occasionally maternity) disputes. It allows parenthood to be established with an extremely high level of certainty, vastly greater than that obtained using conventional genetic marker tests. DNA fingerprints can establish whether the purported relationship is actual by testing whether all of the child's DNA fragments are present in the claimed mother's and/or father's DNA fingerprint. The method was used within a few months of its discovery in an English immigration dispute to prove that a boy from Ghana was the son, and not the nephew or other close relative, of a Ghanian woman resident in England.

The technique has had a revolutionary effect on forensic science and law. Police authorities have hailed its ability to offer positive proof of identity. Whereas other forensic tests, conducted on the biological evidence (such as semen, blood, or hair) that is usually found at the scene of a crime, can identify a suspect with only 90 percent to 95 percent certainty, DNA fingerprinting can rule out everyone else in the world as the possible perpetrator of the crime.

DNA fingerprinting has found wide application in medical genetics. In the search for a cause, diagnostic test, and ultimately treatment of an inherited disease, it is necessary to locate the defective gene on a human chromosome. Gene location is accomplished by a technique called linkage analysis, in which geneticists use marker sections of DNA as reference points to pinpoint the position of a defective gene on a chromosome. The minisatellite DNA probes developed by Jeffreys provide a potent and valuable set of markers that are of great value in locating disease-causing genes. Soon after its discovery, DNA fingerprinting had been used to locate the defective genes responsible for several diseases, including fetal hemoglobin abnormality and Huntington's disease.

Genetic fingerprinting also has had a major impact on genetic studies of higher animals. Because DNA sequences are conserved in evolution, humans and other vertebrates have many sequences in common. This commonality enabled Jeffreys to use his probes to human minisatellites to bind to the DNA of many different vertebrates, ranging from mammals to birds, reptiles, amphibians, and fish; he thereby produced DNA fingerprints of these vertebrates. In addition, the technique has been used to discern the mating behavior of birds, to determine paternity in zoo primates, and to detect inbreeding in imperiled wildlife. DNA fingerprinting can also be applied to animal breeding problems, such as the identification of stolen animals, the verification of semen samples for artificial insemination, and the determination of pedigree.

The technique is not foolproof, however, and results may be far from ideal. Especially in the area of forensic science, there was a rush to use the tremendous power of DNA fingerprinting to identify a purported murderer or rapist, and the need for scientific standards was often neglected. Although trial judges admitted DNA fingerprinting as evidence on the grounds that the methods are "generally accepted in the scientific community," some problems arose because forensic DNA fingerprinting in the United States is generally conducted in private, unregulated laboratories. In the absence of good scientific controls, DNA fingerprint bands between two completely unknown samples cannot be matched precisely, and the results may be unreliable. In a 1987 murder case, DNA fingerprinting evidence was discounted on the basis that the analysis had been poorly performed. Problems can also arise in the interpretation of data. These difficulties are not insoluble, however, and DNA fingerprinting is expected to have increasing applications to biochemical genetics and to general problems involving the establishment of human identity.

Bibliography

Baum, Rudy M. "Genetic Screening: Medical Promise Amid Legal and Ethical Questions." *Chemical and Engineering News*, August 7, 1989, 10-16. This general science article discusses legal, ethical, and social issues raised by the acquisition and use of genetic information.

Jeffreys, Alec J. "High Variable Minisatellites and DNA Fingerprints." *Biochemical Society Transactions* 15, no. 3 (1987): 309-317. The text of Jeffreys' Colworth Medal Lecture; provides a comprehensive discussion of DNA fingerprinting and its applications. Geared to the scientifically knowledgeable reader.

Jeffreys, Alec J., Victoria Wilson, and Swee Lay Thein. "Individual-Specific 'Fingerprints' of Human DNA." *Nature* 316 (July 4, 1985): 76-79. This scientific research article is geared for the geneticist or biochemist. Describes the fundamental work of Jeffreys and coworkers in precise, technical language.

Lewin, Roger. "DNA Fingerprints in Health and Disease." *Science* 233 (August 1, 1986): 521-522. Explains in nontechnical terms the scientific background of Jeffreys' technique and its various applications to forensic science, to establishing parenthood, and to mapping the location of defective, disease-causing genes on human chromosomes.

_____. "Limits to DNA Fingerprinting." *Science* 243 (March 14, 1989): 1549-1551. A scientific news article that discusses limitations in the application of DNA fingerprinting to animal behavior research.

Marx, Jean L. "DNA Fingerprinting Takes the Witness Stand." *Science* 240 (June 17, 1988): 1616-1618. This scientific news article written in nontechnical language discusses the technique of DNA fingerprinting and its application to the identification of murderers and rapists.

Neufeld, Peter J., and Neville Colman. "When Science Takes the Witness Stand." *Scientific American* 262 (May, 1990): 46-53. A general science article discusses problems associated with admitting forensic testimony (including DNA fingerprinting) into criminal cases.

Thompson, William C., and Simon Ford. "Is DNA Fingerprinting Ready for the Courts?" *New Scientist* 125 (March 31, 1990): 38-43. Geared to the general reader, this article discusses the scientific background of DNA fingerprinting and problems associated with the interpretation of forensic evidence.

Thornton, John I. "DNA Profiling: New Tool Links Evidence to Suspects with High Certainty." *Chemical and Engineering News*, November 20, 1989, 18-30. This article is geared to the reader having a general background in science. Provides a good discussion of the applications of DNA fingerprinting and other biological techniques to forensic science.

White, Ray, and Jean-Marc Lalouel. "Chromosome Mapping with DNA Markers." *Scientific American* 258 (February, 1988): 40-48. A general science article discussing the use of DNA markers, including Jeffreys' markers, to map chromosomes and trace defective genes.

Maureen S. May

Cross-References

Watson and Crick Develop the Double-Helix Model for DNA (1951), p. 1406; Kornberg and Coworkers Synthesize Biologically Active DNA (1967), p. 1857; Cohen and Boyer Develop Recombinant DNA Technology (1973), p. 1987; Berg, Gilbert, and Sanger Develop Techniques for Genetic Engineering (1980), p. 2115; Erlich Develops DNA Fingerprinting from a Single Hair (1988), p. 2362.

THE TEVATRON PARTICLE ACCELERATOR
BEGINS OPERATION AT FERMILAB

Category of event: Physics
Time: October, 1985
Locale: Fermi National Accelerator Laboratory, Batavia, Illinois

The Tevatron at Fermilab generated collisions between beams of protons and antiprotons at the highest energies ever recorded

Principal personages:
> ROBERT RATHBUN WILSON (1914-   ), an American physicist and director of Fermilab from 1967 to 1978 who supervised the superconducting magnet project that forms the heart of the Tevatron
> JOHN PEOPLES (1933-   ), an American physicist and deputy director of Fermilab (beginning in 1987) who worked on the colliding beam experiment design group and solved the problems of generating intense beams of antiprotons
> CARLO RUBBIA (1934-   ), an Italian physicist and head of the European Laboratory for Particle Physics (CERN) who was involved in the early planning that led to the use of colliding beams in particle physics

Summary of Event

The Tevatron is a particle accelerator, a large electromagnetic device used by high-energy physicists to generate subatomic particles at sufficiently high energies to explore the basic structure of matter. The Tevatron is a circular, tubelike track 6.4 kilometers in circumference that employs a series of superconducting magnets to accelerate beams of protons, which carry the positive charge in the atom, and antiprotons, the proton's negatively charged equivalent, at energies up to 1 trillion electronvolts—equal to 1 teva-electronvolt (TeV), hence the name Tevatron. An electronvolt is the unit of energy that a unit charge, such as an electron, gains through an electrical potential of 1 volt.

The Tevatron is located at the Fermi National Accelerator Laboratory, also known as Fermilab. The laboratory was one of several built in the United States during the 1960's. Most of the accelerator studies prior to World War II were conducted at the University of California's Radiation Laboratory, now the Lawrence Berkeley Radiation Laboratory in honor of Ernest Lawrence's contributions to the study of nuclear physics, which include his design of the cyclotron particle accelerator. After the war, the European Laboratory for Particle Physics (CERN) at Geneva, Switzerland, was organized. CERN has been in existence since 1952 and operates as a consortium of cooperating Western European governments. Fermilab was conceived at least partly in response to competition from CERN, as well as domestic competition from the

Berkeley Laboratory and the Brookhaven National Laboratory, also established after World War II. Fermilab was established in 1967 and operates as a consortium of fifty-one United States and one Canadian university known as the Universities Research Association. Funding for design and construction came largely from the U.S. Department of Energy.

The heart of the original Fermilab was the 6.4-kilometer main accelerator ring. This main ring was capable of accelerating protons to energies approaching 500 billion electronvolts, or 0.5 TeV. The idea to build the Tevatron grew out of a concern for the millions of dollars spent annually on electricity to power the main ring, the need for higher energies to explore the inner depths of the atom and the consequences of new theories of both matter and energy, and the growth of superconductor technology. Planning for a second accelerator ring, the Tevatron, to be installed beneath the main ring began in 1972.

Robert Rathbun Wilson, Fermilab director at that time, realized that the only way the laboratory could achieve the higher energies needed for future experiments without incurring intolerable electricity costs was to design a second accelerator ring that employed magnets made of superconducting material. Extremely powerful magnets are the heart of any particle accelerator; charged particles such as protons are given a "push" as they pass through an electromagnetic field. Each successive push along the path of the circular accelerator track gives the particle more and more energy. The enormous magnetic fields required to accelerate massive particles such as protons to energies approaching 1 trillion electronvolts would require electricity expenditures far beyond Fermilab's operating budget. By using superconducting materials, which have nearly no resistance to electrical current, Wilson estimated that the Tevatron could achieve double the main ring's magnetic field strength, doubling energy output without a significant increase in energy costs.

The superconducting magnets are constructed of twenty-three thin strands of copper wire, within which twenty-one hundred filaments made of a niobium-titanium alloy are embedded. Such alloys are known to exhibit superconducting properties. During construction, Fermilab was the world's largest single consumer of these materials. To construct the magnets for the Tevatron, Fermilab obtained almost 50 tons of this material. This is more than 30 million meters of superconductor, nearly enough to circle the earth. More than one thousand magnets were constructed of this material, 774 to guide the protons along the track and 240 to focus the beams. The magnets are cooled with liquid helium, which is produced by the largest liquid helium plant in the world.

The Tevatron was conceived in three phases: the Energy Saver, Tevatron I, and Tevatron II. In the energy-saving mode, protons are generated in the main ring at an energy of 150 billion electronvolts and subsequently transferred to the Tevatron, where they are accelerated to 500 billion electronvolts. Recall that this is the maximum energy at which the older main ring could accelerate protons. In this new mode, the protons can be accelerated to the same energy in the Tevatron at a greatly reduced cost. From the Tevatron, these proton beams are directed to experimental areas, where they are aimed at a series of targets and detectors. The collision of these beams and the targets generates other, more interesting particles, such as neutrinos, which are investigated by the sophisticated electronics embedded in the detectors.

Most important, however, were Tevatron I and Tevatron II, where the highest energies were to be generated and where it was hoped new experimental findings would emerge. Tevatron II experiments were designed to be very similar to other proton beam experiments, except that in this case, the protons would be accelerated to an energy of 1 trillion electronvolts. At this energy, Fermilab physicists believed that the neutrinos produced by the interaction of the proton beam and suitable targets or more slowly moving proton beams would be far more energetic than naturally occurring neutrinos or those from other, less powerful accelerators. More important still are the proton-antiproton colliding beam experiments of Tevatron I. In this phase, counterrotating beams of protons and antiprotons are induced to collide in the Tevatron, producing a combined, or center-of-mass, energy approaching 2 trillion electronvolts, nearly three times the energy achievable at the largest accelerator at CERN.

John Peoples was faced with the problem of generating a beam of antiprotons of sufficient intensity to collide efficiently with a beam of protons. Knowing that he had the use of a large proton accelerator—the old main ring—Peoples employed the two-ring mode whereby 120 billion electronvolt protons from the main ring are aimed at a fixed tungsten target, generating antiprotons, which scatter from the target. These particles were extracted and accumulated in a smaller storage ring. As in the Energy Saver mode, these particles could be accelerated to relatively low energies. After sufficient numbers of antiprotons were collected, they were injected into the Tevatron, along with a beam of protons for the colliding beam experiments.

One problem of accumulating the antiprotons is the fact that as they scatter from the tungsten target, they do so along highly divergent and irregular paths. In order to inject this beam into the smaller storage ring, the divergent beam had to be corrected. The physics of such storage, known as "beam cooling," was developed by the Soviet physicist Gersh Budker. Low-energy electrons are directed parallel to the antiproton beam, where the higher-energy antiprotons transfer some energy to the slower electrons. This process provides a more coherent beam of antiprotons and is also known as emittance reduction. An alternative method employing electronic sensors and correcting signals, known as stochastic cooling, was developed by CERN physicist Simon van der Meer. CERN director Carlo Rubbia helped to initiate the use of beam cooling for colliding beam experiments at Fermilab which, under Peoples' guidance, employed van der Meer's method. On October 13, 1985, Fermilab scientists reported a proton-antiproton collision with a center-of-mass energy measured at 1.6 trillion electronvolts, the highest energy ever recorded.

Impact of Event

The Tevatron's success at generating high-energy proton-antiproton collisions affected future plans for accelerator development in the United States and offers the potential for important discoveries in high-energy physics of which no other accelerator is capable.

Physics recognizes four forces in nature: the electromagnetic, gravitational, and strong and weak nuclear forces. A major goal of the physics community is to explain all these forces by one theory, the so-called grand unification theory. In 1967, one of the first of the so-called gauge theories was developed that unified the weak nuclear and the electromagnetic forces. One consequence of this theory was that the weak force was carried by massive particles known as "bosons." The search for three of these particles—the intermediate vector bosons W^+, W^-, and Z^0—led to the rush to conduct colliding beam experiments in the early 1970's. The higher energies of the colliding beams are necessary for such searches because of the instability and large mass of these particles. Because the Tevatron was in the planning phase at this time, Rubbia's team at CERN initiated a multi-million-dollar renovation of their own accelerator. The effort culminated in the awarding of the 1984 Nobel Prize in Physics to Rubbia and van der Meer for their discovery of the particles. In 1989, Tevatron physicists reported the most accurate measure to date of the Z^0's mass.

It is widely believed that the Tevatron is the only particle accelerator in the world with the power to conduct further searches for the more elusive Higgs boson, a particle attributed to weak interactions by University of Edinburgh physicist Peter Higgs in order to account for the large masses of the intermediate vector bosons. In addition, the Tevatron has the ability to search for the so-called top quark. Quarks are believed to be the constituent particles of protons and neutrons. Evidence has been gathered for five of the six quarks believed to exist: up, down, charm, strange, and bottom. Physicists have yet to detect evidence of the most massive quark, the top quark. Although Rubbia's intermediate vector boson data had traces of what he and many others believe may point to the top quark, most physicists believe that only the Tevatron has enough energy to isolate one of these particles.

The success of the Tevatron from a technological perspective, with its superconducting magnets accelerating protons to energies only imagined in the early years of American high-energy physics, led the Department of Energy to abandon its funding for the half-completed proton beam collider "Isabelle" at the Brookhaven Laboratory. Isabelle would make possible proton-proton collisions at 200 billion electronvolts. In 1983, the Department of Energy focused its attention on the design and construction of a gigantic collider that came to be known as the Superconducting Supercollider. The Superconducting Supercollider is planned to be 87 kilometers in circumference and will be located in Waxahachie, Texas, south of Dallas. The ten thousand superconducting magnets of the Superconducting Supercollider will accelerate protons to energies forty times those attainable by the Tevatron. If the Tevatron cannot find the Higgs boson and the top quark, then the task will fall to the Superconducting Supercollider when it is completed.

Bibliography

Robinson, Arthur. "Fermilab Tests Its Antiproton Factory." *Science* 229 (September, 1985): 1374-1376. This brief article discusses the antiproton production capabilities of the Tevatron, as well as its proton-antiproton collision capacity. It stresses the technical problems of producing the antiprotons and discusses the various techniques of beam cooling. The article is not technical, yet it offers sufficient detail for the reader who is interested in a general introduction to collider experiments.

_____. "Proton-Antiproton Collisions at Fermilab." *Science* 230 (November, 1985): 529. A brief introduction to the first proton-antiproton collision at Fermilab. The article is entirely nontechnical but describes some of the logistical problems of operating a particle accelerator that is partially under construction and renovation.

Wilson, Robert. "The Next Generation of Particle Accelerators." *Scientific American* 242 (January, 1980): 42-57. This relatively technical and copiously illustrated article discusses the rationale behind the new generation of particle accelerators, of which the Tevatron is one. It discusses both scientific and engineering details and the role of multinational and public sponsorship of accelerator construction. Offers an especially useful section that illustrates the physics of beam cooling.

_____. "The Tevatron." *Physics Today* 30 (October, 1977): 23-27. Wilson, director of Fermilab during the Tevatron's design and construction, discusses the technical and scientific details of the Tevatron with a special emphasis on the superconducting magnet technology. This is the best article for the individual interested in learning more about the technical details of how the Tevatron works; the article presents ample technical detail and is accompanied by an extensive bibliography.

_____. "U.S. Particle Accelerators at Age 50." *Physics Today* 34 (November, 1981): 86-103. This article puts the development of the Tevatron into historical context by showing its place in the development of American accelerator technology. Discusses almost every aspect of accelerator technology, from funding to engineering. Wilson acknowledges that this is not a definitive history; the bibliography directs the reader to more detailed sources.

William J. McKinney

Cross-References

Lawrence Develops the Cyclotron (1931), p. 953; Cockcroft and Walton Split the Atom with a Particle Accelerator (1932), p. 978; University of California Physicists Develop the First Synchrocyclotron (1946), p. 1282; Gell-Man Formulates the Theory of Quantum Chromodynamics (QCD) (1972), p. 1966; Georgi and Glashow Develop the First Grand Unified Theory (1974), p. 2014; Rubbia and van der Meer Isolate the Intermediate Vector Bosons (1983), p. 2230; The Superconducting Supercollider Is Under Construction in Texas (1988), p. 2372.

TULLY DISCOVERS THE PISCES-CETUS
SUPERCLUSTER COMPLEX

Category of event: Astronomy
Time: 1986-1987
Locale: Hawaii

Tully mapped a complex of galaxy superclusters more than 1 billion light-years in diameter, possibly the largest structure in the observable universe

Principal personages:

R. BRENT TULLY (1943-   ), the optical astronomer at the University of Hawaii's Institute of Astronomy who proposed the phenomenon of galactic supercluster complexes

J. RICHARD FISHER (1943-   ), the radio astronomer at the National Radio Astronomy Observatory in Greenbank, West Virginia, who collaborated with Tully in mapping clusters and superclusters

GÉRARD HENRI DE VAUCOULEURS (1918-   ), a specialist in large-scale structure in the universe who compiled catalogs of galaxies

GEORGE O. ABELL (1927-1983), the astronomer who systematically surveyed groups of galaxies

Summary of Event

R. Brent Tully announced in December, 1987, that he had found a plane of galaxies occupying about one-tenth of the visible universe, a structure larger than any previously suspected. This Pisces-Cetus supercluster complex posed a major problem for the principal theory about the origin and development of the cosmos: It seemed to conflict with a major piece of evidence supporting the theory, which suggested that the distribution of galaxies in the universe should be fairly uniform. Tully's analyses showed otherwise; the universe, he said, is "lumpy."

Twentieth century astronomers have observed ever larger groupings of stars, which theorists have struggled to explain. In the 1920's Edwin Powell Hubble proved that certain "spiral nebulas" lie outside the Milky Way, can be larger than our galaxy, and are moving away at very high velocities. His observations and calculations revealed that the universe not only is much larger than thought at the time but also is expanding—to the surprise of cosmologists, who assumed the universe to be static. To account for this expansion, many astronomers accepted various versions of the big bang theory that Georges Lamaître formulated in 1927. Lemaître proposed that an explosion of the "primordial atom" sent matter hurling in all directions, matter that coalesced into the galaxies of the modern universe. Logically, such an explosion would fill the universe evenly in all directions.

Evidence began to mount even in the 1930's that a perfectly even distribution of matter is not the case. Fritz Zwicky's observations first demonstrated that galaxies tend to group together, and subsequent optical and radio astronomy surveys of the sky have discerned ever larger collections of galaxies bound together by gravity. Astronomers now widely accept that galaxies gather in clusters of about 30 million light-years in diameter. (A light-year is the distance light travels in one year, or about 10 trillion kilometers.) Clusters, in turn, often form structures a hundred million light-years in diameter, called superclusters, or line up in long chains. The Milky Way, for example, is part of a modest group of galaxies near the Virgo cluster, which is the center for the local supercluster. Furthermore, in 1981, scientists announced the discovery of voids as large as superclusters between groups of galaxies.

Tully was among the astronomers who investigated large-scale structures in the universe during the 1970's and 1980's and was a leading advocate for the existence of the local supercluster, along with Gérard Henri de Vaucouleurs and Antoinette de Vaucouleurs, despite general resistance to the idea. Identification of such structures depended upon mapping the vast region beyond the Milky Way, and Tully's interest in the project began in 1972 as he and J. Richard Fisher were completing their doctoral work at the University of Maryland. They collaborated in a project to survey all nearby galaxies, and in the course of their work developed a method to measure distances to spiral galaxies (one of several distinct galaxy classes) based upon a correlation between a galaxy's brightness and the width of the neutral hydrogen line in the spectrum of its light. This "Tully-Fisher relation" is now a widely accepted intergalactic yardstick.

Tully prepared the first prototypes of ten maps while working at the Observatoire de Marseille in France, and revised prototypes were made at the National Radio Astronomy Observatory's drafting department in Charlottesville, Virginia. Meanwhile, Tully and Fisher continued to map the skies, using earlier surveys of galaxy groups by George O. Abell and de Vaucouleurs and drawing data from radio observatories in both the northern and southern hemispheres. Tully moved to the Institute for Astronomy at the University of Hawaii in 1982 and enlisted the help of cartographer Jane Eckelman of Manoa Mapworks in Honolulu in compiling accurate visual representations of the data that were accumulating. This was difficult, Tully and Fisher later wrote, because they still were not sure what they were trying to map; additionally, they wanted to produce complicated three-dimensional graphics to correspond to sectors of the sky.

Their work culminated in 1987 with the publication of the *Nearby Galaxies Catalog* and *Nearby Galaxies Atlas*. While preparing the maps for publication, Tully searched for the edge of the local supercluster. It became increasingly apparent that the supercluster was far larger than had been suspected. He examined densely populated (or "rich") clusters in the Northern Hemisphere, their motion relative to one another, and their distances from one another, and then used a supercomputer to plot their distribution from various perspectives. He found that the clusters appeared to lie in a plane and that this extension of the local supercluster corresponded to a plane of rich clusters in the southern galactic hemisphere centered about 650 million light-years from the Milky Way in the Pisces-Cetus region of the sky. The chance that this correlation of flattened distribution was an accident of random motion, his analyses suggested, was statistically small. Tully concluded that the planes of clusters must be elements of the same structure. In the *Nearby Galaxies Atlas,* Tully and Fisher call this structure the Pisces-Cetus supercluster complex, comprising the Pisces-Cetus supercluster, the Perseus-Pegasus chain, the Virgo-Hydra-Centarus supercluster, the Pegasus-Pisces chain, and the Sculptor region. The entire complex is about 1 billion light-years in length and 150 million light-years in width, making it the largest single structure in the observable universe. Additionally, Tully and Fisher list four other supercluster complexes: in the Hercules-Corona Borealis, Aquarius, Ursa Major, and Leo sectors.

Tully published his findings in two articles in *The Astrophysical Journal* (April 1, 1986, and December 1, 1987). By the time the second article appeared, science writers in newspapers and magazines already had heralded the discovery, emphasizing that it did not seem to fit established concepts of the universe. Tully concluded the December article by pointing out that his discovery posed an obvious challenge to most popular theories of galaxy formation.

Nevertheless, the supercluster complex was not an established fact. Tully acknowledged that the evidence from his surveys was suggestive but not conclusive. The center of the Pisces-Cetus supercluster complex, for example, lies in the southern galactic hemisphere, and knowledge of clusters in that area is still relatively sketchy; furthermore, the center of the Milky Way blocks direct observation of a section of the structure. Therefore, Tully's map of the Pisces-Cetus plane, as he puts it, is full of holes. He had detected an alignment of planes rather than an outright connection. Still, he insisted that the alignment looks too exact to be dismissed as chance.

Astronomers and astrophysicists reacted with interested skepticism to Tully's conclusions. Well-known deficiencies exist in some of the data on which he depended (for example, Abell's catalog of clusters), and some insisted that there simply was not enough knowledge about very large-scale structures in the universe to distinguish actual structures from the illusory effects of imperfect observation. Yet, none was willing to dismiss Tully's work, and further surveys of galaxies have also suggested structures larger than superclusters.

Impact of Event

By the mid-1980's, the big bang theory had undergone major modifications, but a central feature of the theory still predicted that matter in the modern universe should be homogenous in distribution because matter in the early universe expanded at a rapid, uniform rate. Radio astronomers uncovered dramatic supporting evidence for the theory in the 1960's, when they detected background radiation left over from the big bang; the radiation reaches Earth with almost exactly the same intensity from all directions, as would be expected if it were produced by a single explosion. Cosmologists were able to accommodate structures the size of clusters of galaxies in the big bang theory, but when voids and superclusters were found, they had increasing difficulty explaining how the cosmic background radiation could be uniform, while matter was clumping in large, intricate structures. Tully's supercluster complexes presented a new order of difficulty and further observational evidence that the universe is not homogenous. These discoveries inspired theorists to hypothesize novel, sometimes bizarre physical phenomena—based upon both astrophysical and particle physics theories—that could cause galaxies and their groupings to form.

The most widely accepted theory was that soon after the big bang, the universe underwent a sudden, temporary inflation, often called the "cosmic burp." This sent gravity rippling through space that created density fluctuations in the universe's matter and fostered the formation of structures. Another theoretical addition suggested that the light-producing and light-reflecting matter that is observed accounts for only about 10 percent of the total matter of the universe. The "missing matter" is cold and dark—invisible to contemporary instruments—and spread evenly through the cosmos: Visible matter is like froth on top of dark beer, one astrophysicist remarked. Yet, a third theory proposed the existence of extremely thin, high-density, high-energy tubes of space-time that drift through the universe. These "cosmic strings" can be either infinitely long or closed in loops, and their strong gravity could draw matter into galaxies and clusters. Like most astronomers, Tully found aspects of all these theories useful in explaining the results of his observations and measurements.

The proliferation of theories to reconcile observed phenomena to physical laws, as they are understood, often precedes a breakthrough in science. The discovery of increasingly large structures in the universe—Tully's supercluster complexes among them—led some observers to speculate that science may be on the verge of a new basic understanding of the environment as revolutionary as Albert Einstein's theory of general relativity (1916) or Edwin Powell Hubble's proof that the universe is expanding (1929).

Bibliography

Bartusiak, Marcia. *Thursday's Universe: A Report from the Frontier on the Origin, Nature, and Destiny of the Universe.* New York: Times Books, 1986. An entertaining, lucidly written account of astronomical knowledge and theory up to 1986. Although Tully's findings on supercluster complexes were published after the release of the book, Bartusiak introduces readers to related phenomena, such as voids and bubbles, thoroughly and comprehensibly. Probably the best book for newcomers to astrophysics.

Cornell, James, ed. *Bubbles, Voids, and Bumps in Time: The New Cosmology.* New York: Cambridge University Press, 1989. Six articles on developments in cosmology and astrophysics. Margaret J. Geller's essay on mapping the universe is particularly appropriate because it discusses a study of galaxies much like Tully's, as is Alan H. Guth's essay on cosmological theories and how they grope with the problem of structure in the universe. The essays are all readable, often amusing, and accessible to readers who are willing to stretch their imaginations.

Greenstein, George. *The Symbiotic Universe: Life and Mind in the Cosmos.* New York: William Morrow, 1988. A speculative, provocative contemplation as well as a book about science. Greenstein attempts to relate contemporary discoveries about the universe to such large questions as Where did life come from? and Is there a plan behind it all? In the course of his ruminations, he also explains in a loose but clear style astrophysical matters, such as theories about galactic clustering.

Gregory, Stephen A. "The Structure of the Visible Universe." *Astronomy* 16 (April, 1988): 42-47. An astronomer whose specialty is the universe's large-scale structure, Gregory writes clearly for the general reader about the methods of mapping voids, superclusters, and complexes of superclusters and the difficulties in obtaining sufficient data. An excellent, concise background article.

Hodge, Paul W. *Galaxies*. Cambridge, Mass.: Harvard University Press, 1986. With a wealth of photographs and illustrations, Hodge explains galactic evolution, structures, types, clustering, and spacing. He also discusses the Milky Way and nearby galaxies, all in a pleasing, informative style. Useful for background information. Unfortunately, supercluster complexes are not mentioned.

Tully, R. Brent. "More About Clustering on a Scale of 0.1 c." *The Astrophysical Journal* 323 (December 1, 1987): 1-18. The primary scientific vehicle for Tully's conclusions concerning supercluster complexes. Only a reader trained in astrophysics will follow the arguments fully, but a college-level knowledge of astronomy is sufficient to understand portions of it.

Tully, R. Brent, and J. Richard Fisher. *Nearby Galaxies Atlas*. Cambridge, England: Cambridge University Press, 1987. A beautiful collection of color-coded, foldout, two- and three-dimensional maps of galaxies and clusters in a very large format. This book is indispensable in helping one to visualize the structures of superclusters and supercluster complexes.

Roger Smith

Cross-References

Leavitt's Study of Variable Stars Unlocks Galactic Distances (1912), p. 496; Slipher Obtains the Spectrum of a Distant Galaxy (1912), p. 502; Hubble Demonstrates That Other Galaxies Are Independent Systems (1924), p. 790; Lemaître Proposes the Big Bang Theory (1927), p. 825; Hubble Confirms the Expanding Universe (1929), p. 878; Gamow and Associates Develop the Big Bang Theory (1948), p. 1309; De Vaucouleurs Identifies the Local Supercluster of Galaxies (1953), p. 1454.

BEDNORZ AND MÜLLER DISCOVER
A HIGH-TEMPERATURE SUPERCONDUCTOR

Category of event: Physics
Time: January, 1986
Locale: IBM Zurich Research Laboratory, Rüschlikon, Switzerland

Bednorz and Müller discovered superconductivity in a new class of compounds at a significantly higher temperature, setting off a flurry of further research

Principal personages:

J. GEORG BEDNORZ (1950-   ), a German-born physicist whose work was supervised by Karl Alexander Müller and cowinner of the 1987 Nobel Prize in Physics

KARL ALEXANDER MÜLLER (1927-   ), a Swiss-born physicist, Fellow at IBM's Zurich Research Laboratory, and cowinner of the 1987 Nobel Prize in Physics

HEIKE KAMERLINGH ONNES (1853-1926), a Dutch physicist and discoverer of the phenomenon of superconductivity who was awarded the 1913 Nobel Prize in Physics

Summary of Event

Superconductivity, the disappearance of all resistance to the flow of electricity in a material, has challenged experimentalists and theoreticians alike since its discovery in 1911. In that year, Heike Kamerlingh Onnes noted that mercury lost its resistance to the conduction of electricity completely when it was cooled to 4 Kelvins. Despite widespread attempts in the years after 1911, experimentalists could not find substances that would superconduct at temperatures more than a few tenths of degrees above absolute zero. By the early 1970's, the highest recorded temperature at which any material was found to superconduct was 23.3 Kelvins. Theoreticians did not come up with a plausible theory for superconductivity until the late 1950's. That theory, developed by John Bardeen, Leon N Cooper, and John Robert Schrieffer, could not predict which compounds would superconduct.

The reason for the search was simple: The promise of technological advancement and economic and social benefits from superconducting devices was great. Superconducting compounds appeared so promising because they carried electrical currents indefinitely, while normal conductors of electricity, such as copper wire, offered some resistance to the flow of electricity and dissipated much of the power they carried in the form of heat. The promise of cheap, efficient, resistanceless power transmission was only one of the reasons that physicists were so keen to find superconductors that operated at higher temperatures than the frigid regions of the bottom end of the temperature scale.

One of the more important reasons that superconductors had not lived up to their potential was that large, very expensive liquid helium cooling systems were required to keep the superconducting material operating. When the superconductor warmed up to a point above a "transition temperature," the material would revert to its normal state of conductivity. Other reasons were that many could not remain superconducting and still carry the large electrical currents or withstand the high magnetic fields required for many applications. Raising the temperature at which materials would superconduct was the major focus of research for decades.

Over the years, physicists tried intermetallic compounds (combinations of several metals), thin films, and organic compounds in their search for superconductors that would operate at temperatures higher than 23.3 Kelvins. In the 1960's, an oxide (a compound containing oxygen and other chemical elements) was shown to superconduct, but its transition temperature (less than 1 Kelvin) was so low that it was of no practical use. The importance of the discovery was that a new class of compounds— the oxides—was shown to have the potential of superconductivity. The early 1970's saw oxide superconductors achieve the respectable transition temperature of 14 Kelvins—still below the record high transition temperature but high enough to dispel the notion that high-temperature oxide superconductors could not exist.

Karl Alexander Müller was convinced that metallic oxides were promising materials to test for high-temperature superconductivity, in part because the IBM Research Laboratories had a long tradition of research in these oxides. In this context, "high temperature" was taken to be any temperature above 23.3 Kelvins. In early 1983, Müller enlisted J. Georg Bednorz's assistance in the search. Bednorz fabricated a variety of samples of a lanthanum, nickel, and oxygen compound, altering the relative amounts of the constituent elements in hopes of creating a compound that would superconduct. Carefully measured amounts of powdery oxides containing the necessary chemical elements were combined with citric acid and ethylene glycol (automobile antifreeze) and then heated to 8,316 degrees Celsius. The black powdery residue was then compressed under high pressure into a pellet and allowed to cool slowly. Neither the chemicals nor the equipment needed to fabricate the samples was particularly costly or difficult to obtain.

Measurement after measurement on sample after sample demonstrated that Bednorz and Müller were not making any real progress toward their goal of achieving superconductivity at temperatures above 23.3 Kelvins. They could only make measurements in the evening because they had to share equipment with another group at the laboratory. Work on their project almost came to an end, but in 1985 they gained sole use of automatic measuring equipment. At the end of 1985, Bednorz read an article by three French physicists concerning the interesting properties of a lanthanum-barium-copper-oxide compound. The French physicists were not looking for superconductivity; they were concerned with other properties of the compound. Bednorz realized that including copper into their compound theoretically should increase the chances that the compound would superconduct.

Bednorz quickly fabricated a sample of the material. Measurements made on the mixtures of the lanthanum-barium-copper-oxide made in late January of 1986 showed the onset of superconductivity at around 10 Kelvins. Slight alterations in the relative amount of the constituents of the compound yielded transition temperatures of 35 Kelvins. Bednorz and Müller repeated the measurements and confirmed that their superconductor had surpassed the twelve-year-old record by 12 Kelvins. Bednorz and Müller were excited because they had raised significantly the highest temperature at which superconductivity occurred and demonstrated that a new class of compounds could superconduct.

Knowing that any announcement of a new record for the transition temperature of a superconducting metallic oxide would be greeted with skepticism, Bednorz and Müller wanted to be absolutely sure of their results. There could be no doubt that their compound was superconducting if they could demonstrate the Meissner-Ochsenfeld effect—the expulsion of the magnetic field from a compound. The equipment needed to perform these magnetic measurements was as yet unavailable at the IBM Research Laboratory. They decided to publish a cautiously titled paper ("Possible High Superconductivity in the Ba-La-Cu-O System") to establish their priority. Bednorz and Müller spent the time before the publication of their paper attempting to discover the detailed structure of their lanthanum-barium-copper-oxide compound. They were assisted in this work by Masaaki Takashige, a visiting scientist from Japan. After the equipment needed to demonstrate the Meissner-Ochsenfeld effect arrived, Takashige assisted Bednorz in making the measurements. Their compound exhibited the Meissner-Ochsenfeld effect. Bednorz and Müller's paper appeared in September, 1986.

Impact of Event

Bednorz and Müller were met initially with skepticism as they gave presentations at scientific meetings, despite the compelling evidence of the existence of the Meissner-Ochsenfeld effect. This period of skepticism was very short-lived, however. In late November, 1986, a group of scientists in Japan reported that they were able to reproduce Bednorz and Müller's results. Soon after, Paul C. W. Chu at the University of Houston also confirmed the existence of a 35-Kelvin superconducting oxide compound.

One measure of the excitement that this discovery engendered in the scientific community was the large number of scientific papers that flooded scientific journals. In 1987, the year that Bednorz and Müller's results became widely known in the scientific community, more than three thousand papers appeared in the scientific literature concerning superconductivity from laboratories all over the world.

The race was on to identify the superconductor's structure so that even higher temperatures could be achieved through further altering of the compound. Chu raised the transition temperature to more than 50 Kelvins in December, 1986, by applying intense pressure to the compound. He also substituted the rare-earth element yttrium for the lanthanum in the original compound. The transition temperature reached an amazing 93 Kelvins. Besides the great jump in the record temperature, the significance of Chu's discovery was that liquid nitrogen, which was cheap and easy to use, could be used as a coolant. The dream of room-temperature superconductors now seemed much closer.

The high point of Bednorz and Müller's short-term impact came in March, 1987. Scientists were coming up with so many results so fast that scientific journals could not keep up. A special session of the American Physical Society's annual solid-state physics meeting was arranged so that the scientists could share their results. An hour before the session was scheduled to begin, more than three thousand scientists were waiting to enter the hall. Television monitors were set up outside to broadcast the session to the overflow crowd. Dozens of scientists presented their results; presentations continued until after 3:00 A.M. newspapers dubbed the meeting "the Woodstock of physics."

Bednorz and Müller's discovery envigorated the scientific community's search for compounds that superconduct at higher temperatures: They demonstrated that superconductivity could occur in oxides (which are usually electrical insulators at room temperature) at record high temperatures. They also refuted the growing belief that perhaps superconductivity cannot occur above 23 Kelvins. Chu substituted one chemical element in Bednorz and Müller's compound for another and raised the transition temperature to the point where inexpensive and easy-to-use liquid nitrogen could be used for cooling. As a result, Bednorz and Müller were awarded the 1987 Nobel Prize in Physics for their contributions to the field of superconductivity.

Bibliography

Hazen, Robert. *The Breakthrough: The Race for the Superconductor.* New York: Summit Books, 1988. Presents an "insider's view" of the scientific process. The first part recounts the discovery of the 93-Kelvin superconductor. The second part is a firsthand, day-by-day account of the attempts by Hazen and his colleagues to identify the superconducting compound. Vividly portrays the sense of excitement in the scientific community after the discovery of high-temperature superconductors.

Langone, John. *Superconductivity: The New Alchemy.* Chicago: Contemporary Books, 1989. Readable and nontechnical account of superconductivity's history, applications, and future. Written by a longtime science writer, the book focuses on the technical innovations expected from high-temperature superconductivity. Includes glossary and index.

Mayo, Jonathan L. *Superconductivity: The Threshold of a New Technology.* Blue Ridge Summit, Pa.: TAB Books, 1988. Uses straightforward, nontechnical language to explain how superconductivity works and how superconductors are constructed. Describes the applications of superconductor technology. Many illustrations and a good bibliography.

Prochnow, Dave. *Superconductivity: Experimenting in a New Technology.* Blue Ridge Summit, Pa.: TAB Books, 1989. Attempts to make superconductivity accessible to all readers by thoroughly covering all aspects of superconductivity, yet it claims it is not an elementary treatise on superconductivity. Appendices include directions for setting up a demonstration experiment of high-temperature superconductivity, a list of suppliers of necessary materials for the experiment, and a list of books and articles on superconductivity.

Simon, Randy, and Andrew Smith. *Superconductors: Conquering Technology's New Frontier.* New York: Plenum Press, 1988. Intended for readers who have no prior background in physics, electronics, or other pertinent fields. Explores the nature of superconductivity, its history, and current theoretical understanding of the phenomenon. Surveys a wide variety of practical uses for superconductivity. Discusses the recent breakthroughs in superconductivity and evaluates their impact on future technology.

Roger Sensenbaugh

Cross-References

Bardeen, Cooper, and Schrieffer Explain Superconductivity (1957), p. 1533; Esaki Demonstrates Electron Tunneling in Semiconductors (1957), p. 1551; The Superconducting Supercollider Is Under Construction in Texas (1988), p. 2372.

THE FIRST PERMANENTLY MANNED
SPACE STATION IS LAUNCHED

Category of event: Space and aviation
Time: February 20, 1986
Locale: Tyuratam Cosmodrome, Soviet Union

Building upon progress in long-duration flights on Salyut missions, the Soviet Union demonstrated a permanent manned presence on the Mir space station

Principal personages:
LEONID KIZIM (1941-   ), a veteran cosmonaut who commanded the Soyuz T-15 mission to occupy initially the Mir space station
VLADIMIR SOLOVYEV (1946-   ), a Soyuz T-15 cosmonaut who was instrumental in setting up initial habitation of Mir
YURI ROMANENKO (1944-   ), a Soyuz TM-2 cosmonaut
ALEXANDER LAVEIKIN (1951-   ), a Soyuz TM-2 cosmonaut
VLADIMIR TITOV (1947-   ), a Soyuz TM-4 cosmonaut
MUSA MANAROV, a Soyuz TM-4 cosmonaut
ANATOLI LEVCHENKO (?-1988), a Soyuz TM-4 cosmonaut

Summary of Event

While the American space shuttle program was grounded in the wake of the 1986 *Challenger* accident, the Soviet Union proceeded to a new plateau in space station development. Mir (meaning "peace" or "commune" in Russian) was launched from the Tyuratam Cosmodrome on February 20, 1986. Mir represented a third-generation Soviet space station design. Like Salyut 6 and 7, Mir had a docking port at each end. Unlike the second-generation space stations, Mir also had four extra docking ports arranged radially at the forward end just behind the forward axial docking port. Otherwise, Mir was quite similar to Salyut 6 and 7. Nevertheless, this new space station was an evolutionary design, for the Mir space station module was meant to serve as the core for attachment of specialized modules.

Because the heavy-lift Energia booster was not ready to support Mir operations in the second half of the 1980's, Mir was designed for launch by the Proton booster. As a result, the maximum diameter (4.15 meters), length (13.1 meters), and mass (20,000 kilograms) of Mir were virtually identical to the Salyut 6 and 7 space stations. The heavy multiple-docking adapter at the front of Mir forced Soviet mission planners to launch Mir nearly devoid of scientific equipment and without all essential systems and supplies for long-duration habitation. Mir was designed to be heavily supported by initial manned flights, unmanned Progress freighters, and the addition of specialized modules. When launched, the original plan was to dock four additional large modules to Mir before 1990. Those plans were eventually delayed for technological and economic reasons.

Mir was given a number of improved systems that expanded its capabilities beyond those of Salyut 6 and 7. Among these were a capability for routing communications and telemetry data transmission through a communications satellite system at geosynchronous altitude, a new docking system (Kurs) that did not require the space station to be oriented facing spacecraft on final approach during rendezvous maneuvers, a lightweight manipulator arm designed to maneuver add-on modules from one forward docking port to another, a gallium arsenide solar array system capable of generating 9 kilowatts of usable electrical power, and a new computer control system requiring less human input.

When launched, the majority of Mir's interior consisted of living space. Few scientific experiments could be performed initially, as little equipment and apparatus were on board. There were, however, 90 cubic meters of working and living space available inside Mir. The space station was composed of three major parts: the docking and transfer module, the work module, and the crew compartment.

The transfer module served as a hub for traffic into other modules expected to be added to the core Mir space station, and contained an airlock for exiting on extra-vehicular activities (EVA). The primary purpose of the work module was to provide space for experimentation. The crew compartment represented a vast improvement over Salyut designs. For privacy, there were two individual cabins, each equipped with a chair, table/desk, sleeping bag, and a window for a soothing view of Earth. For food preparation, Mir had a galley complete with a refrigerator and a pair of food warmers. A folding table and removable chairs were available for eating meals. A repair shop was equipped with special zero-gravity tools for station maintenance that could be used on EVAs as well. For personal hygiene, there was a zero-gravity toilet, makeshift shower unit, and wash station. Stowed beneath the crew compartment floor was a bicycle ergometer and treadmill for crew exercise and relaxation. The interior of the space station was organized in such a way as to preserve a natural sense of up and down and was painted in pleasing soft pastel colors.

Mir was inserted into a 172-by-301-kilometer orbit. Maneuvers were executed that placed Mir in the same orbital plane as the abandoned Salyut 7 space station. This action led Western observers to speculate that Salyut 7 and Mir were about to be docked together. Those speculations were dashed when Leonid Kizim and Vladimir Solovyev were launched on March 13, 1986, in Soyuz T-15 and rendezvoused and docked to Mir at the forward axial port. Kizim and Solovyev activated the life-support equipment and set up housekeeping on Mir. Soyuz T-15 carried some experimental equipment that the cosmonauts installed and tested. Further equipment, in addition to propellant, food, water, and air supplies, was delivered by the unmanned Progress 25 freighter that docked at the aft port on March 21. Progress 25 also was used to boost Mir to a more stable 336-by-360-kilometer orbit. A second Progress vehicle was dispatched to Mir a month later to complete the refueling of Mir's propulsion system and alter the space station's orbit, making possible a departure of Soyuz T-15 for a trip to Salyut 7.

Kizim and Solovyev undocked from Mir on May 5 and docked to Salyut 7 the next day. The cosmonauts reactivated the station, performed a pair of EVAs to retrieve space-exposed specimens, and removed several pieces of working scientific equipment for use in Mir. On June 25, Soyuz T-15 undocked from Salyut 7, leaving the space station abandoned in orbit. Kizim and Solovyev returned to Mir late the next day and resumed activities to set up Mir for future experimentation and long-duration spaceflights. After some initial scientific work, they prepared Mir for automatic mode and then returned to Earth on July 16. They had spent 125 days in space and lived on two different space stations.

Mir was next inhabited by Yuri Romanenko and Alexander Laveikin when Soyuz TM-2 docked to Mir in the early morning hours of February 8, 1987. This was the first long-duration cosmonaut team assigned to Mir. During their stay on Mir, Romanenko and Laveikin were visited by seven unmanned Progress freighters, the Soyuz TM-3 international cosmonaut team (Alexander Viktorenko, Alexander Alexandrov, and Mohammed Faris—Syrian) that remained on Mir for six days, and the first expansion module, called Kvant. Kvant ("quantum" in Russian) was a combination astrophysical research laboratory, work area, and space station support module. Kvant was 5.8 meters in length and had a diameter of 4.15 meters. At launch, Kvant carried a total of 1,500 kilograms of support equipment and 2,500 kilograms of consumables and other cargo. Inside the expansion module was an extra 40 cubic meters of working space. Kvant docked to the aft port of Mir after some initial difficulty caused by debris preventing the docking mechanism from latching. Romanenko and Laveikin had to perform an emergency EVA in order to dock successfully Kvant to Mir. Kvant had a duplicate port at its own aft to support docking of both Progress and Soyuz spacecraft. On April 21, Progress 29 docked to the aft port of Kvant and was able to refuel Mir's propellant tanks through special plumbing lines routed through Kvant.

Soyuz TM-3 launched on July 22 in order to pay a visit to Romanenko and Laveikin. Soyuz TM-3 flight engineer Alexandrov replaced Laveikin as Romanenko's companion on the long-duration team. Laveikin, Viktorenko, and Faris returned to Earth in the older Soyuz TM-2 spacecraft, leaving the fresher Soyuz TM-3 for the resident crew.

Cosmonauts Vladimir Titov, Musa Manarov, and Anatoli Levchenko were launched on December 21, 1987, in Soyuz TM-4. Their mission was to replace Romanenko and Alexandrov and then remain on board Mir for an entire year. When this team took over residence of Mir, it marked the first time one space station crew was completely exchanged for another without any lapse in habitation, thus demonstrating a permanent occupation of space.

Impact of Event

One of the National Aeronautics and Space Administration's (NASA) earliest space program design studies included the development of a manned space station in low Earth orbit as a platform for scientific research in a weightless environment. Political considerations in the Cold War technological competition between the United States and the Soviet Union led to the abandonment of space station development in lieu of deep space exploration, including manned lunar landings. When the Soviet Union lost the Moon race, the direction of its space program was changed toward the development of Earth-orbiting space stations, a goal often touted as a more important one than exploration of the Moon. The Salyut program was the test-bed for the development of a permanent presence in low Earth orbit. Initial efforts in the Salyut program were geared toward upstaging the efforts of NASA Skylab astronauts.

Salyut 1 was launched on April 19, 1971, and was home to a trio of cosmonauts for twenty-four days, a record-setting endurance, in June of that year. The mission restored much of the diminished Soviet pride in its space program until the tragic loss of these cosmonauts during reentry, when a valve unexpectedly bled cabin atmosphere rapidly into space. Soyuz 11 landed safely, but its crew died. After a detailed investigation of the Soyuz 11/Salyut 1 accident, the space station program renewed its efforts to be successful prior to the launching of Skylab. Several attempts to orbit a Salyut 2 failed. Salyut 3, the first successful Soviet space station, was launched on June 25, 1974, well after the Skylab program had concluded. This space station was primarily a military version and was home to only a single crew of two cosmonauts who stayed on board for sixteen days.

Salyut 4, a civilian version, was launched on December 26, 1974. This space station was home to a pair of crews that lived for twenty-nine and sixty-three days in space before returning safely to Earth. During this mission, an automated docking of an unmanned Soyuz spacecraft to Salyut 4 was accomplished, setting the stage for the Progress freighters that would resupply second-generation Salyut space stations. Salyut 5, the final military version, was launched on June 22, 1976. This space station was home to two crews whose residence on orbit lasted forty-nine and eighteen days.

Salyuts 1 through 5 were not meant for repeated use and had no significant capability for repair and/or resupply. The Soviet Union's second-generation space stations, Salyut 6 and 7, were equipped with a pair of docking ports, one front and one aft. Salyut 6 was launched on September 29, 1977. Before Salyut 6 was deorbited destructively (in July, 1982), it was home to numerous cosmonaut teams for a total of 676 days. Long-duration flights were incrementally increased from 96 days to a record 185 days. Lessons about physiological and psychological adaptation to long-duration exposure to life in an orbiting space station were learned and a confidence in operations grew. Progress and Soyuz resupply flights restocked consumables and propellants and delivered new scientific equipment, samples, and film. Salyut 7 operations (1982 to 1986) increased endurance to 237 days. When abandoned in orbit, Salyut 7 was replaced by the third-generation Mir, a space station with six docking ports and a capability for module expansion and permanent occupation.

Bibliography

Hooper, Gordon R. *The Soviet Cosmonaut Team: Comprehensive Guide to the Men and Women of the Soviet Manned Space Programme.* San Diego: Univelt, 1986.

The subtitle correctly claims this work to be a comprehensive guide to the men and women of the Soviet manned space program. Highly detailed crew biographies, training assignments, and flight assignments. A must reference for serious Soviet spaceflight researchers.

Johnson, Nicholas L. *The Soviet Year in Space: 1987.* Colorado Springs, Colo.: Teledyne Brown Engineering, 1988. Detailed compendium of all Soviet space programs, both manned and unmanned, for 1987. Detailed diagrams, easily understood graphs and charts, and comparative statistics abound. An indispensable reference for the serious Soviet spaceflight researcher.

_____. *The Soviet Year in Space: 1988.* Colorado Springs, Colo.: Teledyne Brown Engineering, 1989. Detailed compendium of all Soviet space programs, both manned and unmanned, for 1988. Detailed diagrams, easily understood graphs and charts, and comparative statistics are included. A necessary reference for the serious Soviet spaceflight researcher.

_____. *The Soviet Year in Space: 1989.* Colorado Springs, Colo.: Teledyne Brown Engineering, 1990. Detailed compendium of all 1989 Soviet space programs, both manned and unmanned. Contains detailed diagrams, graphs and charts, and comparative statistics. An indispensable reference.

Newkirk, Dennis. *Almanac of Soviet Manned Space Flight.* Houston: Gulf, 1990. Thorough yet concise history of the Soviet space efforts focusing on early manned spectaculars, lunar race difficulties, and the development of a permanent manned presence in low Earth orbit. Well researched, with an easy format for both reading and information retrieval.

Oberg, James E. *The New Race for Space.* Harrisburg, Pa.: Stackpole Books, 1984. Written for the layperson interested in comparing the directions of the United States and Soviet space programs in the 1980's. Heavily illustrated and easy reading, but thought-provoking.

U.S. Congress. Senate. Committee on Commerce, Science, and Transportation. *Soviet Space Programs: 1981-1987.* Report prepared by Congressional Research Service, the Library of Congress. 100th Congress, 1988. Committee Print. A two-part document prepared for the 100th Congress. Traces the development of Soviet capabilities in the arena of manned space flight. Particularly strong in detailing the science programs carried out on the Salyut 6 and 7 space stations.

David G. Fisher

Cross-References

Sputnik 1, the First Artificial Satellite, Is Launched (1957), p. 1545; Gagarin Becomes the First Human to Orbit Earth (1961), p. 1693; The First Space Walk Is Conducted from Voskhod 2 (1965), p. 1787; Skylab Inaugurates a New Era of Space Research (1973), p. 1997; *Columbia*'s Second Flight Proves the Practicality of the Space Shuttle (1981), p. 2180; Spacelab 1 Is Launched Aboard the Space Shuttle (1983), p. 2256.

THE CHERNOBYL NUCLEAR REACTOR EXPLODES

Categories of event: Earth science and physics
Time: April 26, 1986
Locale: Ukraine region of the Soviet Union (near Kiev)

A severe nuclear accident occurred in the Soviet Union, killing many people and threatening thousands in several surrounding countries

Principal personages:
> VALERY A. LEGASOV (1936-1988), a Soviet physicist who was one of the chief investigators of the nuclear accident
> ANATOLY DYATLOV, VIKTOR BRYUKHANOV, and NIKOLAI FOMIN, the engineers and plant directors who were convicted for the errors they made that contributed to the accident

Summary of Event

At 1:23 A.M. on Saturday, April 26, 1986, a major accident began at Unit 4 of the Chernobyl Nuclear Power Station about 100 kilometers northwest of Kiev. Over the next nine days, tremendous amounts of radioactivity and radioactive debris were released, endangering tens of thousands of people in the Soviet Union, Northern and Western Europe, and perhaps the whole Northern Hemisphere. Thirty-one workers and firefighters died immediately and more than 240 others sustained severe radiation sickness. The accident has brought significant changes in the nuclear power industry and the cooperation between nations.

Ironically, the accident occurred as a safety test was being conducted just before the plant was to shut down for routine maintenance. The same test had been conducted at other similar reactors in the Soviet Union but had been inconclusive when tried at this plant about a year earlier. A certain amount of pressure had been building to get the test run, since it would be another year before it could be tried again because of the scheduled maintenance. Furthermore, the test had been postponed from earlier in the day because the plant's power had been needed by the electrical system longer than had been thought. It is partly because the plant was undergoing this test that a large amount of technical data were gathered about exactly how the accident occurred. Human and design errors are to blame. Anatoly Dyatlov, Nikolai Fomin, and Viktor Bryukhanov, who were the plant directors and engineers, were sentenced to ten years' imprisonment for their errors that contributed to the nuclear accident.

The test was conducted to see if, upon a sudden and complete emergency shutdown of the plant, sufficient power from the still-spinning turbine would run the plant until the emergency diesel generators could be turned on a few seconds later. As the workers prepared to perform a sudden shutdown of the plant, they also prepared to repeat the experiment again if necessary. This meant that they disconnected several of the safety features that would have meant a longer shutdown of the plant. In bypassing the safety interlocks, they could restart the whole test quickly, if needed. This action was in clear violation of their most fundamental safety rules, but, as they later explained, it was done because they had become complacent with their good safety record.

This style of reactor, called an RBMK-1000 (1,000 megawatts) in the Soviet system, was one of fifteen in use. Two more were under construction at the Chernobyl station at the time; larger ones were under construction elsewhere in the Soviet Union. This reactor has several unique features that contributed to the accident. It is these features that caused the United States to abandon this design for commercial reactors and that caused the English to stop construction of a similar reactor nine years earlier. England warned the Soviet Union that the "positive void coefficient" of this plant presented a serious problem of control. The positive void coefficient in the RBMK reactor was a critical feature that played a major role in the accident. The reactor power in this case increases as the amount of liquid water going through the core of the reactor decreases. This is serious in such a boiling water reactor where the water is vaporized directly inside the reactor. As the water turns into steam, the reactor gets hotter, turning even more water into steam. This calls for very careful control and complicated safety systems. Although difficult to control, it is very efficient. This is the opposite of the behavior in almost all other reactors, especially those in commercial operation in the United States, which have negative void coefficients.

A second design feature contributing significantly to the accident is the fact that the reactor is constructed of graphite. Graphite has several qualities that are good for the nuclear processes occurring, but it is capable of burning. Ultimately, the graphite core of this reactor caught fire and burned. The final feature that distinguishes this reactor from most of the power reactors in the rest of the world is the lack of a containment building. Both pressurized water reactors and boiling water reactors in the United States, Canada, Western Europe, and Japan make use of massive reinforced concrete domes to cover the reactor and main steam generating equipment. These buildings are made of concrete and steel more than 1.2 meters thick. This was not the case at Unit 4 at Chernobyl.

At the RBMK reactor, there was a 1,000-ton cover on the reactor itself, which was easily blown off, and a cover building to house the overhead crane used in refueling operations. While the heavy cover offered some protection, it existed primarily as a radiation shield for the workers who worked on top of the reactor during refueling operations. These RBMK reactors could be refueled while operating, one of their positive features. The building had a ventilation system designed to keep radioactive dust and air inside the building, but it was by no means a containment building. Some of those analyzing the accident several years later have indicated that when the Soviets started building their reactors, they did not have the ability to construct the huge containment buildings that would be needed for these bulky RBMK reactors. As the Soviets gained some experience with these reactors, they did not believe they needed containment buildings. Most analysts are of the opinion that a containment building could have withstood the explosion that rocked Unit 4.

As the operators performed the test that night, they lost control of the reactor; it took off on what nuclear engineers call a prompt power excursion. Without backup safety systems, they could not reinsert control rods quickly nor turn on the emergency core cooling system water. A steam explosion blew the lid off the reactor, fuel rods ruptured, and uncontrolled fission reactions set the reactor on fire, along with surrounding buildings. Brave firefighters extinguished the building fires that night. The adjoining reactors were shut down within the next two days, while a massive burial of the reactor was attempted with helicopters loaded with lead, fire extinguishing chemicals, and concrete. Four days later, however, a heat buildup inside the smothering reactor caused a reeruption of radioactive debris matching the first outbursts. During this time, 135,000 people who lived within 30 kilometers of the plant were evacuated. Eventually, 150,000 people were relocated, some of whom will likely never be able to occupy their former homes.

It is estimated that 3.5 percent of the reactor's fuel and 10 percent of the graphite reactor itself were emitted into the atmosphere, along with significant quantities of radioactive fission products. Because of varying weather conditions over the nine-day span of the accident, the debris was spread first over western Soviet Union and Scandinavia and then the lower Baltic states. Later, the area hardest hit was centered in Austria. The total radioactivity burden added to the world is estimated to have been 4 exabecquerel, or 100 million curies, resulting in 930,000 person-grey, 97 percent of which is confined to western Soviet Union and Europe. The fallout also is estimated to equal the sum of all past atmospheric testing of nuclear weapons.

The spread of the contamination caused a world outcry of alarm. It was not until two days after the explosion that the Soviet government began acknowledging the explosion publicly. Their discussion became quite detailed when Valery A. Legasov led a team of Soviet scientists to a five-day press conference in Vienna, Austria, in August, 1986, to analyze the accident. Their frankness and openness seemed to astound most Western observers.

Impact of Event

For months after the accident, cautious Europeans did their grocery shopping equipped with radiation counters. For years, Laplanders were forbidden to harvest their reindeer herds. Crops in the area around Chernobyl were plowed under. Topsoil was buried near the entombed reactor. By 1988, the English government had reimbursed their farmers $10 million for crops that they could not market because of contamination. Estimates put the cost to members of the European Economic Community as high as $100 billion. Anxiety about future health problems lingers.

The relationship between radiation exposure, especially low levels of radiation, and the likelihood of cancer appearing eventually is very difficult to investigate. Thus, it is not surprising that estimates of the possible death toll vary widely. References that follow testify to the difficulty of making definitive estimates of future deaths. Estimates of the consequences of this accident vary from hundreds of new cancer deaths to half a million. Some believe that more of the deaths will occur in Europe than in the Soviet Union. An authoritative source quoted in *The New York Times* on August 27, 1986, using data supplied by the Soviets in Vienna, noted that most likely twenty-four thousand new cancers would occur in the Soviet Union in the next seventy years as a result of this accident. They also noted that this would mean less than a 0.3 percent increase in the number of cancer deaths that would occur ordinarily and, therefore, would likely go unnoticed.

The accident at Chernobyl solidified increasing distrust of nuclear power in the United States, a distrust that followed the highly publicized incident at the Three Mile Island Power Plant (TMI) in Pennsylvania in 1979. There have been no new nuclear plants ordered in the United States since the TMI reactor core meltdown. Many comparisons have been made between these two accidents. In both, there had been operator complacency, as demonstrated by operators overruling automatic safety systems. Both accidents have resulted in new safeguards and exchanges of technical advice, but now the cooperation is truly international. A worldwide, interconnected system of three hundred atmospheric monitoring stations has been established, meaning a breakthrough in Soviet participation.

Inside the Soviet Union and Eastern bloc countries, the changes in attitudes toward nuclear power are mixed. Several citizen protests have occurred and have been reported widely. Existing RBMK reactors have been retrofitted with operator-proof safety systems or have been closed, especially those near earthquake faults. Yet, new construction of "western-style" reactors occurs as fast as sufficient skilled labor can work. The central Soviet government still maintains that less harm is done to the environment with nuclear power than the burning of fossil fuels. The exportation of coal is important to the Soviet economy, also.

The controversial nature of nuclear power in the Soviet Union will remain so with news that in 1990 nearly a fifth of the radioactivity supposedly trapped in the concrete-entombed reactor rubble had leaked out. Also, the evidence of a controversy is found in the fact that the Soviet Union has designated the Chernobyl area as an international center for the study of radiological effects and that $26 billion would be available for recovery expenses. This occurred at the same time that the Republic of Byelorussia declared 12.3 million acres of the central Ukraine as uninhabitable and estimates that $137 billion more will be needed before twenty-seven hundred towns and villages will be declared secure. In addition, the Ukrainian parliament has demanded that the other three RBMK reactors at Chernobyl be shut down forever.

Bibliography

Ahearne, John F. "Nuclear Power After Chernobyl." *Science* 236 (May 8, 1987): 673-679. Short, balanced, and clear, this article calls for waste handling action, better management, and a major overhaul of RBMK-style reactors.

Gale, Robert P., and Thomas Hauser. *Final Warning: The Legacy of Chernobyl*. New York: Warner Books, 1988. An interesting personal look through the eyes of an American physician called to Chernobyl to help with bone marrow transplants; a hard-line antinuclear stance.

Gould, Jay M., and Benjamin Goldman. *Deadly Deceit: Low-Level Radiation, High-Level Cover-Up*. New York: Four Walls Eight Windows, 1990. A startling book because it attributes everything from AIDS to cancer to low-level radiation. Although much of it has been refuted by various government authorities, it presents its case well.

Gubaryev, Vladimir. *Sarcophagus*. Translated by Michael Glenny. Harmondsworth, Middlesex, England: Penguin Books, 1987. A Soviet play that takes place in the hospital that treats Chernobyl victims. Very moving and thought-provoking.

Haynes, Victor, and Marko Bojoun. *The Chernobyl Disaster*. London: Hogarth Press, 1988. This objective book indicts nuclear power by detailing human failings with this technology.

Medvedev, Zhores. *Legacy of Chernobyl*. New York: W. W. Norton, 1990. An alarmist's point of view from a Soviet émigré who believes that Chernobyl was the worst human disaster.

Mould, Richard. *Chernobyl: The Real Story*. Elmsford, N.Y: Pergamon Press, 1988. This book contains 160 photographs and was done with the full cooperation of the Soviet authorities. Contains no analysis but pulls the facts together well.

Sweet, William. "Chernobyl, What Really Happened." *Technology Review* 90 (July, 1989): 43-52. This short and easily understood article suggests that RBMK reactors cannot be made safe.

Wilson, Richard. "A Visit to Chernobyl." *Science* 236 (June 26, 1987): 1636-1640. A short, clear look at the abandoned city through the eyes of a Harvard University physics professor.

Donald H. Williams

Cross-References

Hahn Splits an Atom of Uranium (1938), p. 1135; Fermi Creates the First Controlled Nuclear Fission Chain Reaction (1942), p. 1198; The World's First Nuclear Reactor Is Activated (1943), p. 1230; The World's First Breeder Reactor Produces Electricity While Generating New Fuel (1951), p. 1419; The United States Opens the First Commercial Nuclear Power Plant (1957), p. 1557.

A GENETICALLY ENGINEERED VACCINE
FOR HEPATITIS B IS APPROVED FOR USE

Category of event: Medicine
Time: July, 1986
Locale: United States

Approval for the vaccine by the FDA was granted to allow widespread use in the United States of an artificial vaccine for hepatitis B that had been produced in yeast by genetic engineering

Principal personages:
PABLO VALENZUELA, a biochemist who led the team that first developed the genetically engineered hepatitis B surface antigen produced in yeast cultures
MAURICE R. HILLEMAN (1919-   ), a scientist who was leader of the group that developed the hepatitis B vaccine
ARIE J. ZUCKERMAN, a physician-scientist who developed vaccines for hepatitis B that contained only subunit sections of the viral surface antigen
GEOFFREY L. SMITH, a scientist who led the group that first genetically engineered the gene for hepatitis B surface antigen

Summary of Event

In July, 1986, the United States Food and Drug Administration (FDA) licensed Merck Sharp & Dohme, a major pharmaceutical corporation, to market a new vaccine against infection with the hepatitis B virus (HBV). The use of this vaccine—called Recombivax-HB—marked a milestone in medical history because it was the first vaccine produced by genetic engineering. After approximately four years of research and testing (led by Maurice R. Hilleman), the new product was to be placed in general use. It became an alternative to another vaccine made by Merck, Heptavax-B, that was produced from the plasma of humans who had recovered from the disease. The Heptavax-B vaccine had been licensed in 1981 and had proved to be both effective and safe, with three levels of treatment during processing to inactivate any viruses, living cells, or harmful chemicals in the plasma. Unfortunately, many of the plasma donors for production of Heptavax-B were homosexuals, who are often infected with the acquired immunodeficiency syndrome (AIDS) virus as well as the virus that causes hepatitis B. Thus, beginning about 1983, there was widespread apprehension about the possibility of AIDS contamination of the pooled plasma-derived HBV vaccine.

The concern about possible contamination of Heptavax-B, as well as other considerations, led to the development of so-called second- and third-generation hepatitis B vaccines. Both second- and third-generation vaccines are based on recombinant

DNA (genetic engineering) technology, in which the genetic information for a protein, called an antigen, on the hepatitis virus is clipped out chemically and inserted into another virus' deoxyribonucleic acid (DNA) or into a cell's DNA. As the host virus replicates, or as the host cell functions, the hepatitis antigen is produced. Because no hepatitis DNA is present, however, no complete infectious hepatitis virus can be constructed, and the vaccine is considered safe for use in humans.

The technique used for production of Recombivax-HB was one of several being developed using genetic engineering. All the experimental vaccines depend upon antigens (chemicals that cause antibody formation against them in an immunized person) that are found in or on the virus. The most commonly used antigen in the vaccines is on the surface of the HBV, called HBsAg. Other antigens are also found inside the virus, and are called HBcAg (for core antigen) or HBeAg (for an enzyme antigen). The antigens in the virus are produced according to genetic information carried in viral DNA. That viral DNA had been taken apart in the laboratory, and the piece that coded for the HBsAg was separated from all the other viral DNA. The piece of DNA specific for HBsAg was then spliced into a kind of carrier DNA molecule—a plasmid—and then moved into a cell or another large virus. Pablo Valenzuela led the team that developed the vaccine that was finally marketed, which was produced in yeast cells, single cells of a fungus with the scientific name *Saccharomyces cerevisiae*, commonly used in baking bread.

Besides the antigens grown in yeast cells, other second-generation vaccine experiments were being performed in other kinds of cells. At the National Institutes of Health in Bethesda, Maryland, a group led by Geoffrey L. Smith joined the HBV gene for HBsAg to the DNA of the vaccinia virus. This is the virus that causes cowpox and was used as the vaccine to eradicate smallpox. The reason for joining the genetic information of these two viruses was to allow formation of an infectious virus, the vaccinia virus, that could infect people naturally and cause them to produce antibodies against hepatitis B as well. It was considered somewhat dangerous, however, because nonvirulent viruses can mutate to cause serious disease.

The HBsAg DNA had also been inserted into live cells of *Escherichia coli* (*E. coli*), the bacterium that is the most studied and most well-understood organism on Earth. It was believed that infections might arise as a result of the use of this vaccine, although work continues on it. Human cells of a continuously growing cell line derived from a liver cell cancer were also used as the host for the gene for hepatitis B surface antigens. The antigens produced by this cell line were excellent, but the cell line was thought to be less safe for vaccine production than yeast because of the possible presence of genes or chemicals in the cells that might cause cancer.

A third-generation vaccine has been developed in the laboratory as well, with protein antigens produced by direct, cell-free chemical synthesis. Arie J. Zuckerman in London has been involved in this work, which is also continuing as a possible source of a vaccine to be marketed in the future.

In the development of Recombivax-B, the viral DNA that coded for HBsAg was inserted into the yeast DNA in such a manner that the antigen, a protein, could be produced within the yeast cell. Large numbers of yeast cells were grown easily in the laboratory and induced to produce the surface antigens of the hepatitis B virus, which were then collected by breaking apart the yeast cells and separating out the specific molecules of antigen. These were purified, treated to kill any contaminants, and combined with other chemicals to form the vaccine.

Tests of the resultant vaccine were conducted in mice, rabbits, chimpanzees, and finally humans. About two years of testing, with thousands of volunteers, occurred in Africa, China, Greece, England, the United States, and other areas of the world. All types of people were tested, from homosexuals and drug abusers at high risk to the "average American" with low risk for exposure to the hepatitis B virus. Infants of infected mothers, health care workers, and immunosuppressed patients such as those on hemodialysis for kidney failure were tested also. In most cases, the vaccine provided immunity to the virus, with the least immunity seen in the immunosuppressed patients. As a result of this widespread testing, it was shown that the new yeast-derived vaccine was essentially similar in its properties to the previous plasma-derived vaccine. It produced no damaging side effects and caused no injury to those who were vaccinated. Merck Sharp & Dohme was therefore given approval by the FDA to market the vaccine.

Impact of Event

The development of this genetically engineered, yeast-grown vaccine against hepatitis B was of worldwide importance in that it made possible the production of much larger quantities of a vaccine of consistently high quality. The vaccine produced by this method was also free of any possible contaminants from human plasma. Licensing of the vaccine for widespread utilization was a major advance in the new technology of genetic engineering through the use of recombinant DNA technology.

Hepatitis B is a viral disease that causes liver damage, jaundice, sometimes death, and possibly liver cell cancer. It is present in about 2 million chronically infected carriers worldwide, mostly in the poorer countries of Africa, Asia, and Latin America. The disease generally is passed from these carriers to others by body fluids or close contact, or through food and water under unsanitary conditions. Specific smaller populations in American and Western European countries are also at high risk, such as health care workers, drug abusers, promiscuous homosexuals and heterosexuals, and persons undergoing dialysis. The best way to prevent continued spread of hepatitis B is to vaccinate those at risk for the disease, producing natural immunity to the virus in each individual.

The first hepatitis B vaccine, Heptavax-B, was shown by extensive testing to be safe and effective at producing immunity to the disease in chimpanzees and humans, the only species in which the virus produces disease. Heptavax-B could be produced only in limited quantities, however, because of the requirement for human plasma from individuals who had recovered from the disease. The vaccine required extensive treatment and testing to ensure that no other viral contaminants were present; this caused it to be quite expensive to produce. As a result of the cost and the reluctance of some potential recipients to trust that they would not be exposed to AIDS with the use of the vaccine, Heptavax-B had a limited impact in the United States and Europe on infection levels of hepatitis B. Also, it had almost no effect on the rate of infection and illness in other parts of the world because it could not be widely used.

The development and release of Recombivax-HB—the vaccine produced in yeast—removed the need for concern about possible AIDS infection, but the cost of treatment did not decrease. In fact, because the new dosage was half of that required with the plasma-derived vaccine, at the same price to the patient, it could be considered to have increased. Discussion in the medical and scientific journals of 1986 and 1987 suggested that the price might be reduced eventually as more vaccine was produced, since the cost of development and testing would be recovered early, but this has not occurred. Because of continuing high cost, the vaccine has had little impact on the Third World nations, where infection frequencies are highest.

The acceptance of the genetically engineered vaccine has led to further work on other forms of hepatitis B vaccines, as well as on other potential vaccines that will allow immunization against viral diseases for which there are no vaccines. Work is continuing on vaccines against herpes, hemorrhagic fever, viruses associated with cancers, and others. In addition, studies are being conducted on alternative forms of genetically engineered viral vaccines that will be more easily produced, leading to wider distribution and greatly reduced cost. The ultimate goal is to immunize every possible victim and eradicate not only the hepatitis B virus but also all other disease viruses from all natural populations, as was done in the 1970's with the smallpox virus. With this first step taken in that direction, the general use of a vaccine produced by the use of recombinant DNA technology, the goal is now in sight.

Bibliography

Brown, Fred, Robert M. Chanock, Harold S. Ginsberg, and Richard A. Lerner, eds. *Vaccines 90: Modern Approaches to New Vaccines Including Prevention of AIDS.* Cold Spring Harbor, N.Y.: Cold Spring Harbor Laboratory Press, 1990. Includes papers presented at a conference held in September, 1989. Several papers refer to the hepatitis B vaccines, including two on the core antigens, one on recombinant HBV with vaccinia virus, and one on the production of HBV vaccine in yeast.

Ginsberg, Harold, Fred Brown, Richard A. Lerner, and Robert M. Chanock, eds. *Vaccines 88: New Chemical and Genetic Approaches to Vaccination.* Cold Spring Harbor, N.Y.: Cold Spring Harbor Laboratory Press, 1988. Includes papers presented at a conference in September, 1987. One of the first papers is on a synthetic vaccine for hepatitis B and another discusses the core antigens as vaccine sources.

Hollinger, F. Blaine. "Hepatitis B Vaccines: To Switch or Not to Switch?" *JAMA* 257 (May 15, 1987): 2634-2636. An editorial written for the physician in practice, to give both sides of the issue of whether patients should be given the recombinant DNA vaccine produced in yeast or the older vaccine from human blood plasma. Presents the dilemma very well, and gives the opinion that a higher dose of the newer vaccine may be required.

Laskey, Laurence, ed. *Technological Advances in Vaccine Development*. New York: Alan R. Liss, 1988. Papers presented at a symposium in early 1988; a rather technical resource. Several papers are related directly to hepatitis B viral growth in yeast or mammalian cells and the production of a synthetic peptide vaccine.

McAleer, William J., et al. "Human Hepatitis B Vaccine from Recombinant Yeast." *Nature* 307 (January 12, 1984): 178-180. Written by the research group at Merck Sharp & Dohme that brought the vaccine to the market; covers the science background for its development and testing. Includes a photograph of the antigen particles produced by the yeast.

Talwar, G. P., ed. *Progress in Vaccinology*. Vol. 2. New York: Springer-Verlag, 1989. Papers presented at an international symposium held in 1986; thus not as current as the date of publication suggests. Emphasizes production of vaccines for Third World use; devotes an entire section to hepatitis. Another paper discusses vaccinia virus with recombinant DNA technology to make it a vaccine for other viral diseases.

Valenzuela, Pablo, et al. "Synthesis and Assembly of Hepatitis B Virus Surface Antigen Particles in Yeast." *Nature* 298 (July 22, 1982): 347-350. A report by the team that placed the genetic information for the surface antigen of hepatitis B virus into yeast cells and produced the surface antigen protein; discusses how this was done. Compares protein particles made by yeast and those produced by human cells with viral infection.

Zuckerman, Arie J. "New Hepatitis B Vaccines." *British Medical Journal* 290 (February 16, 1985): 492-496. Written by a major investigator in this field; discusses polypeptide vaccines, recombinant DNA vaccines, and chemically synthesized vaccines, with the properties and uses of each in preventing hepatitis B.

Jean S. Helgeson

Cross-References

Rous Discovers That Some Cancers Are Caused by Viruses (1910), p. 459; Calmette and Guérin Develop the Tuberculosis Vaccine BCG (1921), p. 705; Zinsser Develops an Immunization Against Typhus (1930), p. 921; Theiler Introduces a Vaccine Against Yellow Fever (1937), p. 1091; Kornberg and Coworkers Synthesize Biologically Active DNA (1967), p. 1857; Cohen and Boyer Develop Recombinant DNA Technology (1973), p. 1987; The U.S. Centers for Disease Control Recognizes AIDS for the First Time (1981), p. 2149; A Human Growth Hormone Gene Transferred to a Mouse Creates Giant Mice (1981), p. 2154; Sibley and Ahlquist Discover a Close Human and Chimpanzee Genetic Relationship (1984), p. 2267.

A GENE THAT CAN SUPPRESS THE CANCER RETINOBLASTOMA IS DISCOVERED

Categories of event: Biology and medicine
Time: October, 1986
Locale: Boston, Massachusetts

Dryja and associates identified and isolated the retinoblastoma (Rb) gene and, thus, stimulated renewed interests in retinoblastoma research

Principal personages:

ALFRED G. KNUDSON, JR. (1922-   ), an American geneticist who developed a model explaining how both genetics and environment may cause retinoblastoma and Wilms' tumor, rare childhood cancers

THADDEUS P. DRYJA (1940-   ), a medical scientist who isolated the retinoblastoma (Rb) gene within human chromosome 13q14—a gene required to suppress childhood retinoblastoma

Summary of Event

Retinoblastoma, a malignant tumor that arises from immature retina, occurs in 1 in 15,000 to 1 in 30,000 live births and represents approximately 2 percent of childhood malignancies. The disease may be inherited or the result of a new germinal mutation. About 10 percent of patients have a family history of retinoblastoma and another 30 percent have bilateral disease. All of these (that is, 40 percent of the patients) will pass the trait to their children as an autosomal dominant. The remaining 60 percent of patients have unilateral and nonheritable disease. A small portion of the cases have a deletion involving chromosome 13q14, but all heritable cases carry a mutant gene.

In 1971, Alfred G. Knudson, Jr., developed a model explaining how both genetics and environment may cause retinoblastoma and Wilms' tumor, rare childhood cancers. Subsequent studies have verified Dr. Knudson's theory that there exist certain anticancer genes, or antioncogenes, that protect against disease. (Genes that play a role in the development of cancer are called oncogenes.) The lack of, or destruction or damage of, antioncogenes can cause cancers such as retinoblastoma, which is a tumor of the eye. Knudson's work also includes an understanding of why some families seem to be prone to cancer and the identification of genetic factors that may relate to other forms of cancer.

In December, 1988, it was reported that researchers had for the first time succeeded in causing cancer cells growing in the laboratory to revert to normal by replacing a defective gene with a healthy one. The gene involved was named Rb. When it is defective, retinoblastoma can develop. The Rb gene discovered in October, 1986, by researchers at the Massachusetts Eye and Ear Infirmary in Boston, was the first of the anticancer genes to be found. The genes known as antioncogenes apparently protect against cancer by preventing adult cells from proliferating.

Normally, an individual receives two Rb genes at conception, one from the father and one from the mother. If both are healthy, a tumor can arise later only if both genes in one cell are disabled by a chemical or virus. Yet, if a child inherits one defective Rb gene, only the one healthy gene must be disabled for cancer to occur. Many children who inherit the defective gene suffer from multiple eye tumors.

The isolation of the Rb gene by Thaddeus P. Dryja and associates required creating a lambdaphage library that contained inserted fragments from human chromosome 13. One of the inserts detected a corresponding fragment that was deleted in two of thirty-seven retinoblastoma tumor DNAs (deoxyribonucleic acid). This suggested that the probed segment was linked to the Rb genes. A nearby probe detected not only the human sequence but also a mouse homolog to a somatic cell hybrid carrying human chromosome 13. The latter probe was used for RNA analysis to detect any transcripts in a retinal cell line. The analysis showed a 4.7-kilobase transcript in the tumor cell line but not in several retinoblastomas. The latter probe also was used to analyze DNA from a large group of retinoblastomas and osteosarcomas (secondary tumors of mesenchymal origin) and found gross changes in genetic structure in approximately 30 percent of the tumors' DNAs. The boundaries of homozygously deleted fragments were mapped. In the analysis of an osteosarcoma and a retinoblastoma, it was discovered that the endpoints of the deletion were within the confines of the genetic unit defined by the probe. This indicated that the target of inactivation was the segment under study and not a neighboring DNA sequence.

Gene probes have been found for identification of potentially affected newborns but until they are widely available, careful and repeated eye examinations of subsequent offspring are mandatory. With early diagnosis and treatment, survival and preservation of sight can be anticipated.

Three groups of researchers reported in mid-1986 that a defective Rb gene also is present in cells from about 20 percent of breast tumors, more than half of small cell cancers, and most bone tumors. The successful conversion of cancerous cells into normal ones was carried out by scientists at the University of California at San Diego. They used a specially engineered virus to insert the healthy Rb gene into retinoblastoma and bone cancer cells. Once the gene had been inserted into the cells, the cells immediately stopped proliferating. When the engineered cells were injected into special laboratory mice that had no immune system, no tumors formed, indicating that the cells were no longer malignant. The researchers predicted that the technique could be tested in humans within five years and might be useful in treating genetically linked conditions such as cancers and Alzheimer's disease (a decline in intellectual function that interferes with an individual's daily living). Researchers have postulated that one or more genes are responsible for familial Alzheimer's, which is characterized by early onset and rapid progress, and could account for 10 to 20 percent of all Alzheimer's cases. The genetic defect found on chromosome 21 suggested a relationship between Alzheimer's and Down's syndrome, a genetic disorder caused by an extra copy of chromosome 21. Alzheimer's may affect both sides of the brain symmetrically or only one side, depending upon which part of the brain is affected; patients may have different behavioral patterns and disorders in language or movement.

In 1987, scientists discovered that many individuals with colon cancer had a defective gene that, like the Rb gene, seems to protect against cancer when it is healthy; this gene was located on chromosome 5.

Impact of Event

The impact of Dryja's research with retinoblastoma was almost immediate. A computer search revealed that more than three hundred papers have been published on eye research since 1986. In 1986, Emil Bogen Mann studied retinoblastoma cell differentiation in culture. Since little is known about the biology of retinoblastoma in vitro because of the lack of adequate culture systems, and only a few retinoblastoma cell lines have been established in the past, a culture system was developed using rat smooth muscle cells as feeder layers that allowed for routine growth of primary retinoblastomas and/or their metastases. These cells were routinely grown up to passages and, in some cultures, spontaneous formulation of Flexner-Wintersteiner rosettes were observed.

In April, 1986, Jack Rootman and associates studied the ocular penetration, toxicity, and radiosensitizing properties of two new nitroimidazoles. To determine the highest level of the drug that is relatively nontoxic for subconjunctival administration, the investigators assessed the effect of varying concentrations of both nitroimidazoles. A dose of 100 milligrams of each drug in 0.5 milliliter of normal saline was the maximum acceptable toxicity as determined by these studies. There was no significant difference in response to injection of either drug. High-pressure liquid chromatographic analysis was performed on test samples from plasma, urine, and ocular compartments. Chinese hamster ovary cells were irradiated (X rays) at a dose of 150 rads per minute. After radiation, cells were washed free of the drug and plated. After seven days of incubation, colonies were stained and scored. Colonies containing fifty cells or more were scored as survivors.

In May, 1986, Yenyun Wang and associates compared diploid fibroblast cell lines derived from two hereditary retinoblastoma (Rb) patients with those of three normal persons of comparable ages for their sensitivity to ionizing radiation, induced transformation to anchorage independence. The target cells were exposed to cobalt 60, allowed to undergo an expression period, and assayed for ability to form colonies in 0.33 percent agar. There was no detectable difference between the Rb cells and the normal cells' response to the transforming action of cobalt 60. Concomitantly, the gene for esterase D (ESD) is known to be tightly linked to the retinoblastoma locus (Rbl) in the q14.1 band of chromosome 13. Jeremy Squire and colleagues were able to clone the ESD gene from a human cDNA library by using oligonucleotides specific for a partial amino acid sequence of the purified enzyme to provide a genetic marker for further studies on retinoblastoma. The putative ESD gene coded for a message of 1.2 kilobases, which was present on all cell types examined and mapped to 13q14.1, thus confirming that it was the ESD gene.

In October, 1987, Wen-hwa Lee discovered that a null allele of esterase D was a marker for genetic events in retinoblastoma formation. Using a rabbit anti-esterase D antibody and the esterase D cDNA probe, Lee and associates found that low but detectable quantities of esterase D protein and enzymatic activity were present in tumor cells from a patient with bilateral retinoblastoma; fibroblasts from this patient contained two copies of the esterase D gene, indicated by heterozygosity at a polymorphic site within this gene; and tumor cells from the same patient were homozygous at this site, indicating loss and reduplication of the esterase D locus. The results demonstrated that one of the two esterase D alleles in this patient acted as a "null," or silent, allele.

Studies on genetic, cytogenetic, and molecular genetic approaches have improved the understanding of the biological events leading to the occurrence of retinoblastoma and have provided tools for the enhanced assessment of risk(s) for some individuals.

Bibliography

Buchanan, Janet, and Cavenee Webster. "Genetic Markers for Assessment of Retinoblastoma Predisposition." *Disease Markers* 5 (February, 1987): 141-152. A review article that provides a conceptual framework for the application of current approaches to the analysis of retinoblastoma in the clinical setting. Progress in elucidating the basis of retinoblastoma is discussed.

Cowell, John, and Jon Pritchard. "The Molecular Genetics of Retinoblastoma and Wilms' Tumor." *CRC Critical Reviews in Oncology/Hematology* 7, no. 2 (1987): 153-168. A description of cancers in adults is presented with a discussion of probable causes of certain types of cancer, such as exposure to environmental carcinogens.

Sturtevant, Alfred H. *A History of Genetics*. New York: Harper & Row, 1965. The text gives a panoramic view of the history of genetics. The publication of Gregor Mendel's paper in 1866 was the outstanding event in the history of genetics; but as is well known, the paper was overlooked until 1900. Many topics are treated separately rather than in chronological order: genes and chromosomes; linkage, mutations, and the cytology of crossing over; genetics and immunology; and biochemical genetics.

Watson, James. *Molecular Biology of the Gene*. New York: W. A. Benjamin, 1965. Developments in molecular biology and biochemistry not only have become a part of the undergraduate biology curriculum but also are increasingly necessary for a general understanding of scientific culture. This book is designed to give interested readers the rigor and the perspective needed to bridge the gap between the single cell and the complexities of higher organisms. Fundamentals of biochemistry, molecular genetics, and cytology are elucidated. Salient topics include the cell theory, chromosomal theory of heredity, the concept of intermediary metabolism, the double helix, the concept of free energy, the concept of template surfaces; chromosome mapping, and the principle of self-assembly and the replication of viruses.

Winter, Jens. "In Vitro and In Vivo Growth of an Intraocular Retinoblastoma-Like Tumour in F-344 Rats." *Acta Ophthalmologia* 64 (April, 1986): 657-663. The growth of a transplantable retinoblastoma-like tumor in the eyes of rats and in cell cultures is described. A valuable article for those interested in retinoblastoma.

Nathaniel Boggs

Cross-References

Morgan Develops the Gene-Chromosome Theory (1908), p. 407; Sturtevant Produces the First Chromosome Map (1911), p. 486; Avery, MacLeod, and McCarty Determine That DNA Carries Hereditary Information (1943), p. 1203; Watson and Crick Develop the Double-Helix Model for DNA (1951), p. 1406; Lasers Are Used in Eye Surgery for the First Time (1962), p. 1714; Murray and Szostak Create the First Artificial Chromosome (1983), p. 2251; Erlich Develops DNA Fingerprinting from a Single Hair (1988), p. 2362.

RUTAN AND YEAGER PILOT THE *VOYAGER* AROUND THE WORLD WITHOUT REFUELING

Category of event: Space and aviation
Time: December 14-23, 1986
Locale: Edwards Air Force Base, California

Rutan and Yeager set a new flight distance record, piloting an aircraft of unusual design around the world without landing or refueling

Principal personages:

BURT RUTAN (1943-   ), an American designer of unconventional aircraft of high efficiency including the *Voyager*

DICK RUTAN (1938-   ), the brother of Burt Rutan, a former Air Force jet pilot, and copilot of the *Voyager*

JEANA YEAGER (1952-   ), an American pilot, holder of five world records in speed and distance in the piston weight class, and copilot of the *Voyager*

Summary of Event

To fly around the world is not an easy task, and standards used for records have not always been well defined. In 1924, four Army open cockpit biplanes started the flight, and after sixty-nine stopovers and 175 days, only two completed the trip. Wiley Post flew around the world twice in a Lockheed Vega that he named *Winnie Mae.* In 1933, two years after his first trip, Post became the first man to accomplish the journey solo in seven days and eighteen hours with stopovers. On an attempted around-the-world flight in 1937, Amelia Earhart vanished in the Pacific Ocean near Howland Island. Howard Hughes, flying his Lockheed 14, reduced the world record to three days and nineteen hours in 1938 utilizing staging facilities.

Since many aviators were flying around the world faster but not always covering the same distance, new distance standards were defined. The Fédération Aéronautique Internationale (FAI) established the minimum distance that would qualify for a world record as the circumference of the Tropic of Cancer or Capricorn equal to 36,800 kilometers. This distance is somewhat smaller than the earth's circumference, which is 40,000 kilometers.

In 1949, the first nonstop flight around the world was accomplished by an Air Force B-50, which was refueled in air and required ninety-three hours for the trip. An Air Force B-52 in 1962 set the longest nonrefueled flight at 20,180 kilometers.

The idea of creating an aircraft to attain aviation's "last milestone" of a non-refueled global flight started in 1980 with a sketch that Burt Rutan outlined on a napkin at a restaurant in Mojave, California, in the presence of Dick Rutan and Jeana Yeager. *Voyager* was the result of the boom in home-built experimental aircraft involving high-technology concepts. These home-built airplanes pioneered the use of composite materials such as carbon fiber spars and Nomex honeycomb sandwich, used to cover wings. The latter is a resin-soaked epoxy fiber coated with layers of carbon fiber cloth that is oven-cured in a pressurized chamber. The use of such exotic construction materials made the airframe flexible and very difficult to fly in turbulent air. The problem was partially solved by specially designed autopilots that would assist the pilots in conditions of marginal stability.

Actual construction on the *Voyager* commenced in 1981, and upon completion, it became the largest composite airplane ever built, with a wingspan of 33.8 meters. Placing the horizontal stabilizer forward of the wing made it of canard design, and with two large outrigger booms connecting both of these flight surfaces, *Voyager* became a "flying fuel tank." Altogether, seventeen fuel tanks were built into the aircraft's interior, leaving only a very small space, 2.1-by-1.0-meters, for the pilots to occupy.

The aircraft was powered by two American-built Continental engines; the front engine was air cooled with four cylinders, but the rear was a water-cooled design. These were high-efficiency engines utilizing fine wire spark plugs to minimize carbon buildup during the long journey. Even the propellers were of a radically different design. Upon completion of sixty-seven test flights, the *Voyager* was readied for its dramatic world flight. The airplane was loaded with fuel until it weighed 4,497 kilograms, of which 3,180 kilograms was fuel, giving an incredible 3:1 ratio of fuel weight to aircraft weight.

The takeoff for the *Voyager* world flight commenced at 8:00 A.M. on Sunday, December 14, 1986, and almost ended in disaster. The aircraft had difficulty becoming airborne because of the downward torsional forces exerted on the wing-canard forward flight surfaces by the fuel weight distribution. *Voyager* required all but 240 meters of a 4,600 meter runway to lift-off, setting another world record of sorts for the longest takeoff roll. The long takeoff damaged both wingtips when both dragged on the ground and were lost early in the flight. The loss of the wingtips did not greatly affect the overall performance, increasing drag by an estimated 6 percent.

The rate of climb, initially at 60 to 100 meters per second, was better than expected, and the crew encountered no problems early into the flight as they climbed to altitudes of between 2,000 to 3,000 meters and maintained speeds of only 180 kilometers per hour. Their first rendezvous with a spotter aircraft was uneventful and occurred over Hawaii on the first night. Late on the second day, *Voyager* was on a collision course with Typhoon Marge coming up from the south and a severe low pressure system converging from the north. The crew, with help from a ground control weather operator, was able to weave a flight path between the two storm systems.

When Dick Rutan flew the first fifty-five out of sixty hours, Jeana Yeager had much work to do monitoring flight systems as well as carefully recording flight speeds, positions, winds, engine speeds, and fuel supplies in the appropriate logs. From time to time, Rutan and Yeager would interchange their positions, with Yeager doing the flying and Rutan resting; such movement was difficult in this limited living space. The crew took about 60 kilograms of food and water provisions with them, eating their meals and drinking from prepared samples placed in small plastic pouches.

Hot meals were prepared by placing the meal pouch around the heating duct.

On the sixth day, the crew noticed that the oil-warning light for the rear engine flashed. Dick had taken a quart of oil out of the engine instead of putting it in by turning the hand crank the wrong direction. The problem was diagnosed quickly, and the oil was cranked back in. Immediately, the airplane was placed in a nose-down attitude to increase the air flow into the cooling ramps. After two hours, the temperature of the engine returned to the normal range. On day seven, severe turbulence from a thunderstorm in the South Atlantic forced the *Voyager* into a ninety degree bank, and both pilots experienced considerable discomfort. A short distance from home on day nine, off the west coast of Baja California, the rear engine ceased running. The front engine, which was not running at the time and kept as a backup system, needed starting, but the aircraft lost more than 1,500 meters of altitude before this could be done. The problem was aggravated by the gravity-activated fuel flow when levels of fuel remaining in the tanks were low. A specially designed eight-to-one valve was used to control the fuel tank switching and proved essential at this stage of the flight.

Finally, on December 23, 1986, at 8:05 A.M., after nine days, three minutes, and forty-four seconds of elapsed time, the *Voyager* landed at Edwards Air Force Base before thousands of well-wishers. It was a magnificent achievement, but only 48 kilograms of fuel remained on board.

Impact of Event

The *Voyager* was the first airplane to fly around the world without landing or refueling and traveling an official distance of 40,212 kilometers. Other notable achievements were made, such as flying in nine days, flying in a variety of extreme weather conditions in perhaps the lowest-powered aircraft for its weight ever constructed, and flying by two pilots cramped in a small horizontal phone booth-sized cabin.

To achieve this flight, the designers had to build the most highly loaded (overall weight divided by wing area) and lightest wing ever built. Since the airplane was designed to fly slowly, the aspect ratio (wingspan divided by average chord or width) was designed to minimize the lift drag. Exotic building materials of carbon fiber and epoxy sandwich skins used on the wing and airframe greatly lessened the overall weight. The resulting wing was touted as the strongest of its type ever built, able to bend about thirty degrees from the horizontal. The experimental canard-boom, another unique design of the aircraft, helped to solve difficult structural problems. The boom fuel weight is situated ahead of the wings as additional lift in the front and is provided by the canard surface. The booms of cylindrical shape had a greater volume to wetted area (exposed to airflow) than a wing, and their presence was less of a hindrance for both drag and structural weight. The booms placed a larger fraction of the aircraft's weight farther out along the wing beam and relieved wing bending moments.

Since the *Voyager* airplane was so radically different in design concept, direct commercial or military spinoffs were not immediately forthcoming, although the carbon-honeycomb materials could be used in the building of long-range transport and reconnaissance aircraft.

Public interest in the flight was greater than expected; there were almost daily reports about the airplane and its crew in the press long before the actual flight. Initially, the project depended upon grass roots for support, soliciting small donations from any airplane fan who was willing to contribute. When the operation grew, so did the number of engineering and technical personnel. This was important in the latter stages of the project, when major aircraft and avionics companies agreed to help. Hartzell, for example, provided the engineering for the new propeller shapes. Each propeller required sixteen different airfoil sections, which were carved from a computer-directed milling machine.

Perhaps the greatest achievement of the project was the dedication of a small group of people to accomplish a very difficult task; it is this kind of spirit that leads to new frontiers of aviation.

Bibliography

Garrison, Peter. "Sketchy Details: Who Really Designed Voyager." *Flying* 114 (July, 1987): 20-22. The *Voyager* was a canard-boom combination design. The reasons for the selection of this design are discussed as well as the initial problems in getting the aircraft airborne.

_____. "Voyager Flight Fantastic: They Made It with 14 Gallons to Spare." *Flying* 114 (March, 1987): 28-35. Highly technical problems had to be solved before the *Voyager* began its world flight. Garrison discusses a number of behind-the-scenes engineering tasks, such as the aircraft's airfoils, ride in turbulence, spark-plug fouling in test flights, fuel efficiency and range, propeller design, and avionics.

Marbach, William D., and Peter McAlevey. "Up, Up and Around." *Newsweek* 153 (December, 1986): 33-44. Dramatic moments of the *Voyager* flight are captured in this news article. Seven color photographs and two color diagrams show the interior of the pilots' cockpit, location of the seventeen fuel tanks, and a map showing the path of the world flight.

Norris, Jack. *Voyager: The World Flight*. Northridge, Calif.: Jack Norris, 1987. This reference on the flight was compiled by the technical director of *Voyager* Mission Control. A brief history of the project is included, with many important statistics of the flight from weights to airspeeds. Tables relate such factors as effect of altitude, engine speed, and coefficient of lift on flight performance.

Yeager, Jeana, and Dick Rutan, with Phil Patton. *Voyager*. New York: Alfred A. Knopf, 1987. Narrative description of the flight by both pilots. Includes memories from their earlier years that led up to their involvement with the *Voyager* project. Eighty-one black-and-white photographs and fourteen full-color photographs are included.

Michael L. Broyles

Cross-References

The Wright Brothers Launch the First Successful Airplane (1903), p. 203; Blériot Makes the First Airplane Flight Across the English Channel (1909), p. 448; Lindbergh Makes the First Nonstop Solo Flight Across the Atlantic Ocean (1927), p. 841; The First Jet Plane Using Whittle's Engine Is Flown (1941), p. 1187; The First Jumbo Jet Service Is Introduced (1969), p. 1897.

SCIENTISTS DATE A *HOMO SAPIENS* FOSSIL
AT NINETY-TWO THOUSAND YEARS

Category of event: Anthropology
Time: 1987-1988
Locale: Gif Sur Yvette, France; Qafzeh cave, near Nazareth, Israel

A team of scientists gave an age of ninety-two thousand years for a modern-looking Homo sapiens, *more than doubling the length of time that modern humans have been known to exist*

Principal personages:
> HÉLÈNE VALLADAS, a French scientist at the Centre des Fables Radioactivités in France and leader of the radiometric team
> BERNARD VANDERMEERSCH, a French physical anthropologist and archaeologist who excavated the Qafzeh site in Israel
> DOROTHY ANNIE ELIZABETH GARROD (1892-1968), an English archaeologist who excavated the caves at Mount Carmel, Israel
> SIR ARTHUR KEITH (1866-1955), an eminent Scottish anatomist and physical anthropologist who analyzed the fossil remains from Mount Carmel
> THEODORE DONEY McCOWN (1908-1969), an American archaeologist and physical archaeologist who assisted Garrod at Mount Carmel and assisted Keith in the analysis of the skeletal material
> RENÉ VICTOR NEUVILLE (1899-1952), a French archaeologist who discovered the cave at Qafzeh

Summary of Event

The origin of modern human beings has been a persistent and vexing problem for prehistorians. Part of their difficulty is widespread disagreement over the relationship between modern man and his closest extinct relative, Neanderthal man. Neanderthals differ from modern and some archaic humans by the extreme robustness of their skeletons and by their heavy brow ridges, extremely large faces, and long and low skull caps. Neanderthal man was once believed to have lived from approximately 100,000 to 50,000 years ago, while modern humans had been thought to have existed for 40,000 to 50,000 years.

Neanderthal skeletons have been found only in Europe and the Middle East, and the robustness and exaggeration of their features increases from east to west, with the most extreme "classic" Neanderthals being found in Western Europe. Although Neanderthals were originally assumed to have been ancestral to modern humans, prehistorians now agree that they were too localized, too extreme, and too recent to have been forerunners of modern people.

Overlapping Neanderthals in time are archaic fossil humans who are more generalized and less robust and who make better candidates as ancestors of modern man. These have been found in sub-Saharan Africa and also in the Middle East, where they overlap both spatially and temporally with Neanderthals. No specimens of this type have been found in Europe.

Arguments regarding the origin of modern humans turn around the issue of one ancestral group as opposed to many ancestral groups. One view assumes that modern man evolved from many local archaic types, including Neanderthal. A variant of this perspective holds that, while the ancestors of modern humans may have originated in one locality, they interbred with local peoples they met as they spread throughout the world. Both these views hold that there has been some degree of regional continuity over many thousands of years, and this continuity explains physical differences between modern populations.

Opposed to this view is the single-origin perspective that maintains that earlier humans were replaced completely by physically and technologically more advanced members of a new group who developed in one locality. Proponents of this view hold that the displaced types contributed no genetic material to the newcomers. The most favored homeland for the new and improved type is sub-Saharan Africa, where there are early examples of possible forerunners of modern humans, but where there are no known examples of Neanderthals.

The region called the Levant, which includes Israel, has been of considerable interest to proponents of both theories, because it is the only land bridge between Africa and the rest of the world. Any population moving from one region to the other had to pass through the Levant. This fact became particularly important when the first Neanderthal found outside Europe was discovered in Galilee in 1925.

Pursuing this lead, English archaeologist Dorothy Annie Elizabeth Garrod excavated a series of caves on Mount Carmel, now in Israel, between 1929 and 1934. She was assisted by a young American, Theodore Doney McCown. In two of these caves, Tabun and Skhūl, McCown discovered human remains associated with the Middle Paleolithic stone tool industry known as Mousterian, an industry previously found with Neanderthal remains. Embedded in rock, the fossils were shipped to England, where McCown, together with the distinguished anatomist Sir Arthur Keith, spent four years prying the bones from their rocky matrices and analyzing them. The resulting description began a controversy that still continues.

McCown and Keith found that there were two types of humans in the Mount Carmel caves. Those at Skhūl, although archaic, resembled modern humans in most of their characteristics. Those at Tabun resembled Neanderthals, although their features were less exaggerated than those of classic Neanderthals. On the basis of similar dentition and technology, plus the fact that the caves were close in both space and time, McCown and Keith believed that the fossils in both caves were part of one single population.

The meaning of these remains was interpreted variously. McCown and Keith believed that the Mount Carmel population was in the process of diverging into two groups from a more generalized ancestor but that neither were ancestral modern humans. Others thought that the fossils represented hybridization between Neanderthals and modern humans. Physical anthropologist F. Clark Howell hypothesized that Tabun and Skhūl were two unrelated populations, that Skhūl was later than Tabun, and that Skhūl might be ancestral to modern humans. A few others thought that Neanderthal had been caught in the act of evolving into modern man.

Overlapping with Garrod's excavations at Mount Carmel were those by René Victor Neuville from the French consulate at Jerusalem. Neuville excavated Qafzeh cave between 1933 and 1935, finding the remains of five individuals. Unfortunately, World War II intervened, followed by the Israeli-Arab conflict of 1947. Neuville died without analyzing his material. As a result, the Qafzeh finds were not taken into account by those attempting to reconstruct human evolution. From 1965 to 1975, the French archaeologist and physical anthropologist Bernard Vandermeersch continued Neuville's excavations, finding the remains of eight individuals, who resembled the non-Neanderthals from Skhūl. By the 1970's, most scholars thought that Qafzeh, like Skhūl, was later than Neanderthal sites in the Levant, and that there had been a replacement of Neanderthals by non-Neanderthals about fifty thousand years ago. Vandermeersch believed that Qafzeh was older than Tabun.

A major problem in making sense of the Levant finds lies in the inaccuracy of dates. While faunal sequences for Europe are accurately established, they are still sketchy in the Levant and are not reliable sources for determining the relative age of sites. Radiocarbon dates do not help because they are inaccurate for sites as old as Qafzeh, Tabun, or Skhūl. Until recently, all that could be known was that humans of some sort had been in the Levant more than sixty thousand years ago and had lived there an undetermined time.

Another method of dating, called thermoluminescence, helped to clarify the sequence. Thermoluminescent dating is used on objects that were heated during the time that they were used. When such objects are heated again, radioactive impurities emit measurable light. Since the first firing of the object had the effect of setting the nuclear clock to zero, the amount of light on the second firing is an indication of how long ago the object was used. Thermoluminescence can be used to date much older material than can radiocarbon methods. Unfortunately, thermoluminescence is not as accurate as radiocarbon. The first objects to be dated by thermoluminescence in the Levant were burnt flints from the Neanderthal burial site at Kebara. The dating was done by a French-Israeli team headed by Hélène Valladas, with results being published in 1987. The level containing the burial was dated at sixty thousand years ago, meaning that if the date is correct, Neanderthals were in the Middle East much later than had been thought previously.

In 1988, a team led by Valladas published a thermoluminescence date of ninety-two thousand years from Qafzeh. If this date is correct, then there were forerunners of modern humans living in the Levant twice as long as had been suspected. Furthermore, these individuals were there either before or at the same time as Neanderthals, confirming Howell's hypothesis that there were two separate populations of the genus *Homo* in the Levant.

Impact of Event

Until there is independent confirmation of the date from Qafzeh, its long-term impact cannot be certain. Inasmuch as thermoluminescence gives only a rough estimate, confirmation by another form of dating is desirable. In the meantime, there have been two dominant reactions by scientists. Those subscribing to the single-origin, "out-of-Africa" model see the Qafzeh date as confirmation of this hypothesis. The English prehistorians Chris Stringer and Peter Andrews believe that this early date for modern *Homo sapiens* is strong evidence that modern man arose in Africa and spread throughout the rest of the world, replacing other archaic humans without any interchange of genes. Most extreme in this position are French scientists such as Vandermeersch and Valladas, who refer to the humans at Qafzeh and Skhūl as "Proto-Cro-Magnons." This term refers to the most famous European fossil type of the succeeding Upper Paleolithic, which is associated with the spectacular cave art of that period.

Others, such as the American Milford Wolpoff, dispute this assessment. Wolpoff believes that there is sufficient anatomical and archaeological evidence to believe that Neanderthals contributed to the genetic makeup of modern Europeans. He points out that dates for presumed early modern humans from sub-Saharan Africa and the Levant have been questioned and that the fossils from sub-Saharan sites are fragmentary and may be poorly reconstructed. Wolpoff correctly notes that Cro-Magnon is a late example of Upper Paleolithic Europeans and quite unlike the earliest post-Neanderthals. He points out that the late Neanderthals in Europe are more like modern Europeans in some respects than are the fossils from Skhūl or Qafzeh.

The date from Qafzeh raises more questions than it answers. Early modern humans in the Levant had exactly the same tool traditions and faunal assemblages as Neanderthals. Consequently, their means of subsistence could not have been significantly superior. A question that anthropologists ponder is how the rest of the world could be invaded and displaced by a human form with a tool assemblage and means of subsistence identical to their own. To complicate matters further, the more efficient tools found about forty thousand years ago, at the start of the Upper Paleolithic, were considered to be a sign of the greater intelligence of modern forms until a Neanderthal skeleton was found in association with an early Upper Paleolithic industry. It needs to be determined if Neanderthals and modern humans, each with similar advanced technologies, coexisted in Europe forty thousand years ago. If this is the case, and if there were no cultural and few genetic differences between modern humans and Neanderthals, it is not known why there has not been genetic interchange. This seems to contradict all that is known about mammals in general and humans in particular. Finally, another perplexing question is, if genetically modern man appeared at least ninety-two thousand years ago, why did it take more than fifty thousand years for the Upper Paleolithic, with its first appearance of modern-appearing culture, including art, to appear? These are difficult questions and will not be answered quickly.

Bibliography

Garrod, D. A. E., and Dorothea Bate. *The Stone Age of Mount Carmel*. Vol. 1. Oxford, England: Clarendon Press, 1937. An archaeological monograph, volume 1 gives details of the ecology of the Mount Carmel site, the excavation methods, and the findings. Garrod's work was outstanding for its time. May be too detailed, but given the importance of the site, it is worthwhile reading. Illustrations.

Holloway, Ralph L. "The Poor Brain of *Homo sapiens neanderthalensis*: See What You Please." In *Ancestors: The Hard Evidence*, edited by Eric Delson. New York: Alan R. Liss, 1985. An amusing article sympathetic to Neanderthal man, this and other contributions to section 7 cover the general issue of the place of Neanderthal in human evolution. Illustrations, diagrams, and bibliography.

Keith, Arthur. *An Autobiography*. London: Watts, 1950. Although the Mount Carmel finds occupy only a small portion of this book, Keith's life history is an interesting and sometimes inadvertently amusing account of English scholarly life in the late nineteenth and early twentieth centuries.

McCown, Theodore D., and Arthur Keith. *The Stone Age of Mount Carmel*. Vol. 2 in *The Fossil Remains from the Levalloiso-Mousterian Levels*. Oxford, England: Clarendon Press, 1939. Discusses the Mount Carmel excavation. For a wide audience. Illustrations.

Stringer, Christopher B., and Peter Andrews. "Genetic and Fossil Evidence for the Origin of Modern Humans." *Science* 239 (March 11, 1988): 1263-1268. Proponents of the "out-of-Africa" hypothesis, the authors survey the evidence and find it supports their views. For the nonspecialist. Illustrations, with bibliography.

Trinkhaus, Eric, ed. *The Emergence of Modern Humans: Biocultural Adaptations in the Later Pleistocene*. Cambridge, England: Cambridge University Press, 1989. The debate between Stringer and Wolpoff is discussed, with the contributions of other scholars. Some articles are technical, but they contain the most current discussion of these issues. Outstanding bibliography.

Valladas, H., et al. "Thermoluminescence Dating of Mousterian 'Proto-Cro-Magnon' Remains from Israel and the Origin of Modern Man." *Nature* 331 (February 18, 1988): 614-616. The article that first announced the early date at Qafzeh. Although the data are technical, the conclusions are easily understandable. Gives the French point of view. Charts, with bibliography.

Lucy Jayne Botscharow

Cross-References

Boule Reconstructs the First Neanderthal Skeleton (1908), p. 428; Zdansky Discovers Peking Man (1923), p. 761; Dart Discovers the First Recognized Australopithecine Fossil (1924), p. 780; Weidenreich Reconstructs the Face of Peking Man (1937), p. 1096; Leakey Finds a 1.75-Million-Year-Old Fossil Hominid (1959), p. 1603; Anthropologists Discover "Lucy," an Early Hominid Skeleton (1974), p. 2037; Hominid Fossils Are Gathered in the Same Place for Concentrated Study (1984), p. 2279.

THE SEARCH CONTINUES FOR THE GENE THAT BEGINS MALE DEVELOPMENT

Categories of event: Biology and medicine
Time: 1987-1990
Locale: Cambridge, Massachusetts, and London

Two attempts to identify the gene on the human Y chromosome that triggers male development failed

> *Principal personages:*
> STEPHEN WACHTEL (1937-   ), an immunologist whose research group was one of the leading proponents of the theory that the protein HY was a key to mammalian sex determination
> DAVID PAGE (1941-   ), a molecular biologist whose interest in studying the chromosomes of males and females led to his discovery of the ZFY gene, a potential trigger for male development
> PAUL S. BURGOYNE, a developmental biologist whose work led him to reexamine Page's conclusions and review the work of others and conclude that the ZFY gene is not the gene that triggers male development

Summary of Event

Ever since the discovery that male development in humans was a result of having a Y chromosome, biologists have looked intensively for the responsible gene(s). As is so often the case, researchers thought on many occasions that they had the answer, only to find out later that their theories were in error. One promising attempt at identifying the gene for "maleness" came in a 1987 announcement by David Page, a molecular biologist at the Massachusetts Institute of Technology (MIT) who thought that a specific gene called ZFY was the key. New evidence concludes that the report was in error, and as of 1990, the gene for male development remained unknown.

Sex determination is controlled by a variety of different genetic and/or environmental mechanisms. In some animals, such as turtles, the temperature at which the eggs are incubated determines the sex of the hatchlings. In certain species of tropical fish, a single dominant animal becomes a male and produces a chemical that keeps all the other members of the school female. Once the dominant male dies or swims away, another adult female from the school changes sex and becomes male to take his place. In most animals, however, sex determination is controlled by the chromosomes rather than by the environment. Flies are female if they have two X chromosomes, while males have only a single X chromosome. In mammals, males have one X chromosome, while females have two X chromosomes. Yet, it is the presence of another chromosome, the Y, rather than possession of one X chromosome, that determines whether an individual will be a male. In 1959, C. E. Ford, W. J. Welshons, and L. B. Russell published papers that implicated the human Y chromosome as critical for male development. No human without a Y chromosome could be male, and all males possessed a Y chromosome.

Over the next thirty years, the role of the Y chromosome in development became clear. Early in human development, the embryo produces gonads, which are indistinguishable between males and females but will eventually become either testes or ovaries. In addition to the gonads, two complete sets of ducts are present: one set destined to become part of the male urogenital system and the other, the functional female system. At the sixth week of development, a major change occurs. If the embryo is XY, the gonads begin to develop into testes, and the male ducts continue to grow. If the embryo is XX, development proceeds to form ovaries with a functional female urogenital system.

Once triggered to develop, the testes of the early embryo produce two critical products which then guide the later development along the male path. These two products include the hormone testosterone, which acts to turn on a number of genes and processes that initiate male development. A second product, the protein AMDF (anti-Müllerian duct factor), causes the degeneration of the superfluous female duct system. The key to male development, then, is the trigger that causes testis development and the production of these two critical products. Supporting this conclusion is the experimental observation that if the gonad is removed from a mammal prior to the time it can produce testosterone and AMDF, the animal develops as an anatomically normal, but sterile, female. Thus, human development follows a female path unless an event occurs which can trigger testis formation.

The evidence for the involvement of the Y chromosome in testis formation was overwhelming. The Y chromosome contains many genes, and the question therefore arose as to whether the entire Y chromosome, or simply one or more genes located on it, would be responsible for testis development. This gene or set of genes was called testis determining factor (TDF).

As geneticists studied the chromosomes of humans, they found certain surprising results. About one in twenty thousand males has two X chromosomes and a similar number of females have an X and a Y. Explaining these unusual individuals gave geneticists the tools to seek out the TDF gene. All the XX males studied had a small portion of the Y chromosome attached to one of their other chromosomes. This rare genetic event is called a translocation. Geneticists have techniques for viewing chromosomes in the microscope and observing translocations. Likewise, all the XY females did not have a complete Y chromosome; in every case, they lacked a part of the Y chromosome. Obviously, the missing part was the same part that was translocated in the XX males. This piece of the Y chromosome must contain TDF, the gene for maleness.

Conceptually, all that remained was to determine the exact boundaries of the piece of the Y chromosome that contained the gene and to search this piece using the powerful methods of molecular biology and recombinant DNA technology to locate and study the TDF gene.

One of the first candidates for TDF was a gene called HY. In the 1970's, Stephen Wachtel and his colleagues at Memorial Sloan Kettering Cancer Research Center led a research group that studied the properties of testis cells and ovary cells. Using the tools of immunology, they discovered a protein called HY, which is found on the surface of testis cells but not on the cells of ovaries. Since the gonad of a six-week-old human embryo requires a signal to develop into a testis, Wachtel and others concluded that the signal was HY, as this protein is produced by a gene that resides on the Y chromosome. Supporting their hypothesis was the observation that in several of the exceptional XX males and XY females, the lack or presence of the HY protein seemed to govern whether development followed a male or a female path.

Contradictory evidence began to accumulate as studies of both humans and experimental animals were done in the late 1970's. Rare XX males were found whose testes were completely devoid of HY protein and rare XY females had functional ovaries that made HY. Obviously, HY was not TDF; now it is believed that HY plays a role in sperm development, an event that occurs much later in male development than the TDF trigger.

The next theory came in 1987, when David Page, at the Whitehead Institute of MIT, published a paper in the journal *Cell*. This paper excited the entire research community because it offered a clear and logical explanation for human sex determination. The candidate gene, ZFY (zinc finger Y chromosome), had molecular attributes which made it a plausible testis determining factor. Page reported an analysis of deoxyribonucleic acid (DNA) from the same groups of exceptional XX males and XY females. He looked at the DNA itself, rather than gonad proteins like HY. Page found in the chromosomes of a twelve-year-old XY girl, 99.8 percent of a normal Y chromosome, yet she had functional ovaries. Thus, the TDF must reside in this small 0.2 percent region of the Y chromosome, a region near the location of HY but distinct from it. Using the methods of recombinant DNA technology and DNA sequencing, Page isolated a candidate gene, and claimed that he had found the trigger for male development, TDF.

ZFY is a gene that controls the production of a protein that can bind to DNA and alter the function of other genes. This is exactly what one would expect a triggering gene would do: throw a switch, turning on ZFY, thus initiating a cascade of other events that would lead the gonad to become a testis. The ZFY protein is a member of a well-studied group of DNA-binding proteins called the zinc fingers. These proteins all possess one or more projections, or fingers, which contain an atom of zinc onto which DNA binds. Other zinc finger proteins are known to have the ability to turn groups of genes on and off. Thus, the ZFY gene was an appealing candidate for TDF.

Puzzling to Page and others in the field, however, was the fact that there was a second identical gene called ZFX on the X chromosome. Thus, Page's theory was confronted with the dilemma that the X chromosome would also have to play a crucial role in testis determination, unless it could be shown that the ZFX gene was nonfunctional.

Two years later, the weight of contradictory evidence to the idea that these zinc finger genes were testis determining factors became overwhelming. Paul S. Burgoyne summarized work in London done in his laboratory and in several others in the journal *Nature*. Burgoyne wrote that further examination of rare exceptional individuals had revealed that males lacking the ZFY gene could be found and that females possessing functional ZFY genes could, nevertheless, make ovaries. Obviously, like HY a decade or so before, ZFY plays some role in male development but clearly is not the triggering gene. The gene that triggers testis determination must still reside in the small portion of the Y chromosome where both ZFY and HY are located. Conventional recombinant DNA technology will ultimately provide the tools for analyzing this piece of DNA completely and should reveal the nature and function of the TDF trigger for male development.

Impact of Event

The testis determining factor (TDF) is one of a number of critical genes that has to function in the correct order and in the correct tissues of an embryo in order for development to proceed normally. This intricately regulated process of gene function is not well understood. One way to learn more about this process is to study genes like the TDF and learn how they make the cells in an embryo different. For example, how genes cause some cells in an embryo to become liver cells while others become skin or kidney or testis. A more complete understanding of how these genes function might enable biologists to intervene in this development should the normal process fail.

The ability to manipulate the TDF and control its function could permit geneticists to alter the sex of a mammalian embryo. Since genes, such as HY and ZFY, have been found in all mammals so far, geneticists expect that TDF also will be found in all mammals that use a Y chromosome to determine maleness. Manipulating the function of TDF would have obvious agricultural economic benefits. Animal breeders might, in some circumstances, prefer female offspring, such as in a dairy herd, while in other cases, they may prefer males for breeding. Control of the sex of their animal's offspring would then become yet another tool to make raising animals more productive.

A second very different result from the search for the TDF is that it shows how science works. Experimental evidence is collected using the best technology available at the time and hypotheses are constructed. Later, as newer procedures or new experimental designs become available, old hypotheses, such as HY or the ZFY being the trigger for testis development, are discarded and the search goes on for new data. The collaborative and competitive efforts of different research groups provide a constant set of checks and balances that ensure that hypotheses receive rigorous reevaluation and continual modification. Scientists can grow quite fond of their own hypotheses, especially those that stem from their own research. They must be willing, however, to abandon those hypotheses in the face of new and conflicting data. Progress in science depends upon the continual refinement of the current view of the world and a major strength of the scientific method is the scrutiny it provides this view.

Bibliography

Burgoyne, Paul S. "Thumbs Down for Zinc Finger?" *Nature* 342 (December, 1989): 860-862. A summary of the arguments against Page's ZFY gene hypothesis. Contains all the references to the relevant research in the original literature but is written as a news summary.

Craig, Ian. "Sex Determination: Zinc Fingers Point in the Wrong Direction." *Trends in Genetics* 6 (May, 1990): 135-137. This is another presentation of the evidence against the ZFY gene being the TDF. Written for a general college-level audience, this article also contains references to the primary research literature.

Gilbert, Scott F. "Sex Determination." In *Developmental Biology.* 2d ed. Sunderland, Mass.: Sinauer Associates, 1988. An excellent college-level description of the various mechanisms by which the sex of an animal is determined. Covers the environmental, genetic, and hormonal controls of this process.

Ohno, Susumu. *Major Sex Determining Genes.* New York: Springer-Verlag, 1979. An advanced college-level text that contains much of the primary research leading to the theory that the HY gene was the trigger for testis development. There is also much basic information about sex determination in mammals that is more readable.

Raven, Peter H., and George B. Johnson. *Biology.* 2d ed. St. Louis: Times-Mirror/ Mosby, 1989. A general college-level text that contains well-illustrated discussions about sex chromosomes and early embryonic development. A complete glossary is included.

Roberts, Leslie. "Zeroing in on the Sex Switch." *Science* 239 (January 1, 1988): 21-23. This article sets out, in a clear and readable fashion, the major events that led to Page's discovery of ZFY.

Joseph G. Pelliccia

Cross-References

McClung Plays a Role in the Discovery of the Sex Chromosome (1902), p. 148; Bevis Describes Amniocentesis as a Method for Disclosing Fetal Genetic Traits (1952), p. 1439; A Human Growth Hormone Gene Transferred to a Mouse Creates Giant Mice (1981), p. 2154; Clewell Corrects Hydrocephalus by Surgery on a Fetus (1981), p. 2174; The First Successful Human Embryo Transfer Is Performed (1983), p. 2235; Murray and Szostak Create the First Artificial Chromosome (1983), p. 2251; Willadsen Clones Sheep Using a Simple Technique (1984), p. 2273.

SUPERNOVA 1987A CORROBORATES
THE THEORIES OF STAR FORMATION

Category of event: Astronomy
Time: February 23, 1987
Locale: Cerro Tololo InterAmerican Observatory, Chile

Observation of supernova explosion and measurement of neutrinos reaching Earth confirmed theoretical physics of star structure and evolution

Principal personages:
TYCHO BRAHE (1546-1601), a Danish astronomer who observed a supernova with the naked eye in 1572
JOHANNES KEPLER (1571-1630), a German astronomer who observed a supernova with the naked eye in 1604
IAN SHELTON, an astronomer who discovered a supernova with the naked eye on February 23, 1987
ALBERT JONES, an amateur skywatcher who observed a supernova at the same time as Shelton

Summary of Event

Supernova 1987A appeared as a bright point of light in the Large Magellanic Cloud, one of a pair of small galaxies some 100,000 light-years from the Milky Way and situated for earthbound observers high in the southern sky. It was bright enough to be seen with the naked eye; an astronomer walking outside Cerro Tololo Observatory in Chile spotted the phenomenon almost by accident, the first supernova visible to the unaided eye in nearly four centuries. Ian Shelton of the University of Toronto Southern Station in Chile observed a supernova on February 23, 1987; Albert Jones, an amateur skywatcher, viewed the supernova with the naked eye almost simultaneously with Shelton. (The Danish astronomer Tycho Brahe was on hand to describe one in 1572, and the Czech astronomer Johannes Kepler observed a supernova in 1604.)

The rarity of supernovas is caused by, in large measure, the physical conditions required. When gravitational forces in a gaseous nebula become sufficient to trigger thermonuclear reactions at the core, a star begins life in what astrophysicists call the "main sequence." This enormous generation of energy balances the tendency of the nebula to collapse under the weight of its own mass. The core reactions begin with the lightest element, hydrogen, fusing hydrogen nuclei to form helium nuclei and releasing energy. The core of a star, thus, is a long-term exploding hydrogen bomb of gargantuan size. Eventually, when the hydrogen in the core has been consumed, the thermonuclear furnace will turn to helium, fusing its nuclei into still heavier elements. Outer portions of the star, still hydrogen-rich, will expand as core temperatures increase, and the star will become a giant.

At this point in the evolution of a star, mass becomes a critical factor. A star with mass comparable to or less than the sun—which is to say the great majority of stars—will expand to become a huge, tenuous, and hot nebula, with the outer layers still consuming light elements, while the core, now contracting and thus generating ever higher temperatures, turns to still heavier ones. By the time the star reaches a stage where the core begins to consume oxygen or carbon, energy generated by thermonuclear reactions will no longer be sufficient to maintain the expanding outer layers against gravitational collapse. After having expanded beyond the orbit of Earth, the sun will experience a lingering death. It will become a white dwarf, with its core matter packed so densely together that its diameter will be only about that of Earth, but it will end its life as a cool, dark, carbonized "cinder." The evolutionary cycle of an "average" star such as the sun covers 10-15 billion years.

Stars that begin life with masses greater than about 1.4 times that of the sun have a foreshortened life cycle and a far more dramatic denouement. Their core reactions proceed at much higher temperatures, and they consume hydrogen more quickly as a result. Stars up to about 10 solar masses may drive off huge amounts of matter into stellar nebulas in their late stages of life, thus reducing their mass. For more massive stars, however, the climax is the unparalleled fury of a supernova explosion.

In their final throes, massive stars develop a core of iron. At this point, thermonuclear reactions, rather than generating energy and fusing nuclei into still heavier elements, merely shatter the nuclear structure of iron and drain energy from the star. Surrounding shells of gas, where lighter elements continue to generate thermonuclear energy, collapse upon the core. Ever-increasing gravitational force actually squeezes electrons into nuclei, forming an almost unimaginably dense neutron core. Neutrinos, weakly interactive and possibly massless particles, are driven away from the star by this process, carrying energy with them and thus hastening gravitational collapse. The supernova explosion occurs as shells of lighter elements carom off the neutron core and collide with other, slower shells of gas collapsing upon the core.

Inasmuch as nearly all theory in nuclear physics and astrophysics developed in the twentieth century, sighting of the first visual supernova in four hundred years offered an unprecedented opportunity to put these models to the test. Within hours of the first sighting of Supernova 1987A (SN1987A), scientists all over the world turned their attention to it. The first problem was to identify which star was involved. Since supernovas are extremely hot in their early stages, they should radiate strongly in ultraviolet wavelengths. The International Ultraviolet Explorer satellite, already in orbit, confirmed quickly that the supernova had been discovered in an early stage but was cooling rapidly. Astronomers identified the exploding star as a blue supergiant of about 50 solar masses. Therefore, it was an extremely strong supernova that had not been able to eject matter into a surrounding nebula before collapsing. It was somewhat unexpected that a blue supergiant should become a supernova. Outer layers of the gas nebula presumably would exist at lower temperatures than required for a bluish luminosity and would show up instead in a reddish color. Red supergiants, in fact, were the common models for supernova events.

Astrophysical theory requires that a supernova event featuring a blue supergiant be particularly energetic, since a much larger mass of matter should be contained in the star. Astrophysicists describe supernovas that are not preceded by ejection of matter into a nebula as Type II events. These are particular rarities among observed supernovas and are of extraordinary importance. (Type I events require that a star receive matter from other stars in a multiple star system.)

Supernovas release nearly all of their explosive energy in the form of neutrinos, weakly interactive particles possessing little or no mass. The upsurge of neutrinos from SN1987A should have been detectable on Earth. Because of their physical characteristics, almost all neutrinos reaching Earth actually pass through the planet as if it does not exist. A few, however, collide with particles. These collisions may be detected in a number of large tanks constructed deep underground to protect their experiments from cosmic radiation and filled with water or chlorine-rich liquids. Occasional neutrino collisions with atoms in the water generate minute flashes of light or, in the case of chlorinated liquids, neutrino impact with a chlorine atom creates radioactive argon. Although these detectors originally were constructed merely to ascertain the existence of neutrinos in the universe, they were excellent facilities for detecting incoming neutrino bursts from SN1987A.

Neutrino detectors installed deep in an abandoned South Dakota gold mine, a salt mine near the southern shore of Lake Erie in Ohio, in Europe, and in Japan, all detected an upsurge in these rare neutrino collisions, confirming crucial aspects of astrophysical theory. Measurements of the energy possessed by neutrinos detected in these underground tanks permitted astrophysicists to reconstruct in considerable detail events in the final stages of the collapse and explosion of the blue supergiant. SN1987A is estimated to have given off more energy when it exploded than an entire galaxy of average size emits in a year. Temperatures in the core at the time of explosion must have exceeded 10 billion degrees.

As the debris of the explosion thinned in its outward rush, intense gamma radiation should have become detectable from the supernova core. Again, theory proved accurate when, in December, 1987, the gamma radiation detector on the Solar Maximum Mission satellite began to register an increase. Theory also predicts that a neutron star will remain at the center of SN1987A. That body could turn out to be a pulsar, a neutron star that rotates at immense speed so that its radio and X-ray emissions sweep through space like a beacon. At the end of 1990, uncertainty remained as to whether a pulsar had been detected or, if not, how much time must elapse before a pulsar signal could be detected through the still thinning gas envelope of the explosion. There remains the possibility that the magnetic field of the pulsar, which focuses the emissions into relatively narrow beams, may be pointed away from Earth.

Impact of Event

As a phenomenon that confirmed a large and critically important matrix of astrophysical theories developed over the last century, SN1987A was an extremely significant and reassuring event. Physical models of the universe, which depended heavily upon these theories, could be called upon now as explanatory paradigms with much greater confidence.

SN1987A represented an unusual opportunity to determine whether the neutrino possesses mass. This question illustrates the many interconnections between theories of stellar evolution and cosmology. Scientists already realized that neutrinos were nearly without mass; an electron carries at least twenty-five thousand times the mass of a neutrino. Nevertheless, theory predicts that neutrinos are about 100 million times as abundant as electrons in the universe, so that if they prove to have any mass at all, neutrinos become a significant portion of the overall mass of the universe. Determining this overall mass could help to settle the most basic debate in cosmology: Is the universe destined to expand forever, or does it contain sufficient mass for gravitational forces eventually to halt the expansion and cause contraction, presumably ending in another singularity or beginning similar to the big bang event thought by most cosmologists to have given birth to the present universe?

Unfortunately, neutrino measurement experiments did not yield conclusive results. Theoretically, if the neutrino does not possess mass, all neutrinos should travel at the speed of light and, according to relativity theory, arrive at a given observation point at the same time. If they possess even a minute mass, those with more energy should arrive slightly ahead of others. All neutrino detections from SN1987A occurred within a span of less than ten seconds. Discrepancies in time of arrival could be caused by differing energy levels for particles with mass, but also by processes of neutrino emission from the supernova which, argues the theory, probably takes about ten seconds. Smaller differences also could have been caused by differing locations of the detectors. SN1987A added greatly to the store of data on neutrinos, but three years after sighting the explosion, physicists still had not solved the mystery of neutrino mass.

One very significant aspect of SN1987A tended to be overlooked in the excitement of the event: The sighting was in the southern celestial hemisphere. Only in the last quarter of the twentieth century have large telescopes been installed in the Southern Hemisphere in sufficient numbers to allow comprehensive observation programs. SN1987A was only about 1 percent as bright as the supernova named for Tycho Brahe in 1572. Had it exploded only one human generation earlier, instrumentation for studying the phenomenon would have been scarce and far less sophisticated. The event might have been missed altogether by astronomers in high latitudes of the Northern Hemisphere. Cave paintings and rock inscriptions by indigenous groups in the Southern Hemisphere suggest a number of important events in the southern skies in recent centuries—some perhaps supernovas—for which there is no record of observation in Europe or North America.

SN1987A quickly became one of the most intensively studied astronomical phenomena in history and will provide an additional wealth of information as later phases of the event unfold. It is likely to remain a focus of considerable research well into the twenty-first century.

Bibliography

Arnett, W. David, John N. Bahcall, Robert P. Kirshner, and Stanford E. Woosley.

"Supernova 1987A." *Annual Review of Astronomy and Astrophysics* 27 (1989): 629-700. Thorough scientific coverage of the event. Portions are technical, but summaries and interpretations are valuable for general readers. Exhaustive scientific bibliography.

Asimov, Isaac. *The Exploding Suns.* New York: E. P. Dutton, 1985. A standard popular treatment of supernovas, relevant theory of their development, and their place in the history of astronomy prior to the 1987 event.

Bahcall, John N. *Neutrino Astrophysics.* Cambridge, England: Cambridge University Press, 1989. This highly technical work discusses the design and execution of experimental techniques in neutrino detection and requires considerable mathematical and scientific background.

Genet, Russell, Donald Hayes, Donald Hall, and David Genet. *Supernova 1987A: Astronomy's Explosive Enigma.* Mesa, Ariz.: Fairborn Press, 1988. Somewhat technical account of the phenomena that assumes some scientific background of the reader. Organized from papers from the June, 1987, meeting of the American Astronomical Society devoted largely to the supernova event.

Kafatos, Minas C., and A. G. Michalitsianos, eds. *Supernova 1987A in the Large Magellanic Cloud.* Cambridge, England: Cambridge University Press, 1988. Highly technical but exceptionally well-organized material from a conference at George Mason University eight months after the event. Contains papers predicting gamma-ray emissions and the light echo phenomenon in advance of their observation.

Marschall, Laurence A. *The Supernova Story.* New York: Plenum Press, 1988. Highly readable account of the history of supernova observation and theory with clear explanations for lay readers. The story of Supernova 1987A is covered in several chapters. Includes an extensive glossary of astronomical and astrophysical terms and a helpful bibliography. The best general work on the topic.

Schorn, Ronald A. "Neutrinos from Hell." *Sky and Telescope* 73 (May, 1987): 477-479. Synopsis of strategies for measuring neutrinos from 1987A and their theoretical significance in reconstructing conditions in the core of the exploding star.

_____. "Supernova 1987A After 200 Days." *Sky and Telescope* 74 (November, 1987): 477-479. Update on supernova observations in the first six months of 1987A. Exemplary of numerous articles on the phenomenon in this widely circulated publication.

Verschuur, L. "The Peculiar Pulsar in Supernova 1987A." *Astronomy* 17 (September, 1989): 20-26. Discussion of the vigil for the emergence of a pulsar at 1987A, and the equivocal early data showing that a pulsar may exist but with highly unusual properties.

Woosley, S. E., and M. M. Phillips. "Supernova 1987A!" *Science* 240 (May 6, 1988): 750-759. Excellent early summary of observational data and comparison with established theory. Also discusses process of identification of a specific star and describes phenomena inconsistent with or not accounted for by theory.

Ronald W. Davis

Cross-References

Russell Announces His Theory of Stellar Evolution (1913), p. 585; Millikan Names Cosmic Rays and Investigates Their Absorption (1920), p. 694; Michelson Measures the Diameter of a Star (1920), p. 700; Chandrasekhar Calculates the Upper Limit of a White Dwarf Star's Mass (1931), p. 948; Davis Constructs a Neutrino Detector (1967), p. 1830; Bell Discovers Pulsars, the Key to Neutron Stars (1967), p. 1862; Cassinelli and Associates Discover R136a, the Most Massive Star Known (1981), p. 2164.

MILLER DISCOVERS A DINOSAUR EGG
CONTAINING THE OLDEST KNOWN EMBRYO

Category of event: Earth science
Time: September, 1987
Locale: Cleveland-Lloyd Dinosaur Quarry, Emery County, Utah

Miller, Hirsch, Stadtman, and Madsen discovered a fossilized dinosaur egg containing the oldest known embryo of any kind, about 150 million years old

Principal personages:
  WADE MILLER (1932-   ), an American geologist from Brigham Young
    University who discovered the complete embryo of a dinosaur in an
    egg in the Cleveland-Lloyd Dinosaur Quarry
  KARL HIRSCH, an American paleontologist who assisted in the study of
    the 150-million-year-old Jurassic egg and embryo
  KENNETH L. STADTMAN, an American geologist from Brigham Young
    University who assisted in the discovery of the 150-million-year-old
    Jurassic egg and embryo
  JAMES H. MADSEN, JR., an American paleontologist who assisted in the
    discovery and study of the 150-million-year-old Jurassic egg and em-
    bryo

Summary of Event

The Cleveland-Lloyd Dinosaur Quarry in Emery County, Utah, is the site in which more than twelve thousand unarticulated, or separated, bones of dinosaurs were found to represent seventy plus individual dinosaurs, or at least twelve genera. The massive amounts of remains found in the area can be explained by the possibility that the lowest member of the fossil-bearing beds of the Brushy Basin Member of the Upper Jurassic Morrison Formation may once have been a marsh or a shallow lake in which the massive reptiles became mired. Two geologists from Brigham Young University, Wade Miller and Kenneth L. Stadtman, along with colleagues from Boulder, Colorado, Karl Hirsch, and Salt Lake City, Utah, James H. Madsen, Jr., are credited with finding a fossilized egg from the 100-million-year gap in the fossil record in geologic time between the Lower Jurassic period and the upper Lower Cretaceous. The egg from the Upper Jurassic period contains an embryo that is approximately 150 million years old—the oldest known embryo yet found of any kind of living creature. The egg was discovered with thousands of other dinosaur bones instead of in the traditional nest.

Discovered by John Horner and Bob Makela, nests are believed to have been made by one or both of the parents of the young dinosaurs. First found in great concentration on a Montana cattle ranch at the eastern boundary of the Two Medicine formation, nests are small hollows, or concave depressions, believed to be the location in which the eggs were possibly laid, hatched, and nurtured for a year or more by the mother and/or father. Horner and Makela found several of these hemisphere-like depressions, each of which contained the bones of baby dinosaurs, all at the same stage of development. Each nest was set off from the red mudstone of the area by the green mudstone, which filled in the cavity of the nest and contained the dinosaur bones. The traditional explanation of bodies found in the same location is that they all died together or that they were moved by natural forces to the same spot after death. In the case of the nests, the identical stages of development indicate the contrary—that these animals were together and not by coincidence. Another inter-esting speculation that has arisen from the idea of nests is that dinosaurs were warm-blooded, not cold-blooded like modern reptiles. There were several indications that the bones were those of babies, such as the size of their bones. On the contrary, there were indications also that the young were quite mature, such as the worn-down quality of their teeth (nearly three-quarters gone) and the hardness of the tendons found on the spine, which are normally soft in hatchlings. The babies were ill-equipped to venture from the nest among adult dinosaurs to compete for food; there-fore, one or both parents must have brought the food. Modern reptiles, which are cold-blooded, take a long time to grow; if the dinosaurs were cold-blooded, the growth process would keep them in the nest for a time as long as a year, and such a lengthy stay in the nest has never been witnessed in modern cold-blooded reptiles. In comparison, warm-blooded creatures, such as birds, grow very quickly and in the nest. In any case, those babies were growing in the nest and at a fast pace.

The fact that the egg from the Utah site containing the embryo was not found in a nest, in addition to two other unusual factors, is a possible indication that the egg was retained in the oviduct, or tube, through which the egg passes from the ovary in the birth process, a phenomenon termed oviducal retention. The egg was found broken into two parts, connected by a hingelike area. The halves were both filled with a fine sediment similar to the shale that enclosed it. Computerized study of the egg revealed that it had fractured on the hingelike structure. The distortion that occurred in other places on the shell and its inverted curvature are both indications that, at some point during the breakage and deformation, the shell was at least semipliable. The only time a normally rigid eggshell is pliable is during the few minutes directly after the beginning of the shelling, the term used for the birth pro-cess of reptiles, or the oviposition. Furthermore, the shell can be pliable only in the oviduct. If the two halves of the egg were put together, it would be roughly 110 by 55 millimeters. The structure of the shell is composed of calcite and calcium carbo-nate, and the primary, or original, shell is preserved perfectly and coated by another layer, the pathological layer, caused by unusual circumstances.

The second unusual factor in the case of the embryo is the multilayering of the shell because of pathological, or disease-related, circumstances at the time of burial. The pathological condition of the eggshell, which is found in modern and fossil reptiles and mammals alike, can be caused by stress, disease, or other environmental conditions. The mother must be alive when the egg shell develops extra layers as a result of retainment in the oviduct; however, the length of time it takes to form the extra pathological layer of the shell is unknown. The embryo may live for a short time and even continue to develop, but the exchange of vital gases is hindered by the misalignment or maldevelopment of pores in the second layer in relation to the first, the end result of which is death by suffocation for the embryo. The hypothesis is that the mother was disturbed before she could lay her egg in the nest. The excellent condition in which the egg was found is perhaps a result of its being held together in the oviduct after fracturing until it was preserved by sediments. The structure of the eggshell itself is that of rigid interlocking units and columns. The secondary pathological eggshell is half the size of the primary eggshell at 0.5 millimeter. Its structure is similar to the primary shell but not fully developed in critical areas, such as the continuation of pores from the primary eggshell to the secondary eggshell, causing poor or restricted facility for gas exchange. The egg cannot be identified either by the known species and genera of dinosaurs or by the other dinosaurs found in the dig. The distinction that makes this eggshell different from any other among the Cleveland-Lloyd fauna is its eggshell structure, which contains a new type of pore canal, and its embryonic remains, which are, as yet, unidentifiable. In addition to finding the oldest known embryo, the team of four paleontologists may have found a new dinosaur.

Impact of Event

The discovery of Miller and his colleagues in the fossil beds of the Cleveland-Lloyd Dinosaur Quarry added to modern knowledge in the field of paleontology. The oldest known and complete embryo preserved perfectly for 150 million years gives researchers a chance to learn more about the process of egg formation and dinosaur embryo development. In addition, the fact that the dinosaur cannot be classed in any of the genera found at the same site is an indication that the egg might be of a new type of dinosaur. Adding to this possibility is the discovery of another specimen, in the Morrison Formation of Colorado in the Jurassic egg site, that has similar structure in the eggshell as well as the pore type. The researchers have mentioned the need to establish a structural morphotype for the eggs found from the Jurassic period.

The state of the embryo from Utah, which is most likely the result of oviducal retention, is a new addition to the study of egg retention and pathological eggshell formation. Because there has been no systematic study of such phenomena, these are areas of process that are still being questioned; however, because oviducal retention still occurs in the mammals and reptiles of the modern era, there is good information available on possible causes of the abnormality. Sometimes, retention causes the movement of the egg backward into the area of its formation, a process called reverse peristalsis. When reverse peristalsis does not occur, the egg develops a second membrane in the oviduct. The Jurassic egg found in Utah is clearly the result of the development of a second membrane and calcareous layer, or one containing calcium carbonate. The Jurassic eggshell is typical of modern pathological eggshells, and so adds to the body of study available on the shelling process in some dinosaurs and, hence, the indirect study of the oviduct. The discovery of dinosaur eggs in their entirety, eggshells that are pathological, and remains of embryos are all extremely rare, especially from the time before the Cretaceous. The discovery in Utah had a great impact on the world of paleontology, especially in view of the possibility of a new dinosaur in the discovered fossilized embryo.

Bibliography

Alexander, R. McNeill. *Dynamics of Dinosaurs and Other Extinct Giants.* New York: Columbus Press, 1989. For everyone who is interested in dinosaurs, including scientists and nonscientists, students and professors. Each chapter has a principal reference source consisting of approximately ten titles. Covers such vital statistics as the weight, footprints, strength, necks and tails, fighting and singing, hot-blooded characteristics, flying marine, and the death of dinosaurs. An informative style with a minimum of scientific language. A few illustrations and a number of drawings and graphs. An excellent reference; highly recommended.

Bakker, Robert T. *The Dinosaur Heresies: New Theories Unlocking the Mystery of the Dinosaurs and Their Extinction.* New York: William Morrow, 1986. Written by an adjunct curator at the University Museum, University of Colorado, a skilled artist and scientist; illustrated with the author's own reconstructions of dinosaurs alive and in action. No scientific jargon; an excellent reference source for the student and layperson. Includes several pages of notes and references for each chapter. Of particular interest are the two parts on warm-blooded dinosaurs.

Hirsch, Karl F., Kenneth L. Stadtman, Wade E. Miller, and James H. Madsen, Jr. "Upper Jurassic Dinosaur Egg from Utah." *Science* 242 (March, 1989): 1711-1713. Featured in a magazine whose contributors are scientists and researchers. The best source of precise knowledge regarding the egg and the state in which it was found. Nonprofessionals might need to consult dictionaries of paleontology to clarify terms and may also want to consult books in which dinosaurs are discussed to learn more specifics on the background of eggs and the birth, or shelling, process.

Horner, John R., and James Gorman. *Digging Dinosaurs: The Search That Unraveled the Mystery of Baby Dinosaurs.* New York: Workman, 1988. An excellent reference to use as a companion to the article by the discoverers of the egg. Discusses the mystery of baby dinosaurs. Written by one of the men who first discovered dinosaur nests, the key to the past that unlocked many secrets of the ancient creatures and opened new doors of insight into other possibilities.

Hotton, Nicholas. *Dinosaurs.* New York: Pyramid, 1963. Covers dinosaurs and paleontology; good for readers who may want to gain general knowledge into the periods of geologic time in which dinosaurs inhabited Earth. Contains famous digging sites and their finds, as well as several illustrations to aid the understanding of fossil finding and other related work.

Matthews, William H. *Fossils: An Introduction to Prehistoric Life.* New York: Barnes & Noble, 1962. A good reference regarding the process of preservation that yields fossils. Includes a discussion on finding and preparing fossils, geologic history, paleontology, and many other topics of related interest. Features a bibliography, photographs, charts, glossary, list of museums, sources of further information, and dealers in fossils.

Paul, Gregory S. *Predatory Dinosaurs of the World: A Complete Illustrated Guide.* New York: Simon & Schuster, 1988. Written by a free-lance dinosaurologist who combines original research in the field with detailed illustrations to express new ideas about dinosaurs. Written for the nonscientific reader. A unique focus on predaceous dinosaurs. Presents skeletal restorations of every predator dinosaur species for which a restoration can be done. An excellent reference. Includes a list of references and notes for each chapter and an extensive bibliography of approximately three hundred references.

Earl G. Hoover

Cross-References

Andrews Discovers the First Fossilized Dinosaur Eggs (1923), p. 735; Barghoorn and Tyler Discover 2-Billion-Year-Old Microfossils (1954), p. 1481; Barghoorn and Coworkers Find Amino Acids in 3-Billion-Year-Old Rocks (1967), p. 1851.

ERLICH DEVELOPS DNA FINGERPRINTING FROM A SINGLE HAIR

Category of event: Biology
Time: 1988
Locale: Cetus Corporation, Emeryville, California

Erlich's DNA fingerprinting technique allowed determination of the identity of an individual from the DNA in a single hair

Principal personages:
> HENRY ERLICH, a molecular geneticist who helped develop the polymerase chain reaction method of DNA amplification and its application to DNA fingerprinting
> KARY B. MULLIS, a molecular geneticist who invented the polymerase chain reaction method of DNA amplification and collaborated on its application to DNA fingerprinting
> ALEC JEFFREYS (1950-   ), a molecular geneticist who first applied the technique of DNA fingerprinting to the identification of individuals

Summary of Event

Deoxyribonucleic acid (DNA) fingerprinting is the identification of an individual from a biochemical description of their unique genetic makeup. Every human has approximately 100,000 genes in the chemical form of DNA. The genetic information coded in these genes is variable enough so that no two humans, except for identical twins, have exactly the same genetic code. The genetic variability among humans is expressed through differences in obvious traits such as eye color and genetic disorders such as hemophilia; however, more genetic variability is hidden from view and can be detected only by direct biochemical analysis of DNA itself. Direct chemical analysis of DNA reveals so much variation in the genetic code from one person to the next that a biochemical description of even a small part of the entire genetic code can identify an individual's unique combination of genetic traits. A description of one's DNA that is detailed enough to distinguish it from any other person's DNA is a genetic "fingerprint."

Genetic fingerprints can be used in place of traditional fingerprints, for purposes of identification, to match samples of dried blood, skin, hair, or semen of suspects in criminal investigations or to determine paternity with virtual certainty. The first DNA fingerprinting technique, developed by Alec Jeffreys in 1985, requires a small sample of blood or tissue that is preserved well enough for large pieces of DNA to remain intact. In 1988, Henry Erlich developed a method of DNA fingerprinting so sensitive that it can be used to obtain a DNA fingerprint from a single hair cell, badly degraded tissue, or less than a millionth of a gram of dried blood that is thousands of years old. Erlich's technique is based on a combination of traditional DNA fingerprinting with a powerful technique called polymerase chain reaction (PCR)—developed by Kary B. Mullis—that can amplify trace amounts of DNA up to a million times to generate quantities large enough for DNA fingerprinting.

Normally, the amount of DNA associated with a single hair is insufficient for DNA fingerprinting. Erlich overcame this limitation by first amplifying the DNA in a single hair with PCR. Erlich mixed the trace amount of DNA in a hair cell with an enzyme, DNA polymerase, which directs the replication of DNA sequences in living cells, plus the raw chemicals the enzyme needs to synthesize new DNA and short DNA sequences called "primers" that the DNA polymerase enzyme needs to get started.

DNA's molecular structure takes the form of a "double helix" composed of two long strands of DNA twisted around each other and connected across the strands, much like a twisted ladder. The two strands of the double helix must be separated, and the rungs broken for the DNA polymerase enzyme to copy the genetic code and duplicate the DNA. The two strands are broken apart in living cells by additional enzymes; but in the PCR reaction, they must be heated to force them apart. Once the strands of the double helix are broken apart and cooled down, the DNA polymerase restores a new second strand to each single separated strand, forming two new DNA molecules, each made up of one new strand and one original strand. Once the reaction has run its course, the amount of DNA has been doubled. The amount of DNA in a single hair, however, is so small that simply doubling the original amount is still insufficient for DNA fingerprinting. The doubling process can be repeated again in the test tube by heating the new DNA until the double strands separate again, allowing DNA polymerase to repeat the reaction. The new DNA, which has been doubled again to four times the original amount, can be redoubled by another round of heating and cooling. The successive rounds of heating, cooling, and doubling constitute the polymerase chain reaction, and the reaction can be run efficiently for approximately twenty-five rounds to amplify the trace amounts of DNA fragments up to about 1 million times the original amount. The PCR technique then, effectively allows the DNA in a single hair to be amplified to an amount equivalent to that found in a million identical strands of hair. The amplified DNA can then be used to obtain a DNA fingerprint.

Erlich and his colleagues used the amplified DNA from a single hair to analyze a histocompatibility gene. Histocompatibility genes code for the tissue-type markers that must be matched in organ transplants because they stimulate attacks from the immune systems of individuals with different tissue types. Histocompatibility genes are highly variable from one person to the next, which is why they are so useful for DNA fingerprinting, since the probability that two unrelated individuals will have the same tissue type is extremely low. DNA fingerprints that are as unique as 1 in 10,000 to 1 in 100,000 can be obtained by analyzing tissue-type genes. Adding other types of genes to the analysis can generate DNA fingerprints that have zero probability of matching with another person, except for an identical twin. Differences in the histocompatibility gene chosen by Erlich for typing were identified by matching DNA probes constructed for this purpose. DNA sequences have the property of self-recognition, and Erlich used this property by preparing samples of the known variants of the histocompatibility gene. Each variant form can recognize matching forms identical to itself in an unknown sample and bind to them but will not bind to any of the other forms. Erlich took the samples of amplified DNA from the hair cells and applied each probe to each unknown sample. The probes—representing the different variants of the histocompatibility gene—stick only to their own form and have a visible stain attached to them so that the high concentrations of probe molecules that stick to a matching sample can be visualized. Erlich was able to identify the differences in histocompatibility genes from the amplified hair cell DNA samples by determining which probes remained attached to each sample.

Erlich and his colleagues showed that the results obtained from single hairs were confirmed by results obtained from blood samples taken from the same people who donated the hair. The technique was also successfully used on seven-month-old single hair samples. One of the first forensic applications of the PCR-DNA fingerprinting technique took place in Pennsylvania. A one-year-old body was exhumed in a homicide case to be examined for evidence, since a previous autopsy was deemed suspicious. The prosecution accused the defendants of tampering with the body by switching some of its internal organs with those of another body to conceal the cause of death. The PCR-DNA fingerprinting technique was used to show that the child's embalmed organs exhumed with the body did in fact match the victim's tissue type, and the defendants were acquitted of the tampering charge.

The DNA fingerprinting technique developed by Erlich and his colleagues at Cetus Corporation was made commercially available in early 1990 in the form of a DNA-typing kit, allowing more widespread application of the polymerase chain reaction to DNA fingerprinting.

Impact of Event

Henry Erlich's refinement of DNA fingerprinting to the point where an individual can be identified from the DNA in a single hair means that one hair, or even microscopic samples of dried blood, skin, or other body fluids found at the scene of a crime, can be analyzed to determine whose body it came from with nearly 100 percent accuracy. Properly used, DNA fingerprinting can be so precise that the margin of error in making a match between biological evidence and a suspect's DNA is less than one in ten thousand, with tests based on tissue-type genes from a single hair, and less than one in a billion with more extensive testing.

Erlich's method of DNA typing from a single hair is a dramatic refinement of DNA fingerprinting. Hair is one of the most common types of biological evidence left behind at crime scenes, so Erlich's improvement over traditional methods of analyzing hair color, shape, and protein composition can be widely applied. Erlich's technique also allows substitution of hair samples for blood or skin when DNA fingerprints are taken; the technique can be automated, making it easier to apply DNA fingerprinting to large populations along with traditional fingerprinting.

The main disadvantage to DNA fingerprinting is the practical difficulty of transferring a new, highly technical procedure from the research laboratory to routine application in the field. While DNA fingerprinting is virtually 100 percent accurate in theory, and works in the hands of highly trained scientists, methods for reliable and economical mass application must be developed and proved before DNA fingerprinting becomes routine.

Evidence based on DNA fingerprinting was introduced for the first time in several dozen court cases in the late 1980's and played a key role in many of them. Widespread adoption of DNA fingerprinting, however, has been hampered by criticism of the reliability of some earlier tests conducted by commercial firms. Most legal experts and forensic scientists, however, expect DNA fingerprinting to become routine since the principles behind it are sound and the technology used for DNA analysis is advancing rapidly.

Erlich's contribution to DNA fingerprinting technology has made this revolutionary technique much more valuable to forensic science. DNA fingerprinting from single hairs and other minute biological samples can provide unequivocal identification in criminal cases where biological evidence is available. The technique can be applied also to paternity testing, medical diagnosis of genetic disorders, and cases involving identification of animals, such as wildlife poaching. The fact that a person's DNA can provide so much information about them, besides simply their personal identity, raises ethical questions about the use of DNA fingerprinting that have not been encountered with the use of traditional fingerprinting. The potential accuracy of DNA fingerprinting, however, and the fact that such absolute certainty can exonerate criminal suspects who are innocent as well as help convict those who are guilty, makes the responsible use of DNA fingerprinting the most important advance in forensic science since the advent of traditional fingerprinting.

Bibliography

Appenzeller, Tim. "Democratizing the DNA Sequence." *Science* 247 (March, 1990): 1030-1032. A nontechnical explanation of the polymerase chain reaction (PCR) technique and its impact on biology and biological applications. This is a useful background article for fully understanding Erlich's DNA fingerprinting technique, since Erlich was also involved in developing the PCR and it is the basis of his DNA fingerprinting method. This article discusses primarily the research and medical applications of PCR, as well as legal disputes between the Cetus Corporation and Du Pont over the patent rights to PCR.

Kirby, Lorne T. *DNA Fingerprinting*. New York: Stockton Press, 1990. This is one of the first books on the topic of DNA fingerprinting that is written for a general audience. The book is written for scientists unfamiliar with the techniques and for lawyers and criminologists who want an authoritative, comprehensive, and practical introduction. The chapters cover topics from basic genetic theory and techniques used to legal and ethical issues and case studies. Includes a glossary, name index, and subject index.

Lewin, Roger. "DNA Fingerprints in Health and Disease." *Science* 233 (August, 1986): 521-522. A nontechnical research news article in the world's most widely read science journal. This article explains the earlier DNA fingerprinting technique developed by Jeffreys. Lewin discusses the impact of DNA fingerprinting on society and describes a case study involving immigration law and maternity. Further references to the scientific literature are included.

_____. "DNA Typing on the Witness Stand." *Science* 244 (June, 1989): 1033-1935. A nontechnical news and comment article describing the legal issues surrounding the initial applications of DNA fingerprinting as evidence in criminal trials. The article focuses on the Castro murder case in New York City, in which expert scientific witnesses from the prosecution and defense met outside the courtroom during the trial, independently of legal counsel, to assess the scientific validity of DNA fingerprinting evidence introduced by the prosecution. The scientists issued a statement that affirmed the principles behind DNA fingerprinting, but questioned the validity of the commercially prepared DNA typing introduced as evidence in this case. Lewin also discusses the status of DNA fingerprinting with respect to the standards for the admissibility of scientific evidence in court.

Marx, Jean L. "DNA Fingerprinting Takes the Witness Stand." *Science* 240 (June, 1988): 1616-1618. A nontechnical research news article describing and comparing Jeffreys' and Erlich's DNA fingerprinting techniques, case studies of each as they have been used in court, and the general impact of DNA fingerprinting on the legal system. Marx also discusses the FBI's program for setting up DNA fingerprinting laboratories.

Rothwell, Norman V. *Understanding Genetics.* 4th ed. New York: Oxford University Press, 1988. A popular college-level introductory genetics textbook that explains all the major areas of genetics in depth. It is clearly written and well illustrated. Each chapter includes problems and questions for analysis, references to the scientific literature, and a summary of key points.

Bernard Possidente, Jr.

Cross-References

Watson and Crick Develop the Double-Helix Model for DNA (1951), p. 1406; Nirenberg Invents an Experimental Technique That Cracks the Genetic Code (1961), p. 1687; Kornberg and Coworkers Synthesize Biologically Active DNA (1967), p. 1857; Cohen and Boyer Develop Recombinant DNA Technology (1973), p. 1987; A Human Growth Hormone Gene Transferred to a Mouse Creates Giant Mice (1981), p. 2154; Jeffreys Discovers the Technique of Genetic Fingerprinting (1985), p. 2296; A Gene That Can Suppress the Cancer Retinoblastoma Is Discovered (1986), p. 2331.

THE OLDEST KNOWN GALAXY IS DISCOVERED

Category of event: Astronomy
Time: March, 1988
Locale: Kitt Peak National Observatory, Tucson, Arizona

A galaxy was discovered with a redshift of 380 percent, which places it among the most distant observable objects, including quasars, in the universe

Principal personages:
> KENNETH CARTER CHAMBERS (1956-   ), an American astronomer who was a graduate student at The Johns Hopkins University at the time of the discovery
> GEORGE KILROY MILEY (1942-   ), an Irish astronomer affiliated with Leiden University
> WILLEM JOHANNES VAN BREUGEL (1948-   ), a Dutch astronomer affiliated with the Institute for Geophysics and Planetary Physics, Lawrence Livermore National Laboratory

Summary of Event

Prior to 1985, the most distant visible objects in the universe were quasars. Because of their great distances, large redshifts were reported. A redshift is the result of a change in wavelength or frequency caused by relative motion away from an observer. This creates a shift in the observed spectral lines toward the red end of the spectrum. A 100 percent shift, for example, would cause a known spectral line at 300 nanometers (one-billionth of a meter) to be observed actually at 600 nanometers. After 1985, enhanced image detectors known as charge-coupled devices (CCDs) were installed on telescopes. The detection of very faint starlight and the viewing of more distant objects was now possible. Astronomers were able to see farther out into the universe, and objects with larger and larger redshifts were discovered. In April of 1988, Kenneth Carter Chambers reported an elongated object with a redshift of 227 percent (2.27). The elongated image ruled it out as a quasar, which is more compact like a star. In August of 1988, Simon Lilly at the University of Hawaii announced that he had obtained the spectrum of a faint source with the 3.6-meter infrared sensitive telescope on Mauna Kea Observatory. Examination of hydrogen and carbon emission lines indicated a redshift of 3.4.

From 1960 on, radio-wave emission from distant sources has been used to pinpoint distant galaxies. Techniques involving the use of more than one radio telescope are called arrays, which improve resolution by offering a larger combined collecting area. An array known as the Very Large Array (VLA) located in New Mexico, links up as many as twenty-seven radio telescopes, each separated by 1 kilometer.

To limit the search for the most suitable sources, a specialized technique was employed. Candidates were selected from the Palomar Sky Survey of radio sources. Radio-emitting sources were selected that had spectra considered ultrasteep. Ultrasteep radio sources are both powerful and distant ones. Various explanations have been advanced to explain such spectra, such as hot regions caused by gaseous jets of material interacting with magnetic fields of the galaxy. Fifty radio sources were selected, and wavelengths of 0.20, 0.06, and 0.02 meter were studied with comparisons made to known quasar radio sources for calibration purposes.

The discovery of the object called 4C41.17 was made in March of 1988 (announced in August of 1988) at the Kitt Peak 4-meter telescope in Tucson, Arizona, by the team of Chambers, George Kilroy Miley, and Willem Johannes van Breugel. The redshift was measured by identifying two emission lines, the brightest one centered at 583.2 nanometers and the fainter one at 743.2 nanometers. The wavelengths agreed with the known Lyman alpha and five times-ionized carbon lines when shifted toward the red by 380 percent of the original values. The actual lines observed in the yellow and red wavelengths would have originated in the ultraviolet. To improve the overall quality of the image, the telescope was used with an enhanced red-sensitive CCD scanned across the image in an incremental fashion. The image obtained was a composite of many separate scans. This served two purposes: to improve the contrast and to eliminate distortion surrounding the image.

It was necessary to obtain much of the imaging in infrared wavelengths because of the high redshifts. At these wavelengths, details may be revealed which are not always discernible at others, such as the gaseous components from stars. For the infrared survey, the 3.5-meter United Kingdom telescope in Hawaii was used. The background sky is ordinarily brighter at the 2.2-micron (one-millionth of a meter) wavelength observed than the distant source. Therefore, to improve the image sharpness, multiple images were taken over the source galaxy, with a total exposure time of 171 minutes.

Spectroscopy was utilized to cover the wavelength range from 480-950 nanometers, including the Lyman alpha line previously identified, as well as the shorter wavelengths down to the threshold of atmospheric filtering in the blue end of the spectrum. These techniques gave the astronomers a wider spectral coverage, which was necessary because of the large redshift. A spectral line of carbon observed at 154.9 nanometers in the laboratory, for example, would appear as shifted all the way over to 743.5 nanometers at a redshift of 3.8. This is truly an incredible shift: The original spectral line observed in the ultraviolet would appear shifted into the infrared.

The optical and infrared emission from 4C41.17 is aligned with the radio axis of the galaxy. The researchers believe this can be explained as resulting from star formation caused by the radio source. The extended source of emission radiation shows spectra identified from stars and is, therefore, probably a distant stellar system. The star formation may have been caused by the radio source compressing the protogalactic clouds. Similarities between the ionized gas, the optical/infrared radiation, and the radio source suggests an interactive mechanism.

From studies of an ideal galactic model, the researchers concluded that the stars would have formed in less than 100 million years, making the galaxy no more than 300 million years old. This would be a very young galaxy, when compared with galactic systems in Earth's solar system, including the Milky Way galaxy.

The galaxy apparently has a very luminous halo out to 100 kiloparsecs, which appears as ionized gas dominated by the Lyman alpha line. The ionization may be caused by hot young stars, or induced by the radio source, or perhaps even be the effect of nuclear emissions.

Distances to objects like 4C41.17 are inferred from a comparison of their observed redshifts to values assumed from Hubble's law relating to the expansion of the universe. Hubble's law states that the velocity of recession of a galaxy or quasar is proportional to its distance. For each megaparsec (3.3 million light-years), the velocity in kilometers per second will increase; this ratio is known as Hubble's constant. Currently accepted values of Hubble's constant range between 50 to 100 kilometers per second per megaparsec. Depending upon the value selected for this constant, 4C41.17 may lie at a distance as great as 15 billion light-years, making it the most distant stellar system known. There are only a few quasars that have greater redshifts and are still farther away.

Impact of Event

The techniques employed by the research team of using Lyman alpha spectral line identification and radio emission from this galaxy would have been detectable out to a redshift of 6.0; at distances greater than this, the spectral lines would no longer be observed optically. Such redshifts are beyond even the most distant quasar presently recorded, which has a redshift of 4.4. Extension of this technique to very faint samples may increase the numbers of known high redshift galaxies and serve as a method of probing the early universe.

A significant discovery of an alignment effect between the optical and infrared radiation from the galaxy extending along the radio source was made by this research team. The radio source may induce star formation and be responsible for this alignment that could occur early in the history of a galaxy. The ultraviolet spectrum (wavelengths from 10 to 100 nanometers) is dominated with starlight, demonstrating that the rate of star formation probably was higher in the past. Most of the radiation emitted by a star consists of invisible wavelengths including ultraviolet, but the atmosphere of Earth filters out nearly all of these wavelengths before they reach the surface. Lilly believes that the visible component of radiation is caused by newly formed blue stars, while the infrared light may be caused by red stars that are from 1 to 2 billion years old.

The discovery of a mature galaxy far out in the visible universe and therefore early in time, perhaps soon after the big bang event, presents some problems for present cosmological theories. Some cosmologists believe that galaxies formed around dense concentrations of invisible particles may compose 90 percent of the material in the universe called cold-dark matter. Models for this theory require millions of years for star systems to form in galaxies, but 4C41.17 probably required much less time than that.

Additional research may give insight on the relationships between quasars and early protogalaxies like 4C41.17. Finding a galaxy at such a great distance was unexpected because, previously, only quasars were known to reside that far away. Quasars were evidently much more abundant in the early universe and have either evolved into galactic systems since then or have disappeared altogether since that epoch.

Bibliography

Chambers, K. C., G. K. Miley, and W. J. M. van Breugel. "4C41.17: A Radio Galaxy at a Redshift of 3.8." *The Astrophysical Journal* 363 (November, 1990): 21-39. The authors report the discovery, observational techniques, and conclusions drawn from the data. Figures show clearly the emission lines detected with evidence for the correct identification using redshifts.

Ferris, Timothy. *Galaxies.* Reprint. New York: Crown, 1987. The author, both a journalist and an editor, has compiled a spectacular large-format book with many full color and black-and-white photographs. Chapters cover the Milky Way galaxy, the local group of galaxies, the form and variety of galaxies, interacting galaxies, clusters of galaxies, and galaxies and the universe. Lookback time is discussed by means of well-illustrated light cone.

Hodge, Paul W., comp. *The Universe of Galaxies.* New York: W. H. Freeman, 1984. A collection of a number of excellent articles that have appeared in *Scientific American.* Some of the topics include dark matter, superclusters, tidal effects between galaxies, and quasars as probes of the past.

Pasachoff, Jay M. *Astronomy: From the Earth to the Universe.* 4th ed. Philadelphia: Saunders College Publishing, 1991. An outstanding chapter on galaxies is included with many photographs of galaxies and clusters of galaxies showing photographic plates imaged with charge-coupled devices (CCDs) and special observatories recording infrared and X-ray radiation. The most distant galaxies are shown in a photograph of 4C41.17; includes a discussion of how it was located.

Time-Life Books, eds. *Galaxies.* Alexandria, Va.: Author, 1988. The reader is taken on a journey through a universe of galaxies. Fantastic color diagrams illustrate the structure of a typical galaxy. Radio galaxies are discussed, showing how the radio lobes evolve from an optical galaxy. A chapter is included on the steps required to transform a protogalaxy into a mature galaxy.

Michael L. Broyles

Cross-References

Kapteyn Discovers Two Star Streams in the Galaxy (1904), p. 218; Slipher Obtains the Spectrum of a Distant Galaxy (1912), p. 502; Slipher Presents Evidence of Redshifts in Galactic Spectra (1920's), p. 689; Hubble Demonstrates That Other Galaxies Are Independent Systems (1924), p. 790; Ryle's Radio Telescope Locates the First Known Radio Galaxy (1946), p. 1271; De Vaucouleurs Identifies the Local Supercluster of Galaxies (1953), p. 1454; Schmidt Makes What Constitutes the First Recognition of a Quasar (1963), p. 1757; Construction of the World's Largest Telescope Begins in Hawaii (1985), p. 2291; Tully Discovers the Pisces-Cetus Supercluster Complex (1986), p. 2306; NASA Launches the Hubble Space Telescope (1990), p. 2377.

THE SUPERCONDUCTING SUPERCOLLIDER
IS UNDER CONSTRUCTION IN TEXAS

Category of event: Physics
Time: November, 1988
Locale: Near Waxahachie, Ellis County, Texas

Construction of a Superconducting Supercollider will allow a new class of particle physics experiments at higher-energy levels, thus testing theories of subatomic structure and nature of matter

Principal personage:
> RONALD W. REAGAN (1911-   ), the 40th President of the United States who supported the Superconducting Supercollider project and approved the construction site

Summary of Event

In June, 1987, President Ronald Reagan approved the recommendations of prominent physicists, leaders in the United States Congress, and other federal officials to construct a gigantic particle accelerator called the Superconducting Supercollider (SSC). After protracted discussion and lobbying by many states, in November, 1988, President Reagan approved a construction site near Waxahachie, Texas, in Ellis County, about 50 kilometers south of Dallas. The SSC—a near-circular tunnel 86 kilometers in circumference housing thousands of superconducting magnets—is designed to generate collisions of proton beams at energies up to 40 trillion electronvolts (40 TeV), energy levels twenty times greater than can be produced by the most powerful accelerator in operation in 1990.

Accelerators are the engines driving research in elementary particle physics—the study of the basic nature of matter, energy, space, and time—and have produced fundamental changes in how these phenomena are viewed and the very nature of the universe. Until about 1960, matter was understood to consist simply of protons, neutrons, and electrons, the building blocks of atoms. As physicists attempted to learn about the forces binding these particles into atoms and nuclei, by driving them to high-energy collisions in small accelerators, or cyclotron rings, they discovered more than one hundred previously unknown particles. Most of these particles had properties similar to those of neutrons and protons, and are collectively classified as hadrons. Hadrons, in turn, appear to be composed of even more elementary particles called quarks. Although electrons appear to be elementary particles themselves, there are about six electron-like particles known as leptons. Quarks and leptons—rather than neutrons, protons, and electrons—appear to be the basic building blocks of all matter.

Physicists also have identified four forces involved in organizing matter. Three of these forces—electromagnetism, the weak force (involved in certain types of radioactive decay), and gravitation—act on both hadrons and leptons. The fourth force, however, known as the strong force, appears to bind quarks together to form hadrons but does not affect leptons in any way.

In late twentieth century physics, attempts to devise a grand unified theory to cover the effects of all four forces in nature have been actively pursued. A unified theory has been developed to cover the weak and electromagnetic forces; by 1990, it was fairly certain that the strong force could be included. The nature of the gravitational force, however, remained elusive. A major component of unified theories has been the discovery of "carrier" particles for three of the four forces. Photons, which carry the electromagnetic force, have been studied since the 1920's. So-called W and Z particles carrying the weak force were discovered in 1983. Strong evidence has emerged from several research centers for the presence of particles, called gluons, as carriers of the strong force. No such discovery has been made for the gravitational force. Discoveries of these myriad new particles have been made through particle collisions under laboratory conditions, at ever higher energy levels; the assumption is that the existence of particles predicted by theoretical physics, but not yet observed in experiments, will be found eventually as energy levels in accelerators increase.

Most early accelerators were linear machines in which magnets drove particle beams in a straight line to collisions with fixed targets. Although linear accelerators can achieve energy levels in the billions of electronvolts, much energy is lost in colliding with a fixed target. Ring colliders, which are essentially circular accelerators, can achieve much higher energy levels and thus reveal the presence of higher energy particles. Rather than focusing a beam on a fixed target, ring colliders focus two beams in opposite directions from a source point. The beams then collide with each other at very high energy levels, just as a head-on collision of two automobiles is more "energetic"—and therefore more destructive—than a collision between an automobile and a tree. Particles are driven to high-energy levels in ring colliders by magnets. The larger the circumference of the ring, and the more powerful the magnets, the higher the energy levels that will be produced in the particle beams. The largest existing rings—at the Enrico Fermi National Laboratory near Chicago (Fermilab) and the European Laboratory for Particle Physics (CERN) near Geneva, Switzerland—can produce energy levels on the order of 1 to 2 TeV, more than enough, for example, to have revealed the existence of the W and Z particles. The SSC, it is hoped, will produce energy levels of around 20 TeV in each of its two proton beams, resulting in collision energies on the order of 40 TeV. This energy level is comparable to those in the first microseconds after the big bang, the physical singularity which most scientists believe to have been the origin of the universe.

The SSC not only will be the largest ring collider facility in the world by a large margin but also will employ about eight thousand sophisticated magnets using superconductivity technology. Superconductivity is a phenomenon in which certain materials, when bathed in liquid helium or otherwise lowered in temperature to a point close to absolute zero, conduct electromagnetic force with almost 100 percent efficiency. Each superconducting magnet in the SSC, as originally designed, was to be about 17 meters in length. The particle beams are driven through one magnet after another until their energy levels become so great that particle velocities approach the speed of light. These magnets are the most costly component of the SSC project, the total initial estimated cost of which was \$4.5 billion, making it second only to the Apollo Manned Lunar Project as the costliest scientific enterprise funded by the federal government. The magnets also proved extremely troublesome and forced numerous delays in the project schedule. Although the advanced cryogenics for supercooling worked well, and shorter versions of the magnet assembly worked perfectly, scaling up the technology to the 17-meter length required in the SSC created difficulties. In effect, the longer magnets introduced a "bend" in the beam, resulting in lower energy levels.

The repeatedly poor performance of prototype magnets led the Department of Energy (DOE) to order a suspension of work on the 17-meter design in August, 1989, just as the DOE was considering bids for the contract to mass-produce the magnets. A DOE panel of physicists and engineers concluded that extensive research and further development of the magnetic design might be necessary and that their production might have to be delayed as long as three years. With the magnet specifications uncertain, plans for other parts of the ring came to a halt.

Just as it had been during the furious lobbying by several states to decide on a site, the magnet problems made the SSC a major political issue once again. The alternatives were to accept a somewhat less powerful SSC, leave the magnet design in place, and proceed with construction or, to redesign the magnets and possibly the ring itself. Physicists throughout the world argued against acceptance of a power loss as unacceptable in the light of the enormous expense involved in building and operating the SSC. At 40 TeV, the ring provided a margin of power of critical importance for certain experiments; even a 10 percent decrease would mean its loss.

The SSC central design group recommended the second alternative: design changes to overcome the problem of power loss in the magnets, including shortened magnets of about 16 meters, a larger number of magnets, possible extension of the ring circumference by about 3 kilometers, and upgrading of the linear booster accelerator to inject protons into the ring at about 1 TeV. The problem with such basic design changes was cost. Some analysts estimated that they will result in a cost overrun of more than \$2 billion, raising the overall cost of the SSC by almost 50 percent. Recommendations for design changes came on the heels of some second thoughts in an austerity-minded Congress. Despite an increased appropriation of \$200 million in fiscal 1990, scrutiny of the SSC project by the administration of President George Bush left everyone involved with no clear idea of how later stages of the project would be received. With the circumference—namely, the precise location—of the ring collider still uncertain, by the end of 1989, federal authorities had not even begun to purchase land for the construction site. Local government units in Texas argued vehemently over who would pay for roads and services in what has been described as almost primeval backcountry around Waxahachie, Texas.

Impact of Event

The economic and financial aspects of the SSC, perhaps inevitably, occupy everyone in the siting, design, and planning stages. The huge cost of the project must be balanced against the economic growth it will create. The SSC will have a community of three thousand scientists and engineers and consume an annual operating budget of nearly $300 million. The staff will require housing and conveniences; the collider itself will have enormous amounts of power. Since one of the criteria in site choice was the presence of a network of universities and laboratories, the SSC will have a synergistic effect on development of high-technology industry and research.

International leadership in physics research also is at stake. In 1988, CERN began planning for an 18-TeV collider. While not as powerful as the SSC, if it is completed while the SSC project languishes, the European community may seize the lead in elementary particle research. Economic prosperity and national pride, however, are not the reasons for building the SSC. It is simply one of the most dramatic projects in history, potentially much more significant than manned space programs in terms of what may be learned about the universe. Physicists are hopeful that the energy levels obtainable in the SSC will reveal the presence of the elusive Higgs boson, a still theoretical particle properties of which may finally succeed in tying force theories together.

The SSC represents a huge step beyond the current frontiers of high-energy physics. Although it is a venture into the unknown, it is difficult not to sense the anticipation of theoretical physicists that the SSC will present humanity with some elemental and final answers about the nature of the universe. The grand unified theory may soon be at hand, and no one can predict its consequences.

Bibliography

Brown, Laurie M., and Lillian Hoddeson, eds. *The Birth of Particle Physics.* Cambridge, England: Cambridge University Press, 1983. Papers from a Fermilab historical symposium on the emergence of particle physics as a field of study in the period from about 1930 to 1950. Owing to the rapid pace of the disciplines, physical scientists are becoming increasingly interested in accurate historical reconstruction of experiments and conditions in earlier periods.

Fisher, Arthur. "The World's Biggest Machine: Superconducting Super Collider." *Popular Science* 230 (June 1987): 56-62. Useful discussion of the principles of particle physics experiments envisioned for the Supercollider and a description of the project for general readers.

Halzen, Francis, and Alan D. Martin. *Quarks and Leptons: An Introductory Course in Modern Particle Physics.* New York: John Wiley & Sons, 1984. Textbook on the main developments and research foci of particle physics. All textbooks in this field require extensive mathematical background. Useful bibliography, including some general works.

Levi, Barbara Goss, and Bertram Schwarzschild. "Super Collider Magnet Program Pushes Toward Prototype." *Physics Today* 41 (April 1988): 17-21. Clearly written, illustrated summary of the technology and manufacturing procedures committed to the controversial first generation of supercollider magnets. Includes discussion of causes of disappointing test results.

National Research Council. Physics Survey Committee. *Physics Through the 1990's.* Washington, D.C.: National Academy Press, 1986. Proceedings and recommendations of the NRC panel that outlined priorities for research in elementary particle physics in the United States through the early twenty-first century. The panel recommended construction of the SSC as the highest national priority, central to the future of American research and leadership in the field.

Ne'eman, Yuval, and Yoram Kirsh. *The Particle Hunters.* Cambridge, England: Cambridge University Press, 1986. Historical account of the process of particle discoveries at ever-increasing energy levels, leading to the need for more powerful accelerators. Particularly good discussion of particle properties and basic forces in language suitable for general readers.

Okun, L. B. *Particle Physics: The Quest for the Substance of Substance.* Translated by V. I. Kisin. New York: Harwood Academic, 1985. Discussion of theoretical issues in particle physics. Rather technical in places but very useful and readable for individuals with a modicum of mathematical and scientific background. The work includes an extensive glossary and one of the most extensive bibliographies on particle physics available outside of specialized publications.

Perkins, Donald H. *Introduction to High Energy Physics.* 2d ed. Reading, Mass.: Addison-Wesley, 1982. Textbook approach, but also contains detailed mechanical and operational descriptions of accelerators and other experimental equipment. Good technical bibliography.

Ronald W. Davis

Cross-References

Lawrence Develops the Cyclotron (1931), p. 953; Cockcroft and Walton Split the Atom with a Particle Accelerator (1932), p. 978; University of California Physicists Develop the First Synchrocyclotron (1946), p. 1282; Hofstadter Discovers That Protons and Neutrons Each Have a Structure (1951), p. 1384; The Liquid Bubble Chamber Is Developed (1953), p. 1470; Gell-Mann Formulates the Theory of Quantum Chromodynamics (QCD) (1972), p. 1966; Georgi and Glashow Develop the First Grand Unified Theory (1974), p. 2014; The J/psi Subatomic Particle Is Discovered (1974), p. 2031; Rubbia and van der Meer Isolate the Intermediate Vector Bosons (1983), p. 2230; The Tevatron Particle Accelerator Begins Operation at Fermilab (1985), p. 2301.

NASA LAUNCHES THE HUBBLE SPACE TELESCOPE

Category of event: Astronomy
Time: April 24, 1990
Locale: Cape Canaveral, Florida

After twenty years of delays, the Hubble Space Telescope was placed in orbit around Earth by the space shuttle Discovery

Principal personages:

CHARLES ROBERT ODELL (1937-), a professor of physics and astronomy at Rice University who was project scientist for the Hubble Space Telescope from 1972 to 1974

HERMANN OBERTH (1894-1989), a German rocket pioneer who proposed that an orbiting observatory was a unique solution to the problem of distortion and absorption of light by the earth's atmosphere in his book *Die Rakete zu den Planetenräumen* (1923; the rocket into interplanetary space)

Summary of Event

When the Hubble Space Telescope (HST) was successsfully placed in orbit, 612 kilometers above the earth by astronauts aboard the space shuttle *Discovery,* one excited astronomer remarked, "At last we are out of the ocean!" His statement summarizes the feelings of astronomers who have had to deal with the frustration caused by the interference of starlight by the earth's atmosphere. No telescope can realize its full potential under 100 kilometers of boiling atmosphere which absorbs much of the visible starlight and particularly the astrophysically important ultraviolet and infrared wavelengths. In 1923, Hermann Oberth, a German rocket pioneer, proposed orbiting a telescope as the solution to peering through the ocean of air.

In 1962, a Large Space Telescope (LST), 305 centimeters in diameter, was proposed in the National Academy of Sciences' report on the "Future of Space Science." Ten years later, the National Aeronautics and Space Administration (NASA) established the LST headquarters at the Marshall Space Flight Center in Huntsville, Alabama, and appointed Charles Robert Odell as lead scientist for the LST project. In 1978, Congress approved the LST project but later changed the name to the Edwin Powell Hubble Space Telescope (HST) and reduced the mirror diameter to 240 centimeters.

Originally scheduled for launch on December 15, 1983, delays, both political and technical, postponed launch until August, 1986. Unfortunately, the space shuttle *Challenger* disaster in January, 1986, further delayed launch while technical problems of the launch vehicle were solved. The HST got as close as the launch pad on March 25, 1990, but again problems delayed launch. Then, on April 24, the hundreds of scientists and engineers associated with the project witnessed twenty years of their work launch into space. The Hubble Space Telescope is by far the most ambitious and expensive scientific tool ever constructed. The price tag of about $1.5 billion has created an instrument whose output will fundamentally alter the way one thinks about the universe.

The technical specifications for the optics and supporting hardware pushed industrial standards of precision, innovations, and designs beyond previous benchmarks of excellence. Both the technical and the organizational problems were complex. For example, it was not known how to keep the telescope (measuring 13.1 meters long, 4.3 meters wide, and weighing approximately 11,600 kilograms) pointing toward one tiny spot in space while orbiting 611 kilometers above the earth at a speed of 27,359 kilometers per hour. Another problem was how to enlist the support of astronomers and engineers for an untried and unproved, long-term project and how to organize them into a coherent team. Therefore, the project spanned two decades in preparation. Among the many technical innovations and the thousands of people who contributed to the success of the HST, the construction of the mirror—the heart of the telescope—is representative of the technological triumph as well as the personal accomplishments involved in the HST program.

The 240-centimeter primary mirror presented the engineers with a major new challenge, requiring more than 4 million man-hours and more than five years to cast, grind, and polish to its demanding specifications. The mirror is actually a glass disk of ultra-low expansion glass (ULE) fabricated from honeycomb cores fused together. Two 2.5-centimeter sheets are fused to the front and back of the 30.5-centimeter honeycomb structure, which weighs less than a ton or about one-tenth of the total telescope's weight. The primary mirror tested virtually optically perfect, deviating from the ideal figure by about 0.000025 millimeter.

Unlike a bathroom mirror, which has a reflective coating on the back side, telescope mirrors are coated on the front side and must be very thin. The HST mirror coating is so thin that if it were peeled off and flung into air, it would float suspended, like smoke. Only 76 millionths of a millimeter of aluminum was deposited onto the glass and then was sealed with 25 millionths of a millimeter of magnesium fluoride to protect against oxidation.

The HST is constructed with the same principles as Sir Isaac Newton's reflector telescope of 1669. The HST is a long tube, open at one end with a concave mirror at the other. In both telescopes, light enters the open end, travels to the primary mirror, and is reflected back toward the open end. In Newton's instrument, a small plane secondary mirror placed near the center of the open end intercepts light and focuses it through a hole in the side of the tube. With the HST, light gathered from the primary mirror strikes a secondary convex mirror and is reflected back toward the center of the primary mirror. A small hole in the center of the primary mirror allows light to pass through the primary mirror and be focused onto a variety of interpretative instruments. Some of these instruments help guide the telescope.

The HST can hold its aiming position for up to twenty-four hours, by using a combination of fine guidance sensors, gyroscopes, and star trackers. This achievement alone is a remarkable technical accomplishment. Other instruments aboard the HST include the wide-field/planetary camera, the faint-object camera, and two instruments designed to analyze the starlight: the faint-object spectrograph and the Goddard high-resolution spectrograph. As the light is focused by the mirror system, it is converted into electrical signals which are transmitted to an orbiting satellite and then to Earth-receiving stations. There they may be stored as digital information or converted to visual images; the problems of atmospheric interference are completely overcome in this manner.

While in orbit, the HST passes through the earth's shadow, where no sunlight can be received on the two solar cell arrays attached to either side of the telescope. At such times, electricity required for the operation of the various electrical components such as telemetry systems, computers, and power supplies, must be supplied by batteries. These nickel hydrogen batteries—six in all—are kept charged by current from the solar cell arrays. When in sunlight, the solar cells provide ample power, approximately 4,000 watts, for the entire spacecraft.

When HST was launched in April, 1990, it was during an intense solar maximum which may result in expansion of the upper atmosphere. To ensure that the HST would not be dragged into a lower orbit by the expanding atmosphere beneath it, the present orbit is about 20 kilometers higher than originally planned. Astronomers saw the first light from the telescope on May 20, 1990. While the images were good, they were not up to expectations. Two problems persisted: There was a micro-wobble and a focusing problem. The wobble was solved with new software radioed to the HST. The focusing problem was attributed to a condition known as spherical aberration. Although both mirrors had been tested and exceeded their individual design specifications, the instrument used to test the curvature, a reflective null corrector (RNC), unfortunately was not accurate nor had the mirrors been tested as a working pair. The flawed RNC fooled the technicians into thinking they were making a near-perfect mirror. The solution to the mirror problem has two parts. Until 1993, scientists can computer-enhance the existing images, resolving some of the focusing problems. It was planned that in 1993, a shuttle crew would replace the scientific instruments onboard HST with their second-generation counterparts. These would have a "correction factor" designed into their optical systems. While this seems disappointing for visible light astronomy, it does not hamper the infrared and ultraviolet research in the same manner.

The beginning of the fifteen-year mission for the HST was witnessed by descendants of Edwin Powell Hubble, for whom the telescope was named. Hubble's work at Mount Wilson Observatory in the 1920's was devoted to determining the nature and scale of the universe. He discovered that the universe is expanding; he also provided an estimate of the rate of expansion, known as the Hubble constant. Although Hubble was never an advocate of an orbiting telescope, which has become his namesake, it is already programmed to make observations that will almost certainly lead to a better understanding of the nature of the expanding universe.

Impact of Event

The disappointing news of spherical aberration seemed consistent with the saga of delays associated with the HST project. These delays had the effect of placing 1970's technology in a 1990's project, and most of the life expectancy of the HST was spent in conditions for which it was not designed: mainly, an atmosphere and gravity. Until 1993, the telescope can devote its time to the infrared and ultraviolet research. It is these frequencies that are the most difficult for ground-based telescopes and will yield the greatest advancement of astronomical knowledge.

When the second generation of instruments are installed in 1993, the potential of the HST may be realized. With the ability to see stars that are fifty times fainter than the faintest ones seen by ground telescopes and at seven times greater distances, the HST will open new avenues of discovery and, consequently, new theories of how the universe is structured. The study of the structure and evolution of the universe is generally accepted as the main scientific innovation for building the Hubble Space Telescope. It is anticipated that the data from HST will resolve mysteries that have perplexed astronomers for the past one hundred years. For example: Will the universe expand forever? The answer depends on measurements of distance and mass. The current distant measuring techniques are believed to be in error by as much as 50 percent. Such uncertainties are disturbing because they mean that the universe's expansion cannot be gauged unambiguously. It is not known if the universe will eventually reverse its expansion and start to contract.

The question of the "missing mass" may be illuminated by HST data as scientists try to determine if the stars and galaxies represent most of the universe or if the matter resides in a yet undetected form. Also, how much time has elapsed since the big bang? Ground-based telescopes have been unable to provide satisfactory answers, but the HST is uniquely qualified to measure distances accurately to a large number of objects in the universe that will lead to significant answers. Current estimates of the rate of expansion of the universe are uncertain by a factor of 2. This rate, known as the Hubble constant, will be improved by several independent techniques with the HST. The study of variable stars, known as Cepheids, will illuminate this question. Bearing heavily on the understanding of the structure and nature of the universe is knowing the nature of individual stars closer to Earth. The capabilities of the HST will allow for more accurate observations of these stars and observing time required will be less.

The observation of known planets and the search for those yet undiscovered planets are high-priority items for HST astronomers. Astronomers think that countless other planets circle stars throughout the universe; the search for these planets is one of the most active areas of research for HST. If found, these other planets will increase the confidence that the solar system is not a cosmic anomaly and would increase astromers' estimates of the probability of the existence of extraterrestrial life. Planets of the solar system are also on HST's agenda. The HST allows scientists to study Earth's planets with a resolution comparable to that of the *Voyager* spacecraft. Best of all is the anticipation of the unknown and the serendipitous discoveries that will lead humans to more fantastic ideas of the universe than can be imagined under the ocean of air.

Bibliography

Chien, Philip. "The Launch of HST." *Astronomy* 18 (July, 1990): 30-37. A very readable account of the events culminated after twenty years of waiting for the launch. Well illustrated with launch and orbit photographs. Chien captures the excitement of the first tense moments and the beginning of its fifteen-year mission.

Dunkle, Terry. "The Big Glass." *Discover* 7 (July, 1989): 69-72. A nontechnical and exciting account of the construction of the telescope optics. Emphasis is on the personalities, skills, and dedication of the people associated with the project.

Field, George, and Donald Goldsmith. *The Space Telescope.* Chicago: Contemporary Books, 1989. An appropriate starting place for the general reader. It is well supported with pictures, illustrations, glossary, and additional readings. Very readable and thorough account.

Maran, Stephen P. "The Promise of the Space Telescope." *Astronomy* 18 (January, 1990): 38-43. A well-written account of the main objectives and expected results of the HST by the senior staff scientist.

Smith, Robert W. *The Space Telescope: A Study of NASA, Science, Technology, and Politics.* New York: Cambridge University Press, 1989. A detailed chronological account of the construction of the telescope from its inception to launch preparation. Perhaps the most complete work on the subject, it includes historical background, involvement with NASA, industry, the scientific community, and politics.

Richard C. Jones

Cross-References

Hale Establishes Mount Wilson Observatory (1903), p. 194; Hale Oversees the Installation of the Hooker Telescope on Mount Wilson (1917), p. 645; Hubble Confirms the Expanding Universe (1929), p. 878; Schmidt Invents the Corrector for the Schmidt Camera and Telescope (1929), p. 884; Lyot Builds the Coronagraph for Telescopically Observing the Sun's Outer Atmosphere (1930), p. 911; Hale Constructs the Largest Telescope of the Time (1948), p. 1325; Voyager 1 and 2 Explore the Planets (1977), p. 2082; *Columbia*'s Second Flight Proves the Practicality of the Space Shuttle (1981), p. 2180; Construction of the World's Largest Telescope Begins in Hawaii (1985), p. 2291.

MEDICAL RESEARCHERS DEVELOP AND TEST PROMISING DRUGS FOR THE TREATMENT OF AIDS PATIENTS

Category of event: Medicine
Time: March, 1991
Locale: United States

As of March, 1991, zidovudine was the only pharmaceutical agent approved for use in the United States for the treatment of patients with AIDS; other drugs are being developed and tested in an attempt to find a cure for this disease

> *Principal personages:*
> SAMUEL BRODER (1945-   ), a physician at the National Cancer Institute involved in researching the efficacy of deoxynucleosides for the treatment of AIDS
> VINCENT MERLUZZI (1949-   ), a microbiologist and section leader of the group at Boehringer Ingelheim responsible for the development of the anti-AIDS drug BI-RG-587
> HIROAKI MITSUYA, a physician at the National Cancer Institute who developed the test used to determine whether dideoxynucleosides are effective anti-AIDS agents
> JOHN SULLIVAN (1946-   ), a professor at the University of Massachusetts Medical School who worked with Merluzzi's group to show that BI-RG-587 has potential as an anti-AIDS drug

Summary of Event

Potential drug therapies for acquired immune deficiency syndrome (AIDS) can be divided into two categories. The first includes pharmaceuticals therapeutic for opportunistic infections that plague AIDS patients. The second includes drugs that interfere with the life cycle of human immunodeficiency virus (HIV), the causative agent of AIDS. A select group of drugs from the latter category are discussed here.

Drugs used against HIV must effectively disrupt the activity of the virus while being nontoxic or, at best, minimally toxic to the patient. Traditionally, candidate drugs undergo a three-phase clinical trial before being approved for use in the United States by the Food and Drug Administration (FDA). Drug safety is tested in phase I, drug efficacy is tested in phase II, and a comparison of the candidate drug with other therapies is studied in phase III. An understanding of HIV infection is essential if one is to understand the mechanism of the action of drugs being used or being tested in AIDS therapy. First, viral components must gain entrance to a human cell; this occurs through binding and subsequent fusion of the two. Second, the genetic information of the virus must be converted into a form that can be stored in a human cell. HIV's genetic information exists as ribonucleic acid (RNA), while human genetic information exists as a slightly different substance, deoxyribonucleic acid (DNA). Fortunately for the virus, HIV contains an enzyme, reverse transcriptase, which copies the virus' genetic information onto newly formed DNA and then destroys the viral RNA. Third, the DNA copy, called provirus, is integrated into human DNA. Once this occurs, the human cell and its progeny are infected with HIV for life.

Candidate anti-AIDS drugs disrupt HIV's life cycle, either by interfering with binding and/or fusion of the virus and the human cell, inhibiting reverse transcriptase, stopping integration, or disrupting events that occur after integration. Four drugs that inhibit reverse transcriptase are zidovudine (formerly called azidothymidine, or AZT, produced by Burroughs Wellcome, Research Triangle Park, North Carolina), dideoxycytidine (ddC; Hoffmann-LaRoche Incorporated, Nutley, New Jersey), dideoxyinosine (ddI, or didanosine; Bristol-Meyers Squibb Company, Princeton, New Jersey), and BI-RG-587 (Boehringer Ingelheim, Danbury, Connecticut).

As of March, 1991, zidovudine was the only anti-AIDS drug approved for use in the United States. Zidovudine, developed in 1964 as a potential (but, unfortunately, ineffective) anticancer agent, was found to have anti-HIV activity in 1985. In 1987, the FDA approved its use for treating people with AIDS or symptomatic HIV infection; in 1990, the FDA extended approval of zidovudine use to include treatment of asymptomatic HIV-positive patients.

Zidovudine belongs to a group of chemicals called dideoxynucleosides. Once inside a cell, these chemicals are phosphorylated (that is, chemical groups called phosphate groups are added to the dideoxynucleosides) to form dideoxynucleoside-triphosphates (ddTPs), which are variants of deoxyribonucleotides, the normal building blocks of DNA. During DNA formation, thousands of deoxyribonucleotides are strung together in a specific order. When phosphorylated, zidovudine (or other ddTPs) are added to a growing DNA chain. They act as chain terminators; that is, no more deoxyribonucleotides can be added to the chain. Thus, after HIV has entered a human cell, zidovudine inhibits infection by interfering with formation of a provirus. Zidovudine does not cure or prevent AIDS; it merely slows down the disease process, buying valuable time for HIV-infected people.

Unfortunately, zidovudine has severe side effects. It is toxic to red bone marrow (the site of white blood cell production), resulting in decreased levels of white blood cells in the body. White blood cells are necessary for proper functioning of body defenses. In addition, lymphomas (a type of cancer) can result from long-term zidovudine therapy; in 1990, it was reported that patients have a 46.4 percent probability of developing lymphoma after thirty-six months of therapy with zidovudine. Another problem is the development of zidovudine-resistant viral strains in patients undergoing zidovudine therapy. Less severe side effects that can result from treatment with this drug include nausea, vomiting, insomnia, abdominal discomfort, diarrhea, malaise (a general feeling of discomfort), myositis (inflammation of muscles), and myalgia (pain in muscles); allergic reactions to zidovudine have also been reported. In addition to adverse side effects, zidovudine does not completely inhibit provirus formation.

Because of existing problems with zidovudine, researchers began studying other dideoxynucleosides to test their efficacy for use as anti-AIDS drugs. Two promising dideoxynucleosides currently undergoing phase I, II, and III trials are ddC and ddI; patients infected with HIV strains resistant to zidovudine respond to therapy with both of these drugs.

After initial treatment with ddC (dideoxycytidine), some patients experience minor side effects, including rashes, fever, malaise, arthralgia (painful joints), diarrhea, edema (swelling caused by increased amounts of tissue fluid) of the ankles, and ulceration of the mouth; these symptoms, however, are transient. More seriously, some patients experience a neuropathy (changes in function or health of nerves) characterized by aching, burning, or tingling sensations in the feet; occasionally, neutropenia (decreased number of neutrophils, a type of white blood cell) or thrombocytopenia (decreased number of blood platelets) are noted. Fortunately, these more severe side effects seem to be dose-related.

Unlike zidovudine or ddC, ddI (dideoxyinosine) is not directly converted to a ddTP in the body. Instead, it undergoes a more complex metabolic pathway, eventually being converted to the ddTP dideoxyadenosine triphosphate. At first, oral administration of ddI was a problem because the drug is sensitive to stomach acid; now, ddI is given in conjunction with an antacid. Mild phlebitis (inflammation of veins) was reported in patients receiving this drug intravenously, but this problem was eliminated by use of more dilute solutions for injection. Mild headaches, diarrhea, and rashes are transient side effects experienced by some patients undergoing treatment with ddI. More seriously, neuropathy, such as that seen in ddC-treated patients, and pancreatitis (a potentially fatal condition in which the pancreas is inflamed) have been reported in some people undergoing ddI therapy; it has not, however, been established whether pancreatitis is caused by ddI or is caused by an unrelated source. In 1990, one patient had grand mal (a form of epilepsy) seizures when treated with ddI. In general, however, severe side effects of ddI, like ddC, seem to be dose-related.

Despite problems, dideoxynucleosides do improve immune functions and prolong the life of HIV-infected people. In an effort to negate side effects caused by these drugs, researchers are studying treatment modalities with more than one type of dideoxynucleoside: for example, therapy in which the patient is alternatively treated with zidovudine and ddC. Combination therapy has three potential advantages over treatment with a single dideoxynucleoside: few side effects are expected; the possibility of development of drug resistance is lessened; and, as each drug seems to enhance the effectiveness of the other drug, treatment efficacy is maximized.

In January, 1991, Boehringer Ingelheim received approval from the FDA to begin phase I clinical trials on a new anti-AIDS drug, BI-RG-587. Studies done jointly by researchers at Boehringer Ingelheim and the University of Massachusetts Medical School (Worcester, Massachusetts) have shown that BI-RG-587, which is not a nucleoside, has anti-HIV activity yet is minimally toxic to human cells in culture. BI-RG-587 specifically inhibits the HIV reverse transcriptase enzyme; it has no effect on reverse transcriptase from related viruses or on several human enzymes that were tested. It is hoped that the negative side effects caused by dideoxynucleoside treatment will not be seen in clinical trials with BI-RG-587.

Impact of Event

From 1981—the year AIDS was first recognized as a disease—to 1990, 100,777 people in the United States were known to have died from AIDS-related complications; 31,196 of these deaths occurred in 1990 alone. Based on current trends, it is estimated that between forty-three thousand and fifty-two thousand additional AIDS-related deaths will occur in the United States in 1991, between forty-nine thousand and sixty-four thousand deaths will occur in 1992, and between fifty-three thousand and seventy-six thousand deaths will occur in 1993. AIDS is responsible for causing the deaths of 14 percent of all males and 4 percent of all females who died in the United States in 1989. Most people who have died of AIDS have been between the ages of twenty-five and forty-four years.

These statistics demonstrate the urgent need for the development of safe and effective drugs for the treatment of HIV-infected people. Although zidovudine can extend the life of AIDS patients, use of this drug is plagued with undesirable side effects. The effectiveness of drugs such as ddC and ddI, of combination drug therapies, of BI-RG-587, and of other potential drug treatments remains to be validated. As of March, 1991, no known drug is available to cure AIDS, and no vaccine is known to prevent HIV infection. Scientific research and AIDS education are the best resources for attempting to prevent the spread of this fascinating yet deadly disease.

Bibliography

Bint, Adrian J., John Oxford, and Philip J. Daly, eds. *AIDS and AIDS-Related Infections: Current Strategies for Prevention and Therapy.* San Diego: Academic Press, 1989. An excellent source for an advanced audience, the articles in this book are based on presentations at a meeting of the British Society for Antimicrobial Chemotherapy, which was held in Stratford-on-Avon in April, 1988. Drugs that interfere with HIV's life cycle and drugs used to treat AIDS-related infections are discussed.

Douglas, Paul Harding, and Laura Pinsky. *The Essential AIDS Fact Book.* New York: Pocket Books, 1989. This book, prepared in cooperation with the Columbia University Health Service, provides basic information for a general audience. The chapter related to drugs discusses health care strategies, approved and experimental drug therapies for AIDS, drugs used to treat opportunistic infections, and psychological factors relating to health care. Each chapter contains references; a resource guide, including phone numbers, is provided.

Gallo, Robert C., and Luc Montegnier, eds. *The Science of AIDS.* New York: W. H. Freeman, 1989. An excellent source for the interested reader. A compilation of ten articles written by AIDS researchers that first appeared in the October, 1988, issue of *Scientific American*, this book provides a wealth of information on various aspects of AIDS. An entire chapter is devoted to AIDS therapies.

Jennings, Chris. *Understanding and Preventing AIDS: A Book for Everyone.* 2d ed. Cambridge, Mass.: Health Alert Press, 1988. This is a readable and informative book for general audiences covering various issues dealing with AIDS. Traditional and experimental therapies, as well as alternative health care measures for AIDS, are discussed. Included are the following from the Centers for Disease Control: the surveillance case definition for AIDS, recommendations for education and foster care of HIV-infected children, and recommendations for health care workers concerning prevention of HIV infection.

Kerrins, Joseph, and George W. Jacobs. *The AIDS File: What We Need to Know About AIDS Now!* 2d ed. Woods Hole, Mass.: Cromlech Books, 1989. This is an easy-to-read, introductory book. Aside from a chapter on AIDS therapies, the book discusses many AIDS-related issues, including a chapter on facts and fallacies. A useful glossary is provided.

Langone, John. *AIDS: The Facts.* Boston: Little, Brown, 1988. Each chapter of this beginning-level book is referenced and designed to answer a specific question about AIDS. The chapter entitled "Can AIDS Be Conquered?" has useful information about potential drug therapies and vaccines for AIDS, as well as a section on modulation of the immune system.

St. Georgiev, Vassil, and John J. McGowan, eds. *AIDS: Anti-HIV Agents, Therapies, and Vaccines.* New York: New York Academy of Sciences, 1990. This an advanced-level book that contains reprints of articles presented at the Second International Conference on Drug Research in Immunologic and Infectious Diseases, Arlington, Virginia, November, 1989. A volume in the Annals of the New York Academy of Sciences.

Susan M. Maskel

Cross-References

Rous Discovers That Some Cancers Are Caused by Viruses (1910), p. 459; Horsfall Announces That Cancer Results from Alterations in the DNA of Cells (1961), p. 1682; The U.S. Centers for Disease Control Recognizes AIDS for the First Time (1981), p. 2149; A Genetically Engineered Vaccine for Hepatitis B Is Approved for Use (1986), p. 2326.

GREAT EVENTS
FROM
HISTORY II

CHRONOLOGICAL LIST OF EVENTS

VOLUME I

CHRONOLOGICAL LIST OF EVENTS

VOLUME II

VOLUME III

VOLUME IV

VOLUME V

ALPHABETICAL LIST OF EVENTS

KEY WORD INDEX

CATEGORY INDEX

CATEGORY INDEX

CXLI

BIOLOGY

CATEGORY INDEX

CXLIII

CATEGORY INDEX

PRINCIPAL PERSONAGES

Abel, John Jacob I-16; II-512; IV-1459
Abell, George Ogden IV-1454; V-2306
Abelson, Philip III-1181
Adams, Ansel III-1331
Adams, Walter Sydney II-645, 815, 948; III-1325; IV-1640
Ader, Clément I-41
Adrian, Edgar Douglas I-243; II-890
Agramonte, Aristides I-73
Ahlquist, Jon V-2267
Aiken, Howard Hathaway III-1396; IV-1593
Albrecht, Andreas V-2047, 2125
Aldrin, Edwin E. ("Buzz"), Jr. V-1907
Allen, William M. IV-1897
Allibone, T. E. III-978
Alpher, Ralph Asher III-1309
Alvarez II-491
Alvarez, Luis W. IV-1470; V-2120
Alvarez, Walter V-2120
Ambrose, James V-1961
Ammann, Othmar H. IV-1782
Anable, Gloria Hollister III-1018
Anderson, Carl David II-532, 694; III-983; IV-1470
Anderson, Herbert L. III-1198
Andersson, Johan Gunnar II-761; III-1096
Andrews, Roy Chapman II-735
Angel, James Roger Prior V-2291
Annis, Martin IV-1708
Anschütz-Kaempfe, Hermann I-303
Appleton, Sir Edward Victor I-174
Ardrey, Robert IV-1808
Aristotle IV-1751
Armstrong, Edwin H. II-939
Armstrong, Neil A. V-1907
Arnold, Harold D. II-615
Arnold, Henry Harley III-1187
Arrhenius, Svante August III-1118; IV-1840
Arteaga, Melchor II-491
Artin, Emil I-438
Arzelà, Cesare I-325
Asaro, Frank V-2120
Aserinsky, Eugene III-1424
Assmann, Richard I-26
Aston, Francis William II-471, 660

Atanasoff, John Vincent III-1213
Audrieth, Ludwig Frederick V-2226
Avery, Oswald III-1203

Baade, Walter II-878, 884; III-1008, 1271; IV-1449
Babakin, G. N. IV-1797, 1819; V-1928, 1950
Babbage, Charles II-846
Backus, John IV-1475
Baekeland, Leo Hendrik I-280
Baeyer, Adolf von I-280; V-1918
Bahcall, John Norris IV-1830
Ballard, Robert D. V-2058
Baltimore, David I-459
Banting, Sir Frederick Grant I-179; II-720
Bardeen, John III-1304; IV-1528, 1533
Bardeen, William V-1966
Bardon, L. I-428
Barghoorn, Elso IV-1481, 1851
Barkla, Charles Glover I-309; II-746
Barnard, Christiaan IV-1866
Barnard, Edward Emerson I-213
Bartels, Julius III-1035
Barthle, Robert C. IV-1598
Barton, Otis III-1018
Barton, Sir Derek H. R. V-1918
Basilevsky, Alexander V-2042
Bateson, William I-61, 270, 314
Baulieu, Étienne-Émile V-2185
Bavolek, Cecelia III-1024
Bayliss, Sir William Maddock I-179
Bean, Alan L. IV-1913; V-1997
Beckers, Jacques V-2291
Becquerel, Antoine-Henri I-93, 199, 412; III-992
Bednorz, J. Georg V-2311
Beebe, William III-1018
Behring, Emil Adolf von I-6; II-567
Bélanger, Alain V-2185
Bell, Alexander Graham II-595; V-2078
Bell, Jocelyn IV-1862
Belyayev, Pavel I. IV-1787
Belzel, George V-2169

WITHDRAWN